ORGANIC CHEMISTRY II LECTURE TEMPLATES

Fourth Edition

Kay I. Kouadio, Ph.D

North Lake College

Name________________________________

www.kendallhunt.com
Send all inquiries to:
4050 Westmark Drive
Dubuque, IA 52004-1840

ISBN 978-1-5249-9687-1

Published in the United States of America

To all my wonderful students (past and current) who have taught me so much how to teach them.

OCHEM II: TEMPLATES - TEXTBOOK RELATIONSHIP

Textbook Chapter #	Template Unit #	Test #
	1	1
	2	1
	3	1
	4	1
	5	2
	6	2
	7	2
	8	3
	9	3
	10	3
	11	4
	12	4
	13	5
	14	5
	15	5

TABLE OF CONTENTS

OCHEM I UNIT 1: REDOX REACTIONS IN OCHEM

A. INTRODUCTION: AN OVERVIEW ON REDOX REACTIONS

1. SOME BASIC DEFINITIONS

- **A reducing agent (*reductant*) is a reactant that loses (or donates) electrons to a second reactant called an oxidizing agent (*oxidant*) in an electron transfer reaction.**

- **Note: The reducing agent becomes oxidized at the completion of the reaction.**

- **Note: A process in which electrons are lost is called an oxidation reaction.**

- **On the other hand, an oxidizing agent (oxidant) is a reactant that gains (or accepts) electrons from another reactant called a reducing agent (reductant) in an electron transfer reaction.**

- **Note: The oxidizing agent becomes reduced at the completion of the reaction.**

- **Note: A process in which electrons are gained is called a reduction reaction.**

- **Note: Oxidation and reduction occur in tandem (meaning happen together) anytime there is an electron transfer during a reaction. Putting half of each word together, the word redox is derived.**

- **So, a redox reaction is a reaction in which one or more electrons are transferred from the reducing agent (most generous one in electrons) to the oxidizing agent (stingiest in electrons).**

- **In short, Reducing agent + Oxidizing agent = Redox Reaction**

- **The following example is a good illustration of a redox reaction:**
  - **$Zn + Cu^{2+} \longrightarrow Zn^{2+} + Cu$**

- **In this reaction:**

- **Zn = reducing agent = losing 2 electrons to Cu^{2+}**

- **Cu^{2+} = oxidizing agent = gaining 2 electrons from Zn**

2. REDOX REACTIONS IN OCHEM

- **Now, let's apply the concepts described above to organic reactions. Unfortunately, redox reactions in OCHEM are not that "*straightforward*". Redox reactions here occurs through a hydride (:H^-) and/or a proton (H^+). In other words, redox reactions occur when two electrons are transferred through a C-H bond. A loss of :H^- ($2e^-$) from a carbon corresponds to a loss of electron density on that carbon. This process is an oxidation (loss of electrons). Conversely, a gain of :H^- ($2e^-$) by a carbon is a gain of electron density for that carbon. This process is a reduction (gain of electrons). The dehydrogenation of an alcohol reaction to give a ketone can be used to illustrate Ochem redox reactions:**

- **The hydrogenation of an alkene to an alkane (addition of H_2) is an example of organic reduction reaction:**

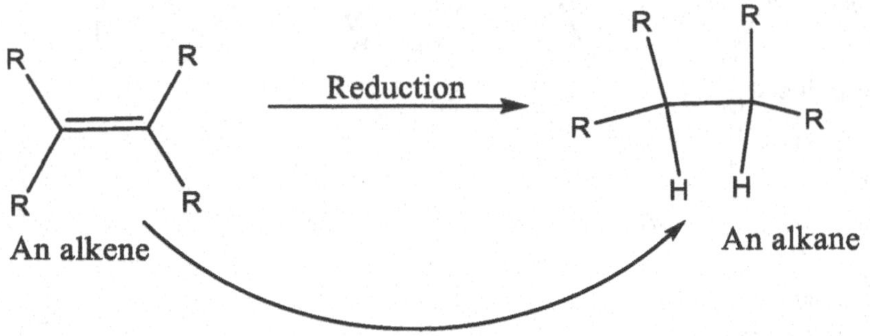

R R R R
Reduction
R R R R
H H
An alkene An alkane

gain of :H^-

- **Indeed, in this reaction, the electron density on one of the carbons of the double bond increases due to the incoming :H^- (hydride) from H_2.**

3. **REDOX REACTIONS IN OCHEM: A SUMMARY**

- **Reduction = Addition of a hydride :H^-(and/or H^+)**

- **Oxidation = Removal of a hydride :H^-(and/or H^+)**

- **Note: Anytime you take away a hydride, oxidation will occur because you have a decrease in electron density.**

- **Note: The more the CO bonds, the more oxidized the compounds. The following compounds are ranked in increasing order of degree of oxidation:**

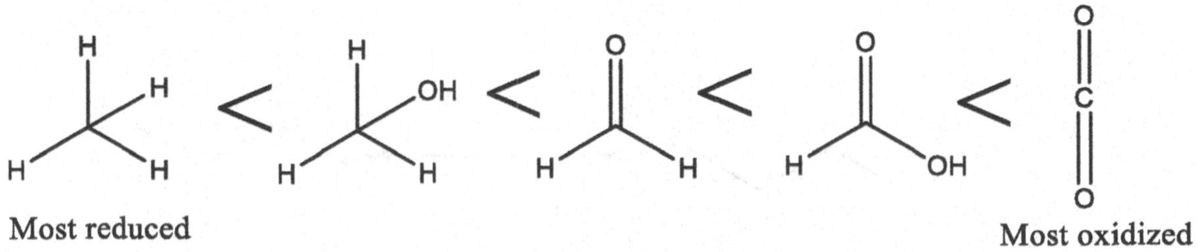

CH_4 < CH_3OH < $H_2C{=}O$ < $HCOOH$ < $O{=}C{=}O$

Most reduced Most oxidized

or

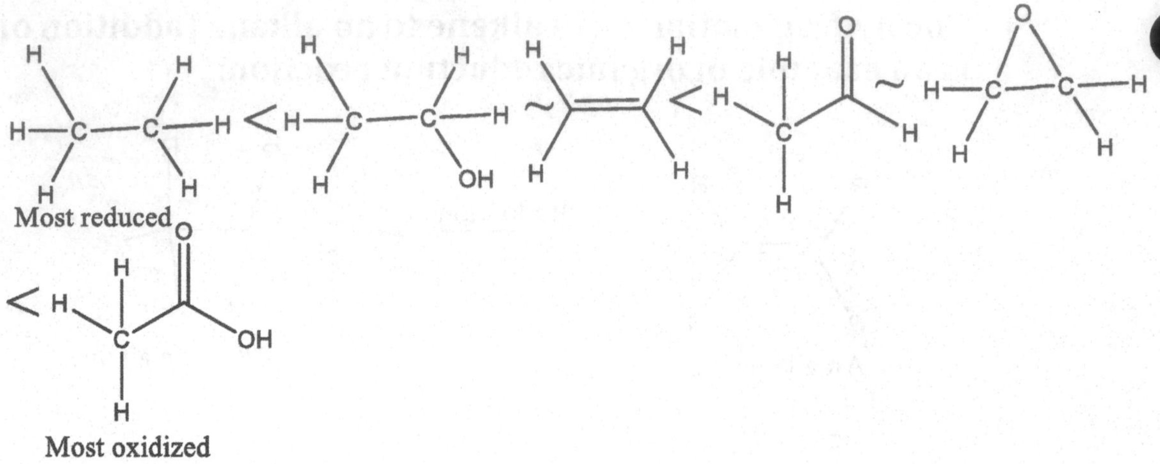

Most reduced

Most oxidized

- **Question: Is the following substitution reaction a redox reaction?**

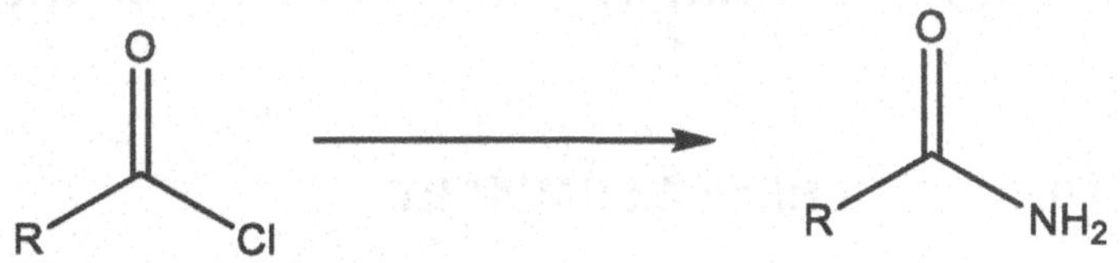

R–C(=O)–Cl → R–C(=O)–NH_2

B. REDOX REACTIONS IN OCHEM

1. OXIDATION

- An **oxidation** reaction is a chemical change in which the number of C-Y bonds (Y = O, halogen) **has increased** or a reaction in which the number of C-H bonds **has decreased.**

Ex:

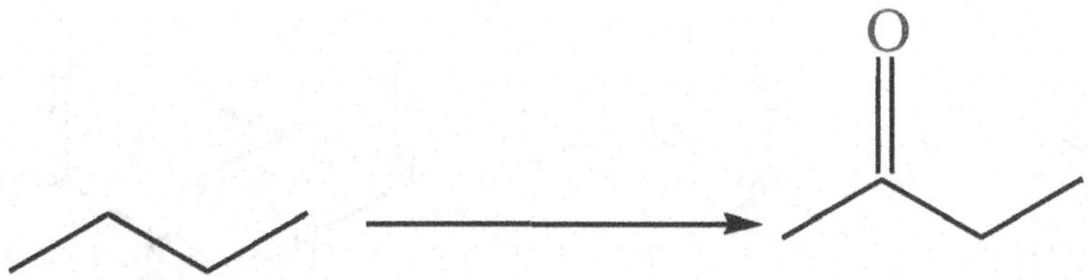

2. REDUCTION

- When the number of C-Y bonds (Y = O, halogen) has **decreased** in a chemical process, a **reduction** reaction has occurred. Reduction also occurs when the number of C-H bonds has **increased.**

Ex:

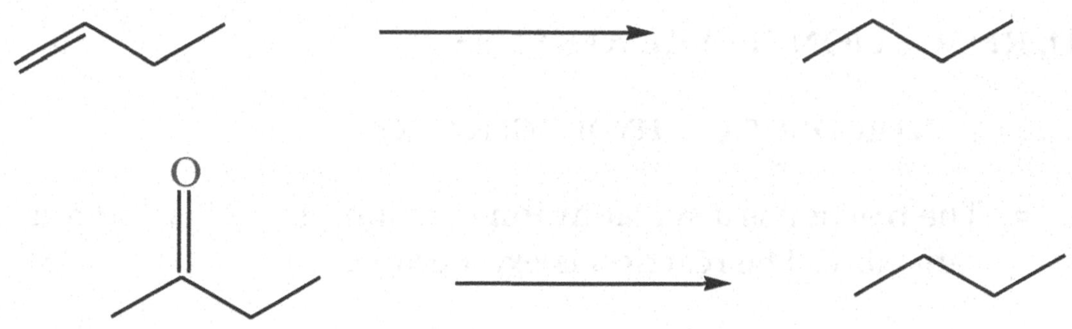

- Do Problem ______, page ______.

C. COMMON REDUCING AGENTS IN OCHEM

1. TYPES OF REDUCING AGENTS: 3

a. H_2/metal

b. $H_2 = 2H^+ + 2e^-$

Ex: Na/NH_3 or Li/NH_3

$$2Na \longrightarrow 2Na^+ + 2e^-$$

$$2NH_3 \longrightarrow 2NH_2^- + 2H^+$$

The overall reaction is:

$$2Na + 2NH_3 \longrightarrow 2Na^+ + 2NH_2^- + 2H^+ + 2e^-$$

c. **Addition of a hydride and a proton**

- There are **2 reducing** agents:
    - **-$LiAlH_4$: Lithium aluminum hydride**
    - **-$NaBH_4$: sodium borohydride**

D. REDUCTION OF THE ALKENES

1. INTRODUCTION: HYDROGENATION

- The reaction is a **syn addition reaction.** The 2 Hs add to the same face. **The reaction is exothermic.**

2. GENERAL REACTION

$$R_2C{=}CH_2 + H_2 \xrightarrow{Pt} R{-}CR(H){-}CH_2{-}H$$

Ex:

$$(H_3C)_2C{=}CH_2 + H_2 \xrightarrow{Pt}$$

- **Mechanism: See page** ______.

3. HYDROGENATION AND STABILITY

- The more stable the alkene, the lower the heat of hydrogenation. **Indeed, the heats of hydrogenation can be used to assess alkene stability as follows:**

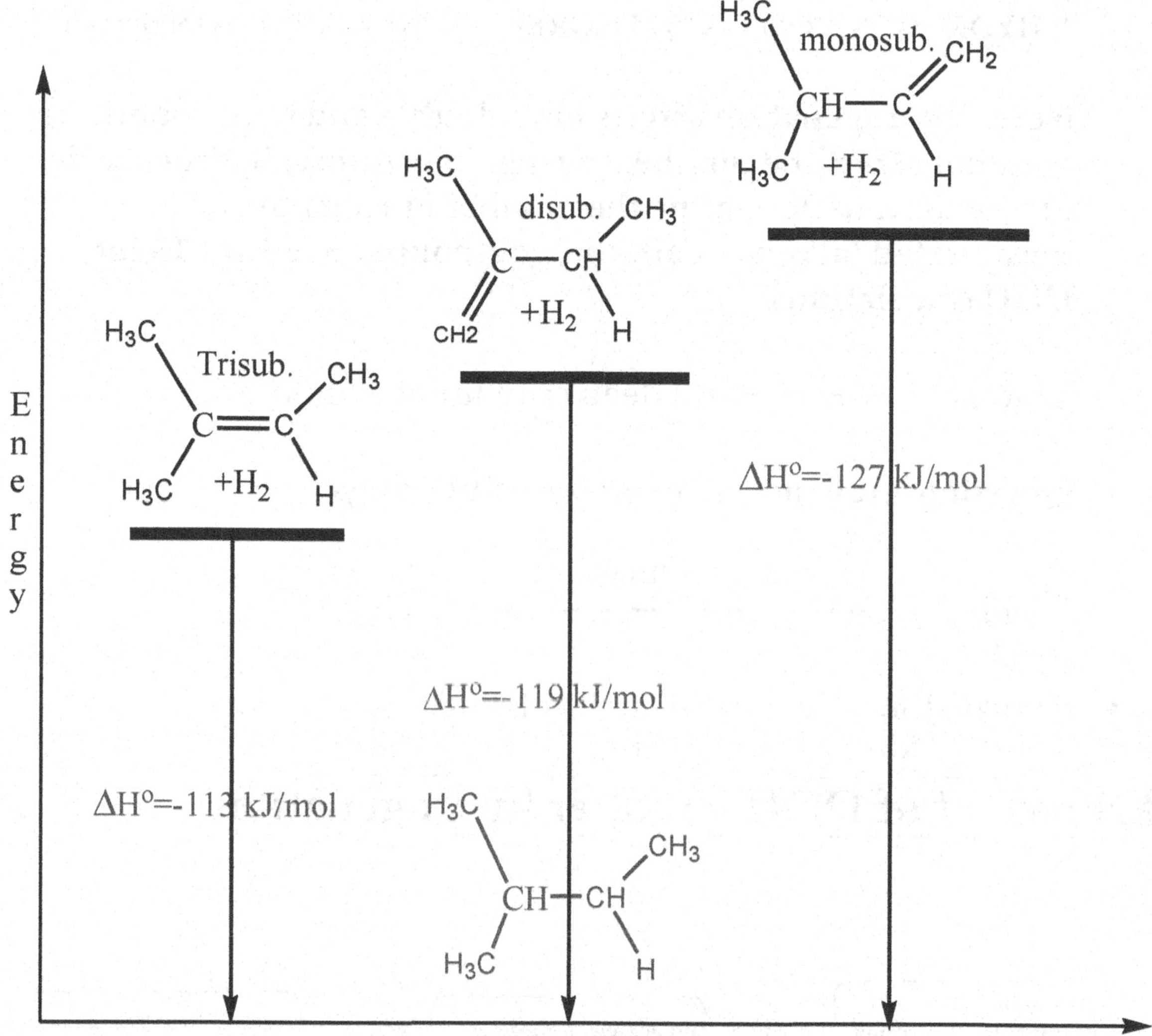

- The rate of the reaction increases as the number of R groups decreases. Indeed, the rate of the hydrogenation reaction follows the following order: monosubstituted alkenes > disubstituted alkenes > trisubs > tetrasub (very hindered). **As a result, the increasing order of stability of the alkenes is: monosubstituted alkenes < disubstituted alkenes < trisubs < tetrasub.**

- **Read pages ______ - ______.**

- **See Fig. ______, page ______.**

- **Do problems on pages ______ - ______.**

4. HYDROGENATION AND DEGREE OF UNSATURATION (DOU)

- **Note: Hydrogenation affects only double and triple bonds. It does not affect the number of rings. Therefore, hydrogenation can be used to determine the number of rings in an unsaturated (alkene or alkyne) compound. See Unit 13 for DOU calculations.**

Ex:

$C_{10}H_{16} \longrightarrow$ 3 degrees of unsaturation

- **Question: How many rings? 3possible rings.**

$$C_{10}H_{16} \xrightarrow{\text{hydrogenation}} C_{10}H_{20}$$

- **Conclusion**: The compound has 1 ring.

 # rings = # of DOU left after hydrogenation

or

actual formula of unsaturated hydrocarbon C_nH_y	$\xrightarrow{+H_2}$	formula after hydrogenation C_nH_z	$\longrightarrow$	corresponding alkane formula C_nH_{2n+2}

$$\text{\# rings} = \frac{(2n + 2) - z}{2}$$

Ex: $C_{10}H_{18} \xrightarrow{\text{hydrogenation}} C_{10}H_{22}$ **#rings =**

- **Read page ______ - ______ and associated problem.**

5. HYDROGENATION OF OILS: HARDENING

$$\textbf{oil} \xrightarrow[\text{Pt}]{H_2} \textbf{fat}$$

- Oils = triacylglycerols with a significant number of unsaturation
- Fats = triacylglycerols with few degrees of unsaturation. They have higher MP due to **increased surface area**. One can go from an oil to a fat through hydrogenation.

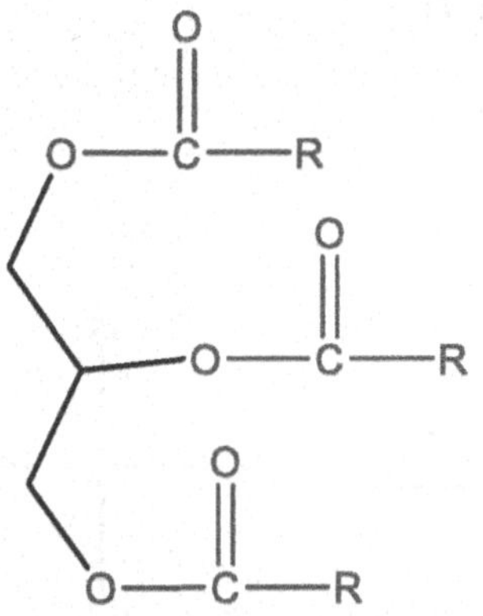

A triacylglycerol
(fat or lipid)

If R, R', and R" have many unsaturations = oil

If R, R', and R" have only few unsaturations = fat

- **Read pages ________ – _______ and associated problem.**

6. OXYMERCURATION-REDUCTION OF THE ALKENES

a. Introduction

- The reaction proceeds through the formation of a cyclic mercurinium ion in the first step with mercuric acetate, $Hg(OAc)_2$.
- The water attacks at the most substituted carbon of the mercurinium intermediate; Markovnikov's rule is followed.
- $NaBH_4$ is used as a reducing agent in the final step.
- An alcohol is produced. OAc = Acetate.

b. The General Reaction

OH H

R R H H

C=C

1.$Hg(OAc)_2$, H_2O/THF

2. $NaBH_4$

R H

R—C—C—H

OH H

An alkene

an alcohol

Ex:

H_3C H

H_3C H

C=C

1.$Hg(OAc)_2$, H_2O/THF

2. $NaBH_4$

CH_3 H

H_3C—C—C—H

OH H

An alkene

an alcohol

c. Mechanism of the Reaction

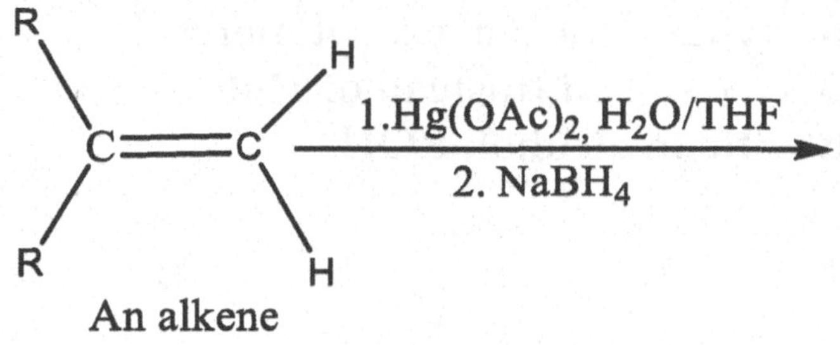

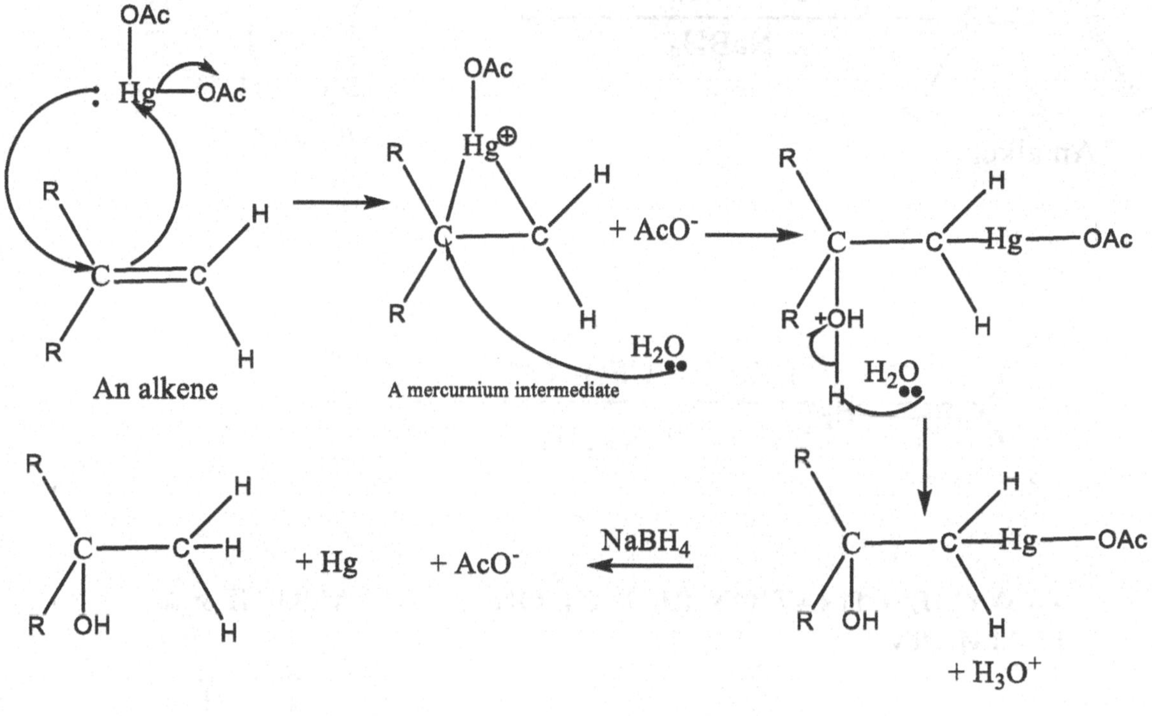

R(R)C=CH(H) — An alkene

1. $Hg(OAc)_2$, H_2O/THF
2. $NaBH_4$

OAc
Hg—OAc
R, R, H, H
C=C
An alkene

A mercurnium intermediate

OAc
$Hg^{\oplus}$
R, R, C—C, H, H
+ AcO^-
H_2O

R, R, C—C—Hg—OAc, H, H
+OH
H
H_2O

R, R OH, C—C—Hg—OAc, H, H
+ H_3O^+

$NaBH_4$

R, R OH, C—C—H, H, H
+ Hg + AcO^-

d. Alkoxymercuration-Reduction of the Alkenes

- **The reaction is similar to oxymercuration, except mercuric trifluoroacetate ($Hg(O_2CCF_3)_2$ is used (instead of $Hg(OAc)_2$) and the water is replaced with an alcohol, ROH.**
- **An ether is produced.**

- **The general reaction is:**

OR H
R H
C=C
R H
An alkene

1.$Hg(O_2CCF_3)_2$, ROH/THF
2. $NaBH_4$

R—C(R)(OR)—C(H)(H)—H
an ether

Ex:

Methylenecyclopentane (=CH_2)

1.$Hg(O_2CCF_3)_2$, CH_3OH/THF
2. $NaBH_4$

7. OXYMERCURATION-REDUCTION OF THE ALKENES: A SUMMARY

R H
C=C
R H
An alkene

1.$Hg(OAc)_2$, H_2O/THF
2. $NaBH_4$

R—C(R)(OH)—C(H)(H)—H
an alcohol (Markovnikov)

1.$Hg(O_2CCF_3)_2$, ROH/THF
2. $NaBH_4$

R—C(R)(OR)—C(H)(H)—H
an ether (Markovnikov)

E. REDUCTION OF THE ALKYNES: HYDROGENATION

1. INTRODUCTION

- There are **3 ways** to hydrogenate an alkyne:
  - **Complete hydrogenation with H_2/Pd or H_2/Pt.**
  - **Partial hydrogenation with Lindlar catalyst.**
  - **Partial reduction using Li/NH_3 or Na/NH_3.**

2. COMPLETE REDUCTION

- Complete hydrogenation requires **2 moles of H_2 (or excess H_2).** The reaction is a **syn addition reaction.**

R—C≡C—R' $\xrightarrow[\text{Pt}]{2H_2}$ R—CH₂—CH₂—R'

Ex:

(cyclopentyl)—C≡C—(phenyl) $\xrightarrow[\text{Pt}]{2H_2}$

3. PARTIAL REDUCTION WITH LINDLAR CATALYST

- The reaction is **stereoselective (syn addition). Only the cis isomer is formed.**
- **Note: Lindlar catalyst: Pd on $CaCO_3$ in $Pb(CH_3COO)_2$ and quinoline. Lindlar catalyst is also called partially deactivated palladium.**
- **The general reaction is:**

R—C≡C—R' $\xrightarrow{H_2/\text{ Lindlar}}$ (R)(H)C=C(R')(H)

A cis alkene

Ex:

(cyclopentyl)—C≡C—(phenyl) $\xrightarrow{H_2/\text{Lindlar}}$

4. PARTIAL REDUCTION USING Li/NH_3 OR Na/NH_3 CATALYST

- The reaction is also **stereoselective (trans addition). Only the trans isomer is observed.**
- **The general reaction is:**

R—C≡C—R' $\xrightarrow{Li/NH_3}$ (R)(H)C=C(H)(R')

A trans alkene

Ex:

(cyclopentyl)—C≡C—(phenyl) $\xrightarrow{Li/NH_3}$

- See mechanism on page ________ and do associated problem.
- See summary on page ________ Fig. _______ and do associated problems.

5. REDUCTION OF THE ALKYNES: A SUMMARY

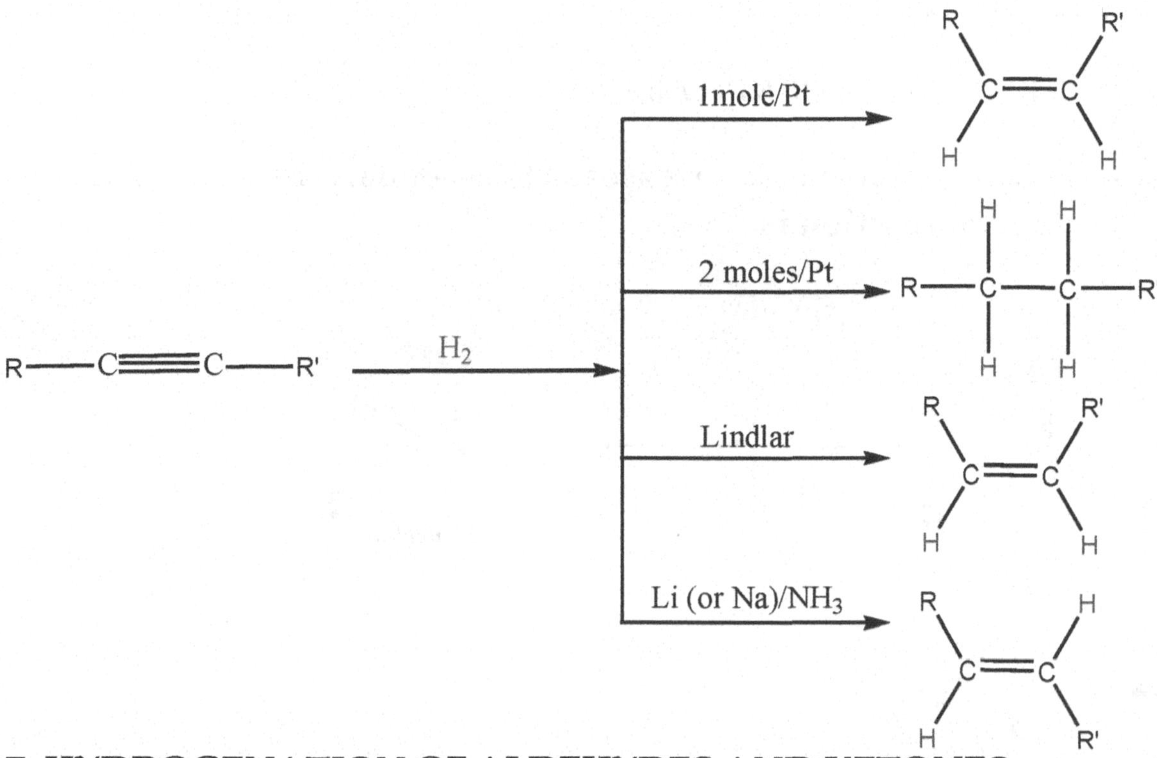

F. HYDROGENATION OF ALDEHYDES AND KETONES

1. REDUCTION OF ALDEHYDES

- Reduction of aldehydes with hydrogen leads to **primary alcohols as follows:**

$$R-\overset{\overset{\displaystyle O}{\|}}{C}-H \xrightarrow[\text{Pd or Raney Ni}]{H_2} R-\overset{\overset{\displaystyle OH}{|}}{C}H_2$$

A 1° alcohol

- **Note: Raney Ni is finely dispersed Ni with adsorbed H_2.**

Ex:

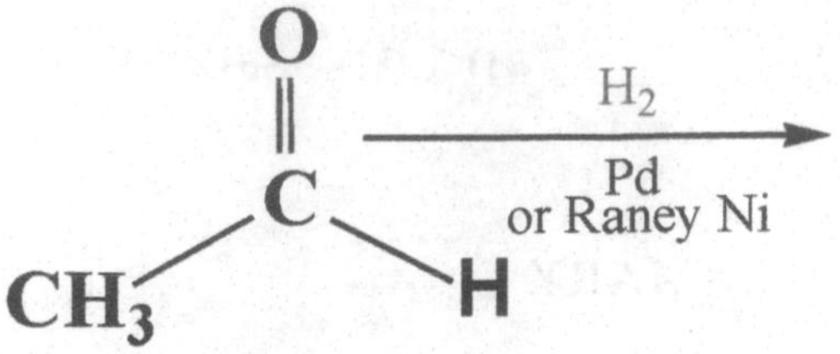

$$CH_3-C(=O)-H \xrightarrow[\text{Pd or Raney Ni}]{H_2}$$

2. REDUCTION OF KETONES

- Hydrogenation of ketones leads to **secondary alcohols. The general reaction is:**

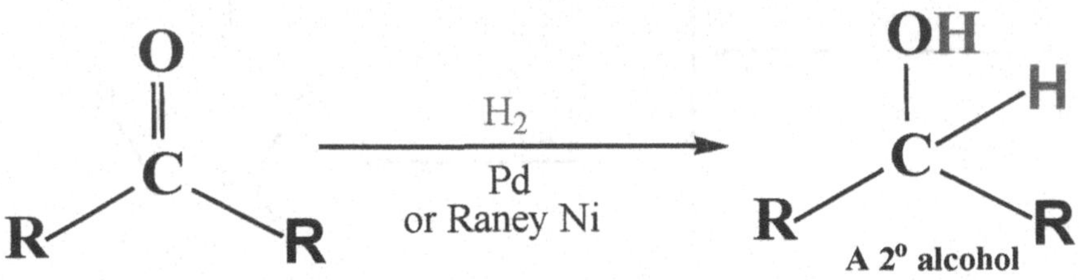

$$R-C(=O)-R \xrightarrow[\text{Pd or Raney Ni}]{H_2} R-CH(OH)-R$$

A 2° alcohol

Ex:

$$CH_3-C(=O)-CH_3 \xrightarrow[\text{Pd or Raney Ni}]{H_2}$$

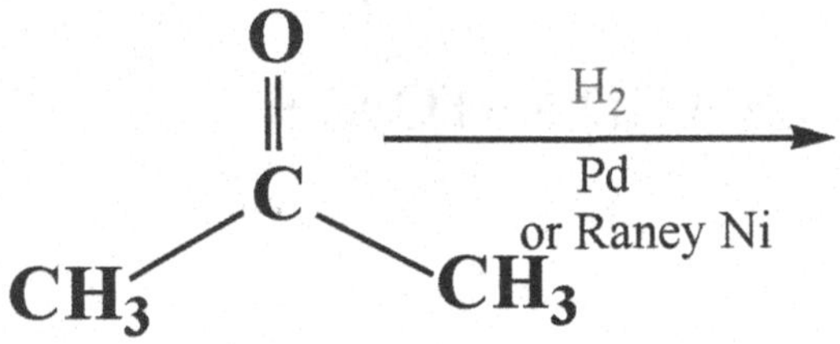

3. REDUCTION OF ALDEHYDES, KETONES, AND ACID CHLORIDES: A SUMMARY

H_2

$R-C(=O)-H$ (an aldehyde) → Pd or Raney Ni → $R-CH(OH)-H$ (a 1° alcohol)

$R-C(=O)-R'$ (a ketone) → Pd or Raney Ni → $R-CH(OH)-R'$ (a 2° alcohol)

$R-C(=O)-Cl$ (an acid chloride) → PD Pd → $R-C(=O)-H$ (an aldehyde)

G. REDUCTION OF ALKYL HALIDES AND ACID CHLORIDES

1. REDUCTION OF ALKYL HALIDES

- The catalyst is $LiAlH_4$. The rate of the reaction decreases from 1° to 3° alkyl halides.

$$R\text{-}X \xrightarrow[2.\ H_2O]{1.\ LiAlH_4} R\text{-}H$$

An alkane

Ex:

$$\text{cyclopentyl}-CH_2Br \xrightarrow[2.\ H_2O]{1.\ LiAlH_4}$$

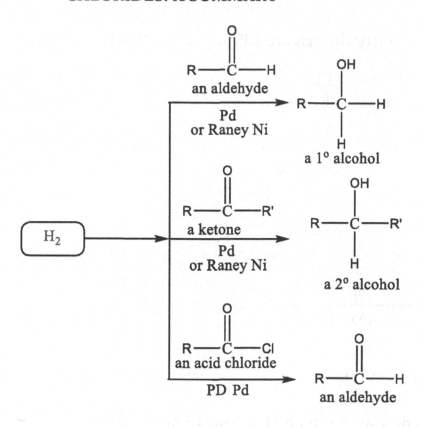

2. REDUCTION OF ACID CHLORIDES: THE ROSENMUND REDUCTION

- The catalyst is **partially deactivated Pd** (aka Lindlard Catalyst).

$$R-\overset{\overset{\displaystyle O}{\|}}{C}-Cl \xrightarrow[\text{PD Pd}]{H_2} R-\overset{\overset{\displaystyle O}{\|}}{C}-H$$

An aldehyde

Ex:

$$\text{cyclobutyl}-\overset{\overset{\displaystyle O}{\|}}{C}-Cl \xrightarrow[\text{PD Pd}]{H_2}$$

H. REDUCTION OF EPOXIDES

- This is an SN2 reaction in which **H^- (from $LiAlH_4$) is the nucleophile.**

- **Note: For unsymmetrical epoxides, the nucleophilic attack occurs at the least substituted carbon.**

- **The general reaction is:**

$$\text{An epoxide} \xrightarrow[\text{2. } H_2O]{\text{1. } LiAlH_4} -\underset{H}{\overset{|}{C}}-\overset{OH}{\overset{|}{C}}-$$

An epoxide

An alcohol

- See Fig. ______, page ______ and associated problems.

Ex:

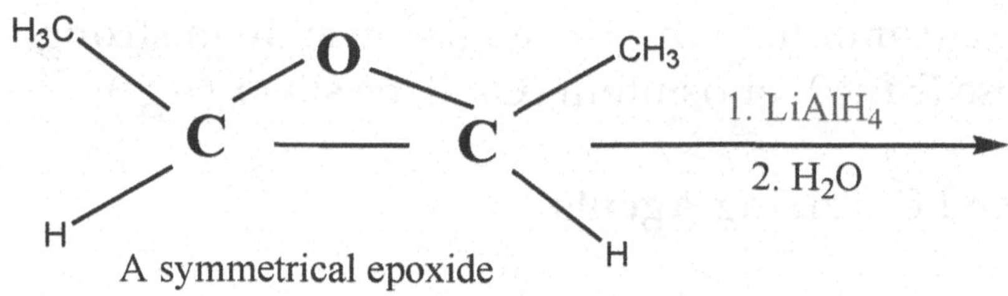

A symmetrical epoxide

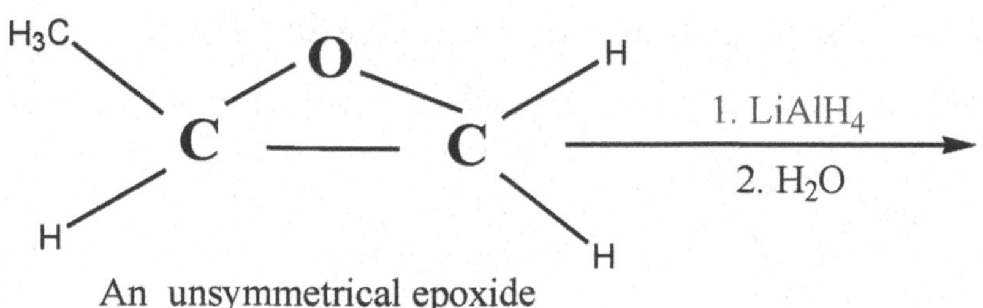

An unsymmetrical epoxide

I. OXIDIZING AGENTS

1. INTRODUCTION

- There are **two types** of oxidizing agents:
  - -The first group consists of reagents having O-O bonds (Ex: peroxides, O_2, O_3).
  - -The other group of reagents contains metal-O bonds (Ex: CrO_4^{2-}).

2. O-O bond oxidizing agents

- Some **are O_2, O_3, H_2O_2, tert-butyl hydroperoxides ($(CH_3)_3COOH$), peroxyacids (RCO_3H).**

- **See Fig. on page ______.**

3. METAL-O BOND OXIDIZING AGENTS

a. **Introduction**

- The metal can be **chromium** in the **+ 6 oxidation state** in **strong acid** or manganese (**Mn^{+7}**), or osmium (**Os^{+8}**), or silver (Ag^{+}) .

b. **Cr-Based Oxidizing Agents**

- CrO_3, $Na_2Cr_2O_7$, $K_2Cr_2O_7$: **strong acid needed.**

- **PCC = pyridinium chlorochromate: no acid needed (mild oxidizing agent).**

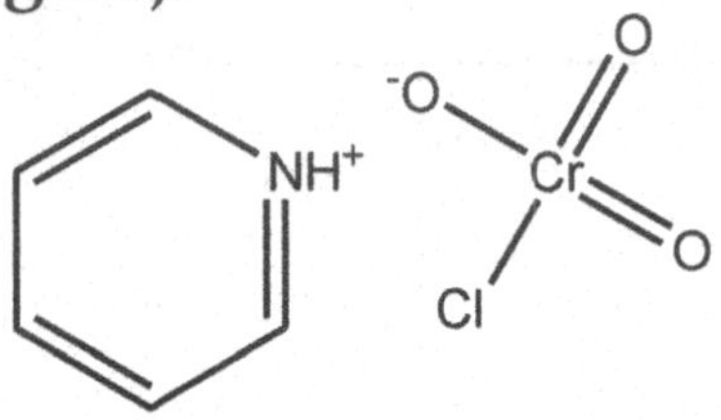

pyridinium chlorochromate

c. Mn-Based Oxidizing Agent = $KMnO_4$

d. Osmium-Based Oxidizing Agent = OsO_4

e. Ag-Based Oxidizing Agent = Ag_2O

J. OXIDATION OF ALKENES

1. EPOXIDATION

a. General Reaction: See Fig. ______, page ______.

- The reaction is carried out with a **peroxyacid (RCO_3H).**

- **The general reaction is:**

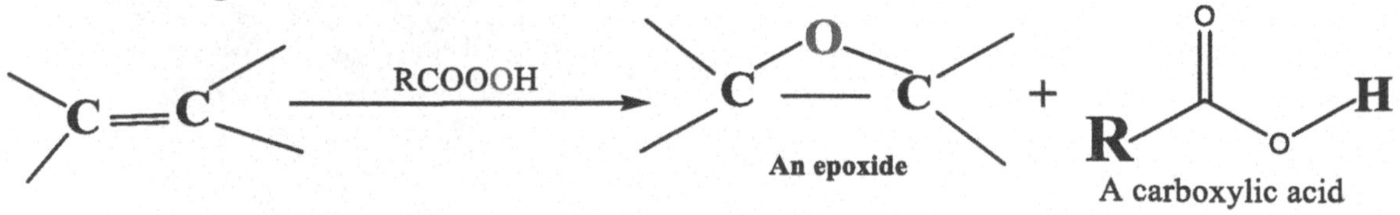

An epoxide

A carboxylic acid

- **Ex:**

H_3C, CH_3, H, H (C=C)

$\xrightarrow{CH_3COOH}$

- **See Fig. on page ________ for the structures of peroxyacids (PAA, mCPBA, MMPP).**

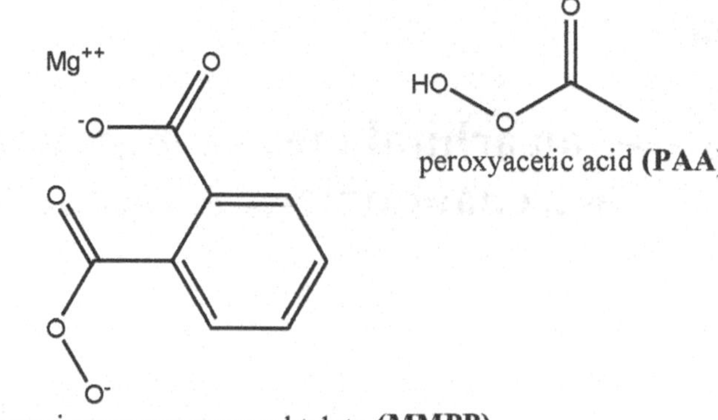

magnesium monoperoxyphtalate **(MMPP)**

HO, O, O

peroxyacetic acid **(PAA)**

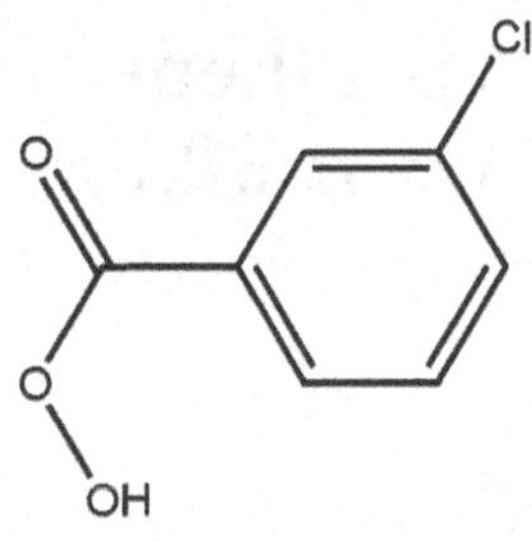

meta-chloroperoxybenzoic acid **(mCPBA)**

- **See mechanism of the reaction on page ______.**
- **See examples on page _______.**
- **Do Problem _______, page ______.**

b. Stereochemistry of Alkene Epoxidation

i. Epoxidation: a Syn Addition

- The reaction occurs through a **syn addition**. The configuration of the alkene is retained.
  - **cis-alkene** ⟶ **cis-epoxide**
  - **trans-alkene** ⟶ **trans-epoxide**

Ex:

Cl Cl C=C H A cis alkene H —mCPBA→ Cl O Cl C* — C* H H A meso compound

ii. Chirality of Epoxidation of the Alkene

- The reaction is **stereospecific.**
  - **cis-alkene → an achiral meso compound**
  - **trans-alkene → 2 enantiomers: A racemate**

Ex:

Cl Cl C=C H A cis alkene H —mCPBA→ Cl O Cl C* — C* H A meso compound H

Cl H C=C H Cl A trans alkene —mCPBA→ Cl O H C* — C* H Cl + Cl H C* — C* H O Cl two enantiomers

- **Do Problem ________, page _______.**

- **Read pages ________ - _______; Synthesis of Disparlure: An Application of Epoxidation on page ______.**

2. OXIDATION OF THE ALKENES: DIHYDROXYLATION

a. Introduction

- **The general reaction is:**

Ex:

R(H)C=C(R)H —catalyst→ HO–C(R)(H)–C(R)(H)–OH

- **Note: The two OH groups can be added on the same side (syn dihydroxylation) or on opposite sides (anti dihydroxylation).**

b. Anti Dihydroxylation of the Alkenes

- The reaction proceeds in **two steps** leading to a **trans-1,2-diol and its enantiomer. A racemic mixture is obtained.** A **peroxyacid catalyst** is used.

An alkene —RCO_3H→ epoxide —H_2O (OH^- or H^+)→ A trans 1,2 -diol and its enantiomer

or

An alkene —1. RCO_3H, 2. H_2O→ A trans 1,2 -diol and its enantiomer

Ex:

cyclohexene —1. RCO_3H, 2. H_2O→

- Do Problem ______ on page ______.

c. Syn Dihydroxylation of the Alkenes

- The catalysts: **alkaline cold dilute $KMnO_4$ or $O.sO_4$/ $NaHSO_3$, H_2O (H_3O^+). The general reaction is:**

R₂C=CR₂ (An alkene) —[$KMnO_4$; H_2O, OH^-]→ A cis-diol

or

R₂C=CR₂ (An alkene) —[1. OsO_4; 2. $NaHSO_3$, H_2O]→ A cis-diol

- **Note: A cis-1,2 diol is produced.**

- **See mechanisms on page ______.**

- **The mechanism of the reaction with OsO_4:**

Alkene + OsO_4 → [A cyclic osmate]‡ —[H_2O; $NaHSO_3$]→ Syn product

A cyclic osmate

- **The mechanism of the reaction with $KMnO_4$:**

Alkene + MnO_4^- → [A cyclic manganate]‡ → product

A cyclic manganate

Ex:

1. OsO_4
2. $NaHSO_3$, H_2O

or

$KMnO_4$
H_2O, HO^-

d. Syn Dihydroxylation of the Alkenes: Using the OsO_4/NMO catalyst

- **OsO_4** is a more **selective oxidizing agent** than $KMnO_4$, but it is toxic and expensive. So a small amount of osmium tetroxide is combined with NMO (**N-methylmorpholine N-oxide**), a good oxidizing agent. In this "tandem" catalysis, Os(VIII) is reduced by the π electrons of the alkene double bond to Os(VI). Then, NMO oxidizes Os(VI) back to Os(VIII) which can be used again. The structure of **NMO** is:

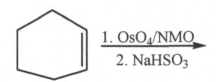

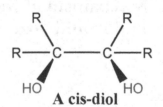

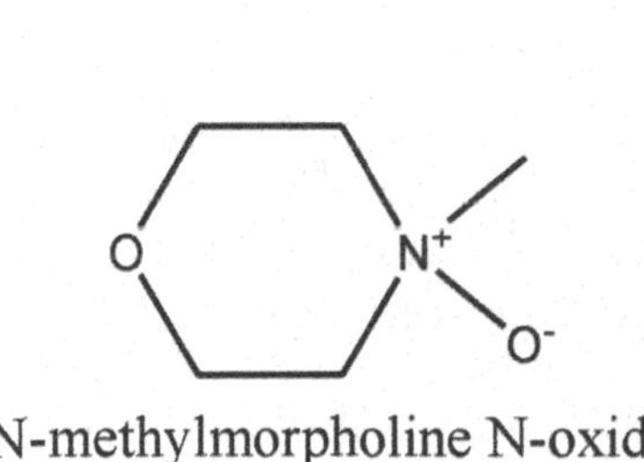

N-methylmorpholine N-oxide

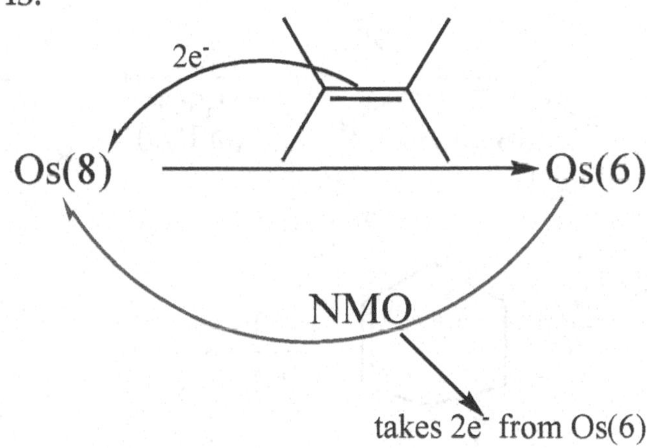

- **The general reaction**

1. OsO_4/NMO
2. $NaHSO_3$

An alkene

A cis-diol

Ex:

1. OsO_4/NMO
2. $NaHSO_3$

- **Read pages ______ – ______. Do problem _______ on page _____.**

K. OXIDATION OF ALKENES: CLEAVAGE REACTIONS

1. INTRODUCTION

- **A cleavage** reaction is the **breaking of the double bond**. The products are **carbonyl compounds**. The **catalysts** are **$O_3/Zn/H_3O^+$, O_3/CH_3SCH_3 (or H_2S) or warm $KMnO_4$.**

2. CLEAVAGE WITH OZONE: OZONOLYSIS

- **The general reaction is:**

$$R_2C{=}CR_2 \xrightarrow[\text{2. } Zn/H_3O^+]{\text{1. } O_3} R_2C{=}O + O{=}CR_2$$

An alkene

or

$$R_2C{=}CR_2 \xrightarrow[\text{2. } CH_3SCH_3 \text{ (or } H_2S)]{\text{1. } O_3} R_2C{=}O + O{=}CR_2$$

An alkene

Ex:

$$\text{cyclohexene} \xrightarrow[\text{2. } CH_3SCH_3]{\text{1. } O_3}$$

- **Mechanism of the reaction: The reaction proceeds through molozonide/ozonide intermediates.** **See page _______.**

A molozonide

An ozonide

product

Ex:

$H_3C-CH=CH-CH_3$ $\xrightarrow[\text{2. } CH_3SCH_3]{\text{1. } O_3}$

- **Note: The carbonyl products are <u>not</u> further oxidized to carboxylic acids.**

- **Do example and problem on pages ________ - _______.**

3. PRODUCTS IN THE CLEAVAGE OF ALKENES WITH ACIDIC OR NEUTRAL $KMnO_4$

- The product of the reaction depends on the alkene:
  - If there are **Hs** on the carbons of the double bond, **2 carboxylic acids** are formed.
  - If there are **2 Hs** on **a** carbon of the double bond, CO_2 is formed from that carbon.
  - If there is **no H** present on the double bond, **two ketones** are formed.

- **The general reaction is:**

$$RHC=CHR' \xrightarrow[H_3O^+]{KMnO_4} R-COOH + HOOC-R'$$

or

$$RHC=CH_2 \xrightarrow[H_3O^+]{KMnO_4} R-COOH + CO_2$$

or

$$R_2C{=}CR'_2 \xrightarrow[H_3O^+]{KMnO_4} R_2C{=}O + O{=}CR'_2$$

Ex:

$$CH_3CH{=}CH_2 \xrightarrow[H_3O^+]{KMnO_4}$$

$$CH_3CH{=}CHCH_3 \xrightarrow[H_3O^+]{KMnO_4}$$

$$(CH_3)_2C{=}CHCH_3 \xrightarrow[H_3O^+]{KMnO_4}$$

$$(CH_3)_2C{=}C(CH_3)_2 \xrightarrow[H_3O^+]{KMnO_4}$$

$$H_2C{=}CH_2 \xrightarrow[H_3O^+]{KMnO_4}$$

4. OXIDATION OF THE ALKENES: A SUMMARY

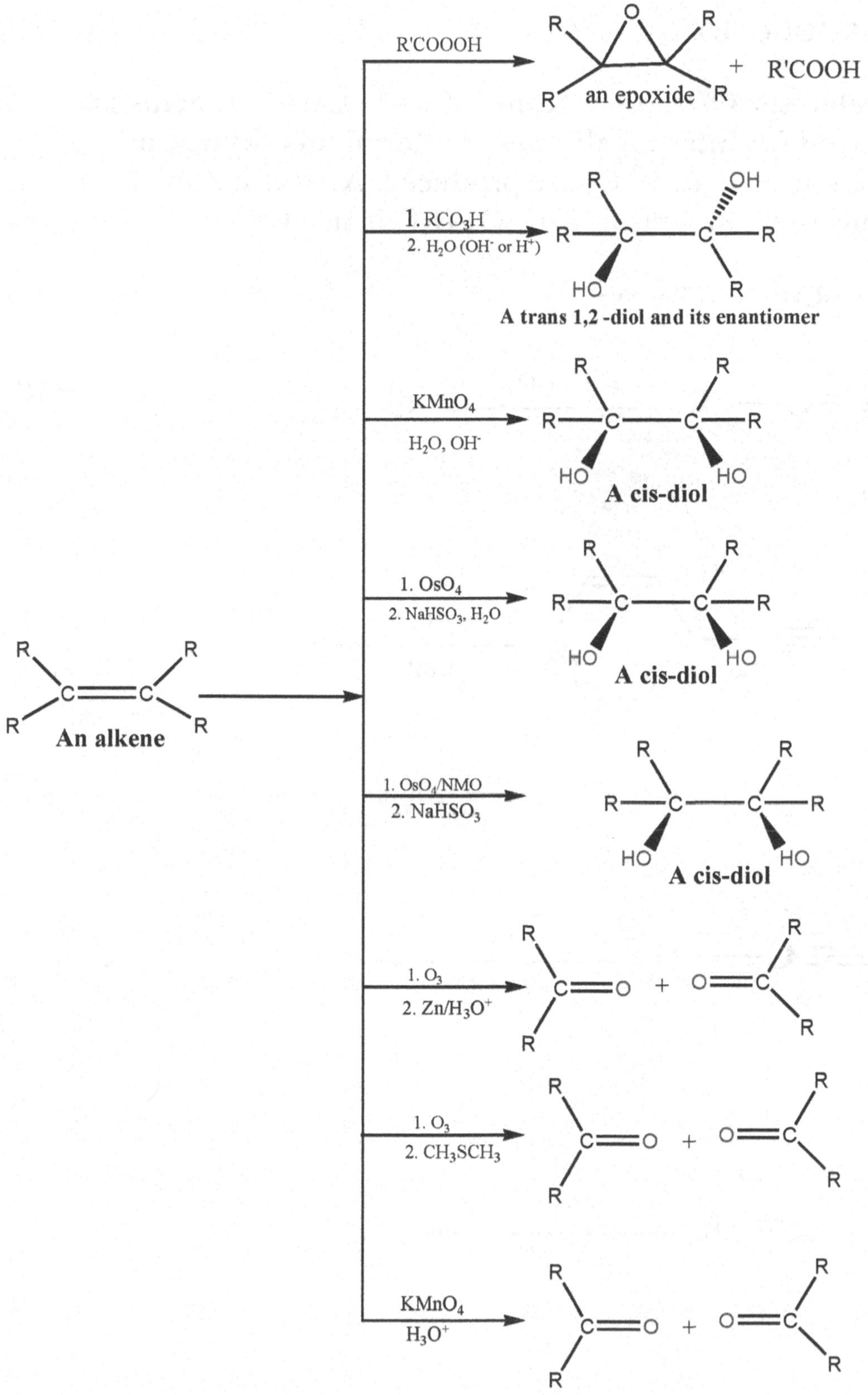

An alkene

$R'COOOH$ → an epoxide + $R'COOH$

1. RCO_3H
2. H_2O (OH^- or H^+)
→ **A trans 1,2 -diol and its enantiomer**

$KMnO_4$, H_2O, OH^- → **A cis-diol**

1. OsO_4
2. $NaHSO_3$, H_2O
→ **A cis-diol**

1. OsO_4/NMO
2. $NaHSO_3$
→ **A cis-diol**

1. O_3
2. Zn/H_3O^+
→ R₂C=O + O=CR₂

1. O_3
2. CH_3SCH_3
→ R₂C=O + O=CR₂

$KMnO_4$, H_3O^+ → R₂C=O + O=CR₂

L. OXIDATION OF ALKYNES: CLEAVAGE REACTIONS

1. INTRODUCTION

- The catalysts: O_3/H_2O or $KMnO_4/H_3O^+$. **Carbonic acids are produced for internal alkynes. For terminal alkynes, a carboxylic acid and CO_2 are produced. An α-diketone is produced when neutral $KMnO_4/H_2O$ is used.**

2. GENERAL REACTIONS

$$R-C\equiv C-R' \xrightarrow[H_3O^+]{KMnO_4} R-C(=O)OH + HO-C(=O)-R'$$

Ex:

$$\text{cyclopentyl}-C\equiv C-\text{phenyl} \xrightarrow[H_3O^+]{KMnO_4}$$

or

$$R-C\equiv C-H \xrightarrow[2.\ H_2O]{1.\ O_3} R-C(=O)OH + CO_2$$

Ex:

$$\text{cyclopentyl}-C\equiv CH \xrightarrow[2.\ H_2O]{1.\ O_3}$$

$$R—C\equiv C—R' \xrightarrow[H_2O,\ neutral]{KMnO_4} R—\overset{O}{\overset{\|}{C}}—\overset{O}{\overset{\|}{C}}—R'$$

an α diketone

Ex:

(cyclopentyl)–C≡C–(phenyl) $\xrightarrow[H_2O,\ neutral]{KMnO_4}$

3. OXIDATION OF THE ALKYNES: A SUMMARY

$R—C\equiv C—R'$

$\xrightarrow[H_3O^+]{KMnO_4}$ R–COOH + HOOC–R'

$\xrightarrow[2.\ H_2O]{1.\ O_3}$ R–COOH + HOOC–R'

$\xrightarrow[H_2O,\ neutral]{KMnO_4}$ $R—\overset{O}{\overset{\|}{C}}—\overset{O}{\overset{\|}{C}}—R'$

an α diketone

- **Note: Oxidation can be used to locate a multiple bond in an unknown unsaturated compound.**

Ex: What is the **structural formula** of the reactant in the following reaction?

$C_{13}H_{22} \xrightarrow[2.\ H_2O]{1.\ O_3}$ (cyclohexyl)–COOH + (pentan-3-yl)–COOH

1 triple bond and 1 ring

- Read page ______ - _______ and do appropriate problems.
- Do example , page ______.

M. OXIDATION OF ALCOHOLS

1. INTRODUCTION

- **Strong, nonselective oxidizing agents such as CrO_3, $Na_2Cr_2O_7$, $K_2Cr_2O_7$ in acid can be used.**
- **PCC (pyridinium chlorochromate) can also be used: no acid needed)**

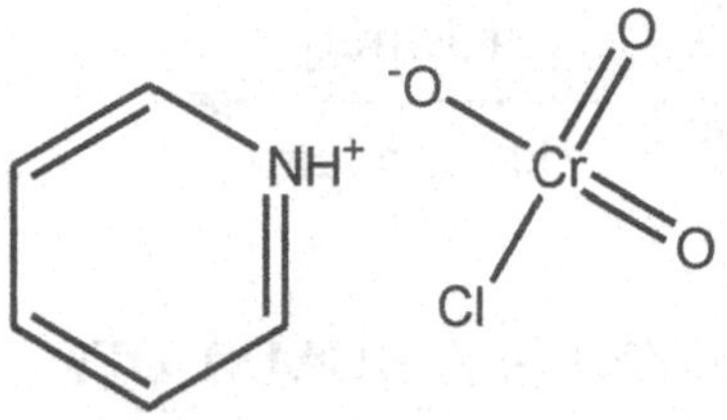

pyridinium chlorochromate

2. OXIDATION OF PRIMARY ALCOHOLS

- **A carboxylic acid is produced. The general reaction is:**

$$\mathbf{RCH_2OH} \xrightarrow{[\,O]} \qquad\qquad \xrightarrow{[\,O]}$$

Ex:

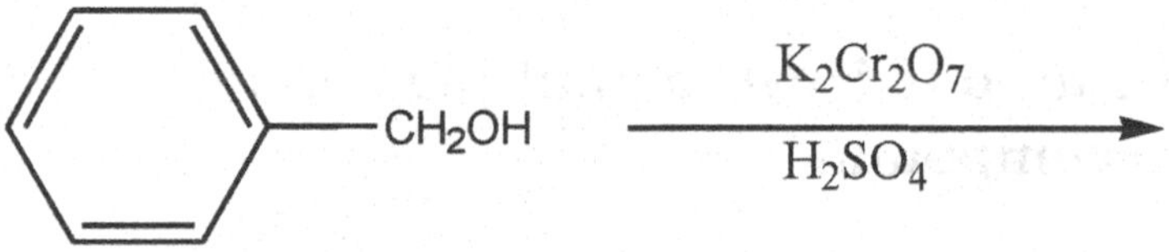

$$C_6H_5\text{-}CH_2OH \xrightarrow[H_2SO_4]{K_2Cr_2O_7}$$

$$\mathbf{RCH_2OH} \xrightarrow{[\,PCC]}$$

Ex:

$$C_6H_5\text{-}CH_2OH \xrightarrow{PCC}$$

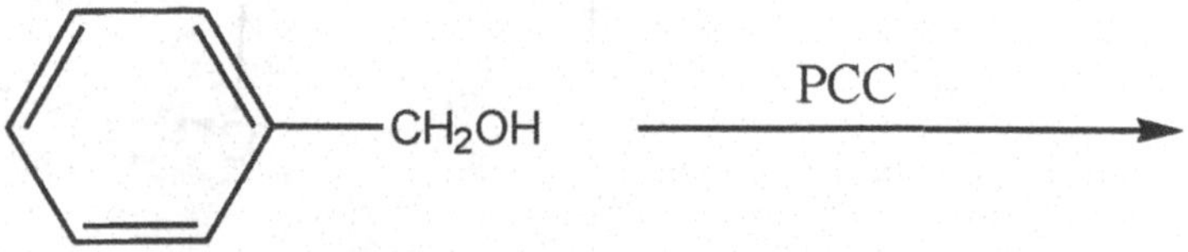

- **See mechanism of reaction on page _______.**

- **Read about the Breathalyzer. See Fig. _______, page ______.**

3. OXIDATION OF 2⁰ ALCOHOLS

- **A ketone is produced. The general reaction is:**

$$R_2CHOH \xrightarrow{[O]}$$

Ex:

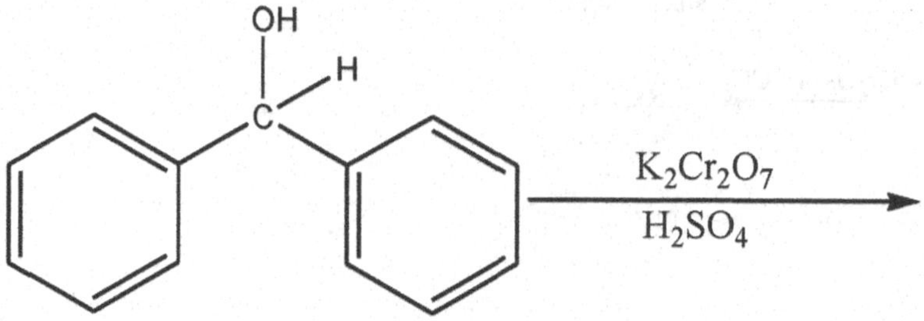

OH

H

C

$K_2Cr_2O_7$

H_2SO_4

4. OXIDATION OF 3⁰ ALCOHOLS

- **No reaction occurs.**

$$R_3COH \xrightarrow{[O]} \text{No Reaction}$$

Ex:

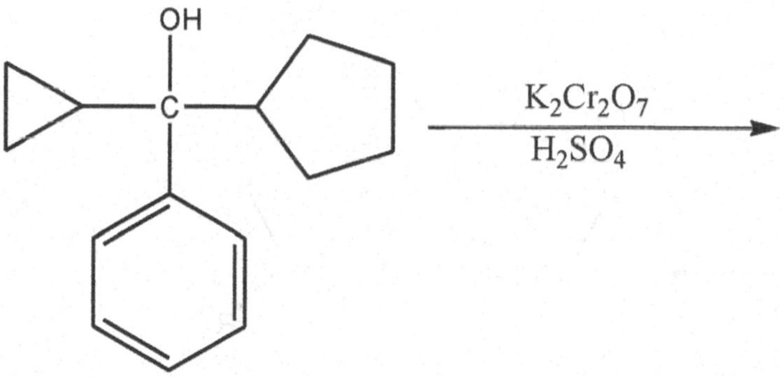

OH

C

$K_2Cr_2O_7$

H_2SO_4

- **Read about the oxidation of ethanol, alcoholism and Antabuse, methanol poisoning and treatment, on pages _____ - _______.**

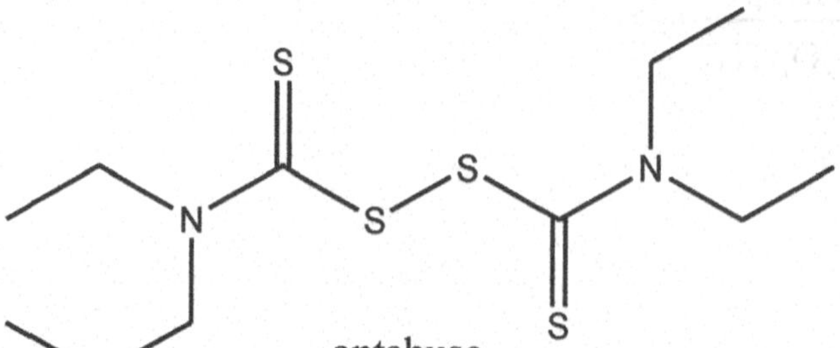

S

S

N

S

N

S

S

antabuse

5. OXIDATION THE ALCOHOLS: A SUMMARY

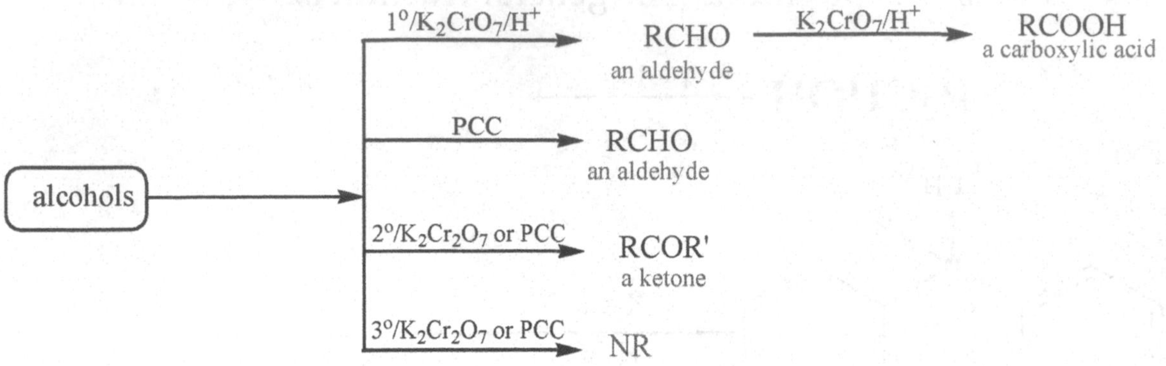

alcohols →
- $1°/K_2CrO_7/H^+$ → RCHO (an aldehyde) → K_2CrO_7/H^+ → RCOOH (a carboxylic acid)
- PCC → RCHO (an aldehyde)
- $2°/K_2Cr_2O_7$ or PCC → RCOR' (a ketone)
- $3°/K_2Cr_2O_7$ or PCC → NR

N. OXIDATION OF 1,2 - DIOLS

1. INTRODUCTION

- 1,2 - diols can be cleaved **using aqueous periodic acid in THF. Two carbonyl compounds are produced. The general reaction is:**

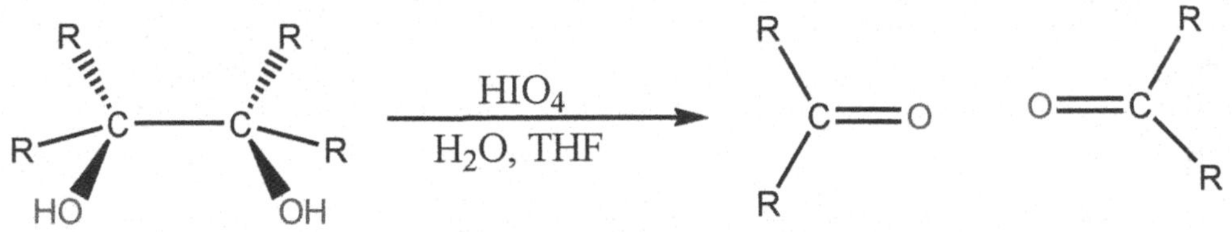

R R
R—C—C—R
HO OH
→ HIO_4 / H_2O, THF →
R₂C=O O=CR₂

Ex:

→ HIO_4 / H_2O, THF →

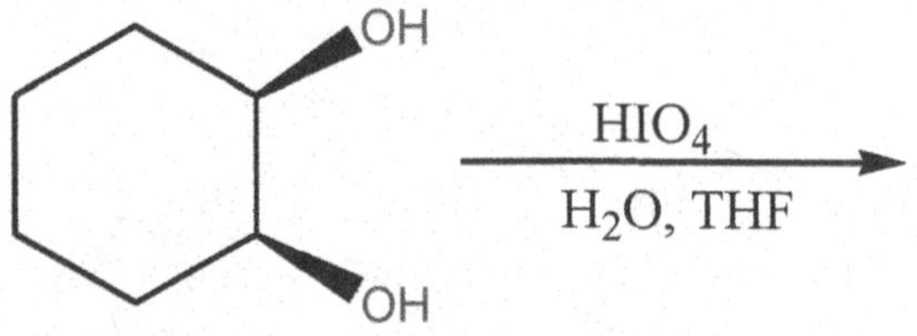

A 1,2 -diol or glycol

- **Mechanism of the reaction:**
- **The reaction proceeds through a cyclic periodate intermediate.**

HIO$_4$ → [cyclic periodate]‡ → carbonyl products

OH OH

OH

A cyclic periodate

Ex:

H$_3$C, CH$_3$, CH—CH, HO, OH

$\xrightarrow[H_2O/THF]{HIO_4}$

2. INDIRECT CLEAVAGE: COUPLING OsO_4 WITH HIO_4

- **The general reaction is:**

R, R, C=C, R, R — **An alkene** $\xrightarrow[NaHSO_3]{OsO_4}$ R, R, C—C, R, R, HO, OH $\xrightarrow[H_2O,\ THF]{HIO_4}$ R, R, C=O O=C, R, R

Ex:

$\xrightarrow[NaHSO_3]{OsO_4}$ $\xrightarrow[H_2O,\ THF]{HIO_4}$

O. SHARPLESS EPOXIDATION

- Named after **K. Barry Sharpless of Scripps Research Institute, Nobel Prize in Chemistry in 2001.**

1. THE PROBLEM

- **Can only one enantiomer be mostly made instead of a racemic mixture in the epoxidation of alkenes?** See page _______.

2. SOME DEFINITIONS

- **Enantioselective reaction = a reaction that gives one predominant enantiomer.**

- **Asymmetric reaction = reaction that converts an achiral reactant to one predominant enantiomer.**

4. CONVERTING ALLYLIC ALCOHOLS (special alkenes) TO EPOXIDES: AN OXIDATION REACTION

- **See page ______.**

- **Read pages _______ - _______.**

- **Sharpless reagent = 3 components: $(CH_3)_3COOH$ + Ti(IV) catalyst + (+ or -) Diethyl Tartrate (+DET or –DET).**

- **See page ________ for the structures of DET.**

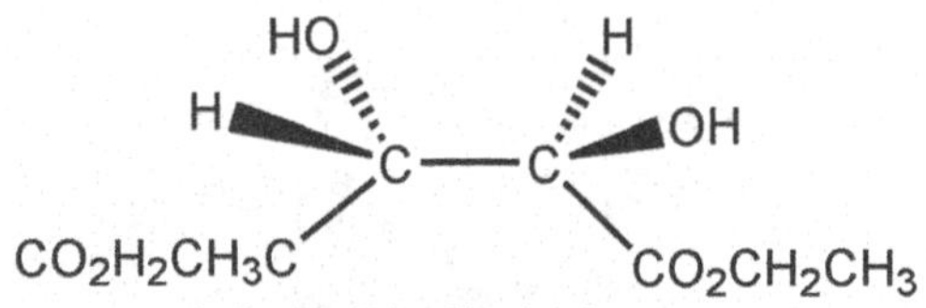

(+)-(R,R)-diethyl tartrate

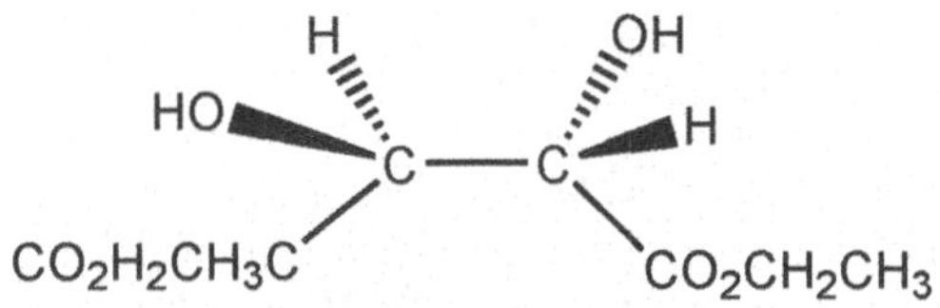

(-)-(S,S)-diethyl tartrate
(-)-DET

- **With (-)-DET, the oxygen atom is added from above the plane. The top epoxide is the major product.**

- **With (+)-DET, the oxygen atom is added from below the plane. The bottom epoxide is the major product.**
- **The general reaction is:**

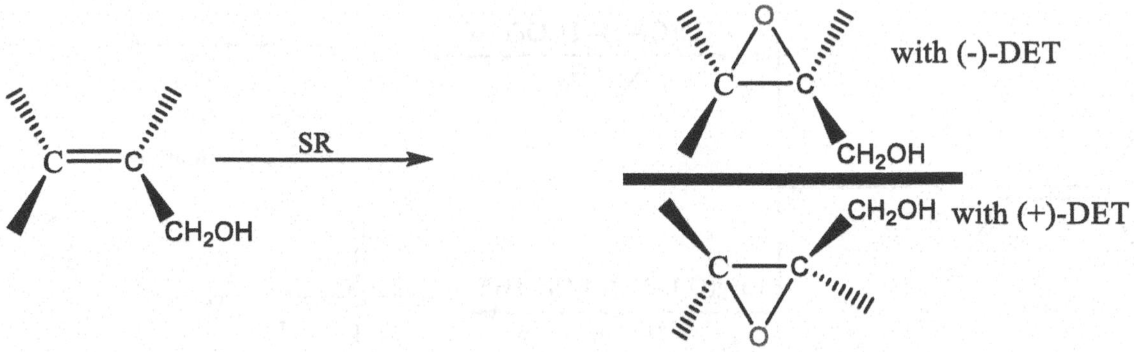

SR

with (-)-DET

CH_2OH

with (+)-DET

Ex:

with (-)-DET

with (+)-DET

allylic alcohol

enantiomers

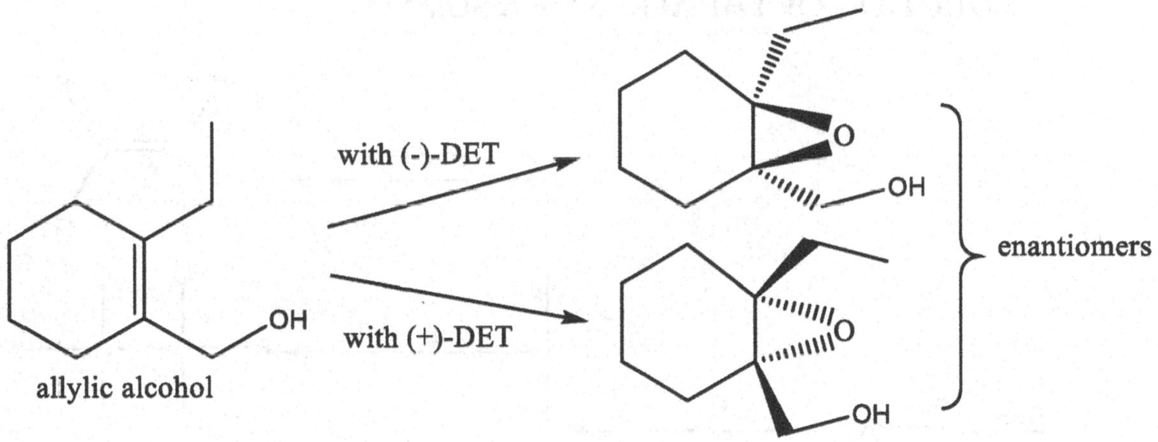

- See examples on pages _____ - _____.

O. SOME SELECTED REACTIONS

1. OXYMERCURATION-REDUCTION OF THE ALKENES: A SUMMARY

An alkene ($R_2C{=}CH_2$)

1. $Hg(OAc)_2$, H_2O/THF
2. $NaBH_4$

→ R–C(R)(OH)–CH$_3$: an alcohol (Markovnikov)

1. $Hg(O_2CCF_3)_2$, ROH/THF
2. $NaBH_4$

→ R–C(R)(OR)–CH$_3$: an ether (Markovnikov)

2. REDUCTION OF THE ALKYNES: A SUMMARY

R–C≡C–R' + H_2 →

- 1mole/Pt → RHC=CHR' (R and R' cis)
- 2 moles/Pt → R–CH$_2$–CH$_2$–R'
- Lindlar → RHC=CHR' (R and R' cis)
- Li (or Na)/NH_3 → RHC=CHR' (R and R' trans)

3. REDUCTION OF ALDEHYDES, KETONES, AND ACID CHLORIDES: A SUMMARY

H_2 →

- $R-C(=O)-H$ (an aldehyde) —Pd or Raney Ni→ $R-CH(OH)-H$ (a 1° alcohol)
- $R-C(=O)-R'$ (a ketone) —Pd or Raney Ni→ $R-C(OH)(H)-R'$ (a 2° alcohol)
- $R-C(=O)-Cl$ (an acid chloride) —PD Pd→ $R-C(=O)-H$ (an aldehyde)

4. OXIDATION OF THE ALKYNES: ASUMMARY

$R-C\equiv C-R'$ →

- $KMnO_4$, H_3O^+ → $R-C(=O)OH$ + $HO-C(=O)-R'$
- 1.O_3 2.H_2O → $R-C(=O)OH$ + $HO-C(=O)-R'$
- $KMnO_4$, H_2O, neutral → $R-C(=O)-C(=O)-R'$ (an α diketone)

5. OXIDATION OF THE ALKENES: A SUMMARY

An alkene ($R_2C{=}CR_2$)

- $R'COOOH$ → an epoxide + $R'COOH$
- 1. RCO_3H; 2. H_2O (OH^- or H^+) → A trans 1,2 -diol and its enantiomer
- $KMnO_4$, H_2O, OH^- → A cis-diol
- 1. OsO_4; 2. $NaHSO_3$, H_2O → A cis-diol
- 1. OsO_4/NMO; 2. $NaHSO_3$ → A cis-diol
- 1. O_3; 2. Zn/H_3O^+ → $R_2C{=}O$ + $O{=}CR_2$
- 1. O_3; 2. CH_3SCH_3 → $R_2C{=}O$ + $O{=}CR_2$
- $KMnO_4$, H_3O^+ → $R_2C{=}O$ + $O{=}CR_2$

6. OXIDATION OF THE ALCOHOLS: A SUMMARY

alcohols →
- $1°/K_2CrO_7/H^+$ → RCHO (an aldehyde) → K_2CrO_7/H^+ → RCOOH (a carboxylic acid)
- PCC → RCHO (an aldehyde)
- $2°/K_2Cr_2O_7$ or PCC → RCOR' (a ketone)
- $3°/K_2Cr_2O_7$ or PCC → NR

7. INDIRECT CLEAVAGE: COUPLING OsO_4 WITH HIO_4

- **The general reaction is:**

R, R, R, R — C=C (An alkene) → OsO_4 / $NaHSO_3$ → R, R, R, R, HO, OH — C—C → HIO_4 / H_2O, THF → R, R C=O + O=C R, R

8. SHARPLESS EPOXIDATION OF ALLYL ALCOHOLS

- **The General Reaction:**

C=C, CH_2OH → SR → C—C, O, CH_2OH with (-)-DET / CH_2OH with (+)-DET, C—C, O

- See Key Concepts on pages _______ - _______.

- **Oils and Fats**

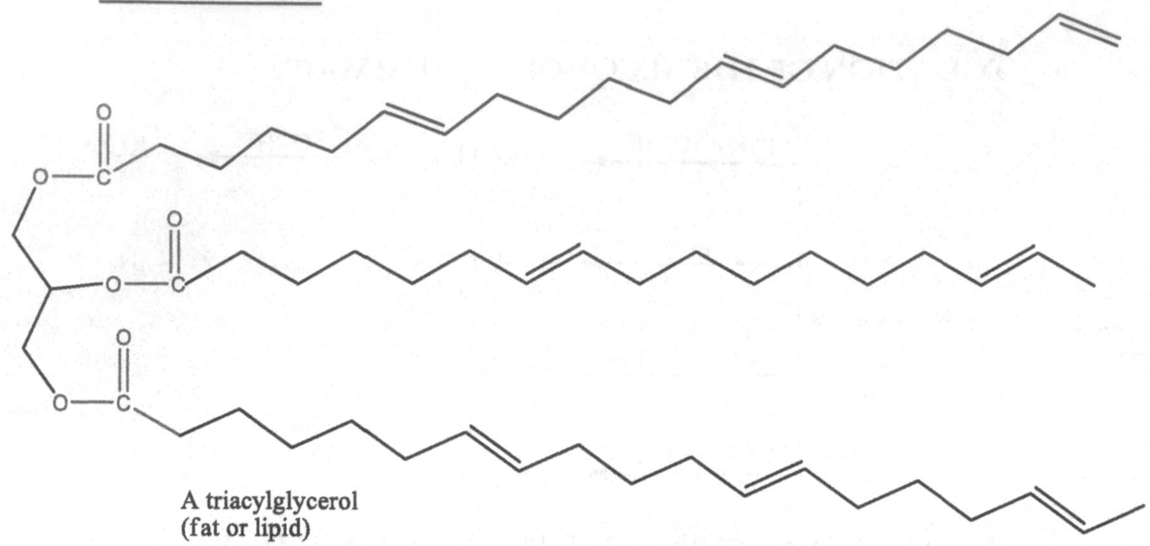

O—C(=O) ... O—C(=O) ... O—C(=O)

A triacylglycerol
(fat or lipid)

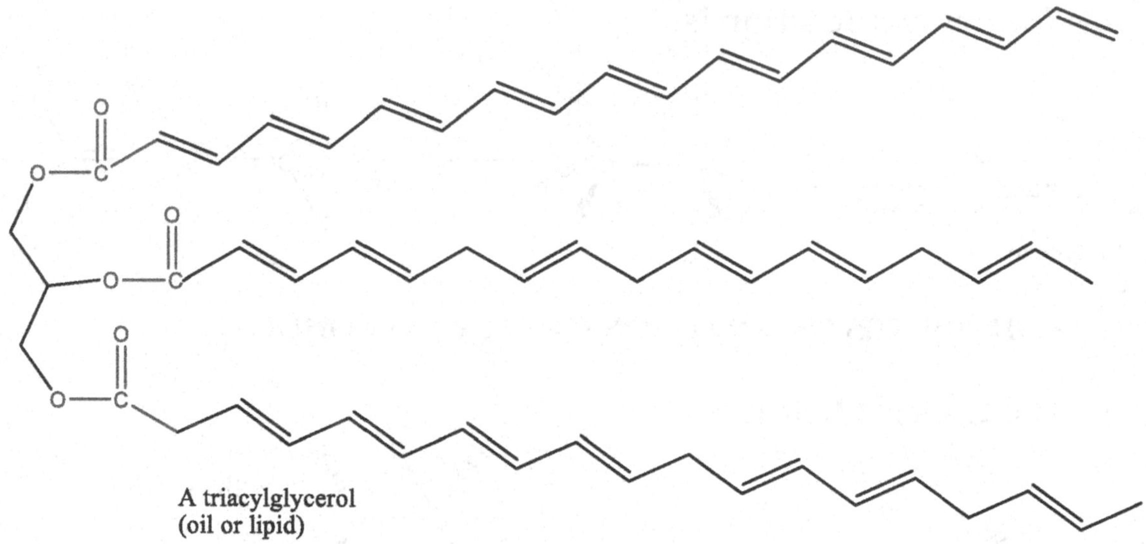

O—C(=O) ... O—C(=O) ... O—C(=O)

A triacylglycerol
(oil or lipid)

OCHEM II UNIT 2: RADICAL REACTIONS

A. INTRODUCTION

1. DEFINITION

a. Definition

- A radical is a very reactive intermediate in which a single electron "resides" **in a p orbital on an sp^2 C.**

- A radical results from a **homolytic (symmetrical) bond breaking.**

b. Structure

- sp^2 C
- **Trigonal planar**

2. CLASSIFICATION OF RADICALS: 4 KINDS

```
     H            H             H              R
     |            |             |              |
H— C•        R— C•        R— C•         R— C•
     |            |             |              |
     H            H             R              R

  methyl         1°            2°             3°
```

Ex:

$$\underset{1^o}{CH_3-\overset{H}{\underset{H}{C}}\bullet} \qquad \underset{2^o}{CH_3-\overset{H}{\underset{CH_3}{C}}\bullet} \qquad \underset{3^o}{CH_3-\overset{CH_3}{\underset{CH_3}{C}}\bullet}$$

3. STABILITY OF RADICALS

- **Recall "Stability Rule": The less stable the reactant, the more stable the product.**

- The stability of a radical depends on the strength of the C-H bond that is broken during radical formation. From bond dissociation energy data (See Table ______, page ______),

Bond	Methyl C-H	1° C-H	2° C-H	3°C-H
BDE(kJ/mol)	435	410	397	381

- The **decreasing order of C-H strength** is:

    - (methyl) C-H > (1°) C-H > (2°) C-H > (3°) C-H

- Since the (3°) C-H is the **weakest** bond, the (3°) C-H is readily broken; therefore its radical forms faster and is **more stable.**

- **Overall, the increasing order of alkyl group radical stability is:**

    - **methyl < 1° < 2°< 3°**

- Read pages ______ - ______.

- Do Problems _______ and ______, page ______.

- **Note: The more R groups attached to the sp² C, the more stable the radical.**

- **See Fig. ________, page ________. Read page ______.**

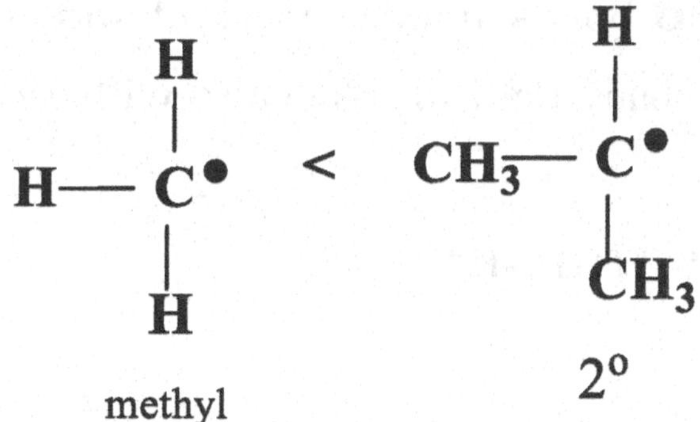

Ex:

- Do problems on page ________.

3. STABILITY OF THE ALLYLIC RADICAL

- The allylic radical is very stable because it is **resonance stabilized.** We have **electron delocalization:**

$$CH_2{=}CH\text{-}\dot{C}H_2 \longleftrightarrow \dot{C}H_2CH{=}CH_2$$

- **The overall increasing order of radical stability is:**

    - **Methyl < 1° < 2° < 3° < allylic**

B. INTRODUCTION TO RADICAL REACTIONS

1. INTRODUCTION

- Radicals are very reactive since they do not have an octet. In general, radicals react with σ bonds in molecules and undergo **addition reactions with π** bonds. Radicals react also with each other.

2. REACTIONS OF RADICALS WITH C-H

- **The general reaction is:**

$$R^{\bullet} + -\overset{|}{\underset{|}{C}}-H \longrightarrow R-H + -\overset{|}{\underset{|}{C}}\bullet$$

Ex:

$$CH_4 + Cl^{\bullet} \longrightarrow \dot{C}H_3 + H\text{-}Cl$$

3. REACTIONS OF RADICALS WITH C=C

$$R^{\bullet} + \;>C=C< \;\longrightarrow\; >C(R)-\dot{C}<$$

Ex:

H3C, H, C=C, H, H + •Br → H3C, Br, C•–C–H, H, H

4. RADICAL-RADICAL REACTIONS

$$R^{\bullet} + R^{\bullet} \longrightarrow R-R$$

Ex:

$$^{\bullet}CH_3 + {}^{\bullet}CH_3 \longrightarrow CH_3-CH_3$$

- Do Problem ______, page _____.
- Read pages _____ - _____.

C. RADICAL INITIATORS-RADICAL INHIBITORS

1. RADICAL INITIATORS: PEROXIDES = RO-OR

$$ROOR \xrightarrow[\text{or heat}]{\text{light}} 2RO^{\bullet}$$

Ex:

$$C_5H_9-CH_2OOCH_2-C_5H_9 \xrightarrow{\text{light}}$$

2. RADICAL INHIBITORS OR RADICAL SCAVENGERS: O_2, VITAMIN E

a. Definition

- Radical **inhibitors** or **scavengers** are compounds that **prevent** a radical to react with other compounds by reacting with it. In other words, they prevent **oxidation** from occurring **(antioxidants)**.

Ex; O_2, Vitamin E, BHT, and related compounds.

- Read pages _____ - _____.

b. General Reaction

Ex:

$$\cdot\ddot{\underset{\cdot\cdot}{O}}-\ddot{\underset{\cdot\cdot}{O}}\cdot + R^{\bullet} \xrightarrow[\text{or heat}]{\text{light}} \cdot\ddot{\underset{\cdot\cdot}{O}}-\ddot{\underset{\cdot\cdot}{O}}-R$$

A diradical

$$\cdot\ddot{\underset{\cdot\cdot}{O}}-\ddot{\underset{\cdot\cdot}{O}}\cdot + \dot{C}H_3 \xrightarrow[\text{or heat}]{\text{light}} \cdot\ddot{\underset{\cdot\cdot}{O}}-\ddot{\underset{\cdot\cdot}{O}}-CH_3$$

A diradical

- **Note: A scavenger can be used to prove whether or not a reaction proceeds via a radical intermediate.**

D. RADICAL HALOGENATION OF ALKANES

1. INTRODUCTION

- Alkanes undergo radical reactions with X_2 (X = Cl, Br) to give **alkyl halides.**

2. GENERAL REACTION = SUBSTITUTION REACTION

$$R-H + X_2 \xrightarrow[\text{or heat}]{\text{light}} R\text{-}X + H\text{-}X$$

- Read pages ______+ - ______.

Ex:

$$CH_4 + Br_2 \xrightarrow[\text{or heat}]{\text{light}} CH_3\text{-}Br + H\text{-}Br$$

3. CHLORINATION OF ALKANES

a. Introduction

- **The reaction is faster, but less selective than bromination. Let's consider the reaction:**

Ex:

$$CH_4 + Cl_2 \xrightarrow[\text{or heat}]{\text{light}} CH_3\text{-}Cl + H\text{-}Cl$$

- This reaction occurs in **3 steps: initiation, propagation, and chain termination. The rate determining step occurs during propagation.**

i. Initiation: 1 Step

$$Cl_2 \xrightarrow[\text{or heat}]{\text{light}} 2Cl^{\bullet} \quad \text{(fast)}$$

ii. Propagation: 2 Steps

$$CH_4 + Cl^{\bullet} \xrightarrow[\text{or heat}]{\text{light}} {}^{\bullet}CH_3 + H\text{-}Cl \quad \text{(slow, exergonic)}$$

$${}^{\bullet}CH_3 + Cl_2 \xrightarrow[\text{or heat}]{\text{light}} CH_3Cl + Cl^{\bullet} \quad \text{(fast, exergonic)}$$

iii. Chain Termination: 3 Steps

$$\dot{C}H_3 + \dot{C}H_3 \longrightarrow CH_3{-}CH_3$$

(fast)

$$Cl^{\bullet} + {}^{\bullet}Cl \longrightarrow Cl_2$$

(fast)

$$\dot{C}H_3 + {}^{\bullet}Cl \longrightarrow CH_3{-}Cl$$

(fast)

- **Note: The reaction continues until it runs out of reactants.**

b. Thermodynamics of Chlorination of Alkanes

- There are 2 **exergonic** steps (**propagation**). The first step is **the rate-determining step**. **The overall process is exergonic. (Ea1>Ea2)**

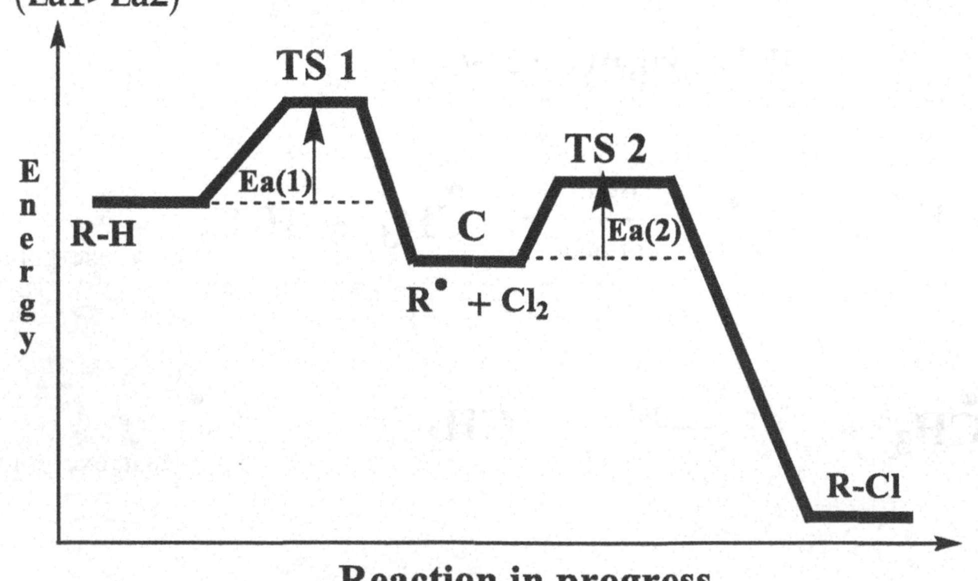

- **See Figs. _______ and ______, page _____: Energy diagram for the chlorination of ethane.**
- **Read page _____ and do all problems on this page.**

c. Products of Chlorination: 2 or More Types of Hydrogens Present on the Alkane

Ex

$$\mathbf{CH_3CH_2CH_3 + Cl_2 \longrightarrow \underset{A}{CH_3CH_2CH_2Cl} + \underset{B}{CH_3\underset{\underset{Cl}{|}}{C}HCH_3}}$$

A mixture of products

- In this alkane, we have **6** primary hydrogens and **2** secondary hydrogens. Therefore, one expects the ratio **A/B** to be **3** to **1**. The **observed ratio** is about 1 to 1, with a little bit more of **B**. This is due to the fact that a 2° radical forms faster than a 1° radical. **Recall that a 1° C-H bond is stronger than a 2° C-H bond. See the following reactions:**

2° Hs

1° Hs

$$CH_3CH_2CH_3 \longrightarrow CH_3CH_2CH_2\cdot \quad (1)$$

1° radical
less stable

$$\mathrm{H{-}\underset{H}{\overset{H}{C}}{-}\underset{H}{\overset{H}{C}}{-}\underset{H}{\overset{H}{C}}{-}H} \longrightarrow \mathrm{H{-}\underset{H}{\overset{H}{C}}{-}\underset{H}{\dot{C}}{-}\underset{H}{\overset{H}{C}}{-}H} \quad (2)$$

2° radical
more stable

- **Read pages ______ - ______.**
- **Do problems on pages ______ - ______.**

4. BROMINATION OF ALKANES

a. Introduction: The overall reaction

$$CH_4 + Br_2 \xrightarrow[\text{or heat}]{\text{light}} CH_3\text{-}Br + H\text{-}Br$$

- The reaction is **slower,** but **more selective** than chlorination. It also proceeds in **3 steps** (initiation, propagation, and chain termination) with the rate determining step occurring **during propagation.**

i. Initiation: 1 Step

$$Br_2 \xrightarrow[\text{or heat}]{\text{light}} 2Br^{\bullet} \quad \text{(fast)}$$

ii. Propagation: 2 Steps

$$CH_4 + Br^{\bullet} \longrightarrow {}^{\bullet}CH_3 + H\text{-}Br \quad \text{(slow,endergonic)}$$

$${}^{\bullet}CH_3 + Br_2 \longrightarrow CH_3Br + Br^{\bullet} \quad \text{(fast,,exergonic)}$$

iii. Chain Termination: 3 Steps

$$\dot{C}H_3 + \dot{C}H_3 \longrightarrow CH_3-CH_3 \text{ (fast)}$$

$$Br^{\bullet} + Br^{\bullet} \longrightarrow Br_2 \text{ (fast)}$$

$$\dot{C}H_3 + Br^{\bullet} \longrightarrow CH_3\text{-}Br \text{ (fast)}$$

- The rate determining step is endergonic. The overall reaction is exergonic.

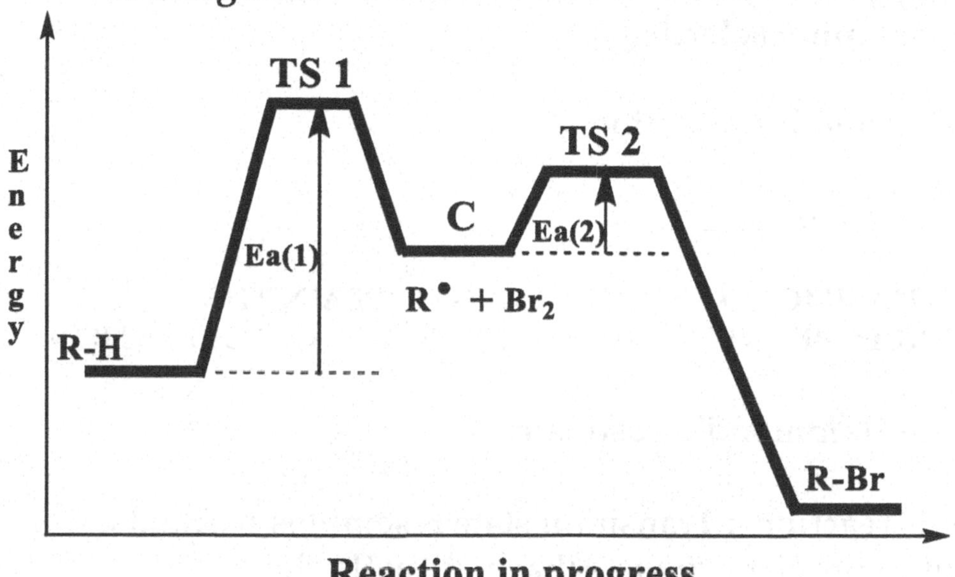

b. Products of Bromination: 2 or More Types of Hydrogens Present on the Alkane

Ex:

$$CH_3CH_2CH_3 + Br_2 \longrightarrow \underset{\substack{A \\ 1\%}}{CH_3CH_2CH_2Br} + \underset{\substack{| \\ Br \\ B \\ 99\%}}{CH_3CHCH_3}$$

- Unlike chlorination (mixture of products), in bromination, the major product comes from the most **stable radical**, the **2° radical.**

5. BROMINATION VS. CHLORINATION

- Bromination is more **selective**: get mostly one product
- Chlorination is less selective: get a mixture of products

Read pages ______ - ______.

- **Selectivity principle: In OCHEM, reagents that are less reactive are more selective.**

Ex: Chlorination and bromination.

Do problems on page ________.

6. USING HAMMOND'S POSTULATE TO EXPLAIN THE DIFFERENCES BETWEEN BROMINATION AND CHLORINATION

- **Recall the Hammond's postulate:**
- **Exergonic reactions: Transition state resembles reactants. Difficult to predict products (have virtually the same activation energy); have a mixture of products.**

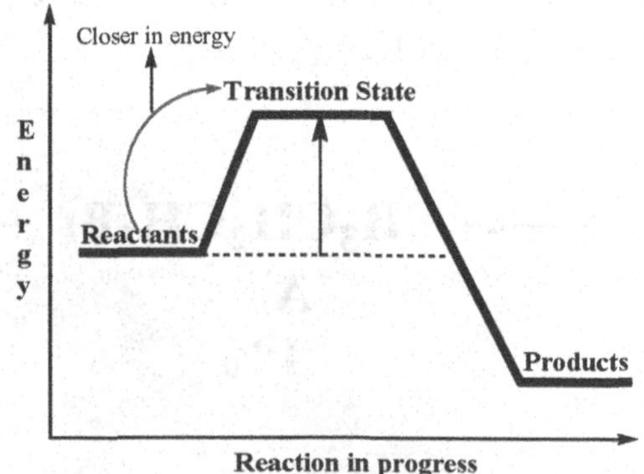

Ex: Chlorination of the alkanes obtained

- **Endergonic reactions: Transition states resembles products. Most stable product (lowest activation energy) forms.**

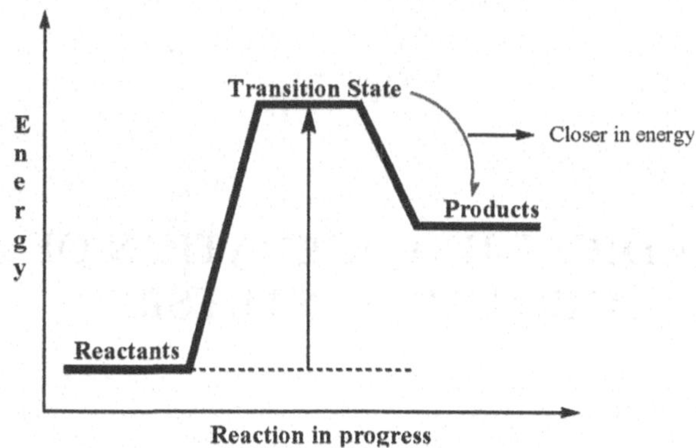

Ex: Bromination of the alkanes

- **Recall: In chlorination, the rate determining step is exergonic.**
- **In bromination, the rate determining step is endergonic.**

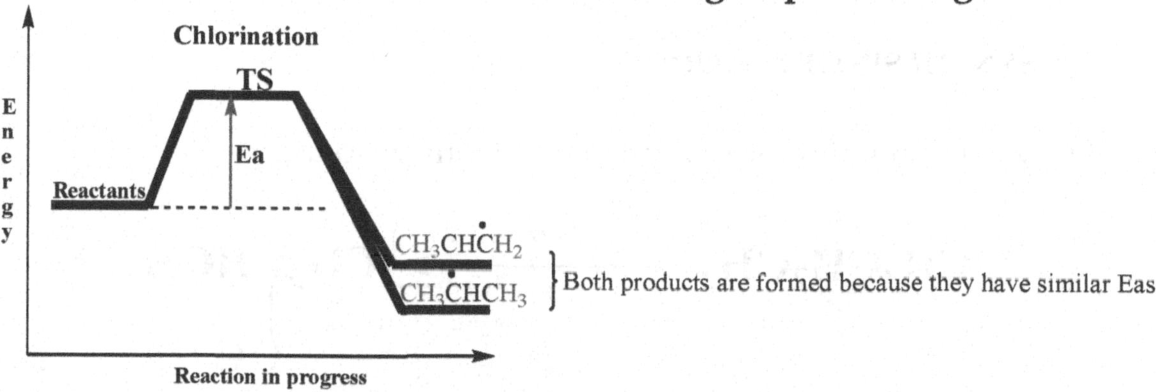

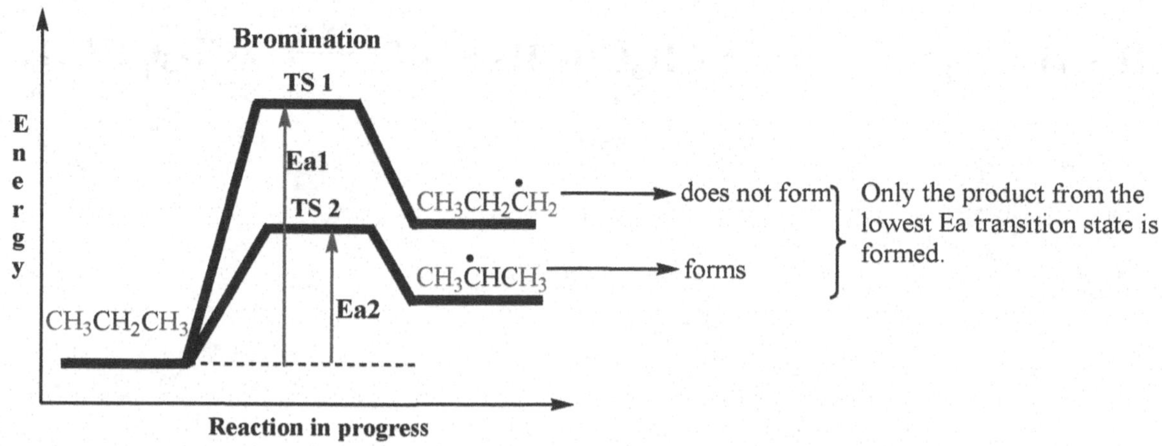

- **Conclusion: In chlorination, one cannot predict the outcome of the reaction; in bromination, the most stable product forms.**
- **Read pages _____ - _____. See Fig. _____ - ______ pages _____ - ______.**

E. APPLYING RADICAL HALOGENATION OF ALKANES TO ORGANIC SYNTHESIS

1. INTRODUCTION

- **Halogenation** can be used to make **alkyl halides** that can be used as **substrates** in substitution synthesis of diverse compounds (ethers, alcohols, etc.) or in elimination synthesis of **alkenes**.

2. SYNTHESIS OF ALCOHOLS

Ex: How do you synthesize 2-propanol from propane?

$$CH_3CH_2CH_3 \xrightarrow{?} CH_3\underset{\displaystyle OH}{\underset{|}{C}}HCH_3$$

- **There are 2 steps**. First, convert the alkane to alkyl bromide. Then, substitute Br with OH^- in an **SN2 reaction**.

$$CH_3CH_2CH_3 + Br_2 \longrightarrow CH_3\underset{\displaystyle Br}{\underset{|}{C}}HCH_3 + OH^- \xrightarrow{SN2} CH_3\underset{\displaystyle OH}{\underset{|}{C}}HCH_3$$

3. SYNTHESIS OF ALKENES

Ex: How would you synthesize propene from propane?

- **There are 2 steps**. First, convert the alkane to alkyl bromide or chloride. Second, eliminate HBr or HCl using a **strong base** in an **E2 reaction.**

$$CH_3CH_2CH_3 + Cl_2 \longrightarrow CH_3\underset{\underset{Cl}{|}}{C}\overset{\overset{H}{|}}{H}CH_2 \xrightarrow[(E2)]{C(CH_3)_3CO^-} CH_3CH{=}CH_2 + HCl$$

- **See examples on pages _____ - _____. Read page _____.**

- **Do problems on page _____.**

F. STEREOCHEMISTRY OF RADICAL HALOGENATION

1. INTRODUCTION

- There are 3 cases depending on the stereochemistry of the alkane reactant. We will examine:

  - **Achiral reactants.**
  - **Chiral reactants in which the reaction does not occur at the stereogenic center.**
  - **Chiral reactants in which the reaction occurs at the stereogenic center.**

2. ACHIRAL REACTANTS

- In general, when an achiral reactant is used, an achiral product and/or two enantiomers are formed.

Ex: The monochlorination of butane: **2 types of hydrogens**

$CH_3CH_2CH_2CH_3 + Cl_2 \longrightarrow CH_3CH_2CH_2CH_2Cl$ + (racemic 2-chlorobutane: H, CH_3, CH_3CH_2, Cl + CH_3, H, CH_2CH_3, Cl)

achiral — A racemate

3. REACTIONS OCCURING AT THE STEREOGENIC CENTER

- In this case, the reaction occurs with racemization.

Ex: The chlorination of (S)-2-bromobutane: Attack at the **stereogenic center (C2).**

H, CH_2CH_3, CH_3, Br + Cl_2 ⟶ Cl, CH_2CH_3, CH_3, Br + CH_2CH_3, Cl, CH_3, Br

Enantiomers (a racemate)

4. REACTIONS OCCURING AT A CARBON OTHER THAN THE STEREOGENIC CENTER

- **Diastereomers** are produced.

Ex: The monochlorination of (S)-2-bromobutane: Attack at the **nonstereogenic center (C3).**

Br, H, C*, CH_3, C, CH_3, H, H + Cl_2 ⟶ Br, H, C*, CH_3, C*, CH_3, H, Cl + Br, H, C*, CH_3, C*, Cl, CH_3, H

Diastereomers

- **Read pages ______ - ______. Do all problems on page ______.**

G. CFCS AND THE OZONE LAYER REVISITED

1. INTRODUCTION

- **Recall: Freons or CFCs are believed to be responsible for the depletion of the ozone layer. See Unit 9, OChem I.**

- **See radical reactions on pages _____ - _____.**

- **See Fig. ______, page _____.**

- **Read pages ______ - ______.**

- **Do Problem ______, page _____.**

2. CFCs and HFCs

- CFCs = chlorofluorocarbons = freons
- used as refrigerants and aerosol propellants

- **$CFCl_3$ = Freon 11**
- **CF_2Cl_2 = Freon 12**
- **CF_3Cl = Freon 13**

- **How Do CFCs Deplete the Ozone Layer?**

- CFCs are believed to be the culprits in the depletion of Earth atmosphere ozone layer. Please see the following diagram:

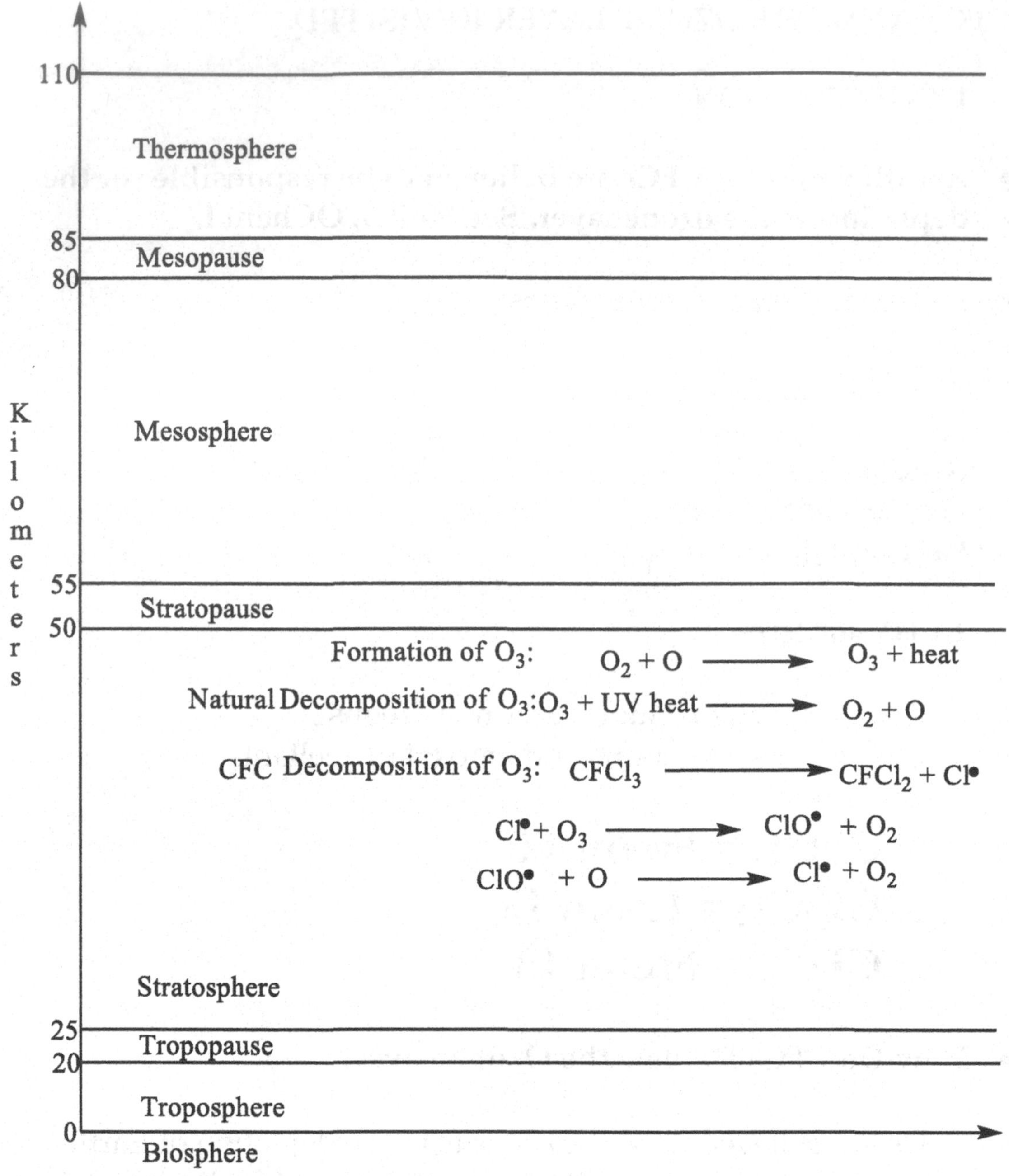

- **Alternatives to CFCs: HFCs**
  - **HFCs = hydrofluorocarbons** = substitutes of CFCs
  - used as refrigerants and aerosol propellants
  - $\mathbf{CFH_3}$
  - $\mathbf{CF_2H_2}$
  - $\mathbf{CF_3H}$

H. RADICAL REACTIONS AT ALLYLIC CARBONS

1. INTRODUCTION

- The allylic radical is very stable because of resonance. Indeed, it is a hybrid of **2 resonance structures:**

$$CH_2{=}CH{-}\dot{C}H_2 \longleftrightarrow \dot{C}H_2CH{=}CH_2$$

2. THE GENERAL REACTION OF ALLYLIC BROMINATION

$$>C{=}C{-}C(H)< \xrightarrow[h\nu \text{ or ROOR}]{\text{NBS}} >C{=}C{-}C(Br)<$$

3. MECHANISM OF THE REACTION

a. Introduction

- The reaction proceeds in **4 steps. A bromine radical** intermediate from **NBS** is formed.

b. Mechanism of the Reaction

i. Formation of a Bromine Radical from NBS

$$\text{N-Br (N-bromosuccinimide)} \xrightarrow{h\nu} \text{N}^\bullet + Br^\bullet$$

N-bromosuccinimide

ii. Propagation

$$\gt C{=}C{-}C(H){-} + Br^{\bullet} \longrightarrow \gt C{=}C{-}\dot{C}{-} + HBr$$

iii. The HBr formed reacts with NBS to generate Br_2 and succinimide as follows:

$$\text{N-bromosuccinimide} + HBr \longrightarrow \text{succinimide (NH)} + Br_2$$

N-bromosuccinimide

succinimide

iv. Then, the allylic radical reacts with the Br_2 generated above:

$$\gt C{=}C{-}\dot{C}{-} + Br{-}Br \longrightarrow \gt C{=}C{-}C(Br){-} + Br^{\bullet}$$

- **Note: The source for both Br_2 (very low concentration) and $Br^{\cdot}$ is NBS.**
- **Read pages ______ - _______.**
- **See example on page _______.**
- **Do problems on pages _______ – ______.**

Ex:

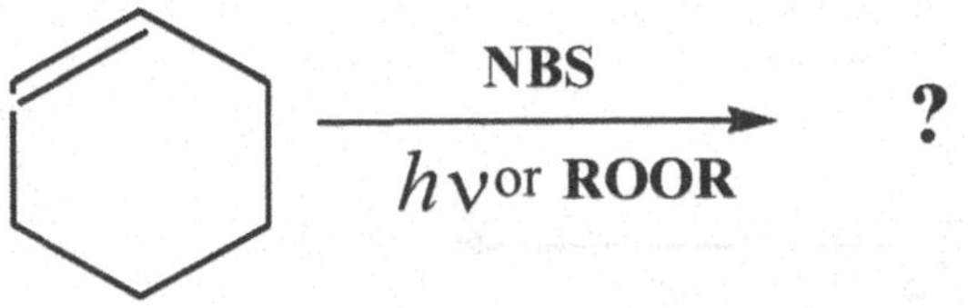

NBS

$h\nu$ or ROOR

?

- **Mechanism:**

4. SUMMARY ON ALLYLIC BROMINATION

- **The outcome of an allylic bromination depends on the catalyst:**
  - **If the catalyst is Br_2 in CCl_4, an addition reaction occurs thru an ionic intermediate.**
- **The general reaction is:**

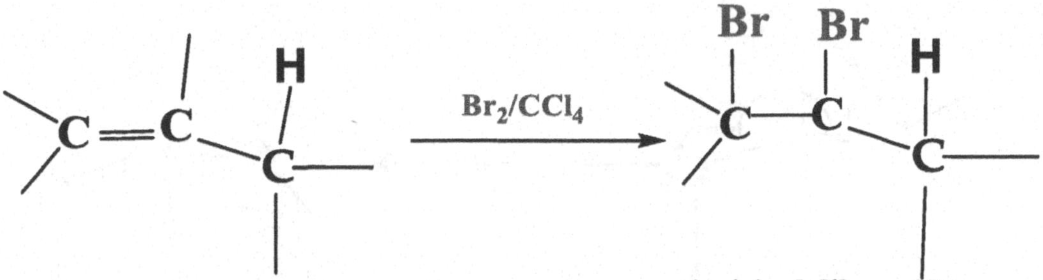

Br_2/CCl_4

A vicinal dibromide

Ex:

$$CH_2{=}CHCH_2CH_3 \xrightarrow[CCl_4]{Br_2} ?$$

- If the catalyst is NBS $h\nu$ or ROOR, then a substitution product results thru a radical intermediate.

$$\gt C{=}C(-)-C(H)(-)- \xrightarrow[h\nu \text{ or ROOR}]{NBS} \gt C{=}C(-)-C(-)(-)-Br$$

Ex:

$$CH_2{=}CHCH_2CH_3 \xrightarrow[h\nu \text{ or ROOR}]{NBS} ?$$

5. PRODUCT ANALYSIS IN ALLYLIC BROMINATION

a. Observation

- Allylic radical bromination often leads to a mixture of 2 products. Why? Because 2 allylic radicals are formed.

$$\gt C{=}C(-)-\dot{C}(-)- \longleftrightarrow \gt \dot{C}-C(-){=}C(-)-$$

Ex:

$$CH_2{=}CHCH_2CH_3 \xrightarrow[h\nu \text{ or ROOR}]{\text{NBS}} ?$$

- **Try this reaction. How many products are possible?**

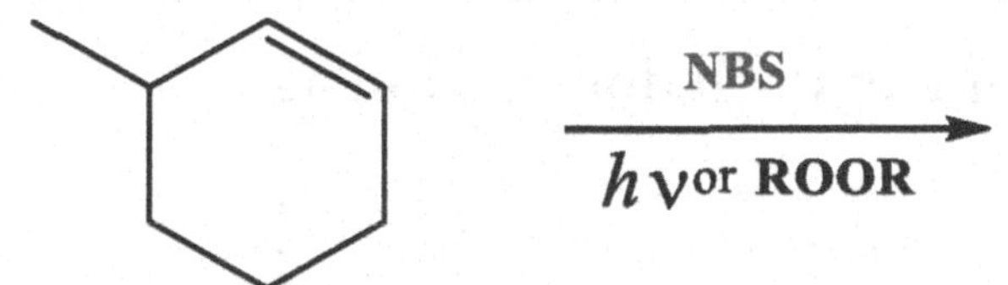

- See mechanisms on page ______.
- Read page ______.

6. OVERALL SUMMARY ON ALLYLIC BROMINATION

allylic carbon

Br_2 / CCl_4

Br

Br

Br

NBS/hν or ROOR

allylic carbon

- Read pages _____ - _____ and do associated problems.
- Read about the oxidation of unsaturated lipids and rancidity on pages ______ - ______.
- Read about antioxidants = radical inhibitors.
  - Vitamin E = body
  - BHT = Butylated Hydroxy Toluene = food preservation

I. RADICAL REACTIONS WITH ALKENES

1. RADICAL HYDROBROMINATION OF ALKENES

a. Reminder

- No light or peroxide, have ionic hydrohalogenation as follows:

$$\text{HC}=\text{C(R)} + \text{HBr} \xrightarrow[\text{no peroxide}]{\text{no light}} \text{H–C(H)–C(Br)–R}$$

- **Recall: The reaction is regioselective since it follows Markovnikov's rule. Indeed the Br goes to the most substituted carbon of the double bond. The H is added first and the Br adds last.**

Ex:

$$H_2C=CH(CH_3) + HBr \xrightarrow[\text{no peroxide}]{\text{no light}} H{-}CH_2{-}CH(Br){-}CH_3$$

- **Mechanism revisited:**

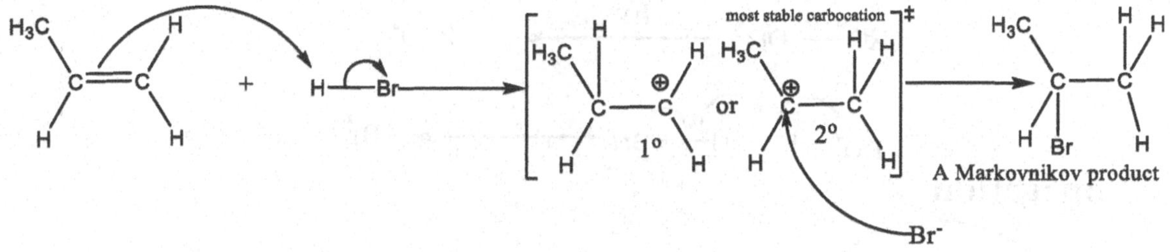

$$H_3C(H)C{=}CH_2 + H{-}Br \longrightarrow \left[H_3C{-}CH_2{-}\overset{\oplus}{C}H_2 \ (1^\circ) \text{ or } H_3C{-}\overset{\oplus}{C}H{-}CH_3 \ (2^\circ) \right]^{\ddagger} \longrightarrow H_3C{-}CH(Br){-}CH_3$$

most stable carbocation

Br^-

A Markovnikov product

b. Presence of Light and Peroxide

- **In the presence of light and peroxide, have radical hydrohalogenation with radical intermediates.**

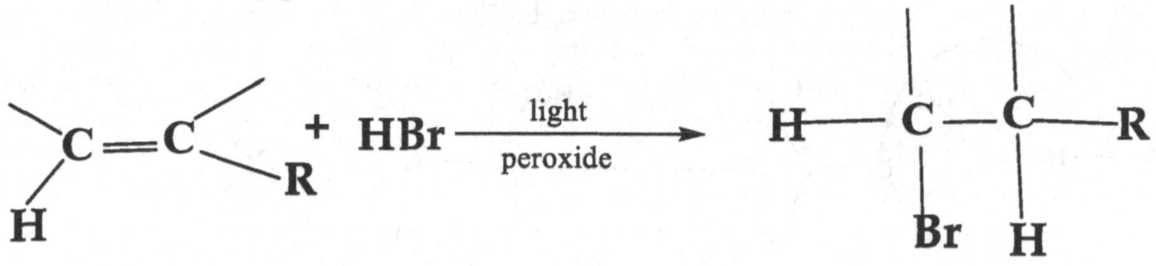

$$\mathbf{>C{=}C(R)- + HBr \xrightarrow[\text{peroxide}]{\text{light}} H{-}C(Br){-}C(H){-}R}$$

- **The reaction is nonMarkovnikov. The Br goes to the <u>least</u> substituted C of the double bond. The Br adds first and the H last.**

- **Note: HCl and HI do not add to alkenes under radical conditions.**

Ex:

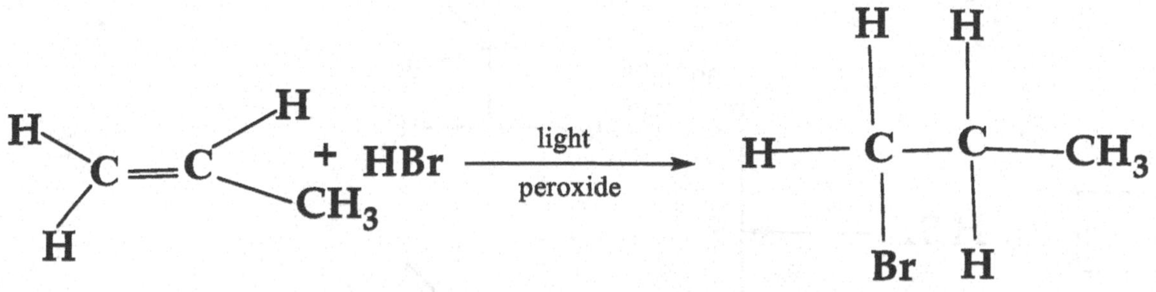

$$\mathbf{H_2C{=}CH{-}CH_3 + HBr \xrightarrow[\text{peroxide}]{\text{light}} H{-}CH(Br){-}CH_2{-}CH_3}$$

- **Mechanism: See page _____.**

1. Initiation

$$RO\!-\!OR \xrightarrow{h\nu} 2RO^{\bullet}$$

$$RO^{\bullet} + H\!-\!Br \longrightarrow Br^{\bullet} + ROH$$

2. Propagation

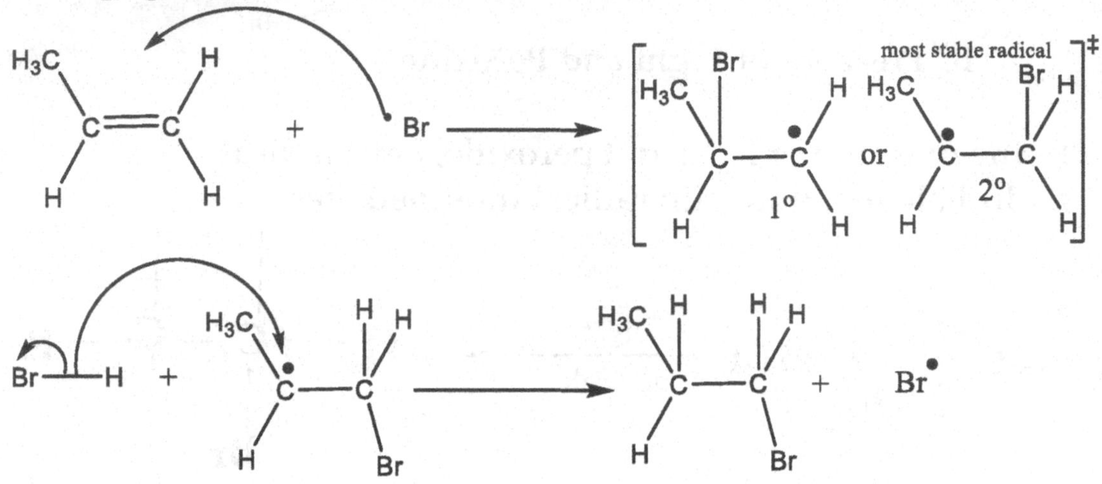

A nonMarkovnikov product

3. Chain termination

$$Br^{\bullet} + Br^{\bullet} \longrightarrow Br\!-\!Br$$

2. HYDROHALOGENATION ALKENES: A SUMMARY

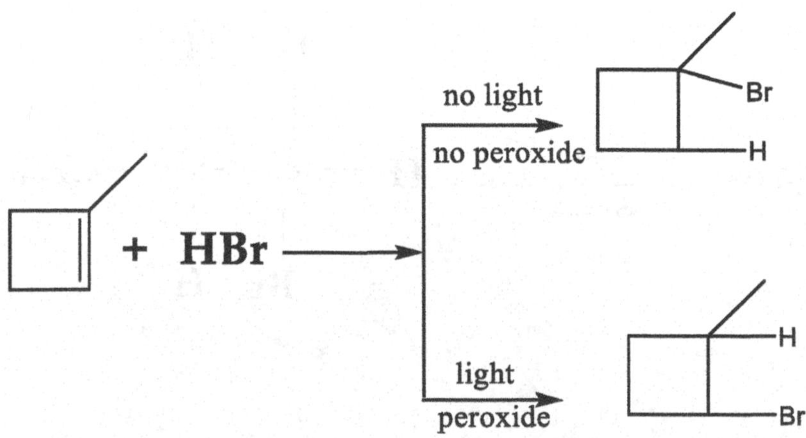

3. THERMOCHEMISTRY OF RADICAL BROMINATION

- From the BDE, it can be shown that the radical addition of HBr to the alkenes is **exothermic.** However, in the cases of HI and HCl, the processes are **endothermic**. This is why HCl and HI do not react.

J. POLYMERIZATION

1. DEFINITION

- **A polymer** is a huge molecule made of a repeating unit called a **monomer.**

Ex: Teflon, PVC, Dacron, Polyethylene, kevlar

2. POLYMERIZATION METHODS: 2

a. Introduction

- There are **2 techniques: radical and ionic.**

b. Radical Polymerization: Requires Peroxide

- **General Reaction:**

RO• + C=C + C=C + C=C + ------ ⟶ RO—C—C—C—C—C—C—

or

n C=C $\xrightarrow{\text{peroxide}}$ [—C—C—]$_n$

c. Ionic (cationic) Polymerization: No Peroxide Required

- **General Reaction:**

$$H^+ + C{=}C + C{=}C + C{=}C + \cdots \longrightarrow H{-}C{-}C{-}C{-}C{-}C{-}C{-}$$

or

$$n\ C{=}C \xrightarrow{H^+} \left[-C-C- \right]_n$$

K. MOLECULAR IONIZATION AND FRAGMENTATION

1. MOLECULAR IONIZATION

- When a molecular substance in the **gaseous state** is bombarded with a beam of high speed electrons, its molecules **ionize** to give **radical ions** called **molecular ions**. The **general process** is:

$$M + e^- \longrightarrow M^{+\bullet} + 2e^-$$

$$CH_4 + e^- \longrightarrow [CH_4]^{+\bullet} \text{ or } [H{-}CH_2{-}H]^{+\bullet} \ (M^{+\bullet}) + 2e^-$$

$$H_3C{-}CH_3 + e^- \longrightarrow [H_3C{-}CH_3]^{+\bullet} \ (M^{+\bullet}) + 2e^-$$

- **Note: The unpaired electron on the free radical is delocalized.**

2. MOLECULAR FRAGMENTATION

- An **ionized** molecule (molecular ion) can fragment to give **neutral** radicals and **positive** ions.

Ex:

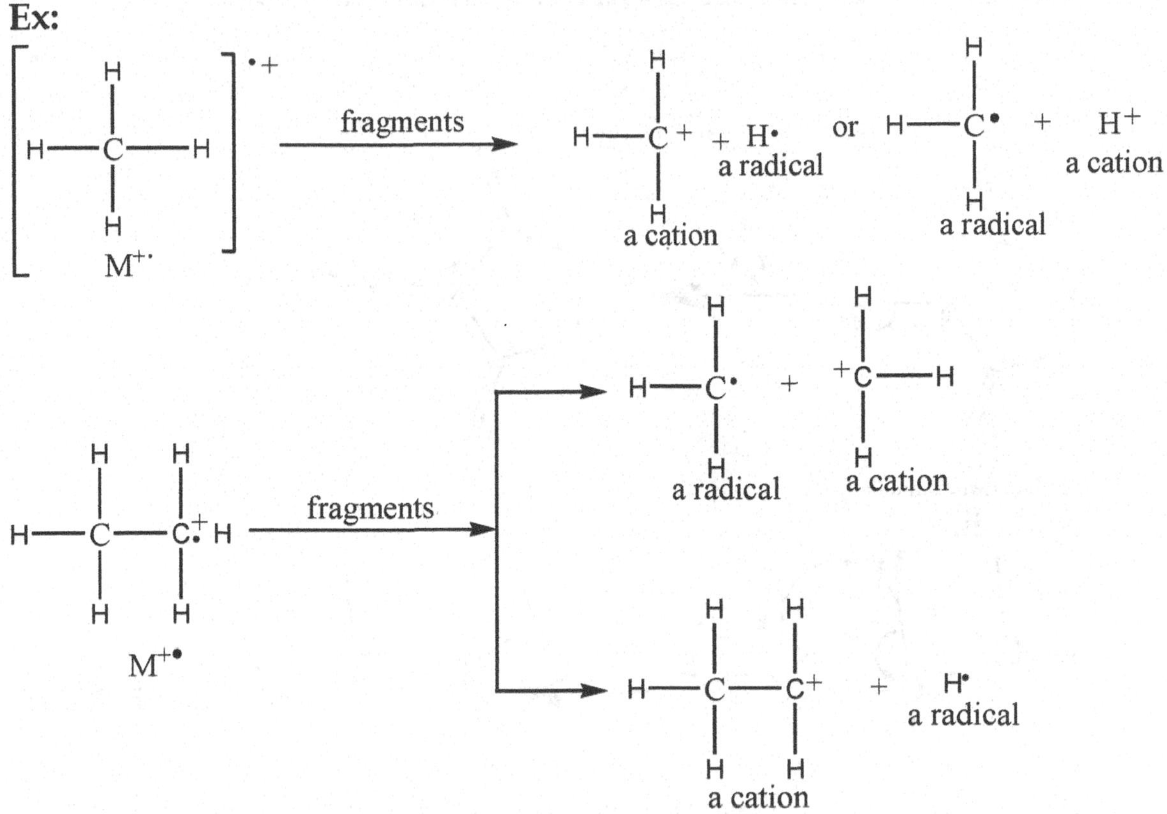

3. M/Z VALUES OF POSITIVE FRAGMENTS

- The **mass to charge ratios (m/z)** of molecular ions and positive fragments obtained from the fragmentation of ionized molecules (molecular ion) can be easily calculated since the charge of each positive fragment is **+ 1**.

Ex:

$M^{+\cdot}$

m/z = 30

a cation
m/z = 29

a cation
m/z = 15

$M^{+\cdot}$
m/z = 16

L. SUMMARY OF SOME SELECTED RADICAL REACTIONS

1. OVERALL SUMMARY ON ALLYLIC BROMINATION

i.

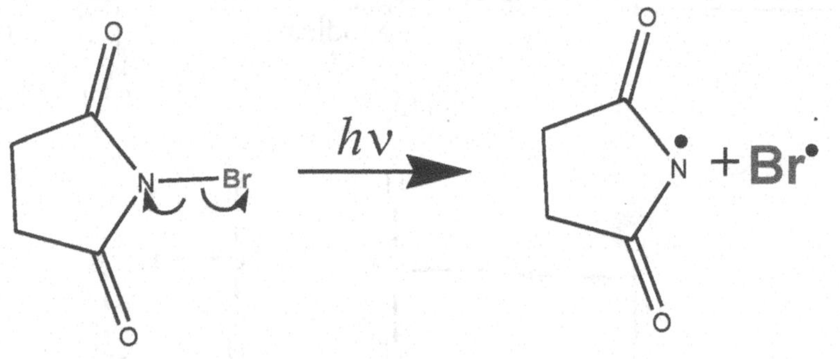

N-bromosuccinimide $\xrightarrow{h\nu}$ succinimidyl radical (N•) + Br•

ii.

>C=C–C(H)< + Br• ⟶ >C=C–Ċ< + HBr

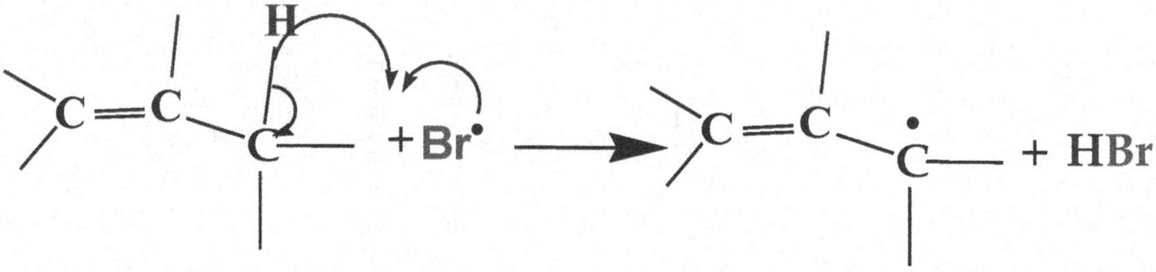

iii.

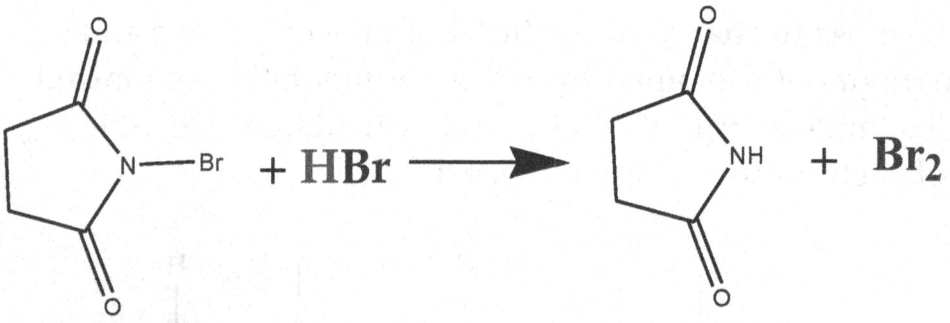

N-bromosuccinimide + HBr ⟶ succinimide (NH) + Br_2

iv.

$$\geq C=C\diagdown \dot{C}- \;+\; Br\text{-}Br \longrightarrow \;\geq C=C\diagdown C(Br)- \;+\; Br^{\bullet}$$

2. OVERALL SUMMARY ON ALLYLIC BROMINATION

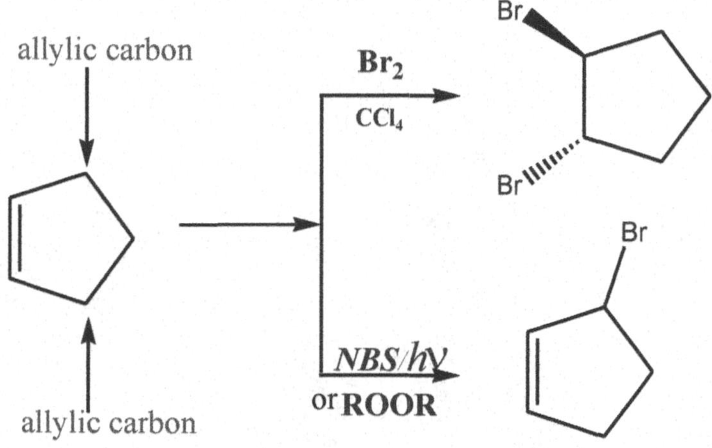

3. HYDROHALOGENATION ALKENES: A SUMMARY

- Read pages _____ - _____.
- Review Key Concepts on page ______ - ______.

OCHEM II UNIT 3: MASS SPECTROMETRY

A. INTRODUCTION TO STRUCTURE DETERMINATION

- **Question: How do chemists know the structures of the products obtained in chemical reactions?**

- **Answer: They rely on techniques that use powerful instruments.** The most useful techniques and their respective uses are in the following table:

technique	use
Mass Spectrometry (ms)	Molecular weight and molecular formula
Infrared Spectroscopy (IR)	Functional group and bond type
Nuclear Magnetic Resonance Spectroscopy (NMR)	Carbon-Hydrogen Framework
Ultraviolet and Visible Spectroscopy (UV-Vis)	Presence of conjugated π electrons

B. MASS SPECTROMETRY

1. USE

- This structural determination of technique can be used to get the molecular weight of a compound and its molecular formula.

2. THE BASIC PRINCIPLE BEHIND MASS SPECTROMETRY

- An unknown **neutral** sample is vaporized with heat **under vacuum.**
- The vaporized sample of the unknown compound to be analyzed is **bombarded** with a stream of high energy electrons (about 70eV or 6700 kJ/mol or 1600 kcal/mol).
- The molecules in sample are thus **ionized** under the collision impact.

- Positive radicals called **cation radicals** are thus produced.
- Most of the cation radicals **fragment** into smaller positive or neutral pieces.
- The **positive** fragments are accelerated toward 2 negative plates.
- From there, the positive fragments are channeled through a curved analyzer in a magnetic field.
- The magnetic field sorts the fragments according to their masses.

3. THE INSTRUMENT: A MASS SPECTROMETER

- **See Fig. ________, page ________. Read page ________.**

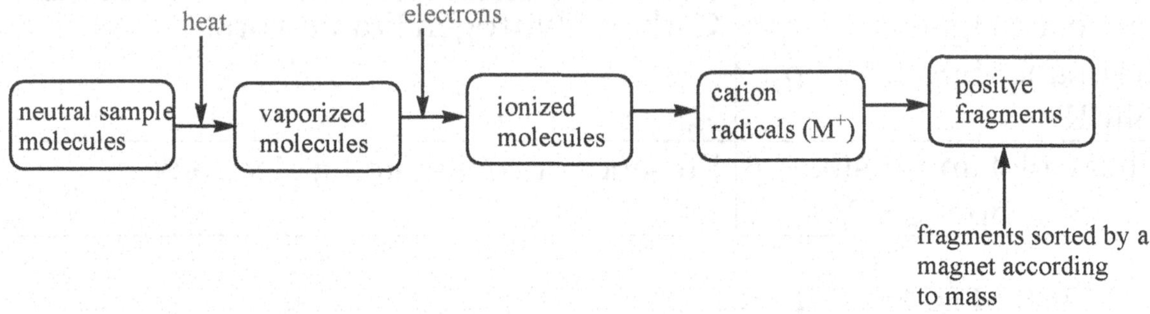

4. MASS-TO-CHARGE RATIO (m/z)

- The ionization process can be summarized as follows:

$$\textbf{organic molecule (M)} \xrightarrow{e^-} \textbf{cation radical } (M^{+\bullet}) + e^-$$

or

$$M \xrightarrow{e^-} M^{+\bullet} + e^-$$

- The cation radical is fragmented as follows:

$$M^{+\bullet} \xrightarrow{\text{fragments}} \textbf{positive and neutral fragments}$$

- The positive fragments are separated according to their mass-to-charge ratios (m/z). **Since most fragments have +1 charge, m/z = m = mass of the fragment.** Using the m/z of the **$M^{+\bullet}$**, one can determine the molecular weight of the unknown.

5. THE MASS SPECTRUM

- A mass spectrum is a **plot of m/z values vs. the intensity** or **relative abundance in %.**

- **The intensity of a fragment or its relative abundance is the number of ions of that fragment (having the same m/z) that strike the detector during the experiment. The more abundant the fragment, the taller its peak.**

Ex: $CH_3CH_2CH_3$

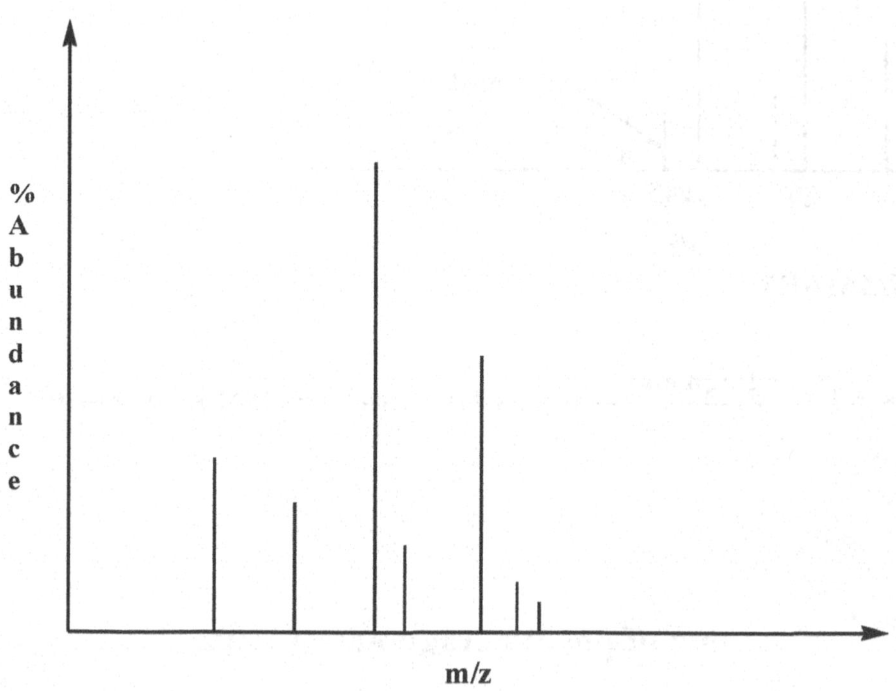

6. SOME DEFINTIONS

- **Base peak = tallest peak in the spectrum; it corresponds to the most abundant fragment.**
- **Parent peak = peak corresponding to the unfragmented cation radical, $M^{+\bullet}$.**
- **$M^{+\bullet}$ is called the molecular ion or parent ion; so, the parent peak is the peak corresponding to the parent ion.**

Ex: $CH_3CH_2CH_3$

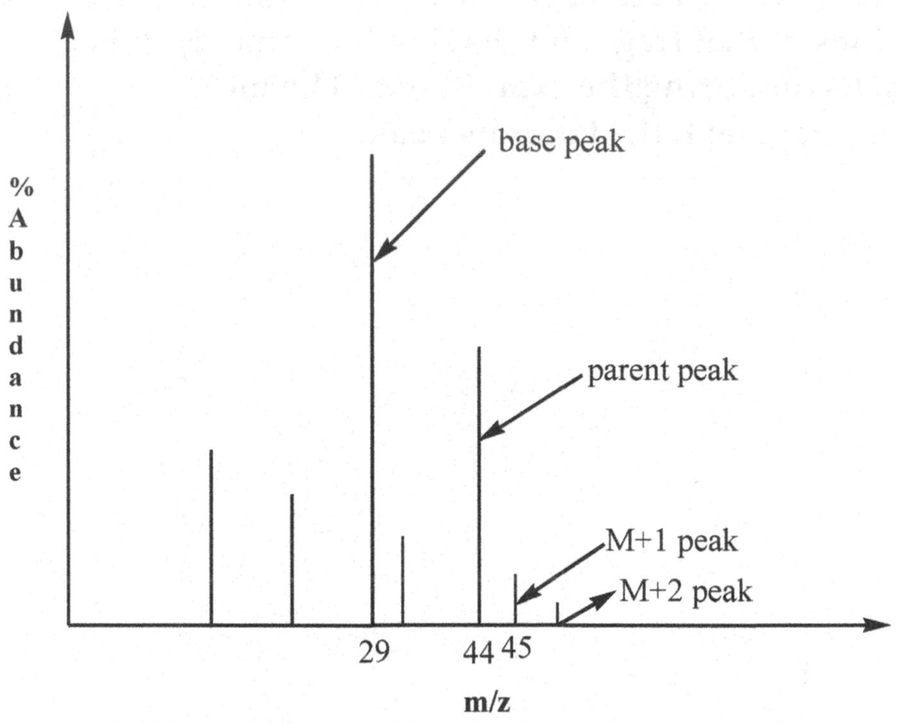

7. MS: A SUMMARY

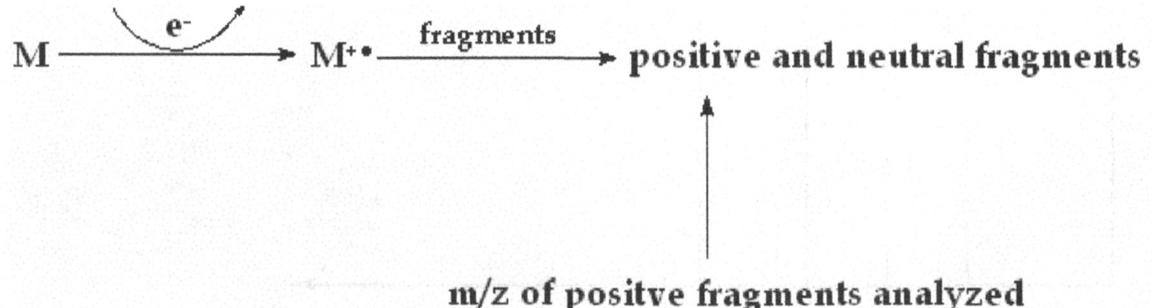

8. AN EXAMPLE: THE MS ANALYSIS OF CH_4

- **Read pages ______ - ______.**
- Consider the MS analysis of methane gas, CH_4:

$$CH_4 \xrightarrow{e^-} CH_4^{+\bullet} \xrightarrow{\text{fragments}} \text{positive fragments}$$

- **Note: The parent ion is: CH_4^+ with m/z = 16.**
- The fragmentation process is:

$$CH_4 \xrightarrow{e^-} \underset{m/z=16}{CH_4^{+\bullet}} \xrightarrow{-H^\bullet} \underset{m/z=15}{CH_3^+} \xrightarrow{-H^\bullet} \underset{m/z=14}{CH_2^{+\bullet}} \xrightarrow{-H^\bullet} \underset{m/z=13}{CH^+} \xrightarrow{-H^\bullet} \underset{m/z=12}{C^{+\bullet}}$$

- **See Fig. on page _____ and Fig. ______ on page ______.**
- **Note: The parent peak can also be the base peak.**

9. INTERPRETING MASS SPECTRA

a. Determination of Molecular Weight

- From the m/z of the molecular or parent ion, one can get the molecular weight of a compound.

Ex. For CH_4, $M^+ = 16 \rightarrow MW = 16$; for C_6H_{14}, $\mathbf{M^+} = 86 \rightarrow MW = 86$.

b. Identification of an Unknown

- One can also use mass spectrometry to identify an unknown substance by comparing its mass spectrum to known mass spectra.

Ex: M^+ for pentane is 72; M^+ for 1-pentene is 70; M^+ for 1-pentyne is 68. Identify each spectrum using your book, pages ______ - ______.

- **Do problem on page ______.**

c. The M+1 Peak

- This usually small peak is due to the presence of a small amount of isotopic (C-13 1.1% of all carbon) and or deuterium (H-2).

- **See pages _______ and _____. For CH_4, the M+1 peak is at m/z = 17. For C_6H_{14}, the M+1 peak is at m/z = 87.**

d. The M+2 Peak

- This peak is due to the presence of Cl or Br. For each of the following atoms C, H, O, N, S, P, F, and I, there is **one major isotope.** However Cl and Br each have **two major isotopes.** Cl occurs naturally as Cl-35 and Cl-37 in a 3:1 ratio. So for alkyl chlorides, 2 peaks are observed for M+ in a 3:1 ratio. The taller **parent** peak is the M^+ (peak for Cl-35) and the smaller peak is called the **M+2** peak (the peak for Cl-37). **In a mass spectrum, when the ratio of M+2 to M is 3:1, Cl is present in the compound.**

- Br occurs naturally as Br-79 and Br-81 in a 1:1 ratio. For alkyl bromides, the 2 peaks are in a 1:1 ratio. The M^+ peak corresponds to Br-79. The M+2 peak is the Br-81 peak. **In a mass spectrum, when the ratio of M+2 to M is 1:1, Br is present in the compound.**

- **Note: Can have a mixture of chlorides and bromides.**

- **See example on pages ______ - ______.**

e. Getting Molecular Formulas from Molecular Weights

i. Introduction

- As mentioned earlier, one can use M+ values to get molecular formulas of substances. First of all, we need to learn some basic rules about organic molecules.

ii. Rules for Determining Molecular Formulas

- **Rule #1:** Hydrocarbons and organic molecules that contain only C, H, and O always have **even** molecular weights.

- **Rule #2: The nitrogen rule:** Compounds that contain an **odd** number of N atoms have **odd** molecular weights. Conversely, compounds that contain an **even** number of N atoms have **even** molecular weights.

- **Problem:** A compound containing C, H, and/or O has M+ = 100. Give the possible molecular formulas for the unknown compound.

- **Note:**
  - **The atomic mass of C = 12; each C ≈ 12 H**
  - **The atomic mass of O = 16; each O ≈ 1 CH_4.**

  - First, find all **possible** "hydrocarbon" formulas (C and H only)

- Next, find all possible C, H, and O containing compounds.

- **Recall**: Each **O ≈ 1 CH_4**

• Do Problems on page ______ and Problems on page ______.

Try $M^+ = 86$.

C. OTHER TYPES OF MASS SPECTROMETRY

1. HIGH RESOLUTION MASS SPECTROMETRY

- **Note: The previous sections in this chapter deal with low-resolution mass spectrometry.**

- In this kind of mass spectrometry, the m/z values are recorded to the nearest whole number: 60, 78, 86 … As a result the M+ obtained can correspond to more than one molecular formula. See the preceding examples.

- In high-resolution mass spectrometry, the m/z values are given with 4 or more decimal places. Since H-1, O-16, and N-14 have **"non-whole"** atomic masses, this can help differentiate between different compounds. The exact masses of C-12, H-1, O-16, and N-14 are in the following table.

isotope	exact mass
H-1	1.00783
C-12	12.0000
O-16	15.9949
N-14	14.0031

Ex: The low-resolution mass spectrum of an unknown compound has an m/z for the molecular ion at 180. Using a high-resolution MS, the parent ion appears at m/z = 180.2064. Which one of the following compounds ($C_9H_{24}O_3$, $C_6H_{12}O_6$, $C_6H_{24}N_6$) is the unknown compound? Explain.

- **See example on page ______. Do problem ________, page _____.**

2. GAS CHROMATOGRAPHY-MASS SPECTROMETRY

a. Gas Chromatography: Basic Principle

- Gas chromatography is an analytical technique used to separate the components of a mixture in the **gaseous state.**

b. Different Parts of a Gas Chromatograph

- A gas chromatograph consists of a sample injecting port or **septum.** Inside a GC, there is a hollow oven that houses a coiled, separatory thin capillary column.

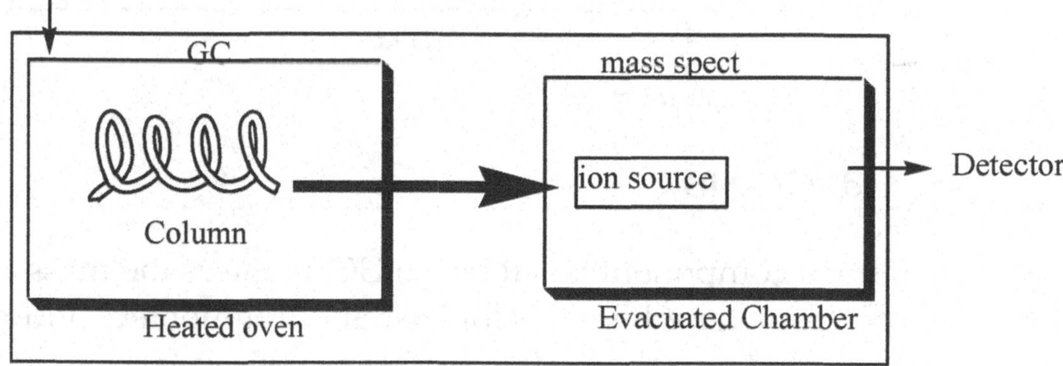

c. How Does a GC Work?

- When a mixture of an unknown liquid sample is injected in the column through the septum, it vaporizes in the oven. There, it is carried through the thin capillary column by an inert gas (He). The components of the mixture are thus separated by boiling points (BP). Low BP components exit first.

- A GC spectrum is the plot of peak intensity (concentration) vs **retention time,** which is the time it takes a given component to travel through the column or to **elute**.

- **See Fig. on page ______.**

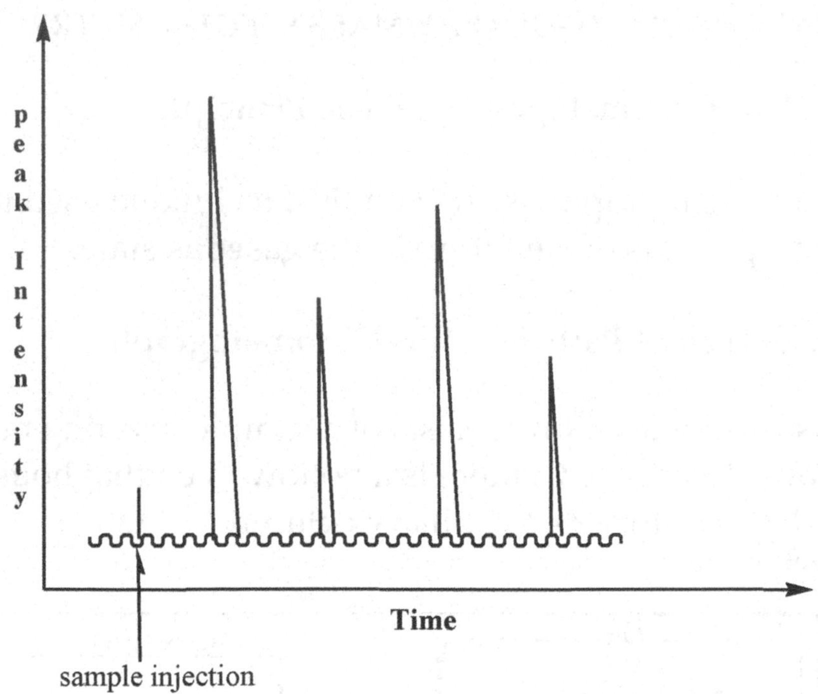

d. GC-MS

- Once a component is out of the GC, it enters the mass spectrometer where it is ionized and fragmented. Fragments are sorted by a magnet according to their m/z.

- **See Fig. ______, page _____.**

- **Do Problem ______, page _____.**

3. ELECTRON SPRAY IONIZATION (ESI) MASS SPECTROMETRY

a. Introduction

- Sample vaporization using heat under vacuum can be done only for samples whose molecular weights are under 800. So, until the 1980s, it was impossible to analyze these samples with an MS. However, in the 1980s, a Yale professor by the name of Dr. John Fenn developed a new vaporization technique that can generate gas phase ions of large molecules. The new technique is called **electron spray ionization or ESI.** This method consists of creating a fine spray of charged droplets in an electric field. When the droplets evaporate, they give gaseous ions that are sorted according to their m/z as in regular mass spectrometry.

- **Read page _______. Do Problem ______, page ______.**

D. INTERPRETING MASS SPECTRAL FRAGMENTATION PATTERNS

1. MOLECULAR FINGERPRINT

- The mass spectrum of a substance can be used as a **molecular fingerprint (like DNA)** since no two molecules have exactly the same fragmentation patterns. Indeed, each molecule fragments in a unique way. There exists a Registry of Mass Spectral Data which is a computerized MS database.

2. STABILITY OF FRAGMENTS

- Tertiary carbocations fragments are usually formed when possible.

Ex:

$$\left[H_3C\text{—}C(CH_3)_2\text{—}CH_3 \right]^{+\bullet} \longrightarrow H_3C\text{—}C^{+}(CH_3)_2 + {}^{\bullet}CH_3$$

$M^{+\bullet}$ — m/z = 57, base peak

- See pages _____ - _____. Do Problems on pages ______ - ____.

3. FRAGMENTATION PATTERNS OF SOME COMMON FUNCTIONAL GROUPS

a. Alcohols: ROH

i. Introduction

- The fragmentation of alcohols in MS follows two possible pathways: α cleavage or dehydration.

ii. α Cleavage

$$RH_2C\text{—}C\text{—}OH \xrightarrow{e^-} \left[RH_2C\text{—}C\text{—}OH \right]^{+\bullet}$$

α

$M^{+\bullet}$

α cleavage

$$\left[>C{=}OH \right]^{+} + RCH_2^{\bullet}$$

Ex:

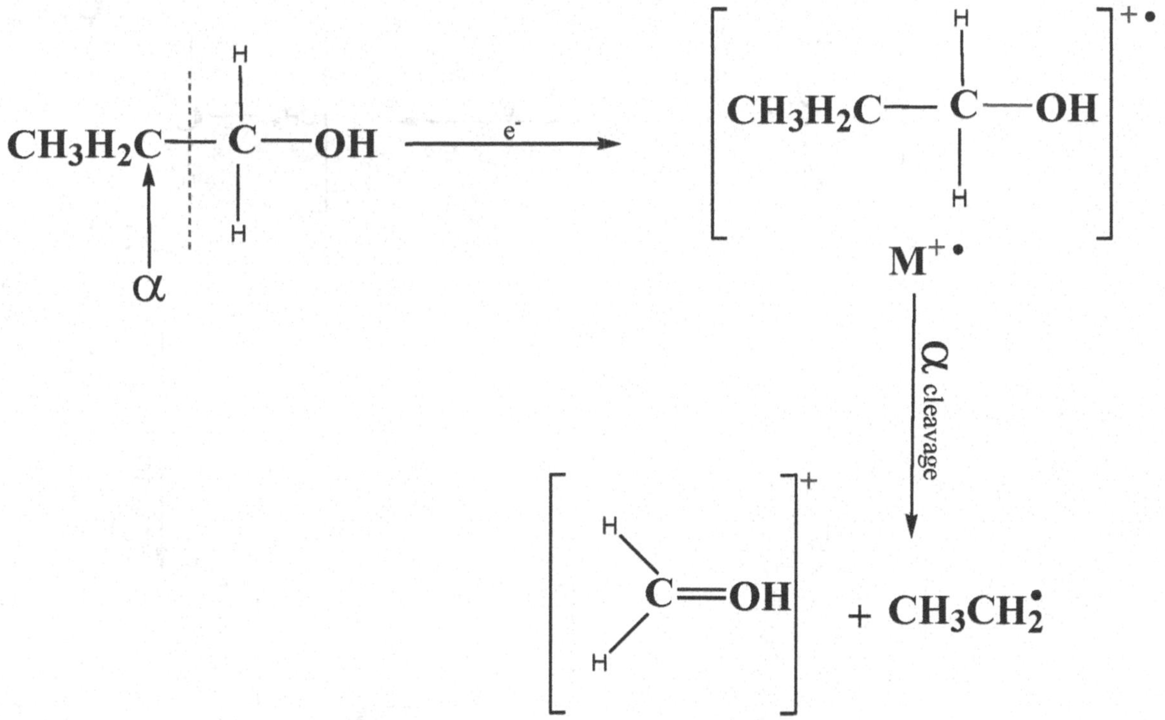

$$CH_3H_2C-CH_2-OH \xrightarrow{e^-} [CH_3H_2C-CH_2-OH]^{+\bullet}$$

α

$M^{+\bullet}$

α cleavage

$$[H_2C{=}OH]^+ + CH_3CH_2^\bullet$$

iii.Dehydration: Elimination of H_2O

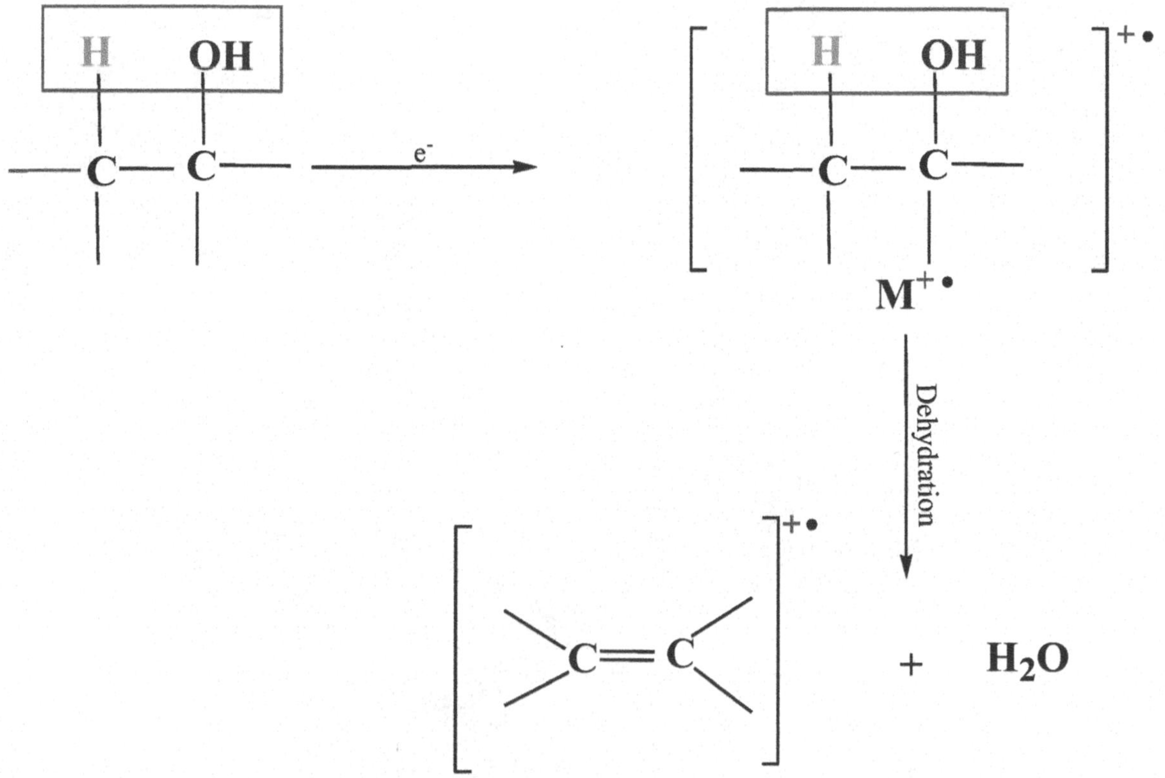

H OH

$$-\overset{|}{\underset{|}{C}}-\overset{|}{\underset{|}{C}}- \xrightarrow{e^-} [-\overset{|}{\underset{|}{C}}-\overset{|}{\underset{|}{C}}-]^{+\bullet}$$

H OH

$M^{+\bullet}$

Dehydration

$$[>C{=}C<]^{+\bullet} + H_2O$$

Ex:

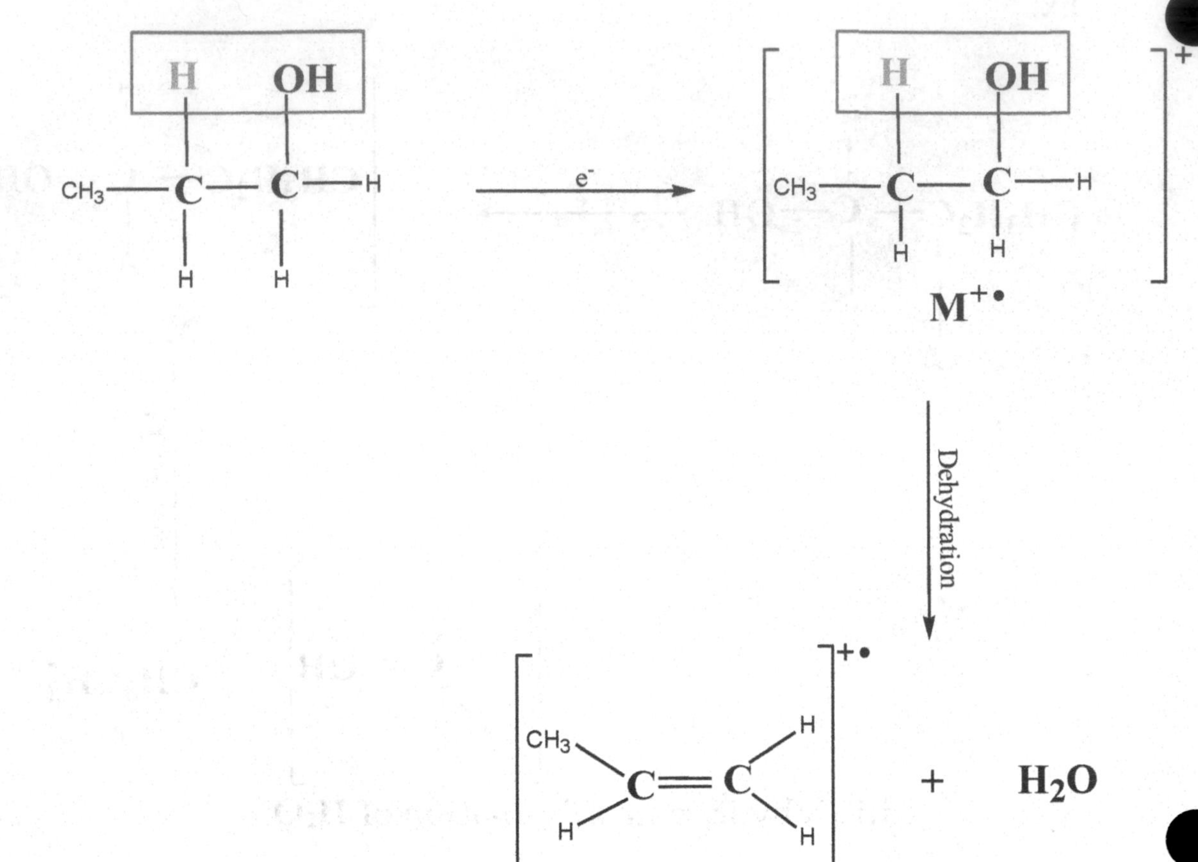

b. Amines: R-NH_2

- Amines undergo α cleavage:

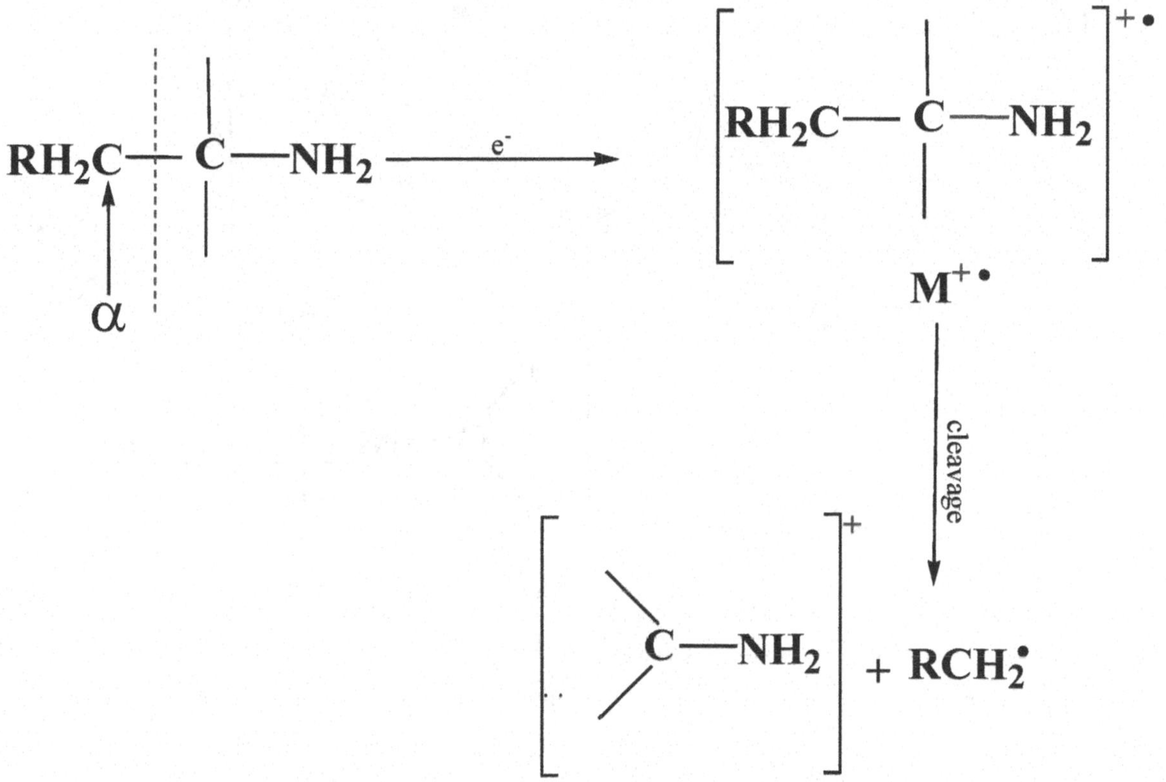

$$RH_2C \,\vdots\, C{-}NH_2 \xrightarrow{e^-} [RH_2C{-}C{-}NH_2]^{+\bullet}$$

α

$M^{+\bullet}$

cleavage

$$[\;C{-}NH_2\;]^+ + RCH_2^{\bullet}$$

Ex:

$$CH_3CH_2-CH_2-NH_2 \xrightarrow{e^-} [CH_3CH_2-CH_2-NH_2]^{+\bullet}$$

α (α-carbon, bond cleaved at dashed line)

$M^{+\bullet}$

↓ cleavage

$$[H_2C-NH_2]^{+} + CH_3CH_2^{\bullet}$$

c. **Aldehydes and Ketones**

i. **Introduction**

- Aldehydes and ketones follow **2 possible** fragmentation pathways in MS: **α** cleavage and the **McLafferty** rearrangement (for long-chain compounds).

ii. **α Cleavage**

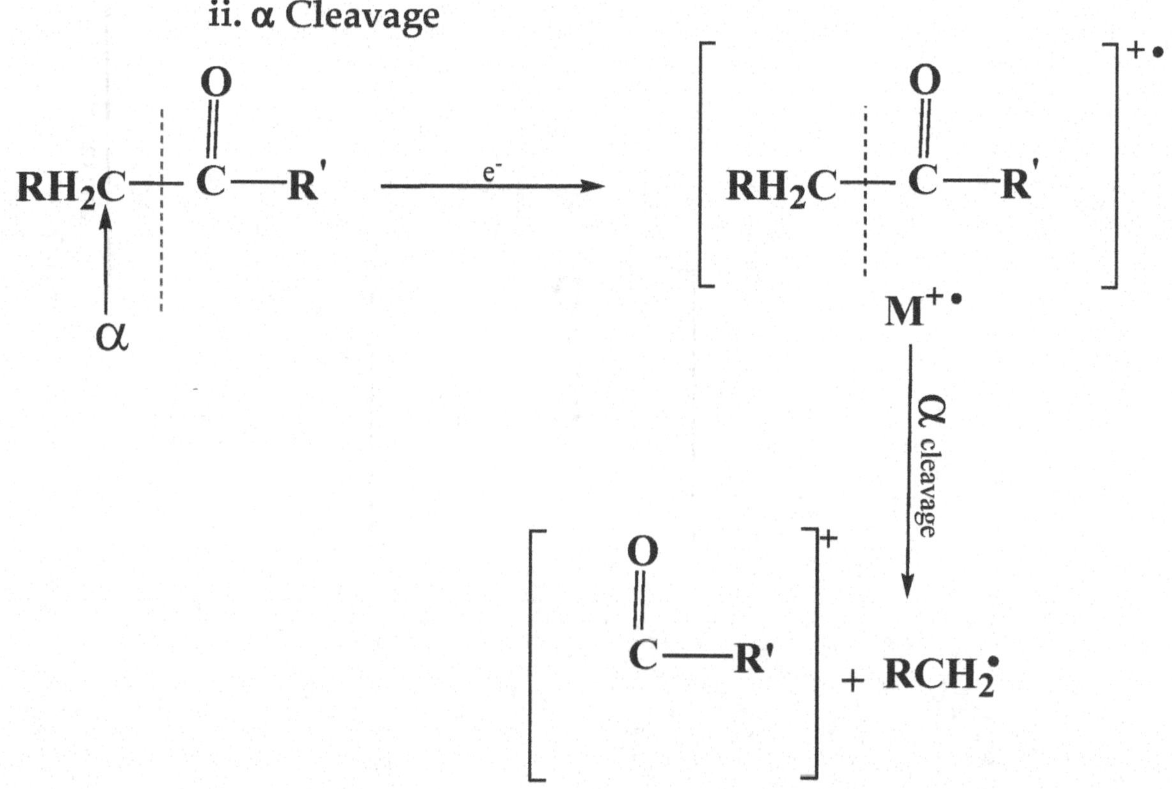

- See pages ______ - ______ and associated problems.

Ex:

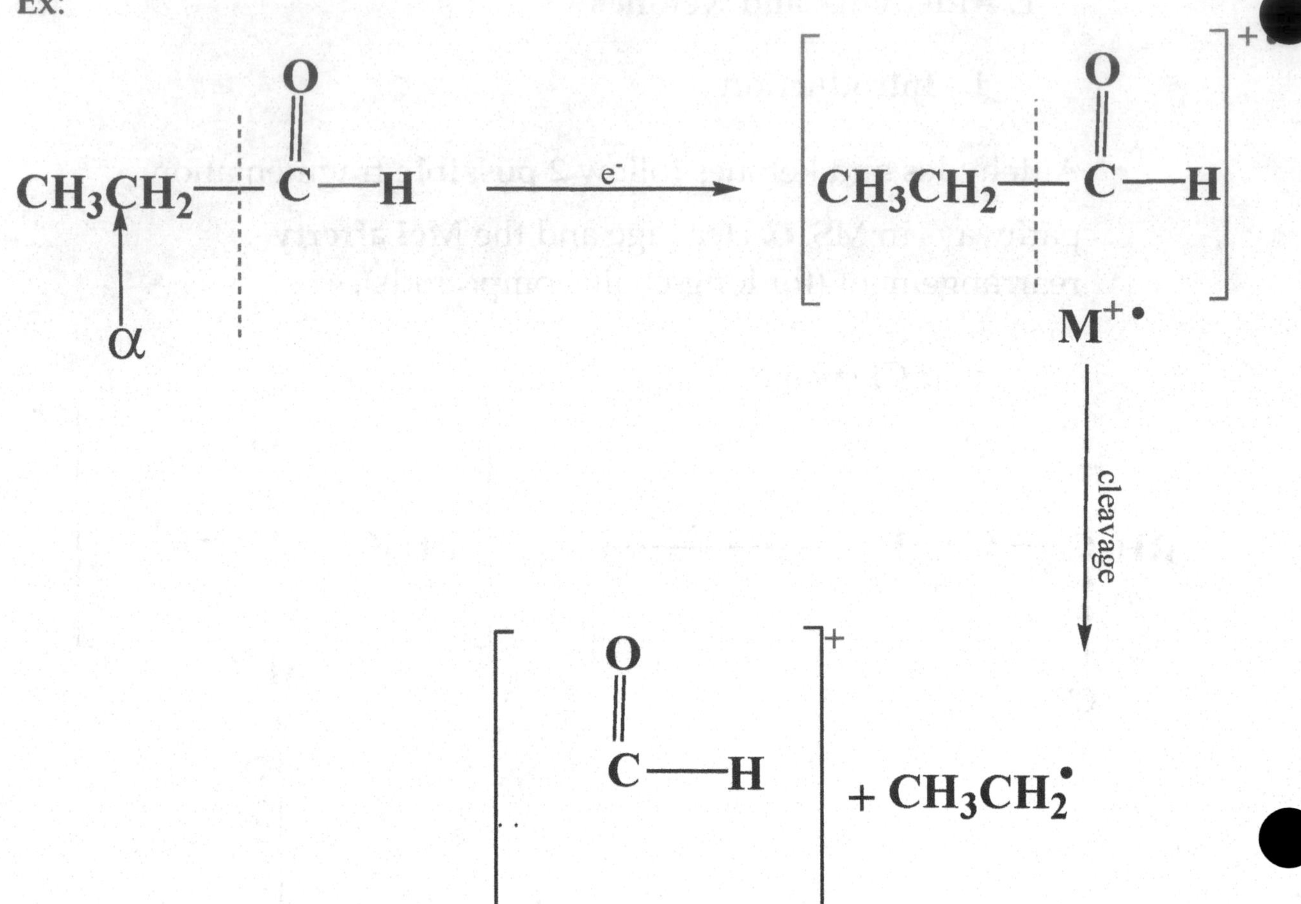
O
CH_3CH_2 — C — H
α
e⁻
O
CH_3CH_2 — C — H
+
$M^{+\bullet}$
cleavage
O
C — H
+
+ $CH_3CH_2^{\bullet}$

iii. McLafferty Rearrangement

H–C–C–C–C(=O)–R →(e⁻)→ [H–C–C–C–C(=O)–R]$^{+\bullet}$ $M^{+\bullet}$

MacLafferty

[R–C(OH)=C]$^{+\bullet}$ + C=C

Ex:

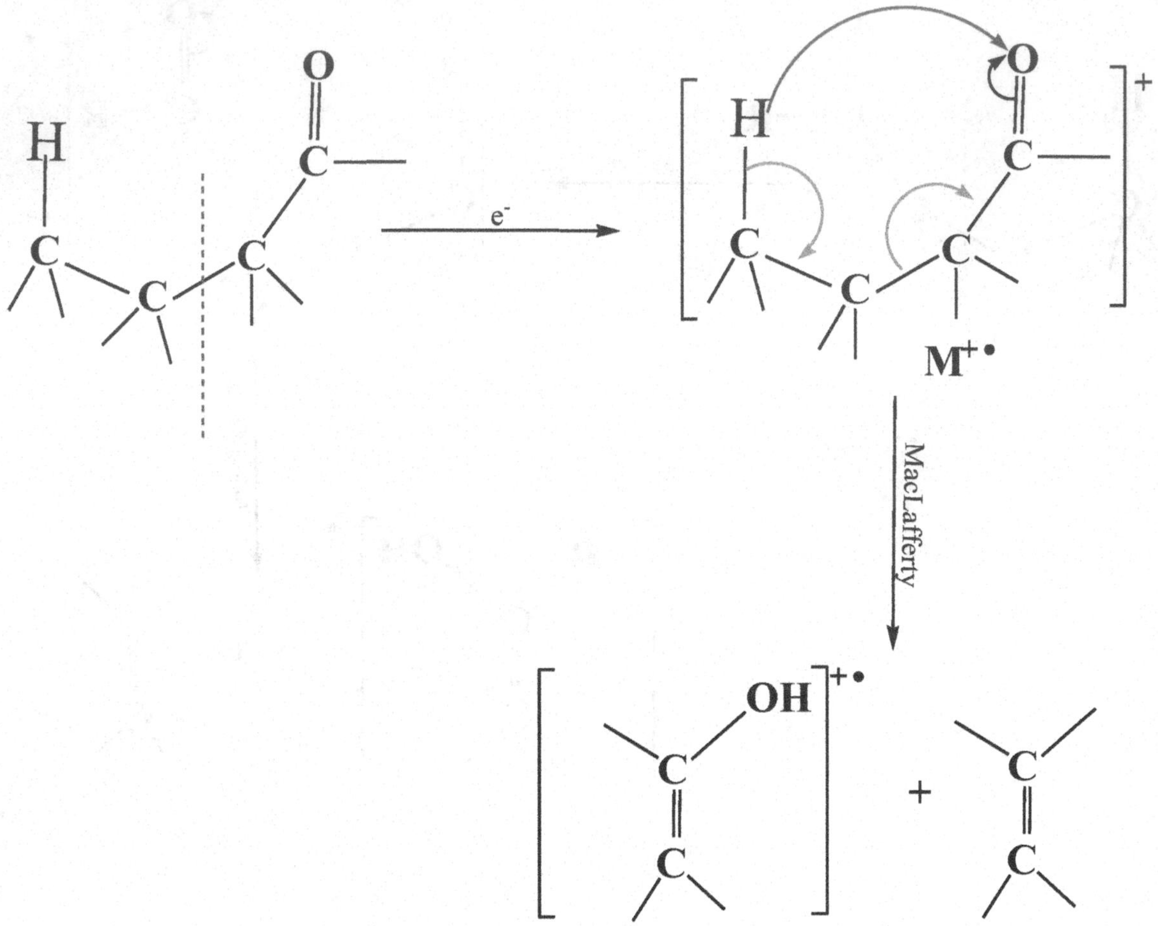

d. Fragmentation Patterns: A Summary

functional group	α cleavage	dehydration	McLafferty
alcohols	√	√	-
amines	√	-	-
aldehydes+ketones	√	-	√

- See Key Concepts on page ________.

OCHEM II UNIT 4: IR SPECTROSCOPY

A. THE ELECTROMAGNETIC SPECTRUM

1. DEFINTION

- The electromagnetic spectrum (EM) is the **complete range** of radiations that move at the speed of 3.00 x 10^8 m/s (speed of light). These radiations have the same speed, but they have different wavelengths and frequencies.

Ex: visible light, X-rays, radio waves, gamma rays, etc.

2. THE EM SPECTRUM

- **See Fig. ______, page ______.**

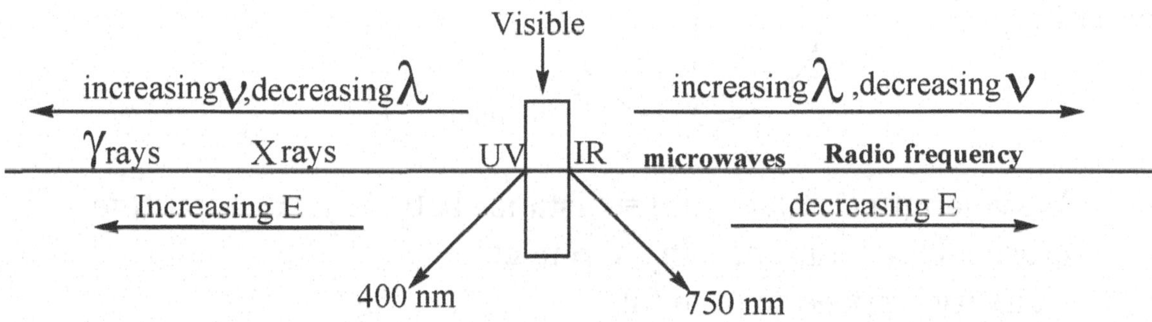

B. CHARACTERISTICS OF LIGHT

1. INTRODUCTION

- Light has a **dual** nature:
  - o Particle nature: **Einstein: light is made of photons; each photon carries a quantum of energy, hv.**
  - o Wave nature: light waves: UV, IR, etc. have different wavelengths and frequencies.
- Here we focus on the **wave nature of light.**

2. **CHARACTERISTICS OF WAVE: 4**

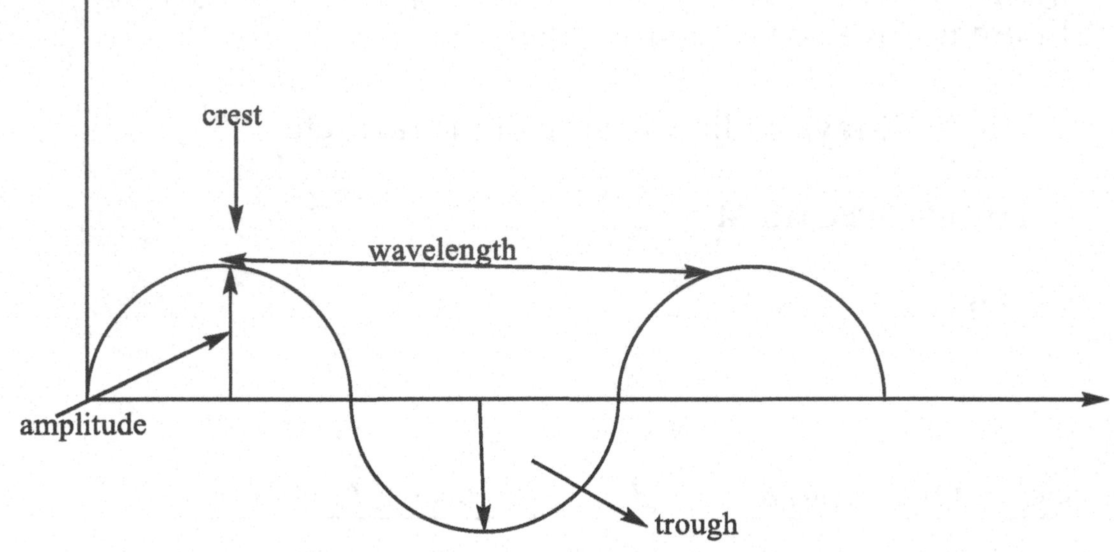

- Wavelength (λ = lambda) = distance between 2 consecutive crests. Units: nm, m, μm, Å, pm, cm.
- Velocity = speed (**c** in m/s)
- Frequency (ν = nu) = number of wave crests that pass a given point in 1 second. Units = cps, 1/s or s^{-1}, Hertz (Hz).
- Amplitude = height of crest.

3. **RELATIONSHIP BETWEEN λ, ν, c, AND ENERGY**

- For a given EM radiation:

$$\lambda = \frac{c}{\nu} \Rightarrow \nu = \frac{c}{\lambda}$$

$$E = h\nu = \frac{hc}{\lambda}; h = 6.63 \text{ x } 10^{-34} \text{J.s}$$

- If ν increases, E also increases, or, if λ increases, ν decreases, and E decreases. From the EM spectrum, X-rays and gamma rays are high- energy radiations. On the other hand, microwaves and radio waves are low-energy waves.

Ex. The wavelength of an EM radiation is 2.3×10^{-6} m: What is the frequency of this radiation? What is the energy of radiation?

- **Do problems on page ______.**

C. GENERAL MOLECULAR SPECTROSCOPY

1. PRINCIPLE OF MOLECULAR SPECTROSCOPY

- **Molecular spectroscopy** is the experimental process of measuring absorbance (or % transmittance) of an EM radiation by a molecule and attempting to correlate absorption patterns with details of molecular structure.

- A molecule is "normally" at rest (**ground state**). So there is an energy difference (gap) between normal state (ground state) and **excited state.** If an external light energy that **matches** the energy difference (or gap) between E_1 (ground state) and E_2 (excited state), the molecule will absorb the energy and go from E_1 to E_2. **See page ______.**

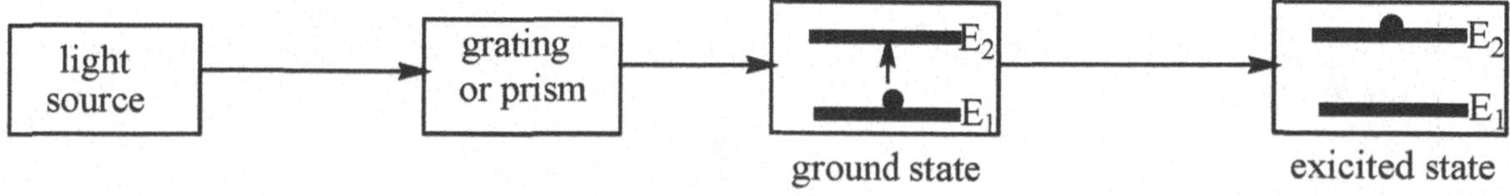

2. GENERAL INSTRUMENTATION IN SPECTROSCOPY

There are 6 parts:

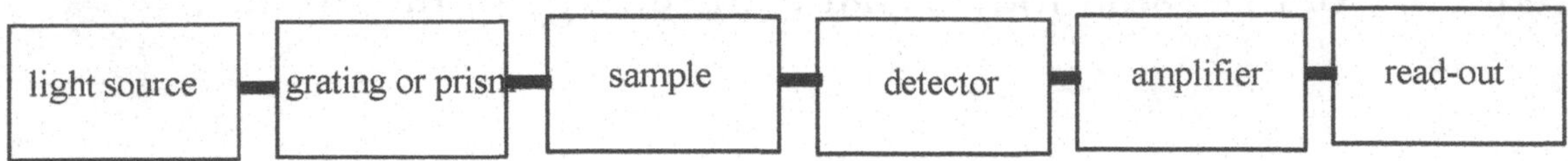

- **Note: One can measure light absorbed by the sample (absorbance) or light transmitted (% transmittance).**

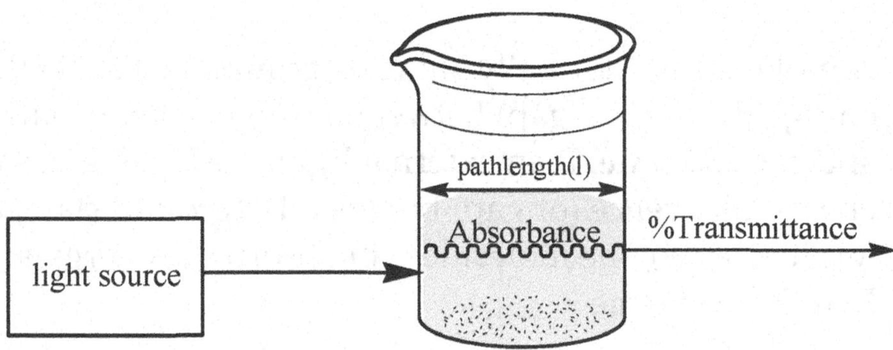

3. DIFFERENT TYPES OF MOLECULAR SPECTROSCOPY

There are 3 types. The following table summarizes the type of EM waves used in each one.

Technique	EM waves used
Infrared Spectroscopy (IR)	Infrared
Nuclear Magnetic Resonance Spectroscopy (NMR)	Radio Frequency
Ultraviolet and Visible Spectroscopy (UV-Vis)	UV-Vis

4. USE OF MOLECULAR SPECTROSCOPY IN STRUCTURE DETERMINATION

- The following table summarizes the use of spectroscopy.

Technique	Use
Infrared Spectroscopy (IR)	Functional group and bond type
Nuclear Magnetic Resonance Spectroscopy (NMR)	Carbon-Hydrogen Framework
Ultraviolet and Visible Spectroscopy (UV-Vis)	Presence of conjugated π electrons

D. INFRARED SPECTRSCOPY OF ORGANIC MOLECULES

1. INTRODUCTION

- IR is based on **vibrational transitions.**

2. IR: BASIC PRINCIPLE

- IR is based on **vibrations of covalent bonds** in molecules. A covalent bond is like a **spring** and vibrates in many ways; it can bend, rotate, wag, stretch, twist, scissor, etc. See page ______ . If the frequency of the EM radiation is the same as that of a given vibration, energy is absorbed (**See general spectroscopy in section C**).

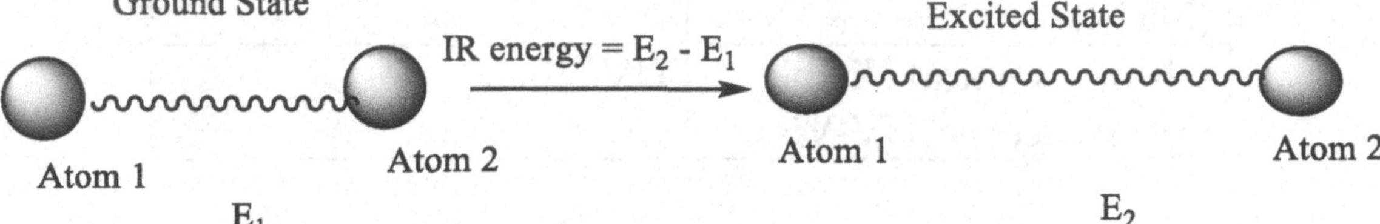

3. GENERAL POTENTIAL ENERGY DIAGRAM OF A COVALENT BOND

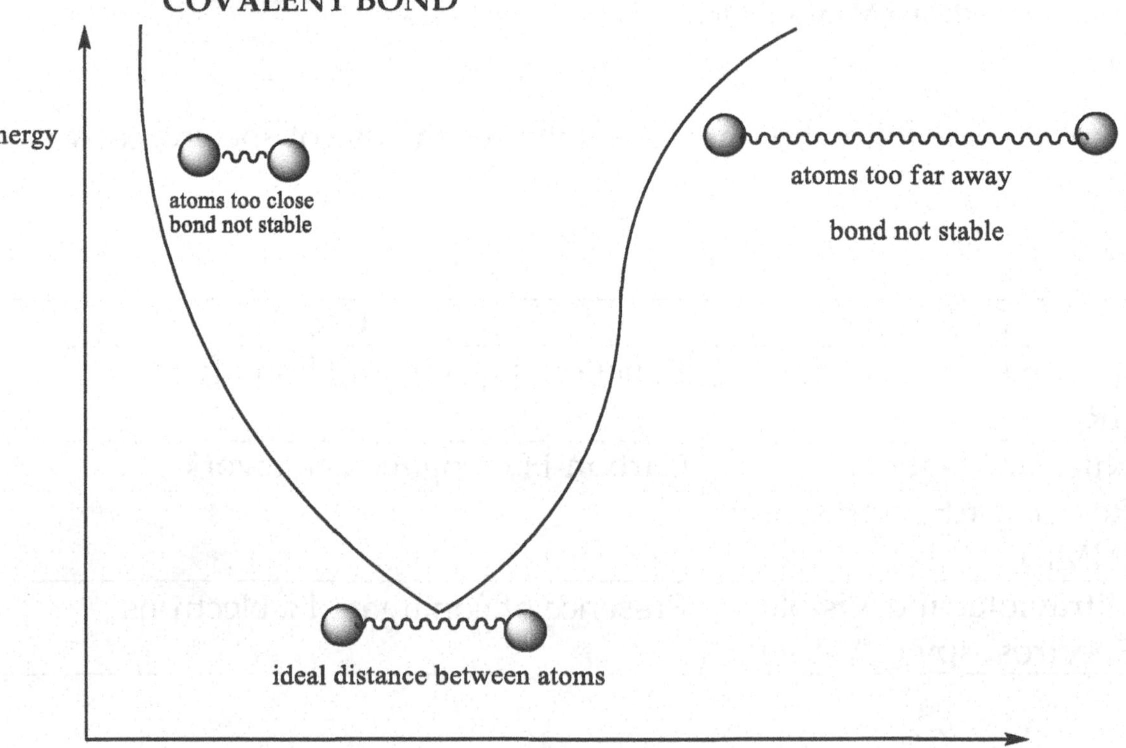

4. **COMMON MODES OF VIBRATION IN IR:**

a. **Stretching: have 2 atoms**

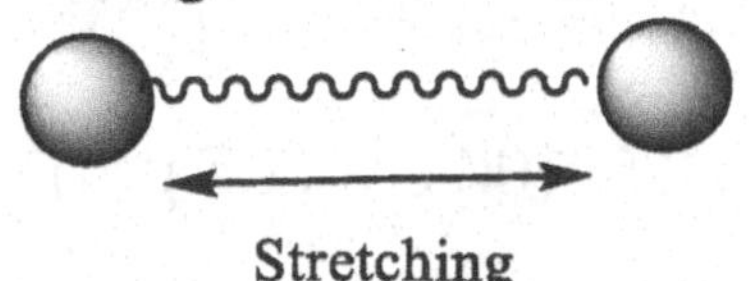

Ex: O-H

b. **Bending: have more than 2 atoms**

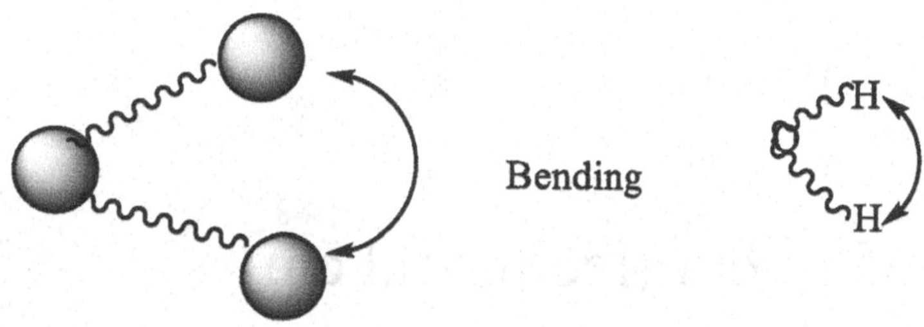

Ex: H_2O

- In essence, IR radiation causes covalent bonds in molecules to be promoted from a **lower vibrational energy level (E_1)** to a **higher vibrational energy level (E_2). In other words, in this technique, IR energy is absorbed by a molecule when the incoming energy matches the energy gap between vibrational excited and ground states. Since different covalent bonds absorb at different frequencies, IR can be used to distinguish different bonds and functional groups.**

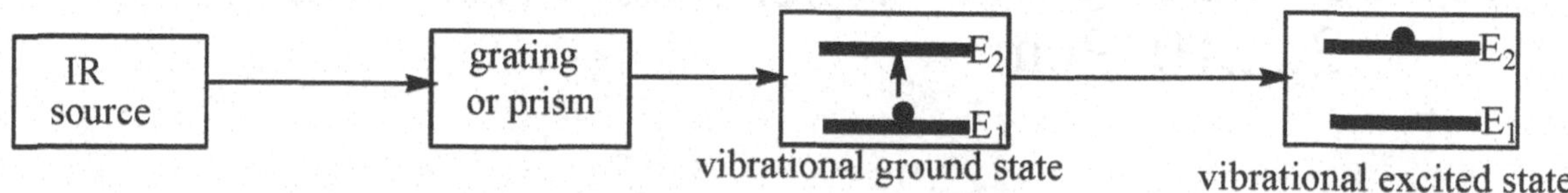

Recall: IR spectroscopy is used to differentiate covalent bonds based on their vibrational patterns. From the types of covalent bonds, one can determine functional groups like -OH, -NH, -C=O, etc.

5. **EM RANGE USED IN IR**

- **See Figure on page _______.**

- Range: from $\lambda_1 = 2.5x10^{-4}$ cm (2.5 μm) → $\lambda_2 = 2.5x10^{-3}$ cm (25μm)

- In IR spectra, use $\bar{\nu}$ = wave number

$$\bar{\nu} = \frac{1}{\lambda}, \lambda \text{ in cm}$$

Unit of $\bar{\nu}$ = cm^{-1}(reciprocal cm).

- Let's express this IR range used in cm^{-1}:

$$\bar{\nu}_1 = \frac{1}{\lambda_1} = \frac{1}{2.5x10^{-4}cm} = 4000cm^{-1}(\text{high frequency})$$

$$\bar{\nu}_2 = \frac{1}{\lambda_2} = \frac{1}{2.5x10^{-3}cm} = 400cm^{-1}(\text{low frequency})$$

- **Note: The IR range used in cm^{-1} (2.5 μm) is from 4000 cm^{-1} → 400 cm^{-1} (25 μm).**

- **Do Problem _______, page _______.**

6. THE IR SPECTRUM

In IR spectroscopy, one can either plot % **Transmittance vs.** $\bar{\nu}$ (in cm^{-1}) or % **Transmittance vs.** λ (in μm). It is a **very complex spectrum.**

- See Fig. on page ______.
- **Note: A peak represents the characteristic signal (or vibrational signature) of a specific bond.**

Ex: O-H

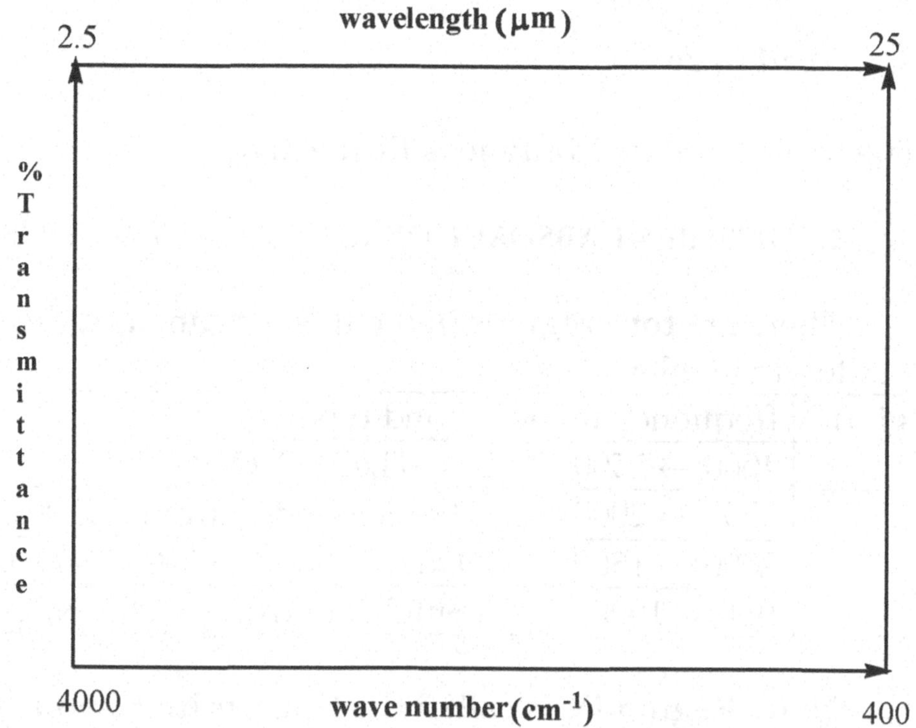

Ex. See Addenda

- The IR instrument is called an **IR spectrophotometer.** **Please, see instrument in our instruments room in C-305.**

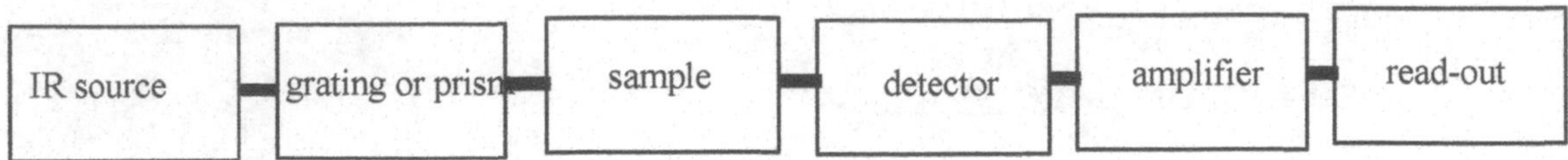

E. INTERPRETING IR SPECTRA

1. INTRODUCTION

- Generally, most organic compounds are **IR active.** ***Why?***
- **Exceptions: Nonpolar, symmetrical bonds are IR inactive because during vibration, a change in the dipole moment must occur.**

- **Read page ________.**

Ex: The triple bond in **2-butyne** is **IR inactive**.

2. DIFFERENT ABSORPTION REGIONS IN AN IR SPECTRUM

- There are **four regions** in the IR spectrum as indicated in the following table:

IR region	frequency range	bond type
I	4000 → 2500	N–H, C–H, O–H
II	2500 → 2000	Triple bonds (strong): C ≡ C, C ≡ N
III	2000 → 1500	Double bonds: C=C, C=N, C=O
IV	Below 1500	Single bonds: C–C, C–N, C–O, C–X

- **Note: Region IV is called the fingerprint region. It is unique for a given substance. It can be helpful in the identification of a compound although it is difficult to analyze.**
- **See Fig. ______, page ______. See Figs. on pages _______ - ______.**

F. EXPLANATION OF ABSORPTION PATTERNS OBSERVED IN IR

1. FACTORS THAT DETERMINE IR ABSORPTION

- There are **4 factors** that determine the IR absorption of a certain bond
  - Bond strength
  - Bond polarity
  - Sizes of bonding atoms
  - %s character

2. BOND STRENGTH

- **Recall: A covalent bond is like a spring!**

- The decreasing order of bond strength for carbon-carbon bond is:

  $C \equiv C > C{=}C > C{-}C$

- The increasing order of bond length for carbon-carbon bonds is:

  $C \equiv C < C{=}C < C{-}C$

- Recall that a shorter and stronger spring vibrates at a higher frequencies. Likewise, a shorter and stronger covalent bond absorbs at higher frequencies in IR.

Ex: $\bar{\nu}$ C ≡ C (2200) **>** $\bar{\nu}$ C=C (1700 cm^{-1})

3. BOND POLARITY

- The intensities of IR absorption bands depend on **bond polarity**. Indeed, the more polar the bond, the more intense the band **because absorption in this area is due to bond stretching.**

Ex: O-H > N-H > C-H

4. SIZES OF BONDING ATOMS

a. Hooke's Law

- It can be shown in physics that a vibrating spring obeys the following:

$$\bar{\nu} = k\sqrt{\frac{f}{m}}$$

- Where f = force, and m = mass. One can see that $\bar{\nu}$ is **inversely proportional to the mass.** So the larger (heavier) the atom bonded to C, the smaller the $\bar{\nu}$. **See Fig. _____, page ____.**

Ex: O is much heavier than H. The following is observed in IR.

$\bar{\nu}$ C–H (3600) > $\bar{\nu}$ C–O (1050 cm^{-1})

5. %s CHARACTER

- A higher %s character means a stronger bond. As previously mentioned, stronger bonds vibrate at **higher frequencies**.

Ex: The %s character of C in ≡ C–H is 50%. On the other hand, the % s character on C in =C–H is 33%. As expected $\bar{\nu}$ ≡ C (3300) > $\bar{\nu}$ =C (3150 cm^{-1} - 3000 cm^{-1}) > C-H (3000).

- **See Table 13.2, page 512. Read pages ______ - ______.**
- **Do Problem on page ______.**

G. IR SPECTRA OF HYDROCARBONS AND SOME FUNCTIONAL GROUPS

1. INTRODUCTION

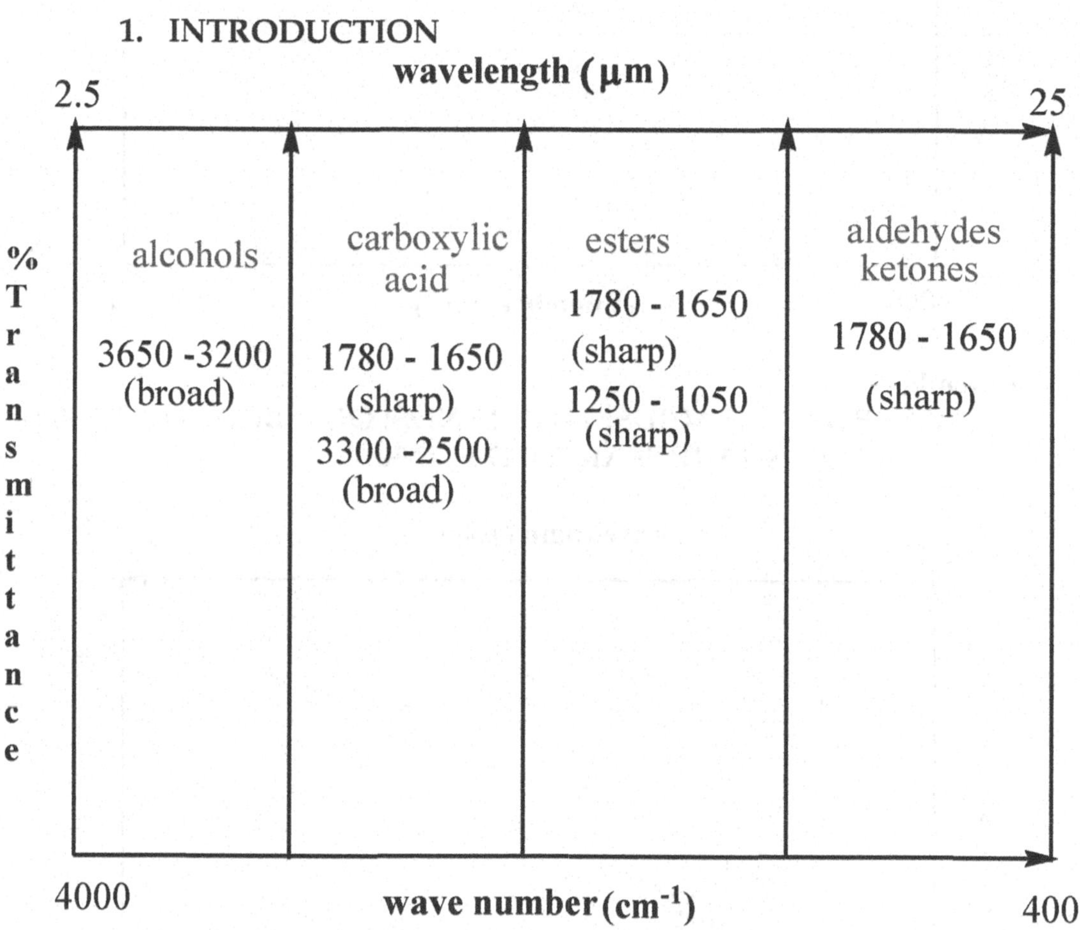

- **For hydrocarbons, read page ______ and associated Figures.**

2. **ALCOHOLS: OH = BROAD BAND AT 3650 - 3200**

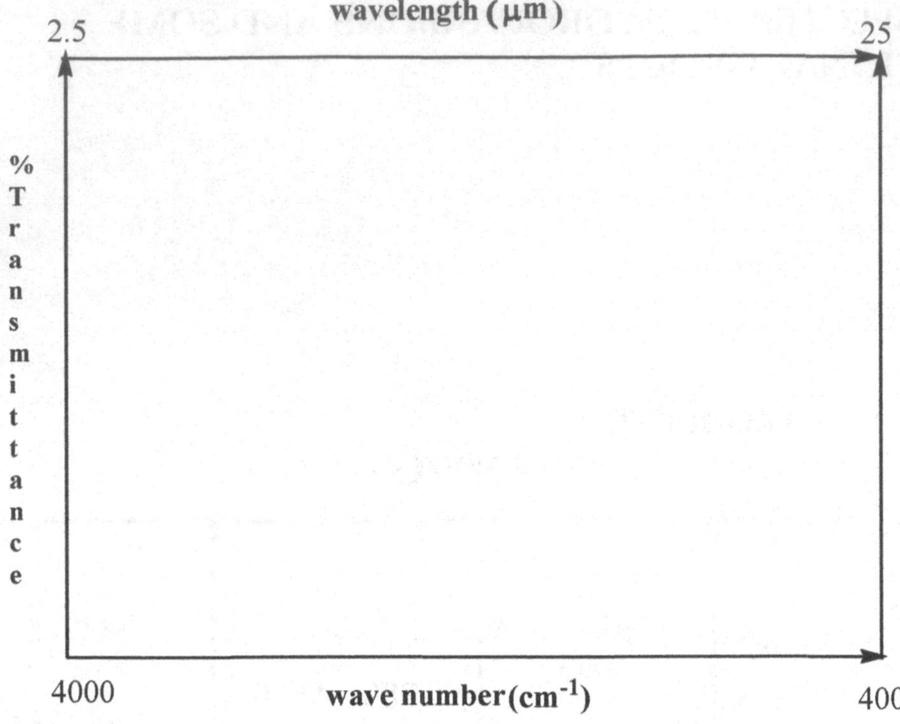

Ex: See Addenda

3. **CARBOXYLIC ACIDS: TWO BANDS: OH = BROAD AT 3300 - 2500 CO = SHARP PEAK AT 1780-1650**

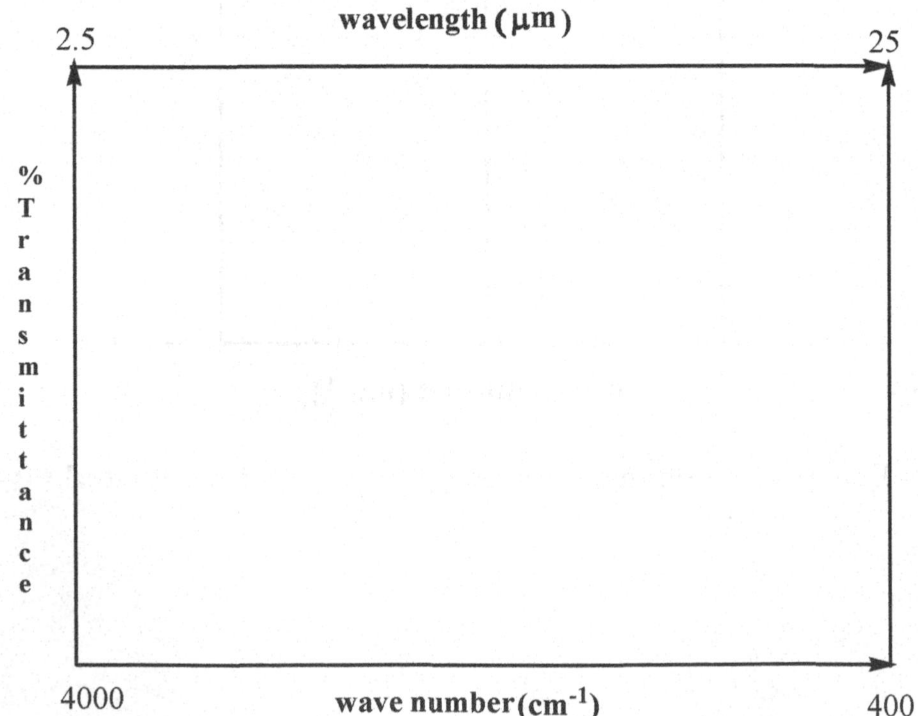

Ex: See Addenda

4. **ALDEHYDES: CO = SHARP PEAK AT 1780 - 1650**

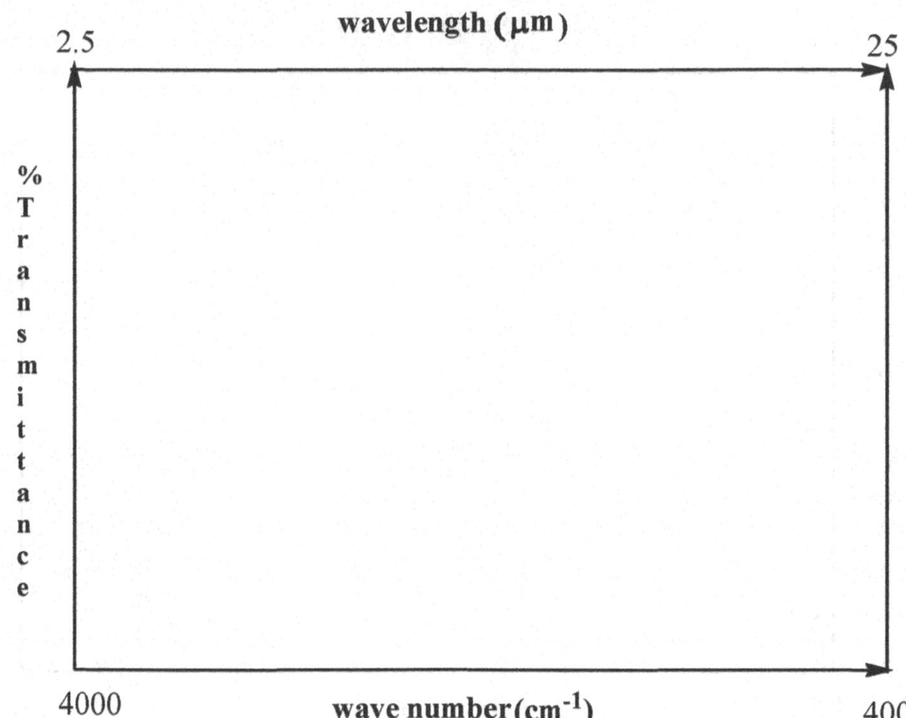

Ex: See Addenda

5. **KETONES: CO = SHARP PEAK AT 178O -1650**

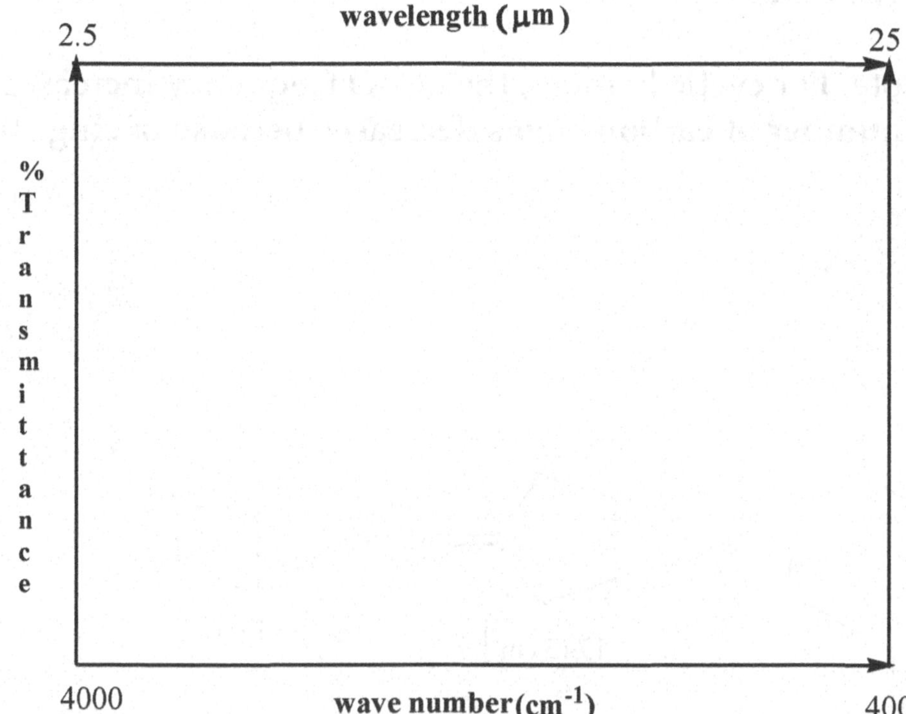

Ex: See Addenda

6. ESTERS: C =O: SHARP PEAK AT 1780-1650

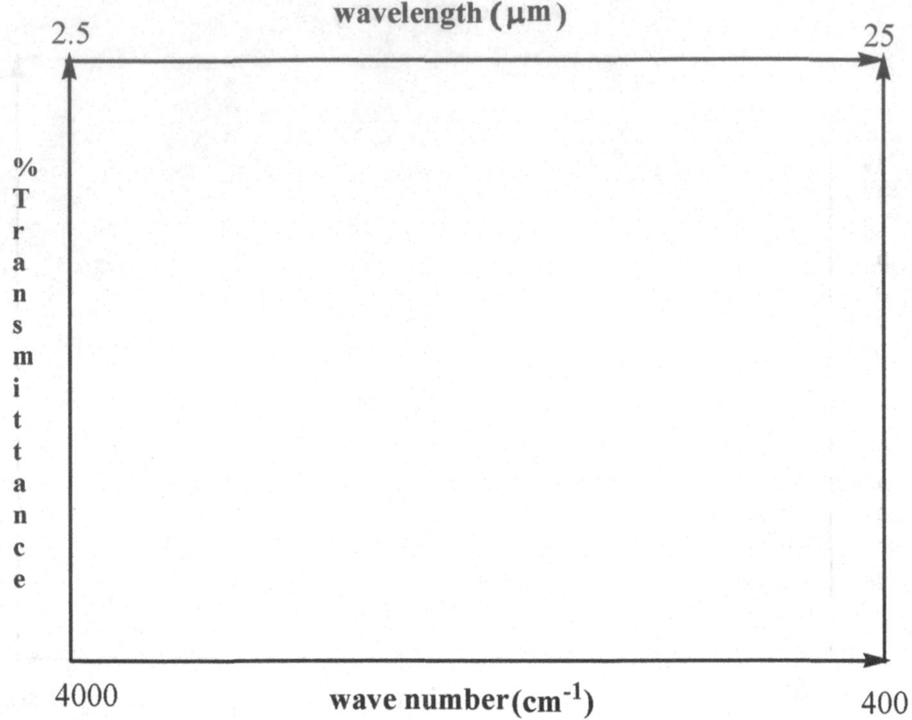

Ex: See Addenda

- **Note: For cyclic ketones, the -C=O frequency increases as the number of carbon atoms decreases because of ring strain.**

Ex:

1715 cm^{-1}

1745 cm^{-1}

1780 cm^{-1}

•Note: For conjugated systems, frequency decreases with increasing conjugation because the -C=O in the conjugated system has a partial single bond character due to resonance. As a result, it is weaker than the nonconjugated C=O. Therefore, it absorbs at a lower frequency.

Ex:

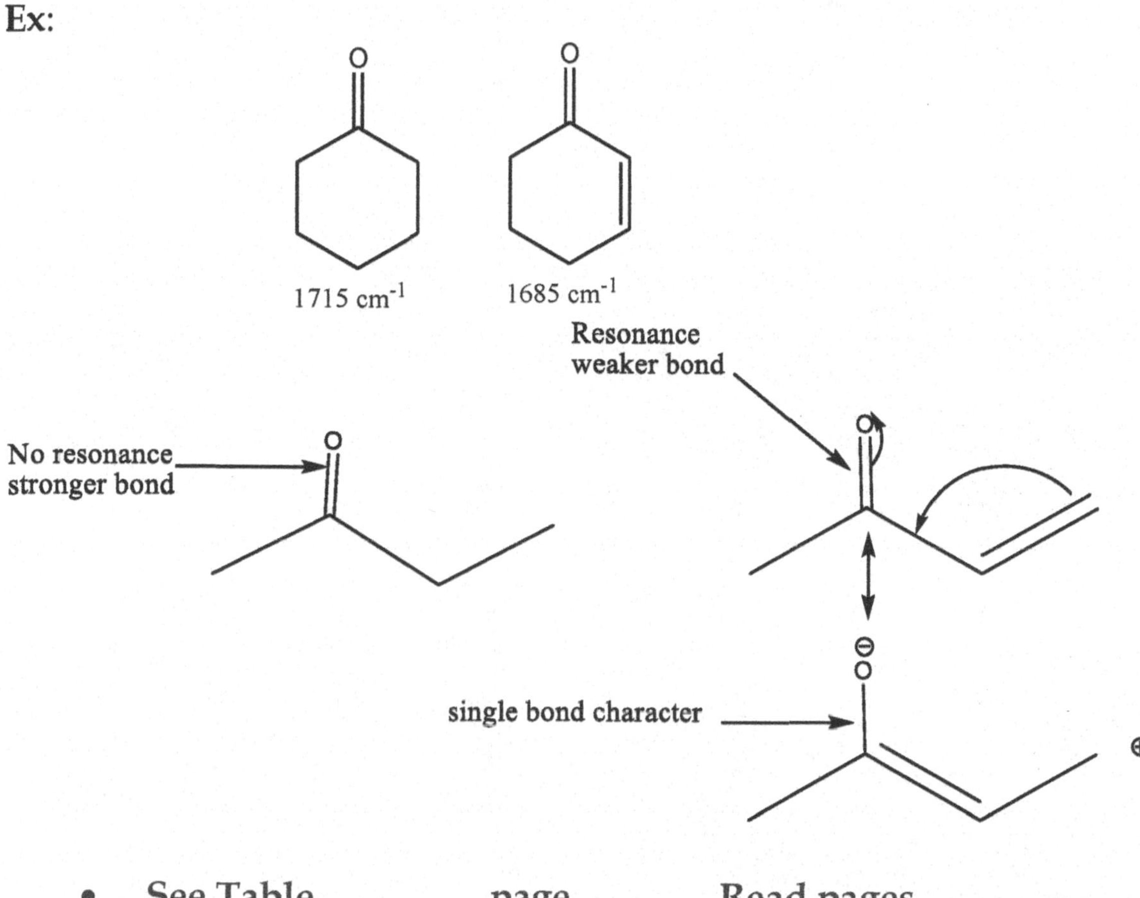

- See Table ________, page ______. Read pages ______ - _______. Take a look at examples and do all problems.

- See Key Concepts on page _______.

OCHEM II UNIT 5: NMR SPECTROSCOPY

A. INTRODUCTION

1. GENERAL REVIEW ON STUCTURE DETERMINATION

- Here are the four types of structure determination.

mass Spectrometry (ms)	molecular weight and molecular formula
Infrared Spectroscopy (IR)	Functional group and bond type
Nuclear Magnetic Resonance Spectroscopy (NMR)	Carbon-Hydrogen Framework
Ultraviolet and Visible Spectroscopy (UV-Vis)	Presence of conjugated π electrons

- We have already covered the first two techniques (**please, see Units 3 and 4).** In this unit, we are covering NMR. This technique **complements** MS and IR by providing a "map" of the C-H (and C-C) framework.

2. TYPES OF NMR

- There are several types of NMR. However, the most common types are **^{1}H NMR** (aka **proton NMR**) and **^{13}C NMR** (aka **carbon NMR**). Whereas ^{1}H NMR allows us to determine the number of types of **hydrogens** in a substance, ^{13}C NMR is used to get the number and kinds of **carbons** that are in an organic compound.

B. BASIC PRINCIPLE BEHIND NMR

1. NUCLEAR SPINS

- **Nuclear spins** stem from the spinning of **positive nuclei** in atoms in molecules. These spinning nuclei act as **tiny bar magnets.** As a result, they can interact with an **external magnetic field, B_o.Ex**: C-13 and H-1.

2. BEHAVIOR OF NUCLEAR SPINS IN THE PRESENCE OF AN EXTERNAL MAGNETIC FIELD

a. No Magnetic Field Applied

- When there is **no external field** acting on them, nuclear spins orient themselves **randomly.**

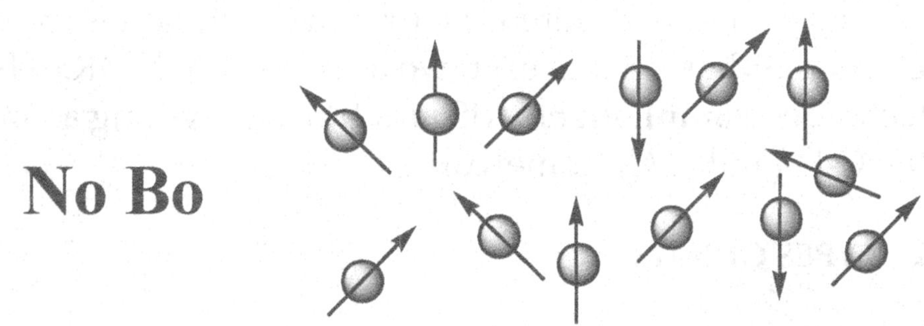

b. Nuclear Spins in an External Magnetic Field

- In the presence of a **strong external magnetic field, Bo,** a spinning H-1 (or C-13) nucleus **aligns itself either with the field or against it.**

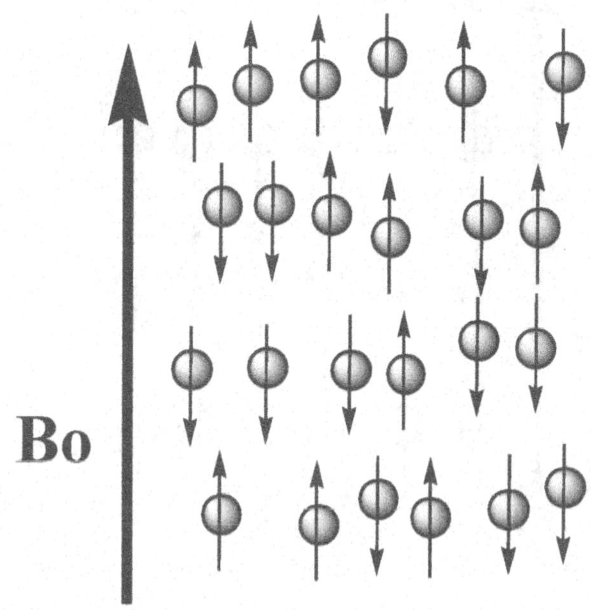

- See Fig. on page _______.

c. Types of Nuclear Spin Orientations in an External Magnetic Field

i. Parallel Orientation

- When nuclear spins orient themselves **with** the applied field, they are said to have adopted a **parallel orientation.** This orientation represents **a lower energy state (or α state) and is more favorable (most stable).**

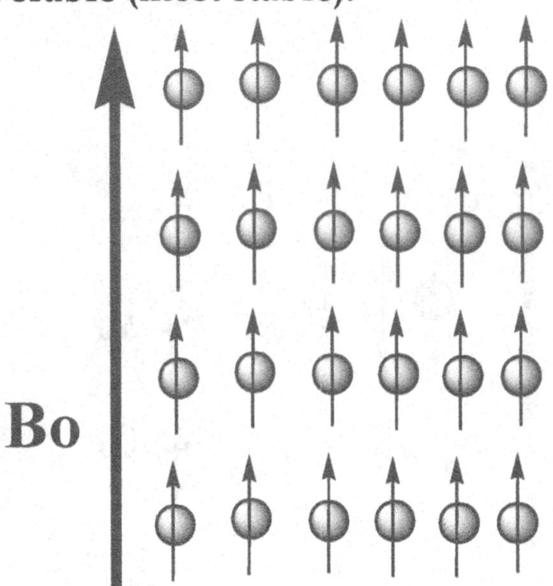

ii. Anti Parallel Orientation

- When nuclear spins align themselves **against** the applied field, they are said to have an **anti parallel orientation.** This orientation is of **higher energy state (or β state) and is less favorable (less stable).**

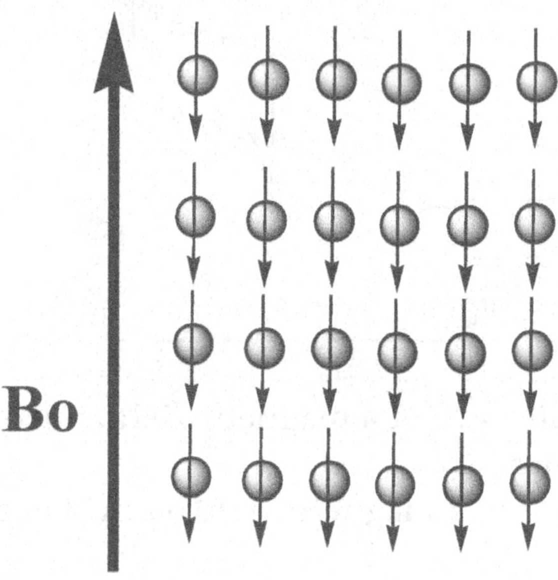

3. RADIO FREQUENCY, SPIN FLIP, AND RESONANCE

- When a **radiofrequency (RF)** having an energy that **matches the energy gap** between the **α state** (low energy, **ground state**) and the **β state** (high energy, **excited state**) is sent through the sample cell in the applied magnetic field, energy is absorbed. The nuclear spins in the **α state** go from this **ground state** to the higher energy level, the **β state.** This phenomenon is known as **"spin flip".** At this point, the nuclei are said to be in **resonance** with the applied field. The name of this technique, **Nuclear Magnetic Resonance or NMR,** comes from this "**resonance**" induced by spinning nuclei in a magnetic field. **By using a technique called Fourrier-Transform, the energy of relaxation of nuclei (returning from β to α state) is measured.**

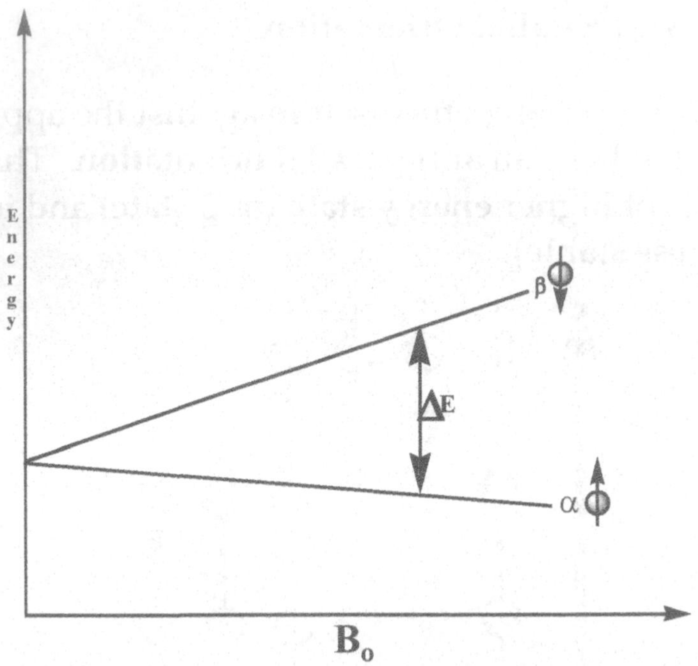

- Note: In the absence of a magnetic field, nuclear spin states have equal energies.
- In the presence of a magnetic field, spin states have unequal energies.
- The higher the applied magnetic field, the greater the energy gap (ΔE) between the spin states. Therefore, strong magnetic fields are used.
- See prior Fig.

- To summarize: NMR is the technique in which spinning nuclei in a strong magnetic field absorb radio frequency to go from a lower energy level to a higher energy level (resonance). See Fig. on page _______.

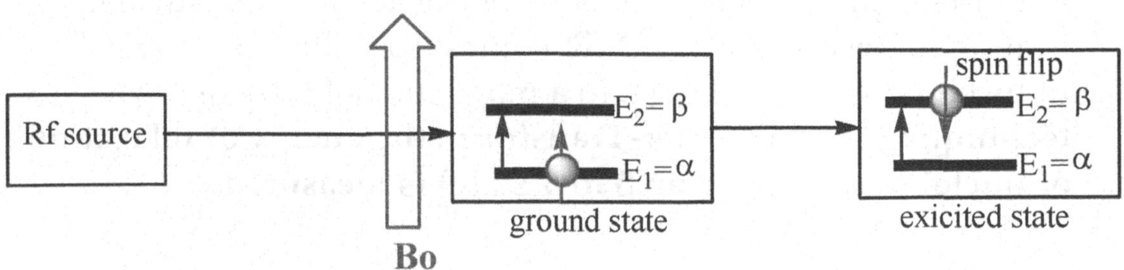

4. THE APPLIED FIELD AND RF

- The RF used in NMR depends on the strength of the applied field and the identity of the **nuclei.** Stronger magnetic fields require higher frequencies and vice versa.

- The applied field, $\mathbf{B_o}$, is directly proportional to the frequency of the RF required to bring the nuclei into resonance. The relationship is:

$$\Delta E = h\nu = \text{constant x } B_o \Rightarrow B_o = \frac{h\nu}{\text{constant}}$$

- To bring ^{1}H nuclei into resonance, RF = 200 MHz, Bo = 4.7 Tesla. A field of 21.6 Tesla (the highest) can also be used with a 920 MHZ RF.

- To bring ^{13}C nuclei into resonance, RF = 50 MHz, Bo = 4.7 Tesla.

5. MAGNETIC-NONMAGNETIC NUCLEI

- All nuclei with an **odd number of protons** (^{1}H, ^{19}F, ^{31}P, ...) and all nuclei with an **odd number of neutrons** (^{2}H, ^{13}C, ^{14}N, ...) exhibit magnetic properties. Nuclei with **even number of both protons and neutrons are non-magnetic. (Sorry, the theory behind this fact is beyond the scope of our study here!)**

- **Read page _______.**

6. THE NMR SPECTROMETER (OR SPECTROPHOTOMETER)

- **There are 6 parts. See Fig. _______, page ______.**

1. RF source.
2. Prism or grating
3. Sample cell wrapped in a magnetic field
4. Amplifier
5. Detector
6. Recorder

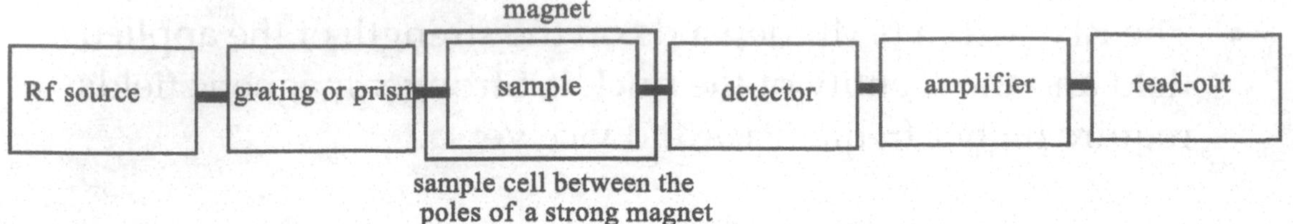

- **Sketch: See Addenda**

- **Note: The sample to be analyzed is dissolved first in deuterated chloroform or $CDCl_3$. Other solvents that can also be used are CD_2Cl_2, $C_6D_6Cl_6$, CD_3CN, and D_2O. Why?**

C. THE NATURE OF NMR ABSORPTION

1. EFFECTIVE MAGNETIC FIELD: B_{eff}

- The **effective magnetic field** (B_{eff}) is the **actual** field (or portion of the applied field) **felt** by a spinning nucleus.

- **Note: In atoms with spinning nuclei, there exist also spinning electrons moving around the nuclei.**
- The electrons act also as **tiny bar magnets.** As a result, they create their own tiny **local magnetic fields.** These local fields act in **opposition** to the applied field, B_o. $B_{effective}$ is given by:

$$B_{effective} = B_{applied} - B_{local}$$

or

$$B_{effective} = B_o - B_{local}$$

- **Note: Nuclei that feel a high B_{eff} require high frequencies to bring about resonance. On the other hand, nuclei that feel a low B_{eff} require low frequencies.**

2. THE SHIELDING EFFECT

- In essence, the local fields created by the spinning electrons **"shield" (or protect) spinning nuclei from the full effect of the applied field.** The nuclei are said to be **"shielded."**
- Depending on the electronic environment, (**electron density around the nucleus**), the effect of the applied field varies. Since different nuclei in different environments absorb differently, one can get **distinct NMR signals** for different nuclei. Structures of substances containing these nuclei are thus determined.

3. THE SHIELDING EFFECT: A SUMMARY

Nucleus	Next to O, N, halogen	Far from O, N, halogen
Status	Deshielded	Shielded
Beff	High, high freq.	Low, low freq.

D. THE NMR SPECTRUM

- An NMR spectrum is obtained by plotting the **effective field strength** vs. **the intensity of absorption** of RF energy or **the intensity of peaks vs. their chemical shifts in ppm (δ).**

- **Note: C-13 and H-1 spectra cannot be recorded on the same chart since each peak represents a chemically distinct ^{1}H or ^{13}C.**

- See Fig. on page ______.

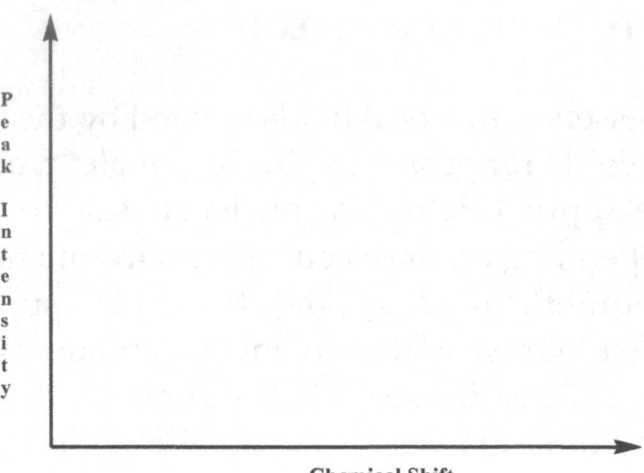

E. CHEMICALLY EQUIVALENT NUCLEI

1. INTRODUCTION

- Nuclei that have the same **electronic environment (same electron density around them)** are said to be **chemically equivalent**. This means that these nuclei are shielded to the same extent by surrounding electrons. As a result, they appear as **a single absorption sharp peak on the NMR spectrum.**

- **Note: If two nuclei are equivalent, they appear in one single peak. Different nuclei appear as distinct peaks.**

Ex: How many different types of H and C are there in the following compound? How many proton NMR signals do you expect? How many carbon NMR peaks do you expect?

- **Note: In identifying equivalent nuclei (either H-1 or C-13), all H atoms must be fully drawn. Do not use skeletal or shorthand notations.**

Ex:

```
   H  O      H  H
   |  ||     |  |
H—C—C—O—C—C—H
   |         |  |
   H         H  H
```

- **Read pages _______ - ______. Do problems on pages ___ - ___.**

2. CHEMICAL SHIFT

a. Downfield - Upfield

- There are two major parts on an NMR spectrum: **downfield (or larger effective magnetic fields) and upfield (or smaller effective magnetic fields). Downfield** refers to the **left-hand side** of the spectrum; it corresponds to **higher frequencies and larger effective magnetic fields. The right-hand side** of the spectrum is called **upfield and corresponds to lower frequencies and smaller effective magnetic fields.**

- **Recall:**

$$B_{effective} = B_o - B_{local}$$

- **The effective magnetic field is directly proportional to the radiofrequency. Therefore, larger effective magnetic fields require higher frequencies. Smaller effective magnetic fields require lower frequencies.**

- Nuclei that absorb **downfield** have little shielding from the surrounding electrons. They are said to be **deshielded.** $\mathbf{B_{local}}$ is small. So they feel a **larger overall effective magnetic field** ($\mathbf{B_{eff}}$). As a result, they appear **downfield (left-hand side of the spectrum) at higher frequencies.** C and H atoms bonded to **electronegative atoms** such as O, N or halogens are highly deshielded and absorb downfield. On the other hand, nuclei that are highly shielded and are not in the vicinity of these electronegative atoms absorb **upfield (right-hand side of the spectrum).** $\mathbf{B_{local}}$ is large. So they feel a **smaller overall effective magnetic field.** As a result, **lower frequencies** are required to bring about resonance. Therefore, they appear **upfield**, closer to **TMS.**

b. Calibration of an NMR Spectrum

- The positions of the peaks on an NMR spectrum are measured relatively to the position of a **reference, tetramethylsilane or TMS,** for both ^{13}C and ^{1}H NMR. TMS is $(CH_3)_4Si$ or:

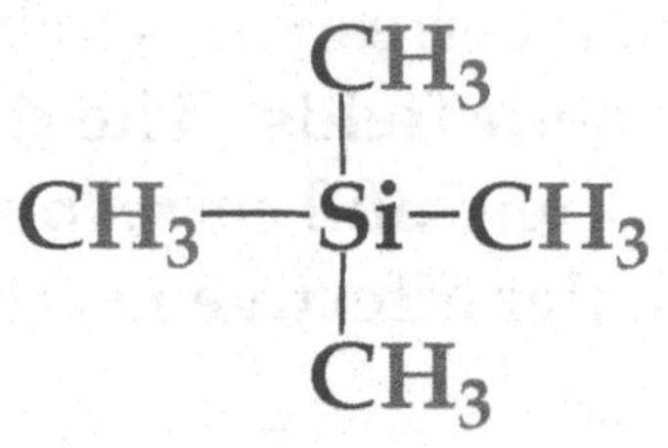

$$
\begin{array}{c}
CH_3 \\
| \\
CH_3{-}Si{-}CH_3 \\
| \\
CH_3
\end{array}
$$

- **Question: Why TMS? TMS is used as a reference because it has (4 carbons) 12 highly shielded equivalent hydrogens that absorb upfield of any known organic compound and give a single peak. Before an NMR experiment, a small amount of TMS is added to the sample to be analyzed.**

c. Chemical Shift

- **The chemical shift** of a nucleus is the **position** of its absorption peak from TMS on the NMR spectrum.

- **Note: The chemical shift of TMS is arbitrarily set to 0 ppm.**

d. The Delta (δ) Scale

The delta (δ) scale is an **arbitrary** scale defined as:

1 δ = 1 ppm of the spectrometer operating frequency

- **Recall: B_0, the magnetic field is directly proportional to the applied frequency.**

- **Note: 1 ppm of a 200-MHz frequency is 200 Hz; therefore, for this spectrometer, 1 δ = 200 Hz. Likewise, in an instrument whose applied frequency is 500 MHz, 1 δ = 500 Hz.**

- In general, the chemical shift of an absorbing nucleus can be calculated as follows.

$$\delta = \frac{\text{observed chemical shift in Hz}}{\text{spectrometer frequency in MHz}}$$

Ex: A 200-MHz spectrometer records a proton that absorbs at a frequency of 2500 Hz downfield from TMS. What is the chemical shift of this proton?

Ex: A proton absorbs at 480 Hz on a 120-MHz spectrometer. The same proton absorbs at 2400 Hz on a 600-MHz spectrometer. What is the chemical shift in **δ (or ppm)**?
From the first spectrometer, δ = 480/120 = 4 δ; from the second spectrophotometer, δ = 2400/600 = 4 δ.

- **Note: The chemical shift (in δ) for a given nucleus is the same for all NMR spectrometers, regardless of the applied frequency.**

- **Analogy: The distance between two cities A (0 mile) and B is always the same (suppose 600 mi), regardless of the speed (or vehicle) of travel. Suppose driver 1 travels the distance in 10 hours at a speed of 60 MPH. It will take 12 hours for driver 2 to travel the same distance at 50 MPH. City B will always be at 600 mi away from city A. In NMR, A is TMS.**

Do problems on pages ______ - ______ .

e. Chemical Shift (in δ) Ranges for ^{1}H and ^{13}C NMR

- **The general chemical shift range for ^{1}H NMR is 0 → 10 δ.**
- **The general chemical shift range for ^{13}C NMR is 0 → 220 δ.**

- **Note: Deshielded H and C nuclei absorb downfield of TMS, at high δ values. On the other hand, shielded H and C nuclei absorb upfield closer to TMS, at low δ values. δ of TMS is 0.**

f. The NMR Spectrum: A Summary

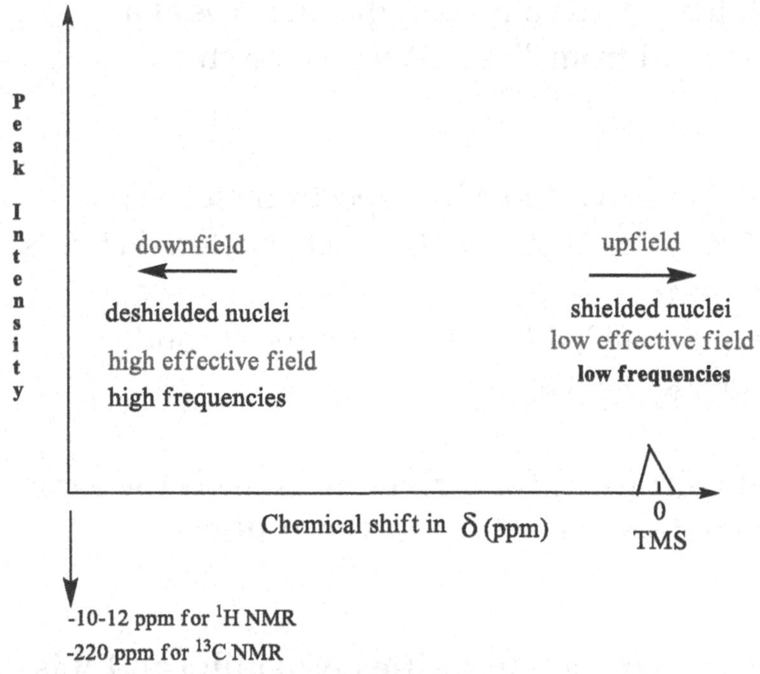

Ex: See Addenda

F. ^{1}H NMR SPECTROSCOPY AND PROTON EQUIVALENCE

1. INTRODUCTION

- Each **electronically distinct H** has a unique sharp peak in the ^{1}H NMR spectrum. A set of equivalent hydrogens gives also a single sharp peak. There are **four different types** of hydrogens or protons:

  i. Unrelated hydrogens or protons
  ii. Homotopic hydrogens or protons
  iii. Enantiotopic hydrogens or protons
  iv. Diastereotopic hydrogens or protons

2. UNRELATED PROTONS

- **Unrelated protons** have **different** chemical environments and therefore give **different** NMR **signals.**

Ex: Pentane

$$CH_3CH_2CH_2CH_2CH_3$$

- **Hint: The monohalogenation of a compound gives two different products from unrelated protons:**

Ex: propane: 2 types of Hs

unrelated Hs

$$CH_3CH_2CH_3 \xrightarrow{Cl_2} CH_3CH_2CH_2Cl + CH_3CHClCH_3$$

2 chlorinated products

3. HOMOTOPIC PROTONS

- **Homotopic protons** have the **same** electronic environment. As a result, they have the **same** ^{1}H NMR signal.

Ex: Pentane

$$CH_3CH_2CH_2CH_2CH_3$$

- **Hint: The monohalogenation of a compound gives only one product from homotopic protons:**

Ex: propane: 6 homotopic 1° Hs and 2 homotopic 2° Hs

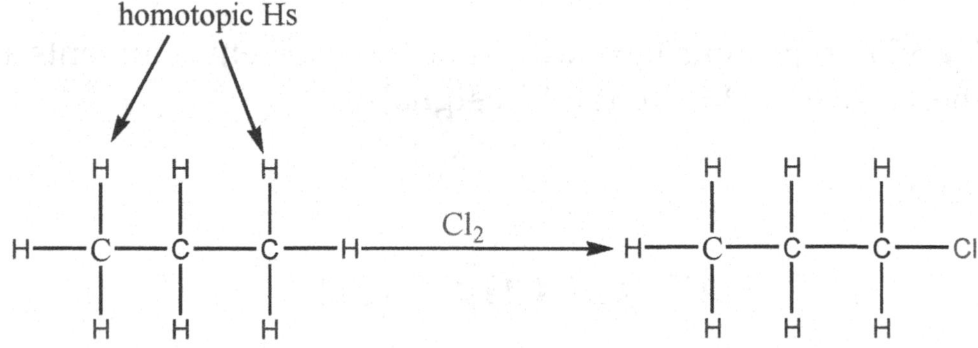

1 chlorinated product

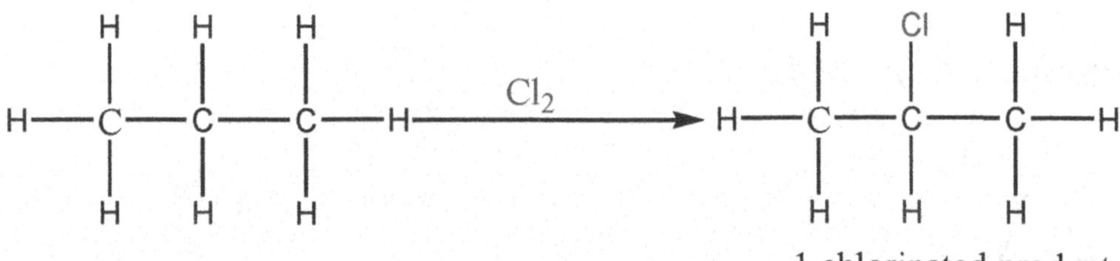

1 chlorinated product

4. ENANTIOTOPIC PROTONS

- **Enantiotopic protons** are **different** but have the **same** electronic environment. Therefore, they have the **same ^{1}H NMR signal**. The replacement of one of the protons creates a new **stereogenic center**.

Ex:

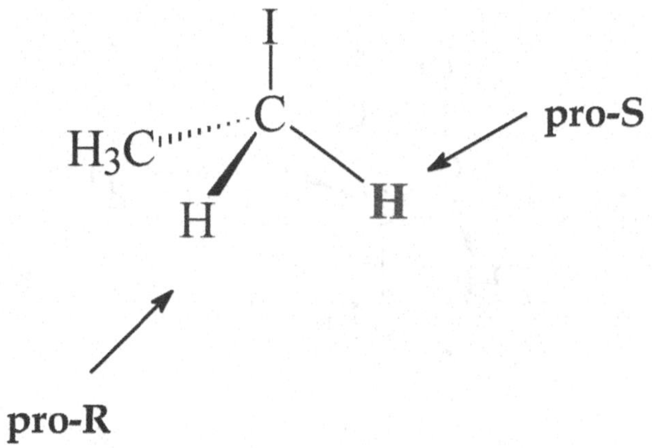

I
H_3C C H H
pro-S
pro-R

- **Hint: The replacement of enantiotopic protons leads to two different enantiomers :**

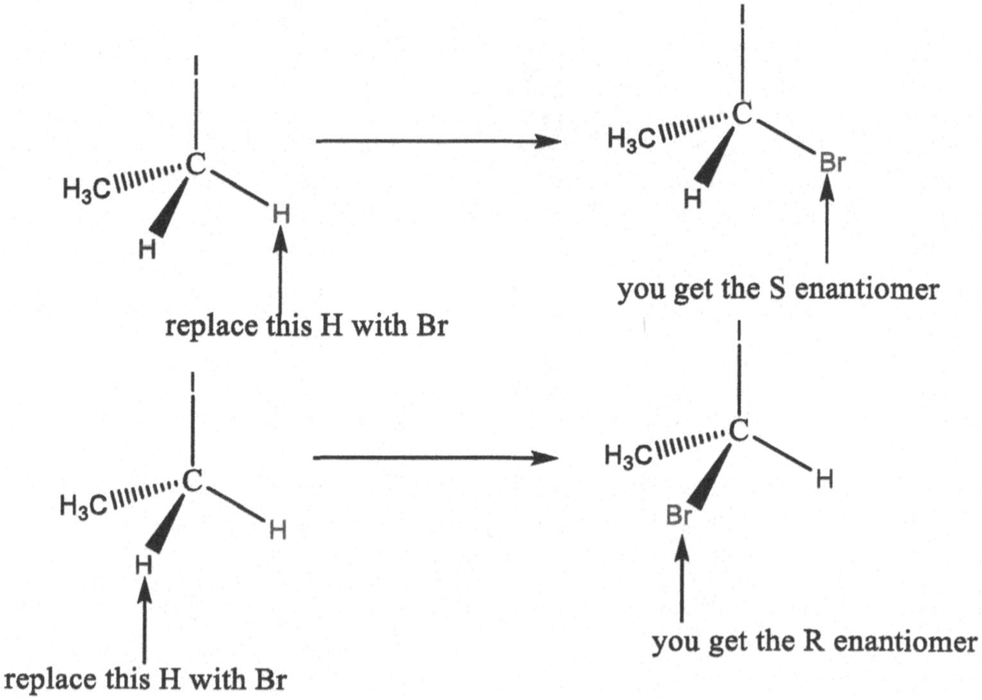

I
H_3C C H H
replace this H with Br

I
H_3C C H Br
you get the S enantiomer

I
H_3C C H H
replace this H with Br

I
H_3C C Br H
you get the R enantiomer

5. DIASTEREOTOPIC PROTONS

- **Diastereotopic protons** are **not** chemically or electronically equivalent. Therefore, they show **different ^{1}H NMR signals**.

Ex:

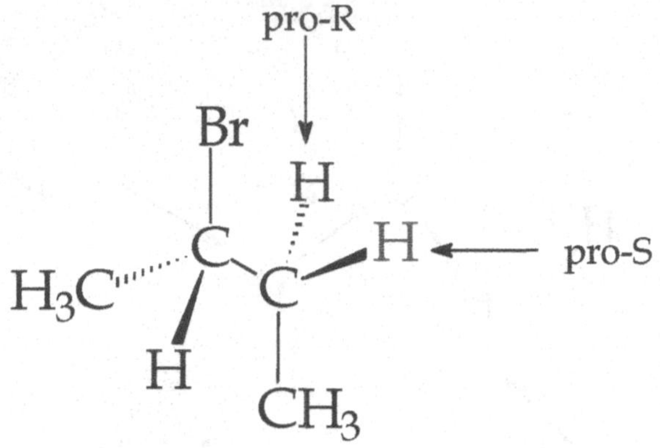

pro-R

Br H

H_3C C C H ← pro-S

H CH_3

- **Hint: The replacement of one of the protons creates a second stereogenic center.**

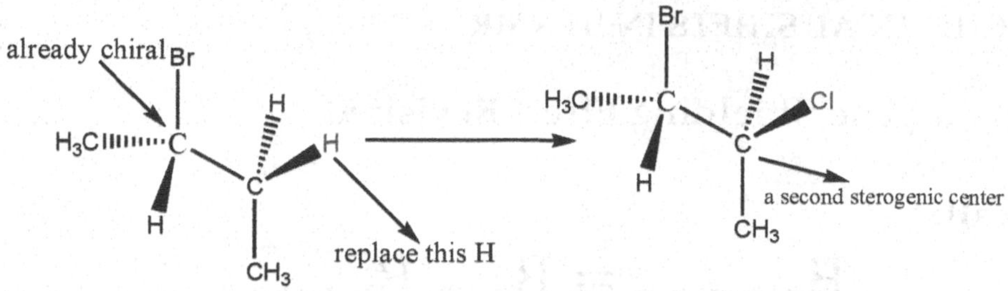

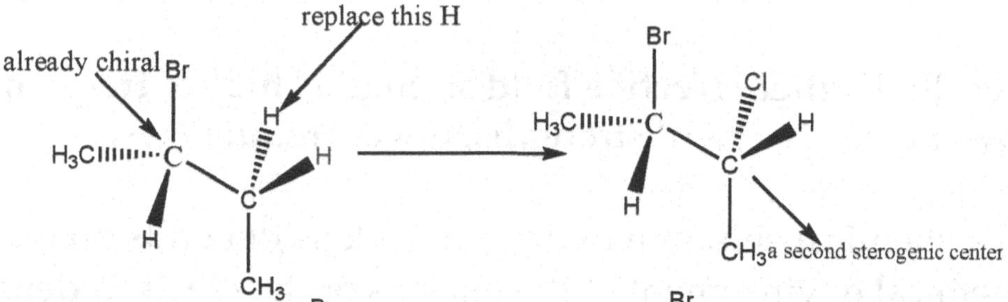

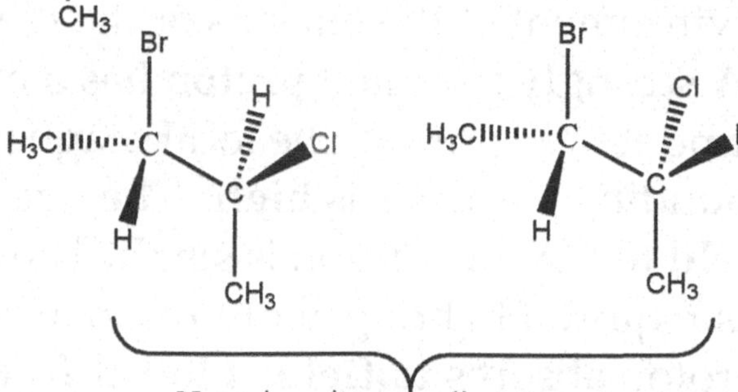

already chiral

replace this H

a second sterogenic center

already chiral

replace this H

a second sterogenic center

Not mirror images: diasteromers

- **Read pages ______ - ______. Do problems on page _____.**

- **Activity: State if each set of shown protons in the following compounds is made of unrelated, homotopic, enantiotopic, or diastereotopic protons.**

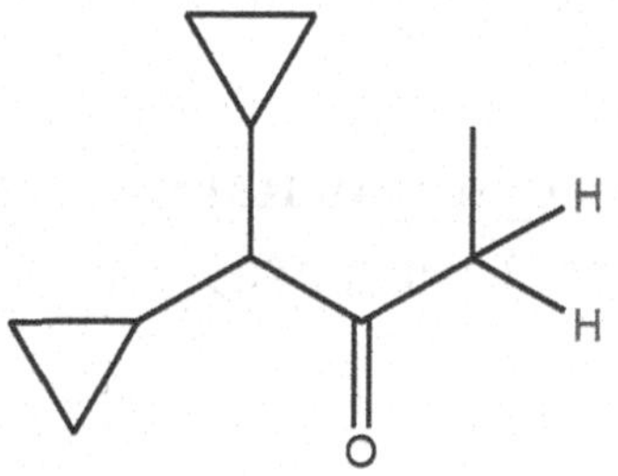

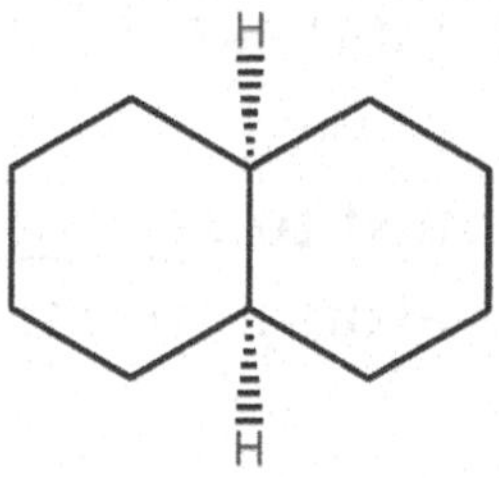

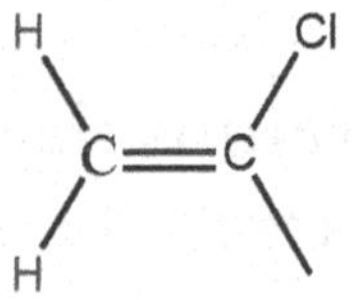

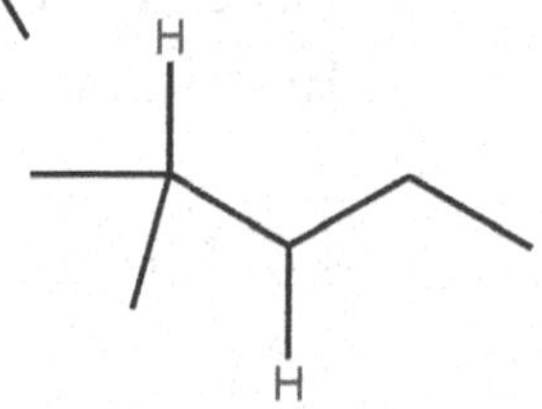

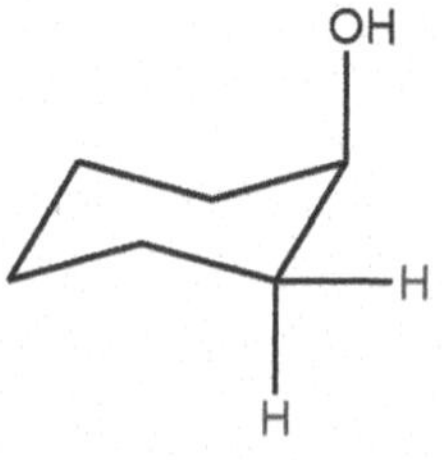

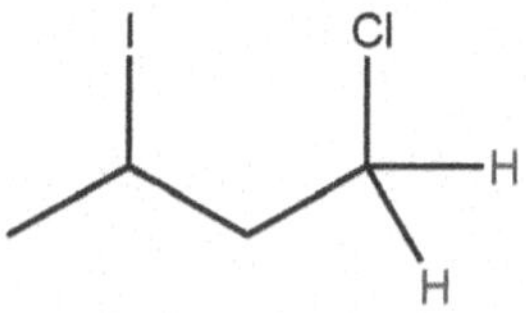

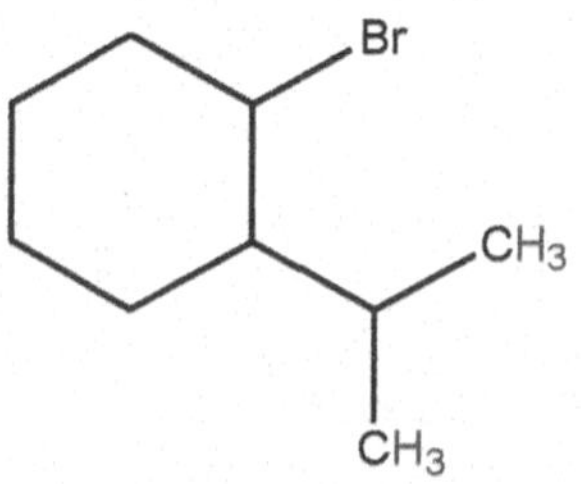

6. CHEMICAL SHIFTS IN ^{1}H NMR

a. The Shielding Effect Revisited

- **Recall:**

$$B_{effective} = B_o - B_{local}$$

- **Recall: Higher effective field strengths, higher frequencies; lower effective field strengths, lower frequencies.**

- The chemical shift of a hydrogen nucleus depends on the chemical environment of the nucleus or the **electron density around it. A strongly shielded proton** has a high electron density around it. As a result, the local magnetic field induced by the surrounding electrons is **high.** The overall **effective magnetic field** felt by the proton is **small.** Therefore, a **small frequency** is required to bring about resonance. The net result is that the proton absorbs **upfield at lower frequencies** (right hand side of the spectrum). On the other hand, a **deshielded proton** has a **low** electron density around it. As a result, the local magnetic field created by the surrounding electrons is **low.** The overall **effective magnetic field** felt by the proton is **high**. Therefore, a **large frequency** is required to achieve resonance. The net result is the proton absorbs **downfield at higher frequencies** (left hand side of the spectrum).

- **Recall: Protons that are next to <u>electronegative atoms</u> have lower electron densities around them and are, therefore, deshielded.**

- **See Fig. ________ page _______.**

Ex: Which protons are the most **deshielded?**

CH_3CH_2F; CH_3CH_2Br; CH_3CH_2Cl

CH_3CH_2F; CH_3CHF_2; CH_3CHCl_2

- **See Fig. ______, page ______.**
- **Read pages ______ - ______. Do all problems on pages ______ - ______.**
- **Recall: The general range of ^1H NMR spectroscopy is:**
  - **$0\ \delta \rightarrow 10\delta$**
  - **^1H NMR Regions**
- There are **6 general regions** in an NMR spectrum. See Fig. ______, page ______.
  - **-Region I: 0 – 1.5 δ**
  - **-Region II: 1.5 – 2.5 δ**
  - **-Region III: 2.5 – 4.5 δ**
  - **-Region IV: 4.5 – 6.5 δ**
  - **-Region V: 6.5 – 8.0 δ**
  - **-Region VI: 8.0 – 12.0 δ**

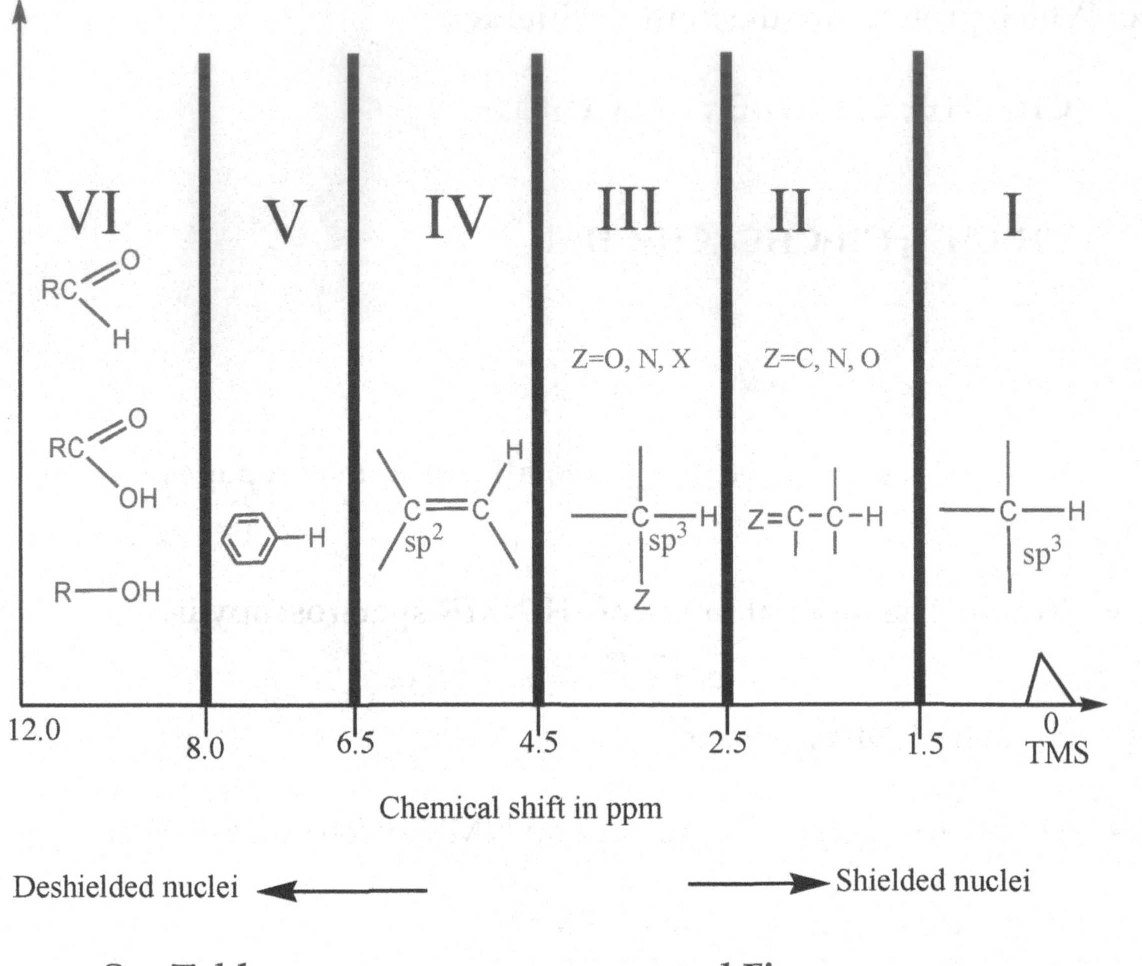

Deshielded nuclei ← → Shielded nuclei

- **See Table _______, page _______, and Fig. _______, page _____.**

Ex: How many ^1H NMR signals do you expect for the following compound? Rank the protons in increasing order of chemical shift.

```
   H O   H H
   | ||  | |
 H-C-C-O-C-C-H
   |     | |
   H     H H
```

- **Read pages ______ - _____. Do problem on page ______.**

b. Special Chemical Shift Values

i. The sp^3 Proton

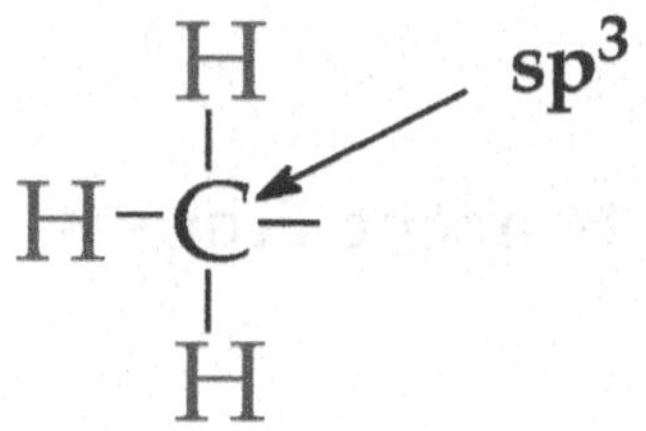

$$\mathrm{H{-}C(H)(H){-}} \quad \leftarrow sp^3$$

- This H is **strongly shielded** when there is no electronegative atom in the vicinity. It absorbs at about **0 – 1.5 δ.**

- **Note: Substituted sp^3 carbon protons have higher chemical shift values than methane because carbon is more electronegative than H. The more the alkyl groups, the higher the chemical shift.**

- **Do Problem ________, page _______.**

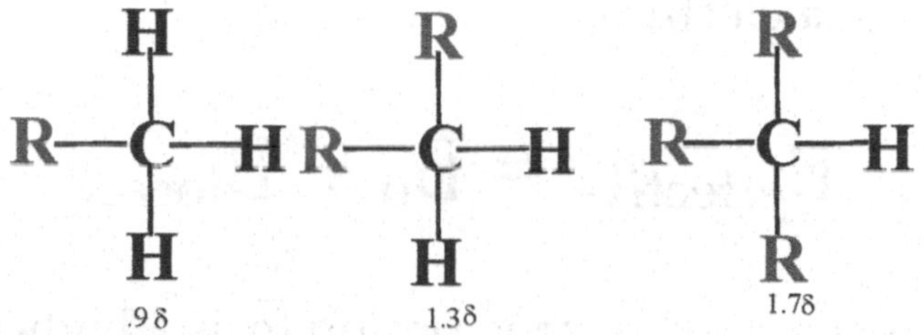

$$\mathrm{R{-}CH_2{-}H} \quad \mathrm{R_2CH{-}H} \quad \mathrm{R_3C{-}H}$$

.9 δ　　1.3 δ　　1.7 δ

ii. The sp^2 Protons

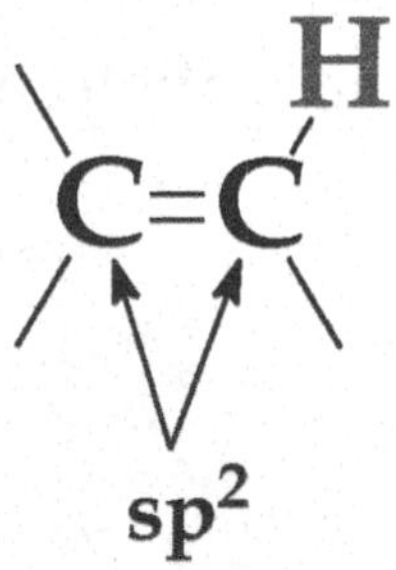

$$\mathrm{C{=}C{-}H}$$

sp^2

- The loosely held π electrons induce a **local** magnetic field that **adds** to the external applied magnetic field, B_o.

$$B_{effective} = B_o + B_{local}$$

- As a result, the proton feels a larger effective magnetic field. So it is **deshielded.** It takes a **higher frequency** to bring it into resonance. This proton absorbs **downfield** at about **4.5–6.0 δ.** **Read page _______.**

iii. Benzene Protons

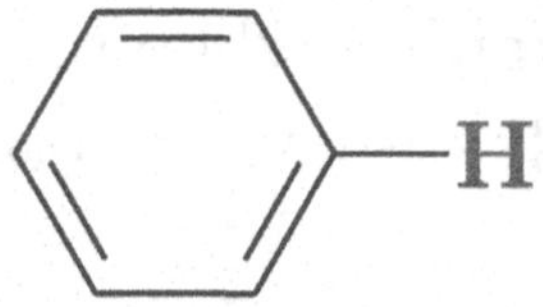

- In the presence of an external magnetic field, the **6** delocalized **π** electrons in the benzene circulate around the ring. This phenomenon creates a **ring current** that induces a **local** magnetic field that **adds** to the external magnetic field as in the case of the alkene **π** electrons.
-

$$B_{effective} = B_o + B_{local}$$

- The net result is that the benzene proton feels a higher effective magnetic field. Therefore, a **higher frequency** is required to bring about resonance. So, a benzene proton is highly deshielded and absorbs downfield at about **6.5–8.0 δ.**

- **Read page _______.**

- **See Fig. on page _______.**

iv. The sp Proton

$$-C \equiv C-H$$

sp

- In this case, the 4 π electrons circulate around the C-C axis and also create a ring current in the presence of an external magnetic field. However, the induced **local** magnetic field that results **does not add to B_o.** It rather acts in **opposition** to the applied field.

$$B_{effective} = B_o - B_{local}$$

- This makes the overall effective magnetic field felt by the **proton smaller.** As a result, a **smaller frequency** is required to achieve resonance. The net result is that the sp proton is **shielded** and absorbs **upfield** at about **2.5 δ.**

- **Read page _______.**

- **See Fig. on page ______.**

7. INTEGRATION OF AN ^{1}H NMR SPECTRUM

a. Introduction

- The height of a peak is proportional to the area **under a peak** which is itself directly proportional to the **number of hydrogens** that give rise to that peak. So by "integrating" the **areas under the peaks in an NMR spectrum,** one can determine the number of hydrogens responsible for the peaks. **It is noteworthy to know that integration is automatically**

done by the spectrometer during analysis. The values of integrals are given in an arbitrary unit.

- **Read pages _______ - ______.**
- **See example on page ______.**
- **Do Problems on page ______.**

b. Determination of the Number of Protons in an NMR Signal from the Molecular Formula of the Analyzed Sample

The following formulas are useful.

$$\text{Total number of units} = \sum \text{integration units}$$

$$\text{Number of units per proton} = \frac{\text{total number of units}}{\text{number of protons}}$$

$$\text{Number of protons per signal} = \frac{\text{integration value}}{\text{number of units per proton}}$$

- **See example on page ______.**

Ex 1: The molecular formula of a compound is $C_9H_{10}O_3$. There are 3 signals with integration values of 54, 23, and 33. Find the number of protons responsible for each signal.

Ex 2: Try $C_5H_{12}O$. In the spectrum, there are 2 signals with integration values of 20 and 60. Find the number of protons that give rise to each signal.

- **Read pages ______ - ______.**
- **Do Problems on page _____.**

8. SPIN-SPIN SPLITTING IN ^{1}H NMR

a. Introduction

- The absorption signal of a proton or equivalent protons is sometimes split into two or more peaks. This phenomenon is called **spin-spin splitting or spin coupling.** It occurs with **nonequivalent protons** on the **same carbon or adjacent carbons.** It is due to the local magnetic fields of neighboring protons. Indeed, the magnetic field of a nucleus affects that of its neighbor nuclei and vice versa.

- **Note: An NMR signal is the total absorption of a proton or a group of equivalent protons. A signal is made of 1 or more peaks.**

b. Explaining Doublets and Triplets

- The NMR spectrum of 1,1-dicholro-2-fluoroethane shows 2 signals as expected since it has 2 types of protons. However, the first signal is split into 2 peaks (**a doublet**). The second signal is split into three peaks (**a triplet**). **Why?**

H_a Cl

F–C–C–H_b

H_a Cl

♣ **Explaining the doublet: absorption signal of the H_a protons**

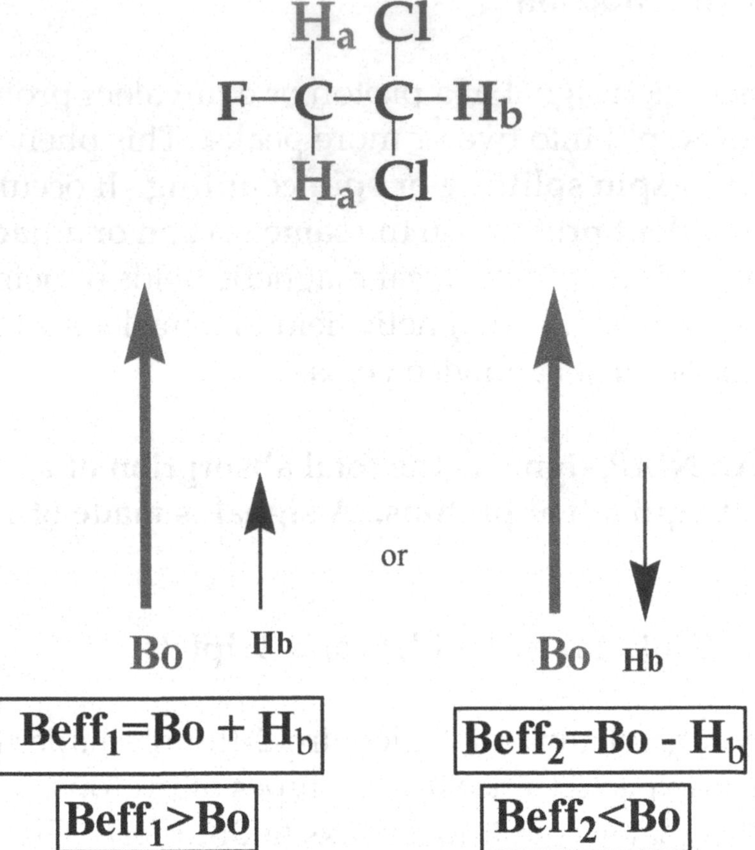

- The doublet represents the absorption signal of both **H_a hydrogens.** If these **equivalent hydrogens** were not adjacent to H_b, their absorption signal would appear as a **single peak.** In other words, the presence of H_b has caused the signal to be split into **two peaks**. This split is due to the fact that in the presence of a magnetic field, the H_b nucleus can either align itself **with** the field or **against it**. This causes the two H_a to feel **two different effective fields**, one slightly higher than the other. The net result is the H_a protons absorb at **two different frequencies yielding a split of their absorbing signal into two peaks. The two peaks in a doublet have about the same area, a 1:1 ratio.**
- See page ______.
- Read pages _______ - ______.

- **Explaining the triplet: absorption signal of the H_b proton**

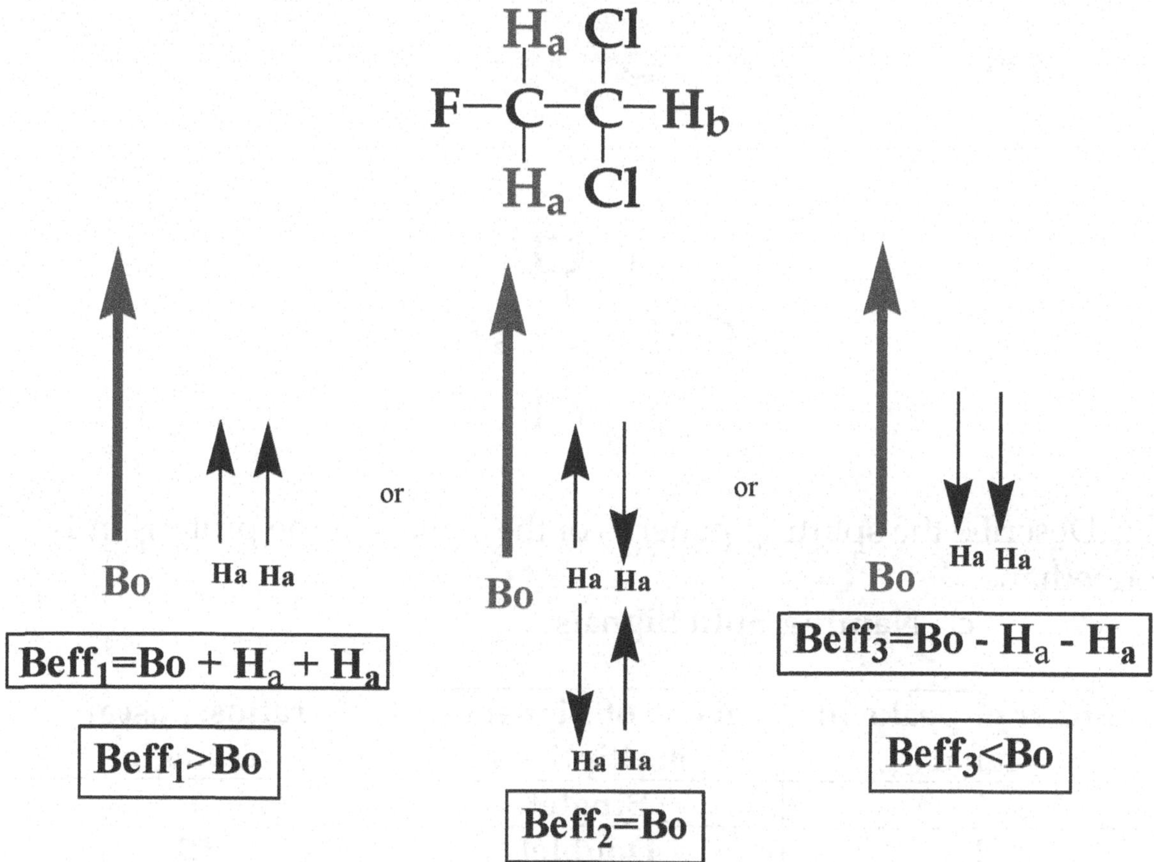

- The triplet is the absorption signal of the H_b hydrogen. **As in the case** of the H_a hydrogens, if this proton were not adjacent to the H_a hydrogens, the absorption signal would appear as a **single peak.** In other words, the presence of the H_a protons has caused the signal of H_b, to be split into three peaks. The split is due to the fact that in the presence of an external magnetic field, the two H_a protons can either align themselves **with** the field or **against** the field in **three ways.** This causes H_b to feel **three different effective fields: one slightly higher than the applied field, one slightly smaller than the applied field, and the third with the same strength of the applied field.** The net result is the H_b proton absorbs at **three different frequencies in the NMR spectrum yielding a split of its absorbing signal into three peaks. The intensity of the middle peak in a triplet is higher than the other two outer peaks because there are 2 ways the 2 H_a protons can arrange themselves. The ratio is 1:2:1.**

- See page ______.
- Read pages ______ - ______.
- Summary:

```
    Ha Cl
    |  |
F—C—C—Hb
    |  |
    Ha Cl
```

Ex: Describe the splitting patterns of the signals of the protons in 1-iodoethane.

c. Naming Split Signals

number of peaks in signal	name of Signal or multiplicity	ratios: Pascal Triangle
1	Singlet	1
2	Doublet	1:1
3	Triplet	1:2:1
4	Quartet	1:3:3:1
5	Quintet	1:4:6:4:1
6	Sextet	1:5:10:10:5:1
7	Septet	1:6:15:20:15:6:1
More than 7	Multiplet	--

- See Table _______, page _____.

d. The Coupling Constant, J

- J is the **distance** between peaks in a ^1H NMR signal. It is expressed in unit of frequency: the **Hertz.** J is the **same** for coupled protons. Furthermore, J is independent of the spectrometer field strength. For open-chain alkanes, J ranges from **6 to 8 Hz.**

e. **Splitting Rules**

- Chemically equivalent **protons do not split each other.**

Ex: Describe the splitting patterns of the protons in 1,2-dibromoethane.

- **The n+1 rule:** If a proton has **n** equivalent neighboring protons, its NMR signal is split into **n+1 peaks.**

Ex: 1-bromo-2-fluoroethane:

$$Br—CH_2CH_2—F$$

- **See Table ________, page _______ for common splitting patterns.**

- **Note: If there are more than 7 peaks in a signal, we have a multiplet.**

- Splitting occurs with **nonequivalent protons on the same carbon or on adjacent carbons.**

Ex: How many ^1H NMR signals (and peaks) do you expect for the following compounds?

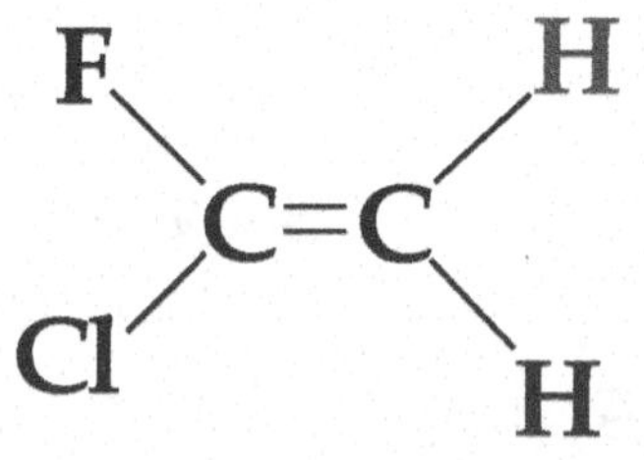

F, Cl on one carbon; C=C; H, H on the other carbon

```
   H  O     H  H
   |  ||    |  |
H—C—C—O—C—C—H
   |        |  |
   H        H  H
```

- **Read pages _______ - ______.**

- **Do problems on page _______.**

- **Activity: Please, use solid lines to represent the following splitting patterns:**

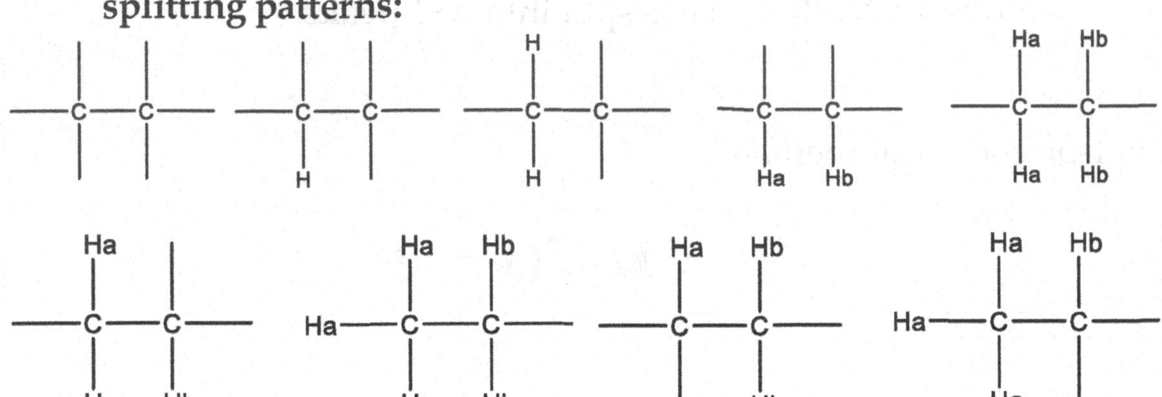

f. More Complex Spin-Spin Splitting Patterns

- **The n and m rule: The signal of a proton or set of equivalent protons "sandwiched" between n and m nonequivalent protons is split into (n+1)(m+1) peaks.**

Ex: Describe the splitting of the signals observed in the analysis of: **$CH_3CH_2CH_2F$.**

- **Read pages ______ - ______. See Fig. ______, page _____.**

- **Do problems on page ______.**

g. Spin-spin Splitting in the Alkenes

i. Introduction

- There are three types of hydrogens to be considered: **geminal, cis, and trans.** In each case, the NMR signal appears as a doublet. However, the coupling constants, J, are different. The increasing order of J is:

$J_{geminal}$ (0-3 Hz) < J_{cis} or J_Z (5 -10 Hz) < J_{trans} or J_E (11-18 Hz)

ii. Spin-Spin Splitting in Monosubstituted Alkenes

Ex: In the NMR spectrum of the following compound, there are 3 doublets of doublets. Explain using a "tree diagram".

H_a H_c
C=C
H_b Cl

- **Hints: H_a splits H_c into a doublet. H_b splits H_c, etc. Read pages _______ - _______. See Fig. ______, page ______.**

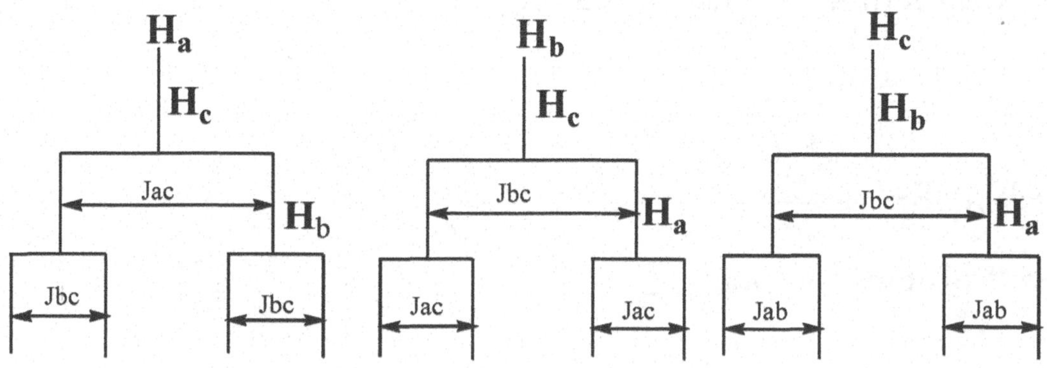

- **Do problems on page ______.**

- **Signals of the OH and NH protons**

- The OH and NH protons are not split by neighboring protons in a magnetic field. They appear as **single peaks or singlets.**

Ex: How many NMR signals (and peaks) do you expect for ethanol?

- **Read page ______. Do problems on pages ______ - ______.**

- **Do Problem , page ______.**

h. Signals of the Protons in Cyclohexane Conformers

- In this case, the axial and equatorial signal is a **singlet because NMR sees an average environment. It cannot distinguish between axial and equatorial hydrogens.**

- **Read pages ______ - ______.**

i. Signals of the Protons in Benzene

- All six protons appear as a **singlet** since they are equivalent, but **deshielded** because of the ring current.

- **Read page ______.**

- **Do Problems, on page ______.**

9. USES of ^{1}H NMR SPECTROSCOPY

- Proton NMR can be used in **two major ways:**
  - To identify the structure of an unknown or the products of a chemical reaction
  - To support a reaction mechanism.

Ex: The non-Markovnikov regiochemistry of alkenes.

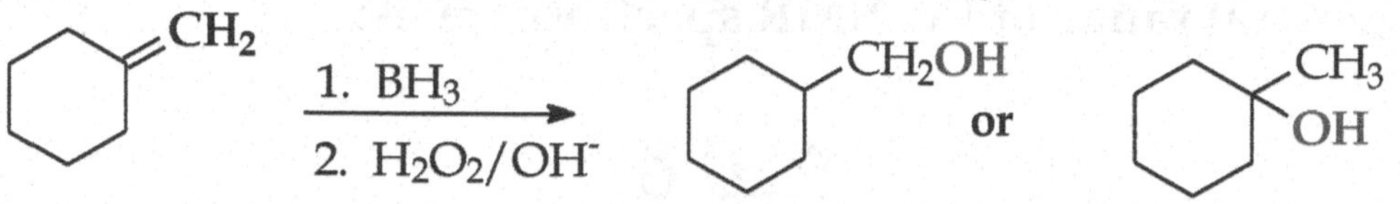

CH_2 1. BH_3 2. H_2O_2/OH^- → CH_2OH or CH_3 OH

non-Markovnikov **Markovnikov (not observed**

Read pages _______ - ______. Do example.

Do all problems on pages ______ - _____.

G. CHARACTERISTICS OF ^{13}C NMR SPECTROSCOPY

1. INTRODUCTION

- Each **electronically distinct** C has a unique sharp peak in the ^{13}C NMR spectrum. A set of carbons that have the same electronic environment gives also a single sharp peak. **Signal averaging** is used to get **better spectrum** (less noise) due to the low abundance of C-13 (**1.1% of all carbon**). In addition, **Fourrier-Transform** is used to increase the **instrument speed.**

Ex: Predict the number of signals in the ^{13}C NMR spectrum of each of the following compounds:

CH_3CH_2OH

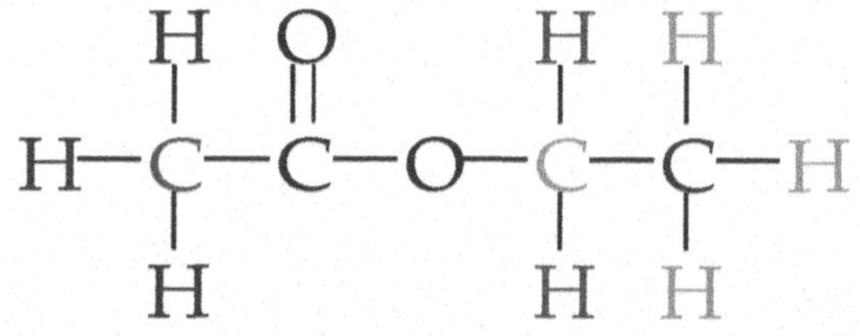

```
   H  O     H  H
   |  ||    |  |
H—C—C—O—C—C—H
   |        |  |
   H        H  H
```

- Read pages ______ - ______.
- See examples on pages ______ - ______.
- Do problems on pages ______ - ______.

2. CHEMICAL SHIFT RANGE IN ^{13}C SPECTROSCOPY

Recall: The general range of ^{13}C NMR spectroscopy is:

$$0\ \delta \rightarrow 220\ \delta$$

- See page ______.

3. SOME OBSERVATIONS ON ^{13}C SPECTROSCOPY

- Carbon atoms bonded to **electronegative atoms (O, N, halogens) are deshielded and absorbed downfield (left-hand side of TMS).**
- **See Fig. on page ______.**
- sp^3 carbons of alkyl groups are shielded and absorb between **0 and 90 δ.**
- sp^2 carbons of alkenes and benzene rings are **deshielded** and absorb between **110 and 220 δ**.
- Carbonyl carbons (C=O) are **highly deshielded** and absorb between **160 and 220 δ**.
- See Table ______, page ______.

Ex:

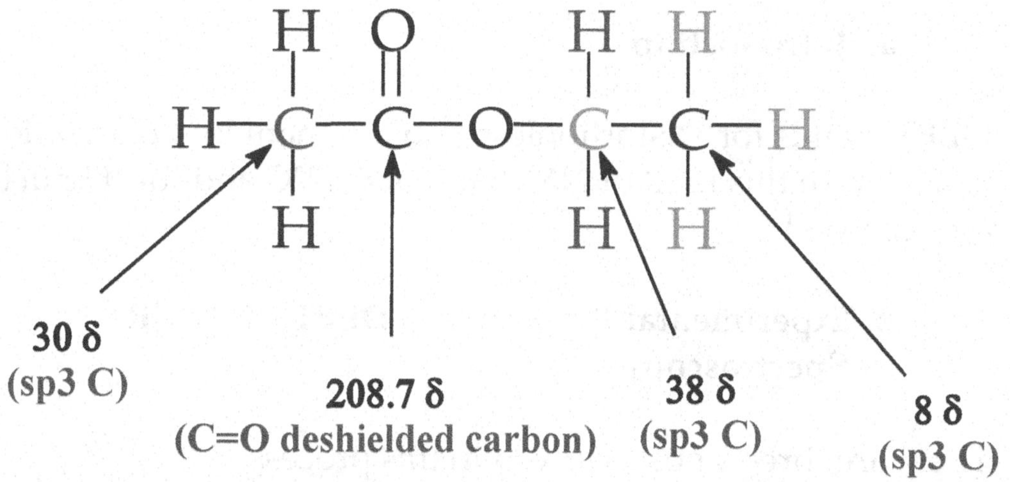

H—C(H)(H)—C(=O)—O—C(H)(H)—C(H)(H)—H

30 δ (sp3 C)

208.7 δ (C=O deshielded carbon)

38 δ (sp3 C)

8 δ (sp3 C)

- **Read pages _______ - _______.**
- **Do problems on pages ______ - ______.**
- Unlike ^{1}H NMR, in ^{13}C NMR, the absorption signals **are not split by neighboring carbon atoms.** As a result, peak intensities **are not proportional to the number of carbon atoms that give rise to them. So ^{13}C NMR signals are not integrated.**

4. USES of ^{13}C NMR SPECTROSCOPY

- ^{13}C NMR spectroscopy can be used in **two major ways:**
  - To identify the structure of an unknown or the products of a chemical reaction
  - To support a reaction mechanism.

Ex: Confirmation of Zaitsev rule.

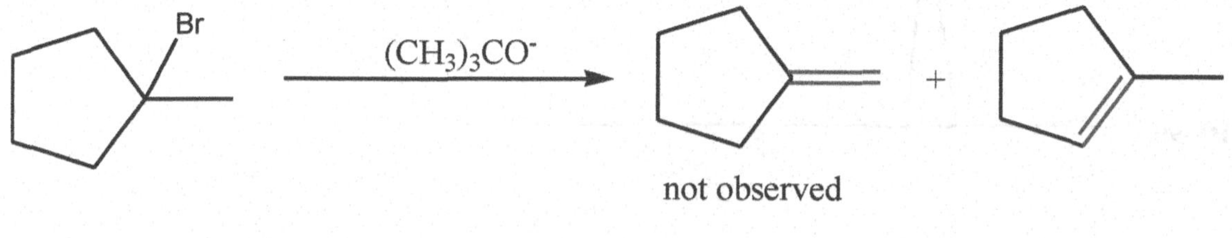

Br

$(CH_3)_3CO^-$

\+

not observed

5. INTRODUCTION to DEPT ^{13}C NMR SPECTROSCOPY

a. Introduction

- **DEPT** stands for Distortionless Enhancement by Polarization Transfer. In this kind of NMR, signals of ^{13}C and/or ^{1}H nuclei are recorded.

b. Experimental Procedure in DEPT ^{13}C NMR Spectroscopy

- There are three steps involved in this process:
    - The first run gives a **Broad Band Decoupled (BBD)** spectrum. In this spectrum, signals of all carbon groups present in the compound appear as **positive** signals.

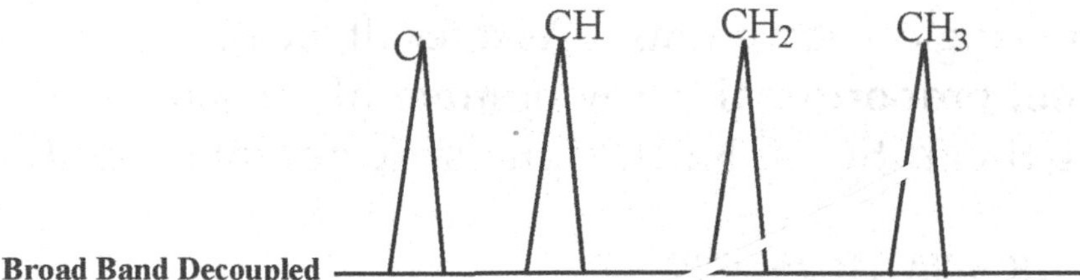

- In the second step, a **DEPT-90** spectrum is obtained. Such a spectrum shows **only** signals of **all -CH-** groups present in the compound.

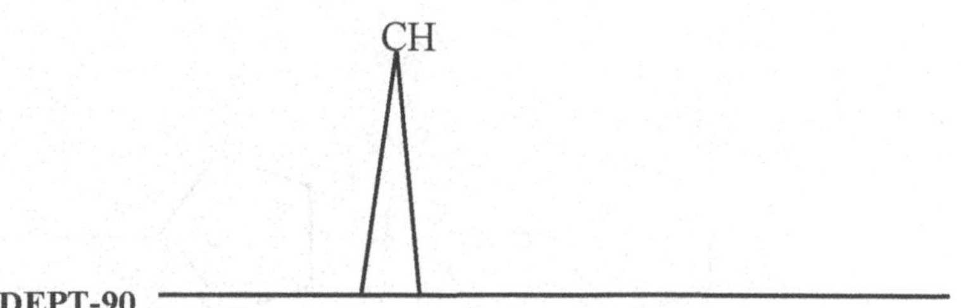

- Finally, a third spectrum called **DEPT-135** shows all **CH-** and **CH_3-** groups present as **positive signals,** and all **$-CH_2$ groups** as **negative signals.**

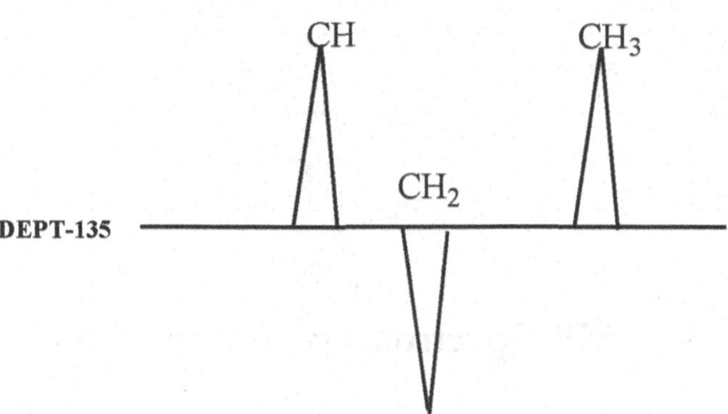

- **#C = BBD – DEPT-135; #CH = DEPT – 90; #CH_2 = negative DEPT-135; #CH_3 = DEPT-135** positive **–DEPT-90**

b. DEPT ^{13}C NMR Spectroscopy: A Summary

Group\Technique	BBD	DEPT-90	DEPT-135
$-CH_3$	√+	-------	√+
$-CH_2$	√+	-------	√-
-CH	√+	√+	√+
-C	√+	-------	------

or

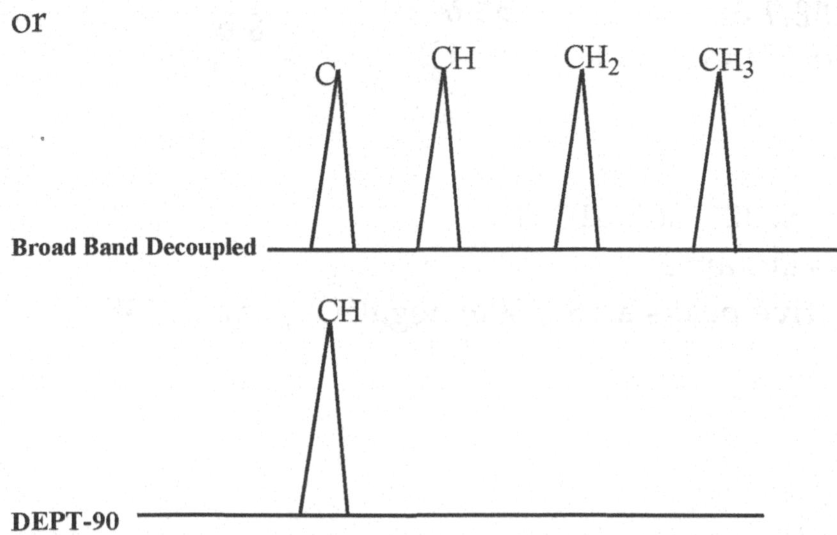

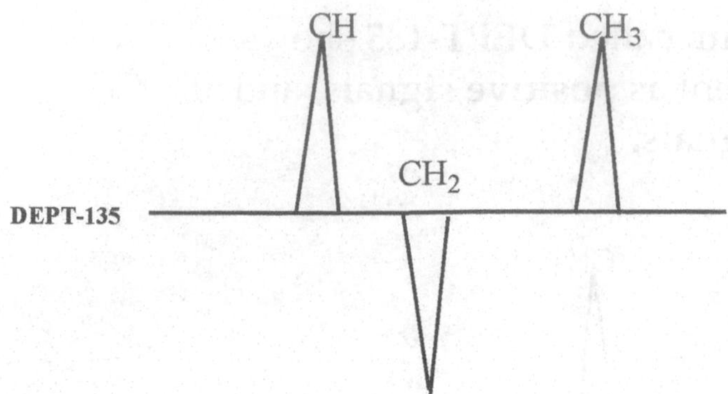

Ex 1: Describe the DEPT ^{13}C NMR Spectroscopy result of the following compound:

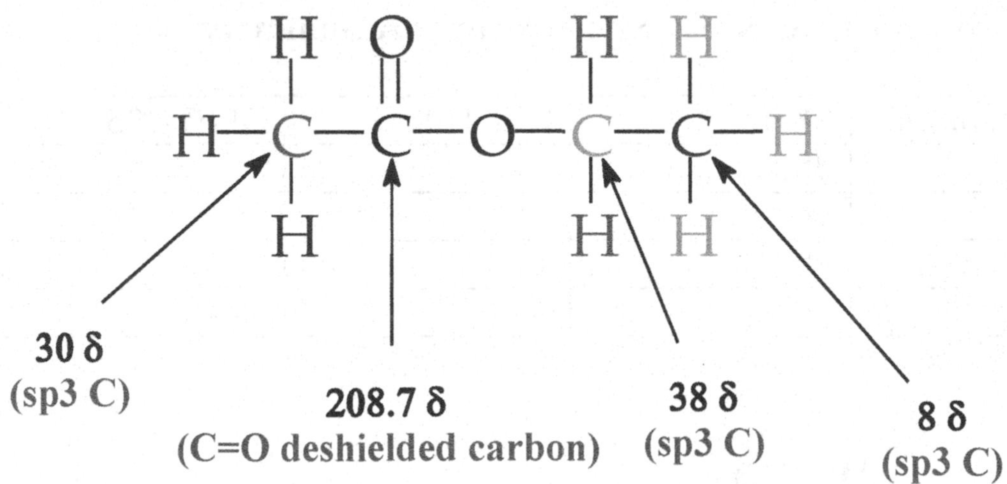

- **BBD: peaks at: 8δ; 30δ; 38δ; 208.7δ**
- **DEPT-90: peaks at: none**
- **DEPT-135: positive peaks at: 8δ; 30δ; negative peak at 38δ.**

Ex 2: Describe the DEPT ^{13}C NMR Spectroscopy result of the following compound:

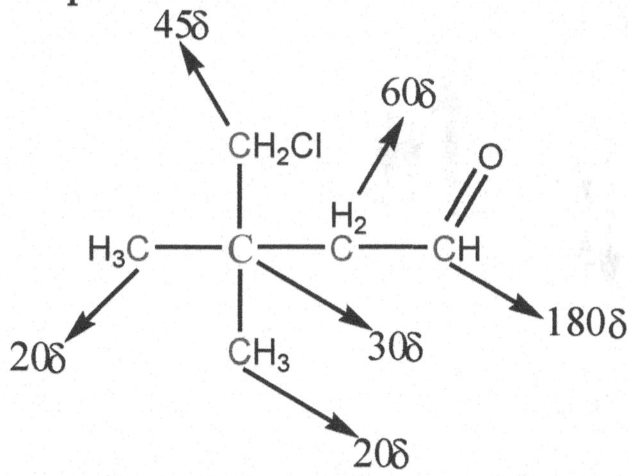

- BBD: peaks at:
- DEPT-90: peaks at:
- DEPT-135: positive peaks at:
  negative peaks at:

6. SOME QUESTIONS

- Why is there no carbon-carbon coupling in ^{13}C NMR?
- Answer: There is no carbon-carbon coupling in ^{13}C NMR because of the low abundance of C-13 (only 1.1% of all carbon).
- Why is there no carbon-hydrogen coupling in ^{13}C NMR?
- Answer: As a matter of fact, there is ^{13}C - ^{1}H coupling. However, the instrument uses "broad band decoupling" to suppress it.
- Read page ______.
- Read about MRI on page ______.
- See Key Concepts page ______.

*Splitting pattern of CH_3CH_2I explained.

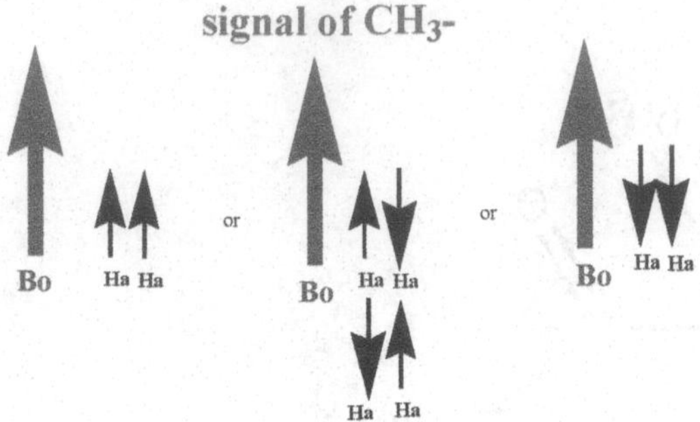

signal of CH_2-

Bo Hb Hb Hb or Bo Hb Hb Hb Hb Hb Hb Hb Hb Hb or Bo Hb Hb Hb Hb Hb Hb Hb Hb Hb or Bo Hb Hb Hb

OCHEM II UNIT 6: CONJUGATED DIENES AND UV-Vis SPECTROSCOPY

A. INTRODUCTION

1. DEFINITION

- **Conjugated unsaturated hydrocarbons** are compounds that contain double bonds that "alternate" with single bonds. They are widespread in nature.

Ex: isoprene, lycopene

1,3-butadiene   isoprene   a polyene

See page ______.

2. DIENES

a. Definition

- **Dienes** are compounds that contain 2 double bonds. There are **three types of dienes:**
  - -Conjugated
  - -Cumulated
  - –Isolated

b. Conjugated Dienes: An Example

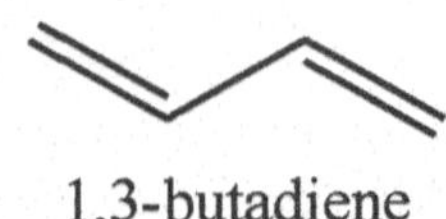

1,3-butadiene

c. Cumulated Dienes: An Example

d. Isolated Dienes

B. 1,3-CONJUGATED DIENES

1. PREPARATION

a. Using Allylic Alkenes

- **The general reaction is:**

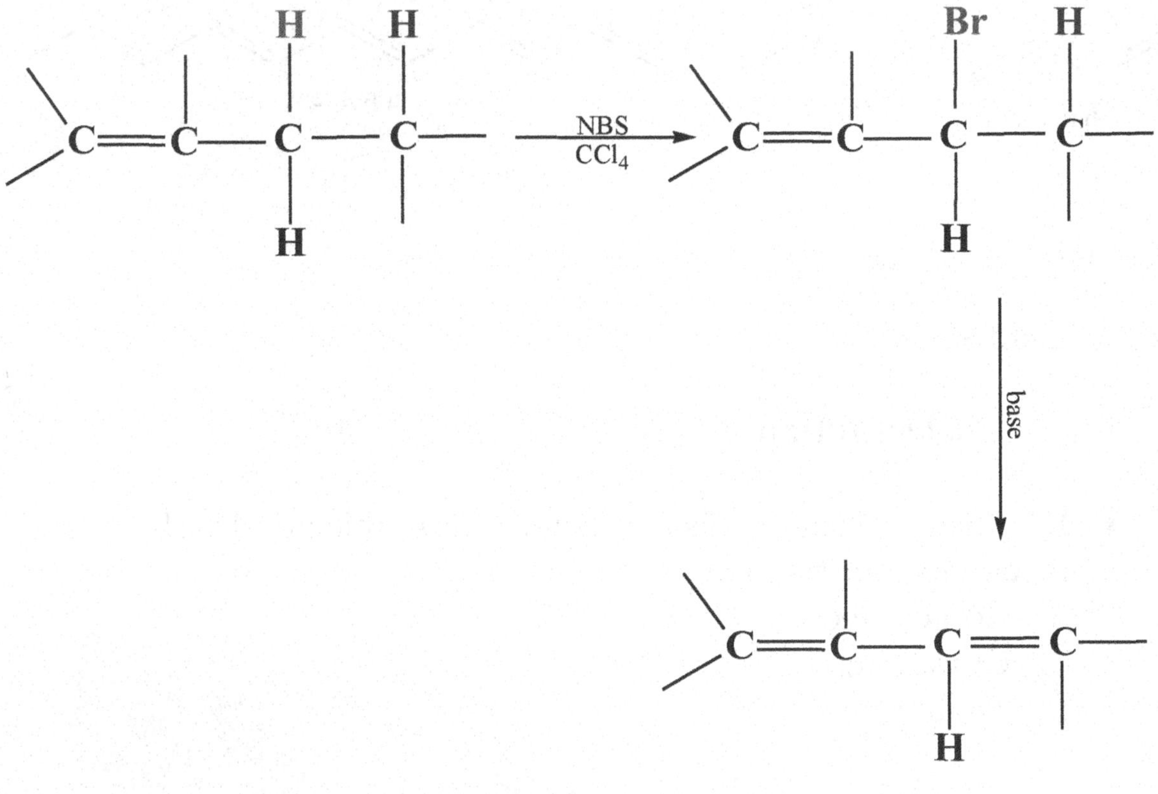

H H

C=C—C—C—

H

NBS

CCl_4

Br H

C=C—C—C—

H

base

C=C—C=C—

H

Ex:

$$\text{cyclohexene} \xrightarrow[\text{2. KOH}]{\text{1. NBS/CCl}_4}$$

$$\text{2-octene} \xrightarrow[\text{2. KOH}]{\text{1. NBS/CCl}_4}$$

b. Industrial Thermal Cracking of Alkanes

Ex:

$$CH_3CH_2CH_2CH_3 \xrightarrow[\text{catalyst}]{600°C} CH_2{=}CHCH{=}CH_2 + 2H_2$$

- **The catalyst is: Cr_2O_3/Al_2O_3.**

c. Acid-Catalyzed Double Dehydration of Diols: Isoprene Preparation

$$CH_3C(CH_3)(OH)CH_2CH_2OH \xrightarrow[\text{Heat}]{Al_2O_3} CH_2{=}C(CH_3)CH{=}CH_2 + 2H_2O$$

2. SPECIAL BONDING PROPERTIES OF CONJUGATED DIENES

- The single bond between the "alternate" double bonds in a conjugated system is shorter than a normal carbon-carbon single bond.

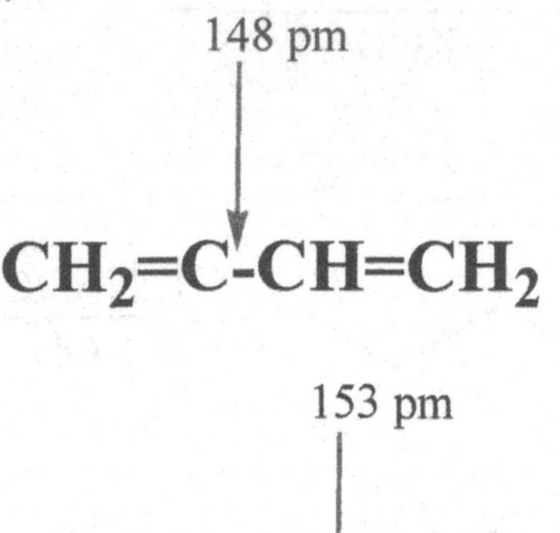

$CH_3CH_2\text{-}CH_2CH_3$

- **Read pages _______ - _______. Do problems on page ______.**

3. STABILITY OF CONJUGATED DIENES

- Numerous experiments (heats of hydrogenation, etc.) have shown that conjugated systems are more stable than nonconjugated ones.

- **Note: The higher the heat of hydrogenation, the least stable the substance, and vice versa.**

Ex:

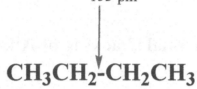

non conjugated — $\Delta H^o = -61$ kcal/mol

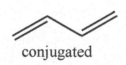

conjugated — $\Delta H^o = -54$ kcal/mol

- **Read pages 6 ______ - _______. Do problems on page _______.**

4. THE VB DESCRIPTION OF 1,3-BUTADIENE

- All carbons are sp^2 hybridized. **Delocalization** of the electron density (**4 pi electrons**) adds more stability to the compound. Indeed, all 4 pi electrons spread over all 4 carbons.

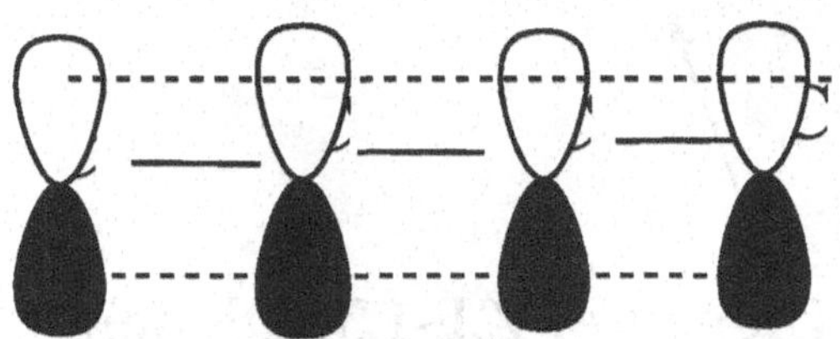

- **Read pages _______ - _______.**

- **Do problems on page _______.**

5. STEREOCHEMISTRY OF CONJUGATED DIENES

a. Introduction

- There are 3 possible stereoisomers for **RCH=CH-CH=CHR′:**
    - Trans-trans-1,3-diene (or E,E-1,3-diene))
    - Cis-cis-1,3-diene (or Z,Z-1,3-diene)
    - Cis-trans-1,3-diene (or Z,E-1,3-diene)

b. Trans-Trans-1,3-Diene ((or E,E-1,3)-Diene):))

- The two double bonds are trans (or E).

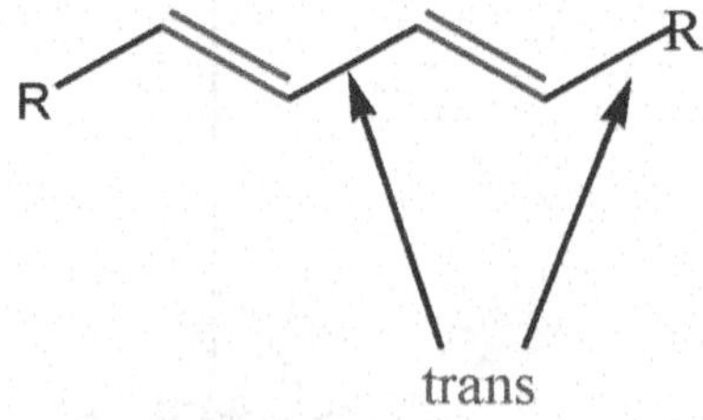

c. **Cis-Cis-1,3-Diene ((or Z,Z-1,3)-Diene):**

- The two double bonds are cis (or Z).

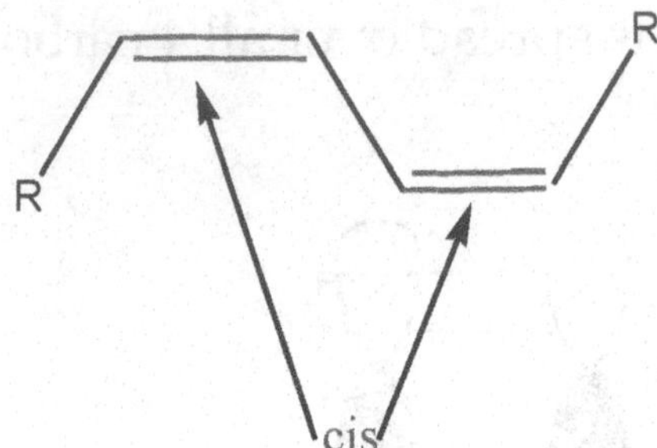

d. **Cis-Trans-1,3-Diene ((or Z,E-1,3)-Diene))**

- One double bond is cis (Z) and the other one is trans (E).

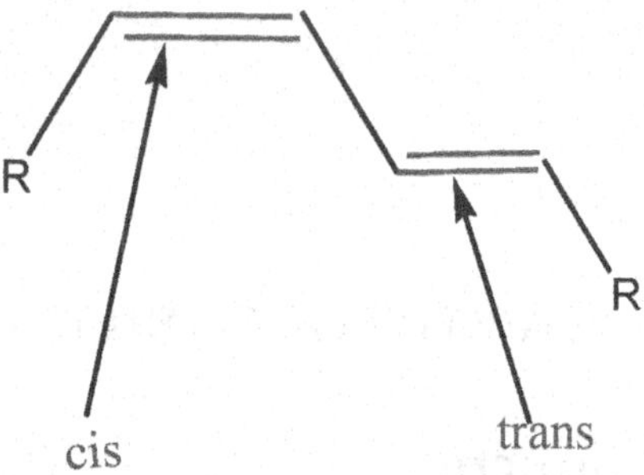

e. **Conformers: 2 for Each Stereoisomer:**

- **Two** conformers result from the rotation around the **single bond** connecting the two double bonds.
  - **s-cis conformer: The two double bonds of the diene are on same side:**

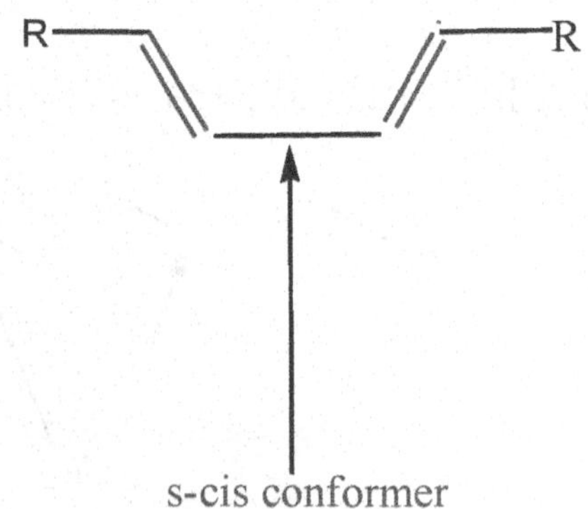

- s-trans conformer: : **The two double bonds of the diene are on opposite sides:**

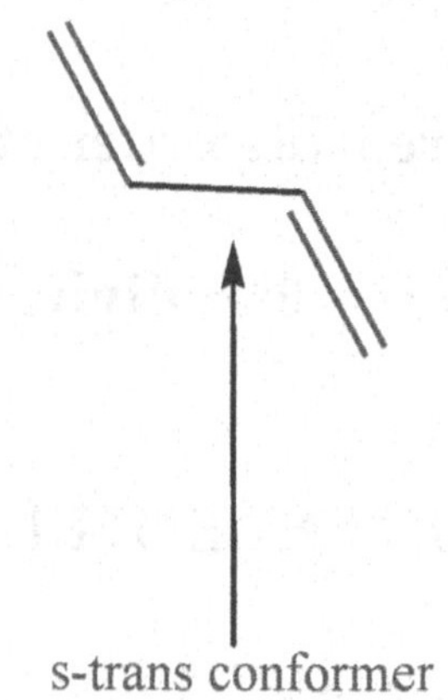

- **Read pages ______ - ______. Do Problems on page ______.**

C. THE ALLYLIC CARBOCATION

- Each carbon is **sp^2** hybridized. The allylic carbocation is **conjugated.**

sp^2

sp^2 sp^2

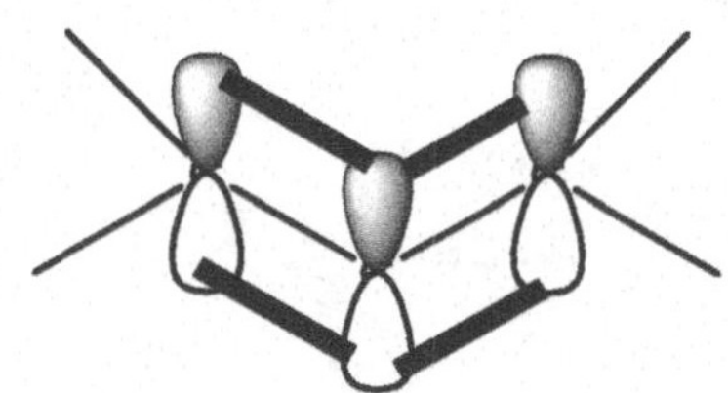

- **The electron density of the two pi electrons spreads over all the three carbons, adding stability to the allylic carbocation (delocalization).**

- **Recall: The increasing order of stability of carbocations is:**

Methyl < 1° < 2° ~allylic~benzilic < 3°

D. COMMON RESONANCE STRUCTURES

1. RESONANCE STRUCTURES

a. Definition

- Resonance structures are **alternative** Lewis structures that do not exist by themselves.

b. Systems Giving Rise to Resonance Structures: 4 Types

- **Allylic systems**
- **Conjugated systems**
- **Species in which a lone pair is adjacent to a positive charge.**
- **Species containing a double bond in which one bonding atom is more electronegative than the other.**

- Read pages ______ - ______.

2. ALLYLIC SYSTEMS:3

- They can be either **radicals, cations, or anions.**

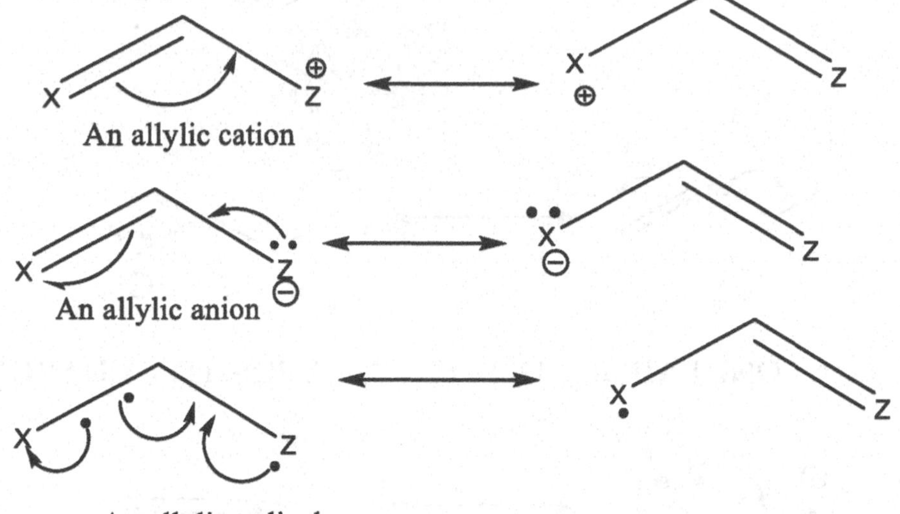

An allylic cation

An allylic anion

An allylic radical

Ex:

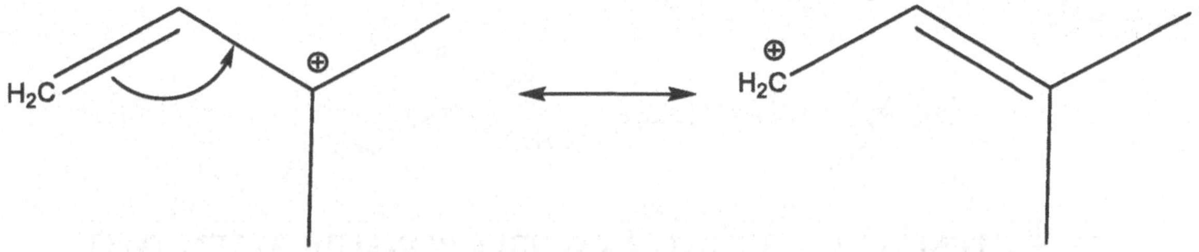

H_2C ⟷ H_2C

An allylic cation An allylic cation

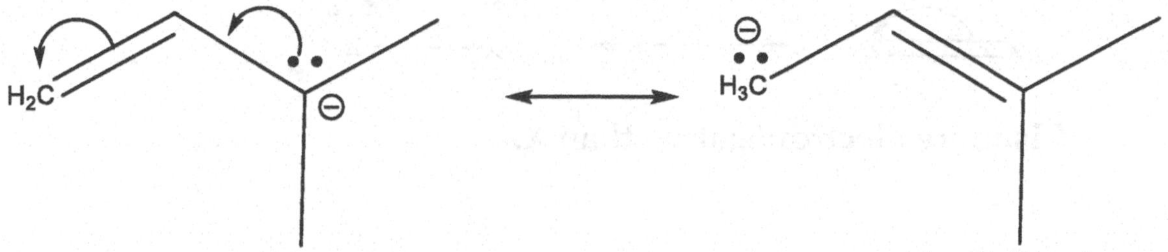

H_2C ⟷ H_3C

An allylic anion An allylic anion

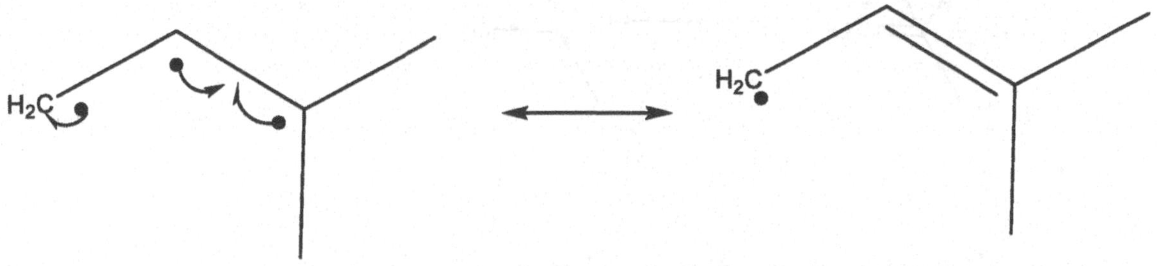

H_2C ⟷ H_2C

An allylic radical An allylic radical

3. CONJUGATED SYSTEMS

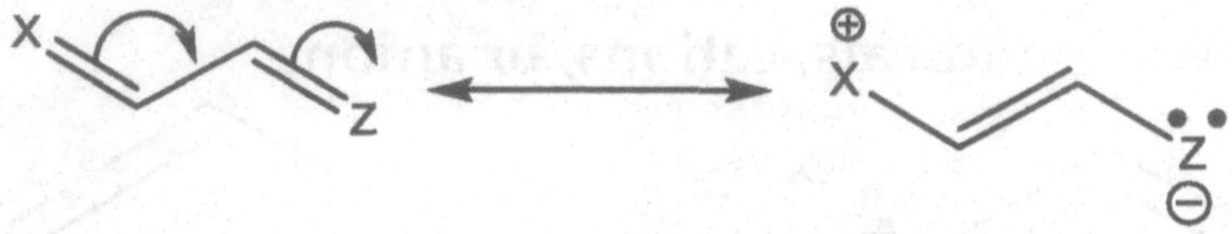

X ... Z ⟷ ⊕X ... Z⊖

Ex:

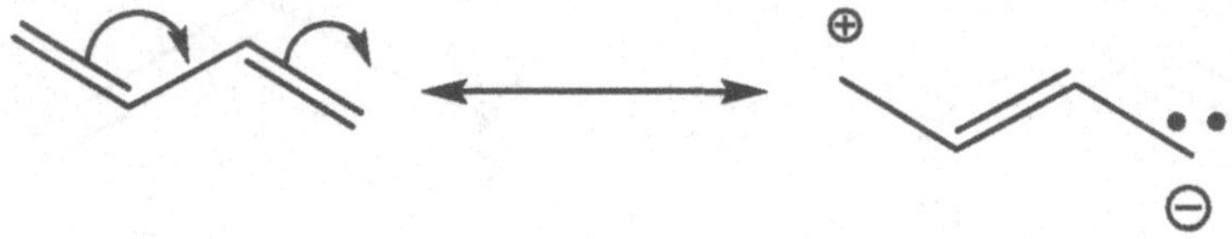

4. A LONE PAIR IS ADJACENT TO A POSITIVE CHARGE

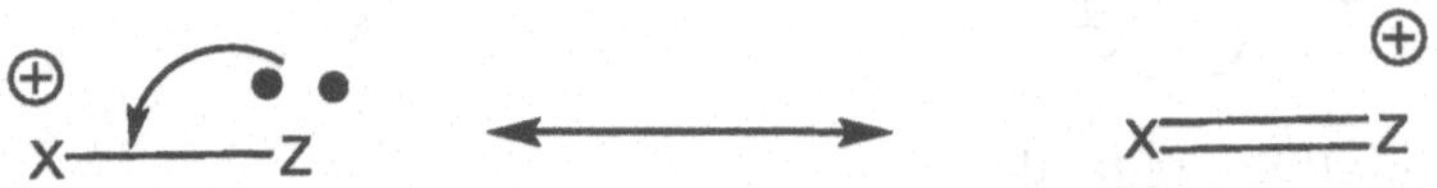

⊕X—Z: ⟷ X=Z⊕

Ex:

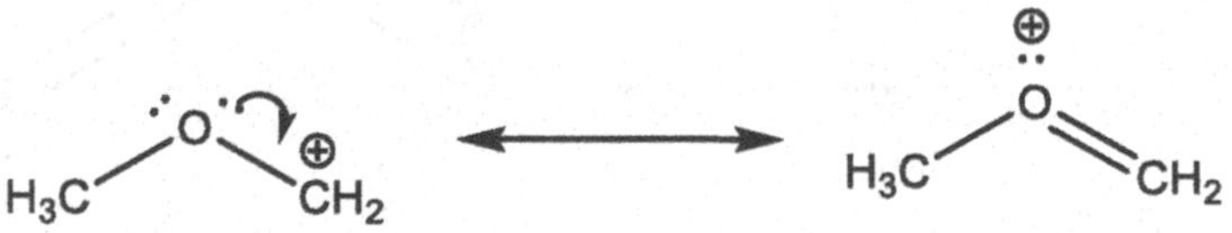

H_3C–Ö–$CH_2^{\oplus}$ ⟷ H_3C–$O^{\oplus}$=CH_2

5. SYSTEMS CONTAINING A DOUBLE BOND BETWEEN TWO ELECTRONEGATIVE ATOMS

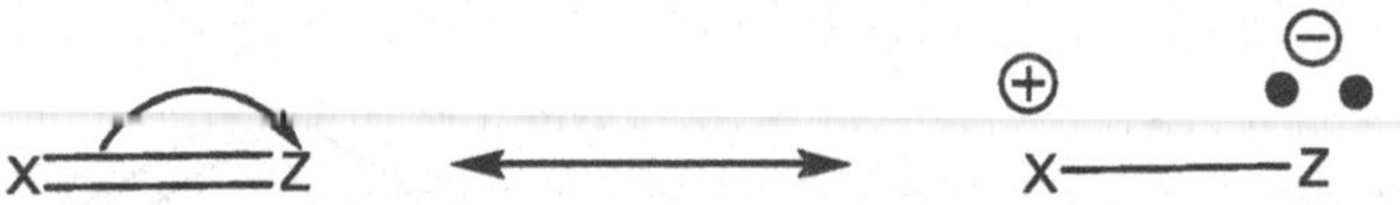

X=Z ⟷ ⊕X—Z:⊖

- Z is more electronegative than X.

Ex:

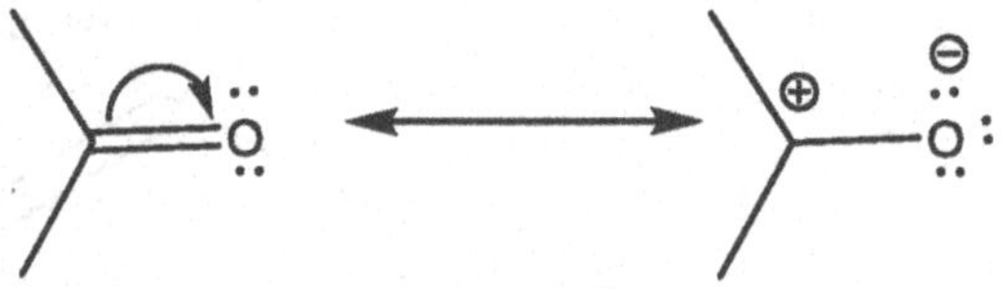

- Read pages ________ - ________.

- Do all problems and examples.

E. THE RESONANCE HYBRID

1. DEFINITION

- A resonance hybrid is the "**combination**" or **average** of all possible resonance structures.

Ex: O_3

Equal contributors

- **Note: The most stable resonance structure is the major contributor to the resonance hybrid. The minor contributor is the least stable resonance structure.**

Ex: CO_2

major contributor

minor contributor

2. RULES FOR IDENTIFYING MAJOR AND MINOR CONTRIBUTORS

- **Rule # 1: The resonance structure with the more bonds and fewer charges is the major contributor.**

  Ex:

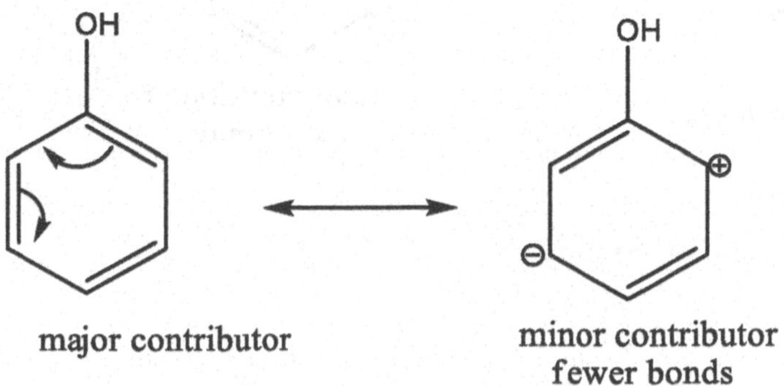

major contributor

minor contributor
fewer bonds

- **Rule # 2: The resonance structure in which all second-row atoms have an octet is the major contributor.**

Ex:

OH / OH

no octet for this carbon

major contributor ⟷ minor contributor

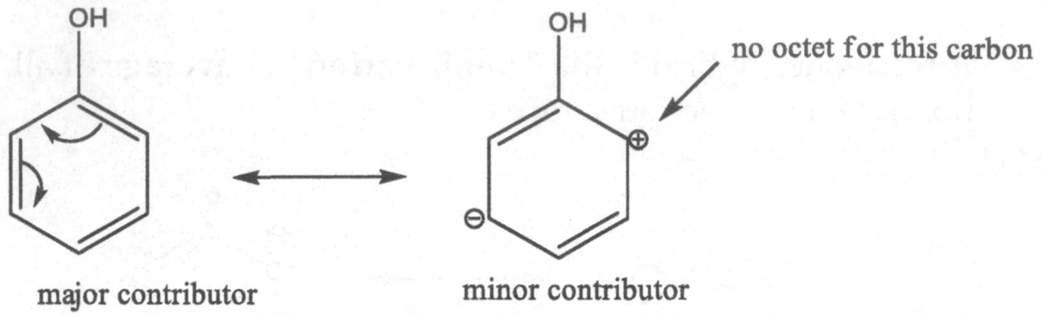

- **Rule # 3: The resonance structure that places a negative charge on a more electronegative atom is the major contributor.**

Ex: CO_2

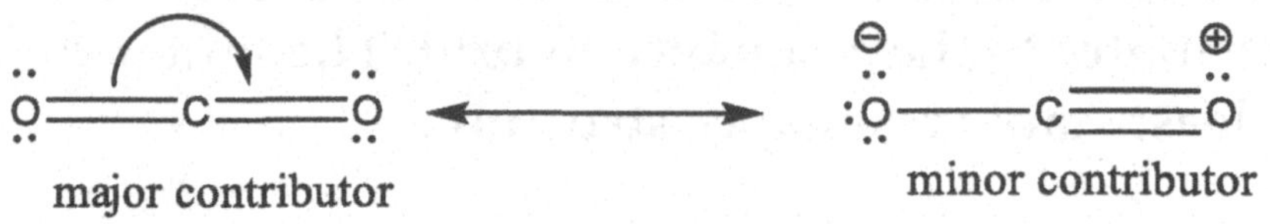

major contributor ⟷ minor contributor

- **Rule # 4: The resonance structure with neutral atoms is the major contributor.**

Ex:

neutral atoms

OH / OH

major contributor ⟷ minor contributor
fewer bonds

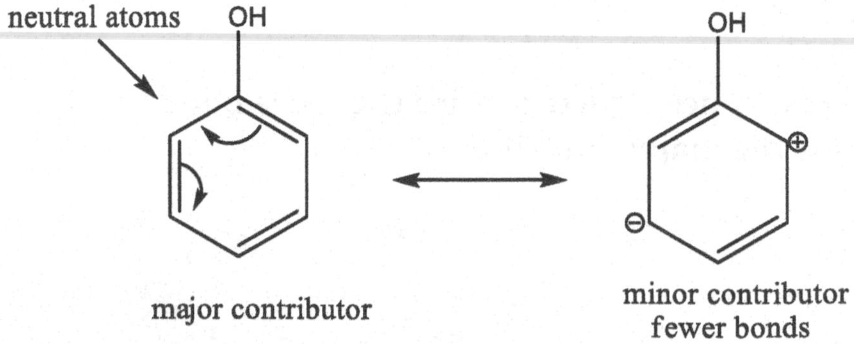

- **Rule # 5: The least stable contributor is the one in which 2 adjacent atoms have positive charges.**

Ex:

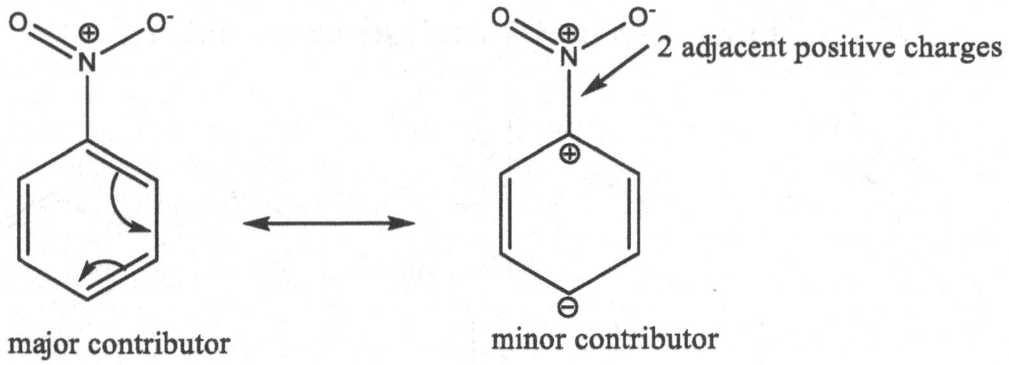

2 adjacent positive charges

major contributor

minor contributor

- **Read pages ________ - _______.**

F. ELECTRON DELOCALIZATION AND HYBRIDIZATION

- For a conjugated system such as X=Y-Z, **z is always** sp^2 hybridized.

Ex:

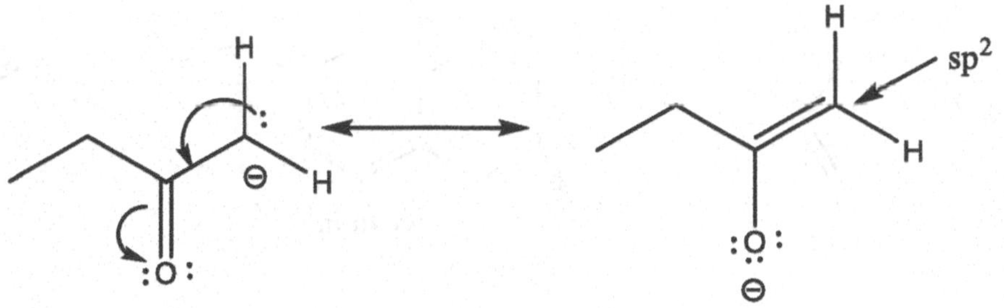

sp^2

- **Read pages ______ - ______.**

- **Do all problems and examples.**

G. IMPORTANT DIENES AND POLYENES

- **Read page _______. See Fig. _______, page ______.**

  - Isoprene
  - Lycopene

- β-carotene
- Vitamin D
- Zocor
- Linearmycin B (**Check structure online**)

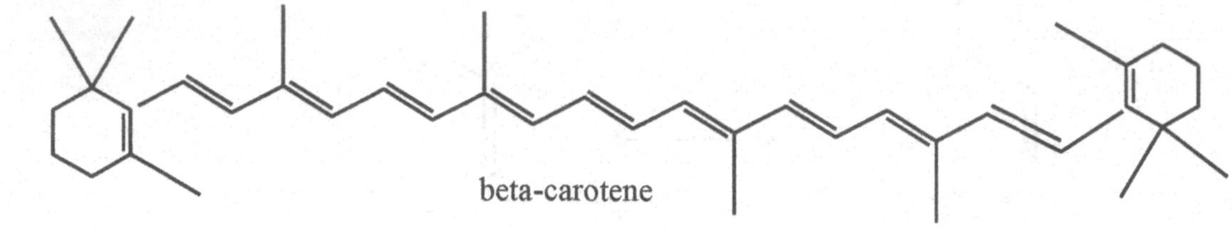

beta-carotene

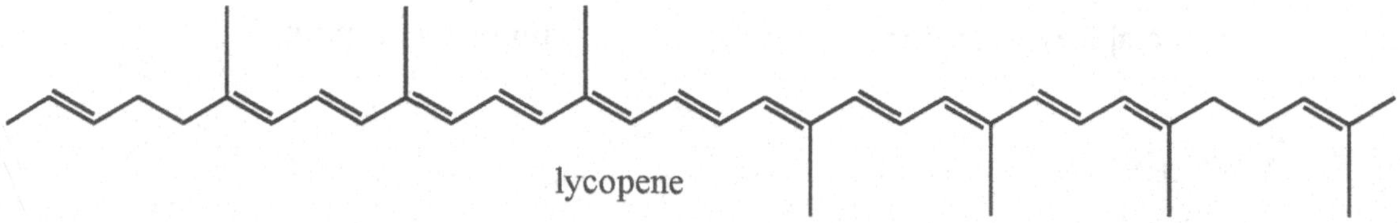

lycopene

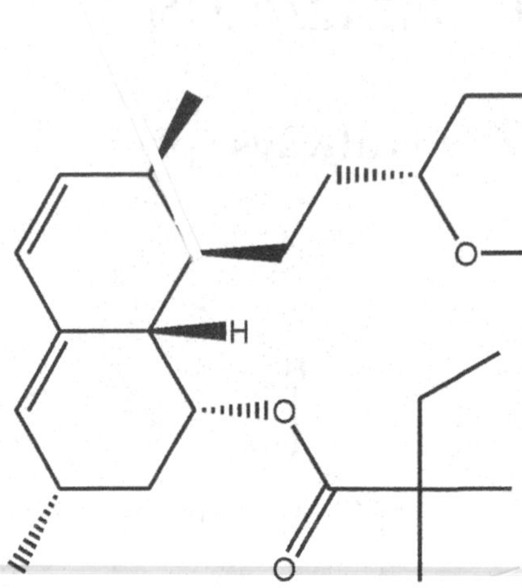

zocor

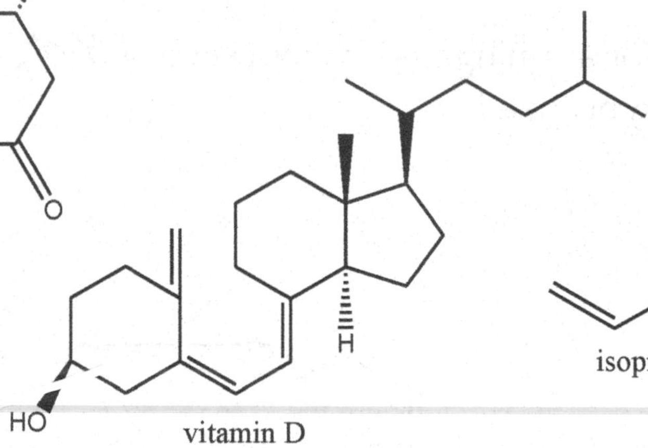

vitamin D

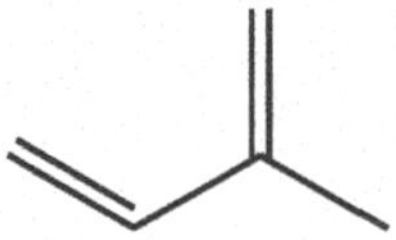

isoprene

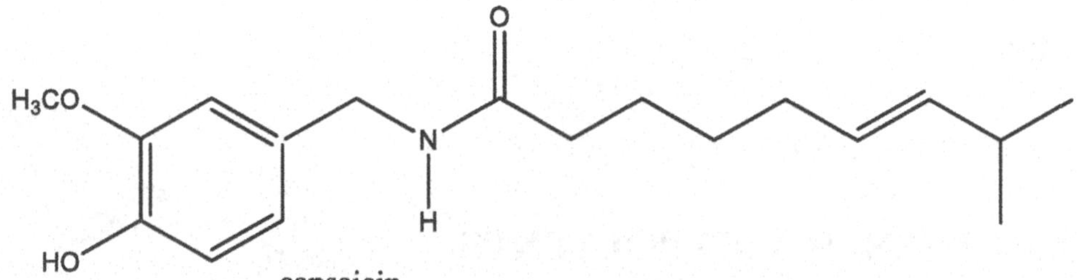

capsaicin

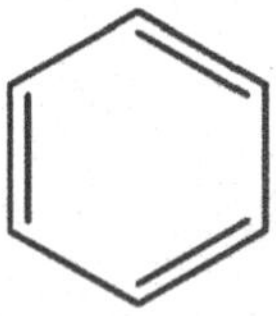

benzene

H. ADDITION OF HX (X = Cl, Br, I) TO NONCONJUGATED AND CONJUGATED DIENES

1. INTRODUCTION

- **HX** can add to both **nonconjugated** alkenes and **conjugated** dienes. We will cover both.

2. ADDITION OF HX TO NONCONJUGATED ALKENES: A REVIEW

- **The reaction follows Markonikov. (See OCHEM II, Unit 2)**
- **The general reaction is:**

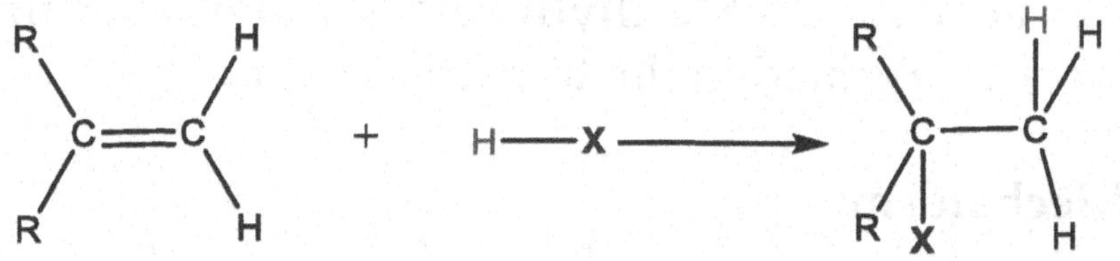

A Markovnikov product
(one product)

- **Mechanism:**

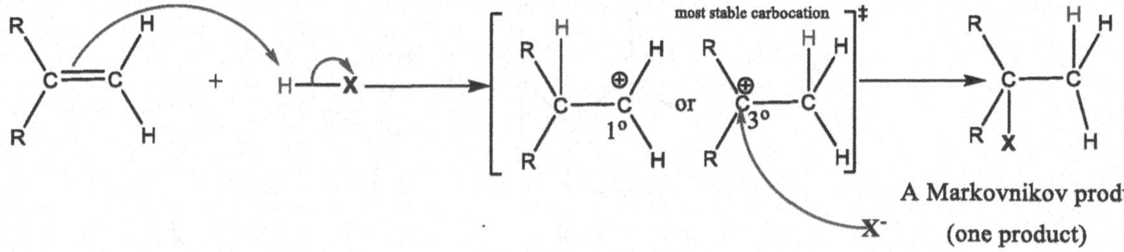

most stable carbocation

A Markovnikov product
(one product)

3. ELECTROPHILIC ADDITION OF HX TO CONJUGATED DIENES

a. Introduction

- The reaction proceeds in two steps. **Two** products are produced.

Ex:

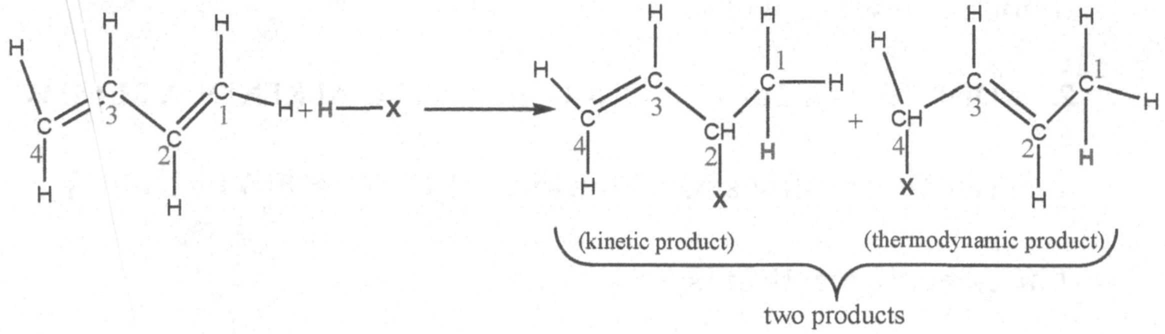

H₂C=CH–CH=CH₂ + H–X → (kinetic product) + (thermodynamic product)

two products

- **Question: Why are two products obtained? Two products are produced because two stable allylic carbocations (stabilized by resonance) are formed in the transition state.**

- **Reaction Mechanism:**

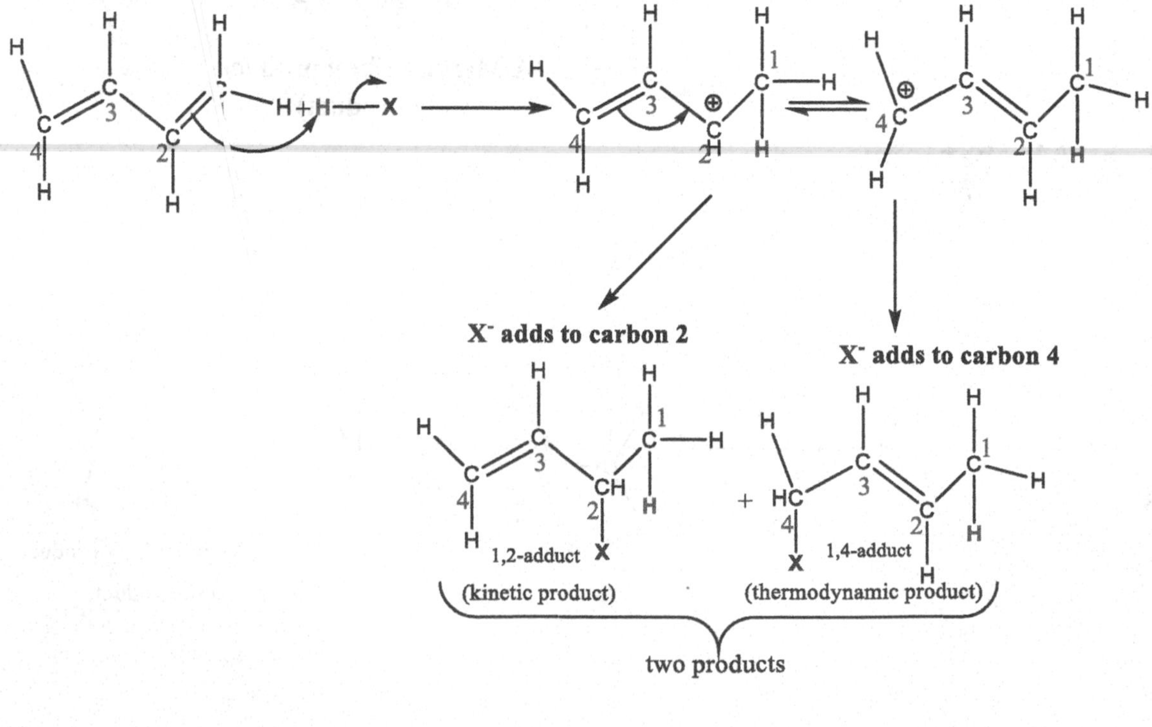

X^- adds to carbon 2

X^- adds to carbon 4

1,2-adduct (kinetic product)

1,4-adduct (thermodynamic product)

two products

- Note: The 1,2-adduct (product) forms faster at lower temperatures (under mild condition = kinetic control). This is called the *kinetic* product and is less stable (least substituted). On the other hand, the 1,4-adduct forms more slowly at higher temperature (under vigorous conditions = thermodynamic control). This product is called the *thermodynamic* product (more substituted). It is more stable and predominates at equilibrium.

b. Energy Diagram: Kinetic Control

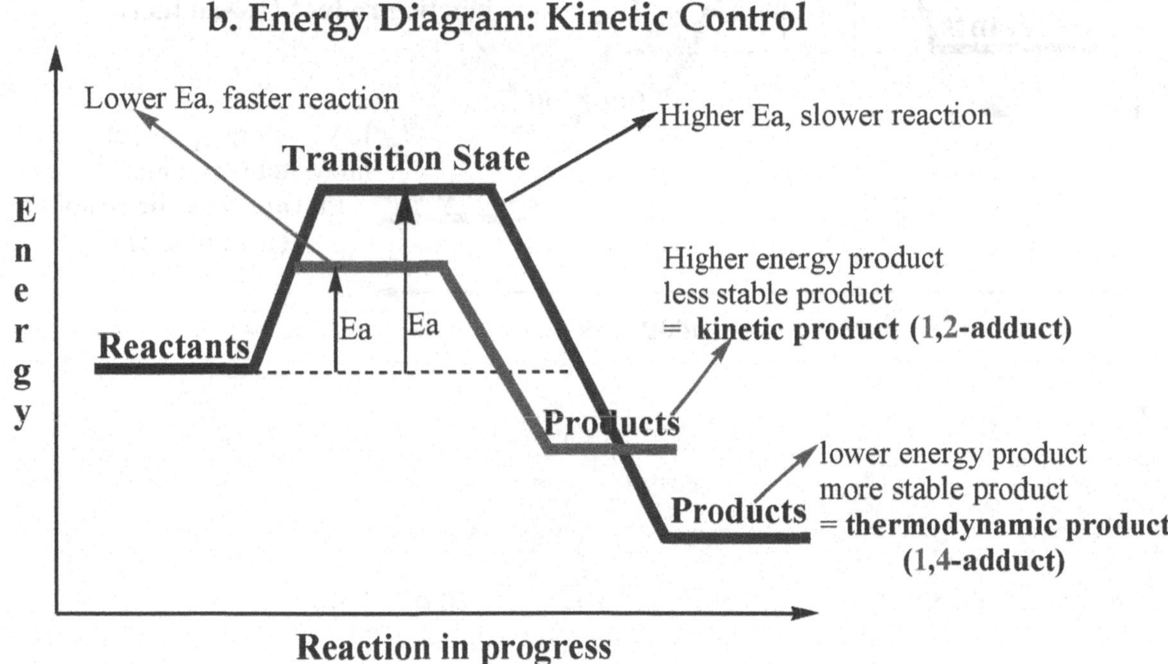

c. Energy Diagram: Thermodynamic Control

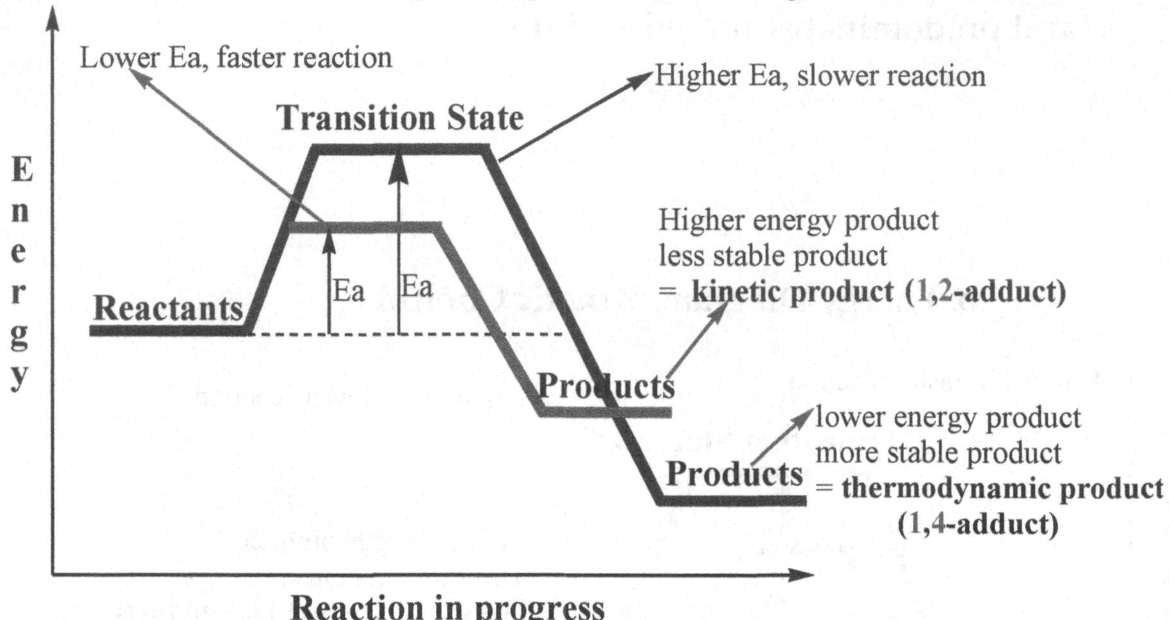

- Read pages ______ - _____. Do all problems.

I. THE DIELS-ALDER CYCLOADDITION REACTION

1. INTRODUCTION

- The Diels-Alder reaction was discovered by Otto Diels and Kurt Alder. This **heat-initiated** reaction (thermal reaction) involves a **1,3- diene** and a **dienophile** (alkene) undergoing a **pericyclic reaction** in which there is a **concerted bond breaking and bond forming**. The reaction is very fast and yields a **cyclic product.**

- **The general reaction is:**

$$\text{Diene} + \text{Dienophile} \xrightarrow[\text{heat}]{\text{benzene}} \text{Cyclic alkene}$$

or

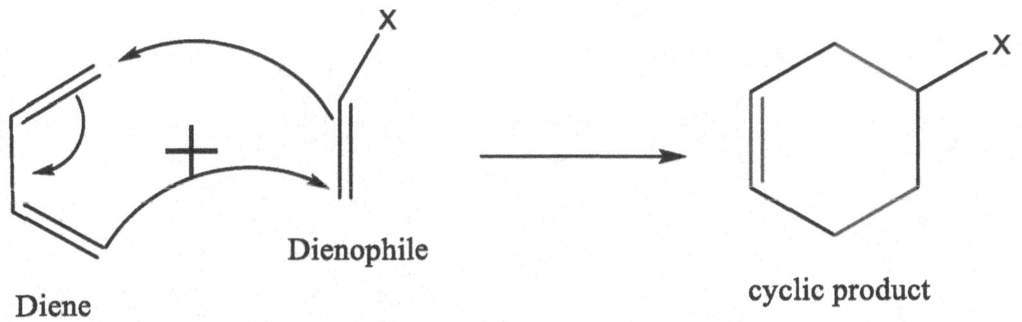

Ex:

- **Read pages _______ - ________. Do all problems on pages _____ - _______.**

2. CHARACTERISTICS OF THE DIELS-ALDER REACTION

a. Introduction

- The reaction
  - o Is initiated by **heat**
  - o A **six-membered** ring is produced
  - o Is **concerted**: bond **breaking** and **forming** happen at the same time.

- The Diels-Alder reaction depends also on the **diene and the dienophile.**

b. The Diene

- The diene must be able to adopt an **s-cis conformation.** It should not be sterically strained and should not be rigidly fixed in an **s-trans** conformation.

Ex:

Read page ________. Do all problems on page ______.

c. The Dienophile

- Good dienophiles have **electron withdrawing groups (EWG).** The more EWG on the dienophile, the more reactive it is.

Ex:

- **See Fig. _______, page _______. Do all problems on page ____.**

d. Stereochemistry of the Cycloalkene Product.

- The Diels-Alder reaction is **stereospecific**. This means that the stereochemistry of the alkene product is the same as the dienophile. In other words:

Diene + **cis**-Dienophile $\xrightarrow[\text{heat}]{\text{benzene}}$ **cis**-Cyclic alkene

Ex:

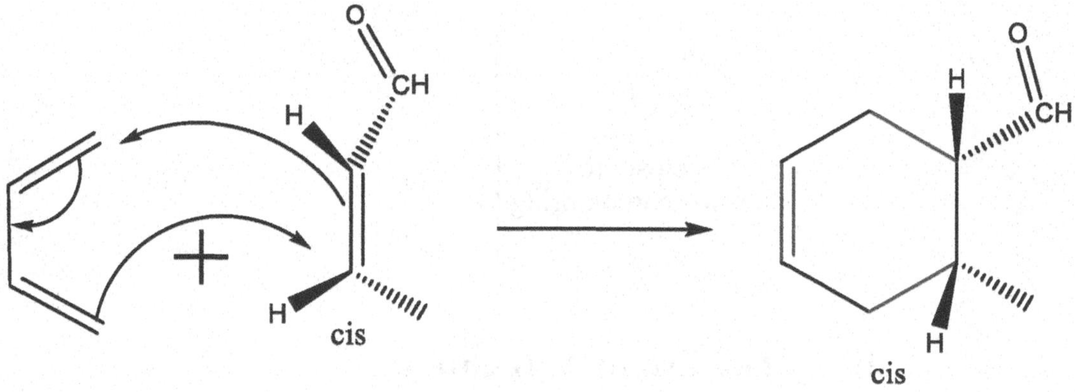

O CH H H cis

cis

Diene + **trans**-Dienophile $\xrightarrow[\text{heat}]{\text{benzene}}$ **trans**-Cyclic alkene

Ex:

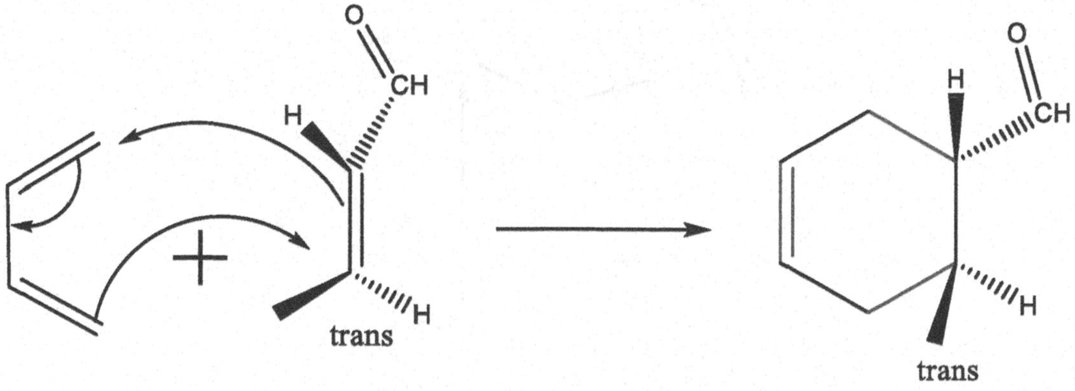

O CH H H trans

trans

- **Read pages _____ - _____. Do problem on page _____.**

e. Endo-exo cyclic compounds

i. The Endo or Syn Group

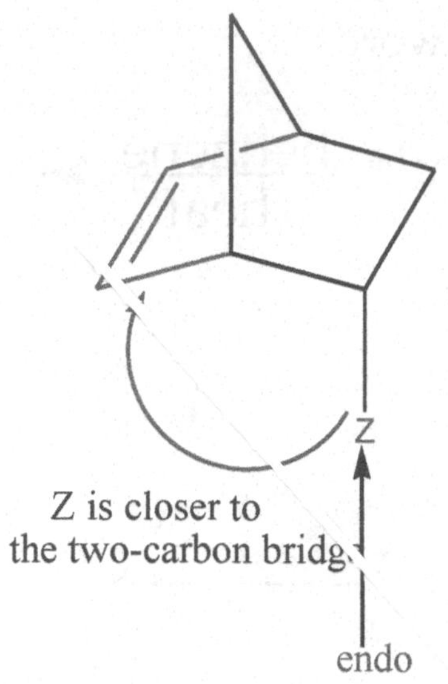

ii. The Exo or Anti Group

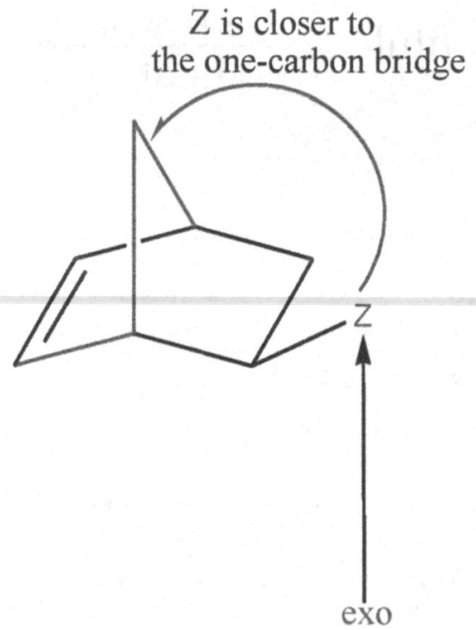

- **Note: In a Diels-Alder reaction,** the major product is the endo product.

Read pages ______ - ______. Do problem ______, page ______.

Ex:

endo (observed)

exo (not observed)

or

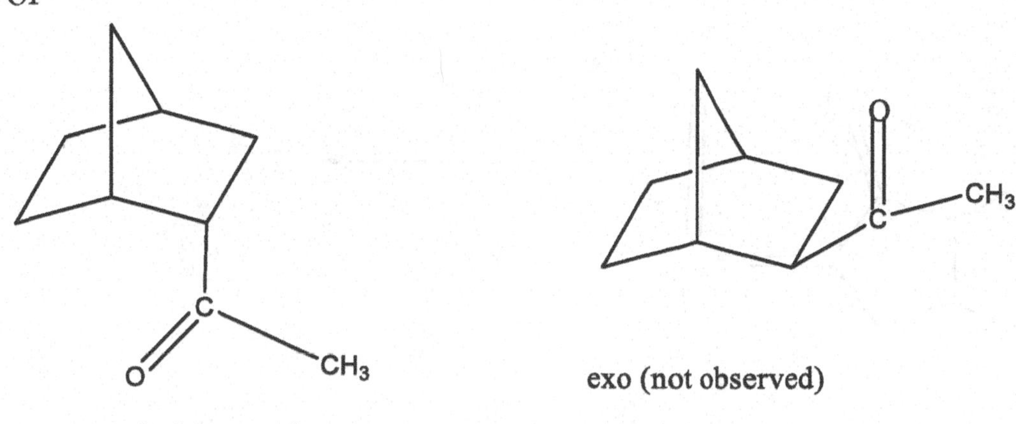

endo (observed)

exo (not observed)

3. **RETROSYNTHETIC DIELS-ALDER REACTION**

- A Diels-Alder reaction can be used to identify the original diene and dienophile. Here is an example. Please, find the diene and the dienophile.

? ⟶

CN

- Read pages _____ - _____.
- See Fig. _______ page ______.
- Do Problem ______, page _______.
- Read pages _______- ______ = Application of the Diels-Alder
- Reaction: Steroid Synthesis.

Ex:

4. SUCCESSIVE DIELS-ALDER REACTION

Ex:

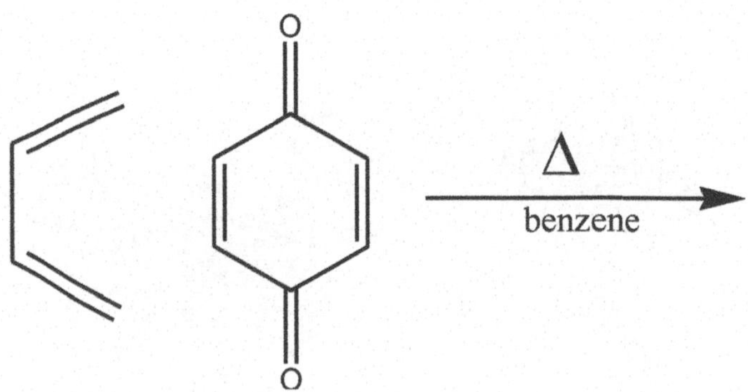

J. OTHER PERICYCLIC OR REARRANGEMENT REACTIONS

1. INTRODUCTION

- These are reactions in which **a sigma** bond moves across the face of one or more **pi bonds.**

2. CLAISEN REARRANGEMENT OF ALLYL PHENOL ETHERS

a. The General Reaction

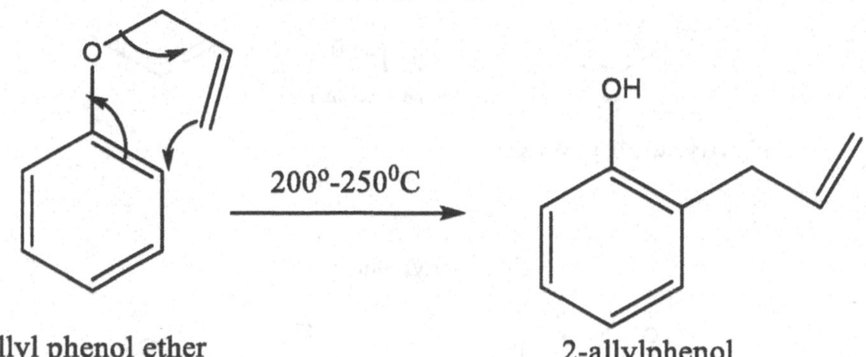

Allyl phenol ether

2-allylphenol

b. Tautomerization

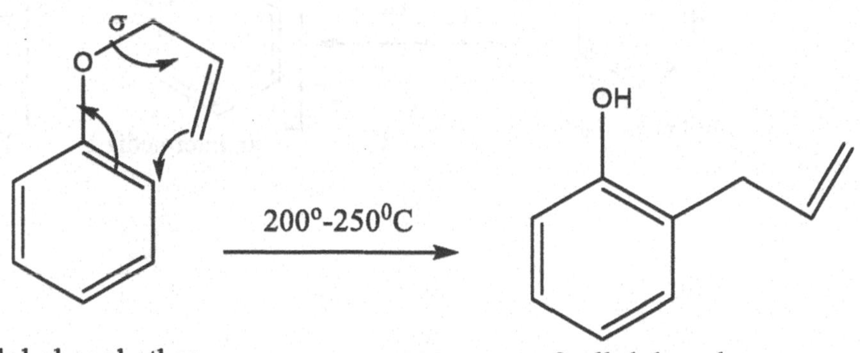

Allyl phenol ether

2-allylphenol

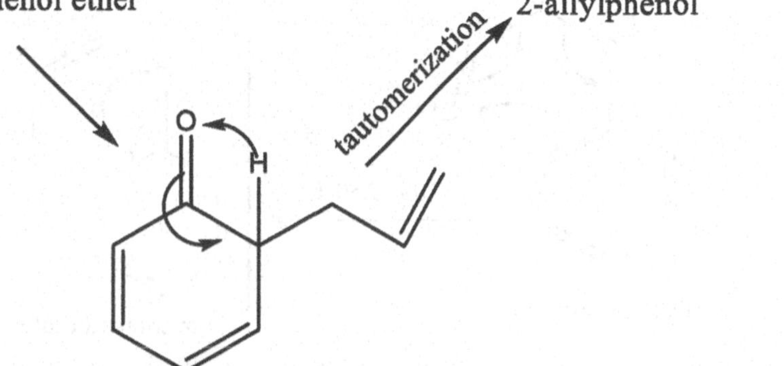

2-allyl-2,4-cyclohexadienone intermediate

Ex:

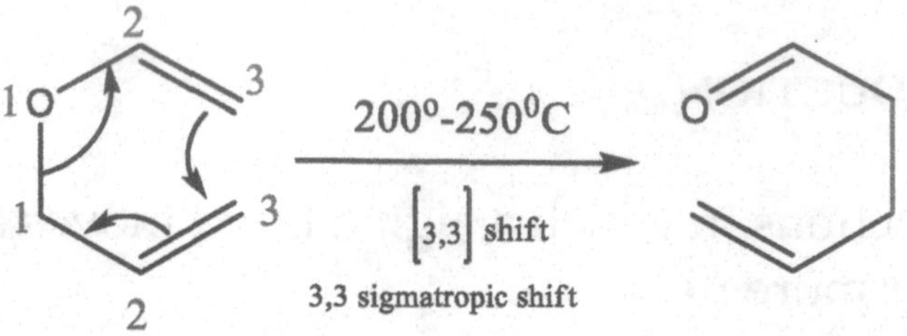

1O 2 3 1 3 2

200°-250°C

[3,3] shift

3,3 sigmatropic shift

Allylic vinyl ether

or

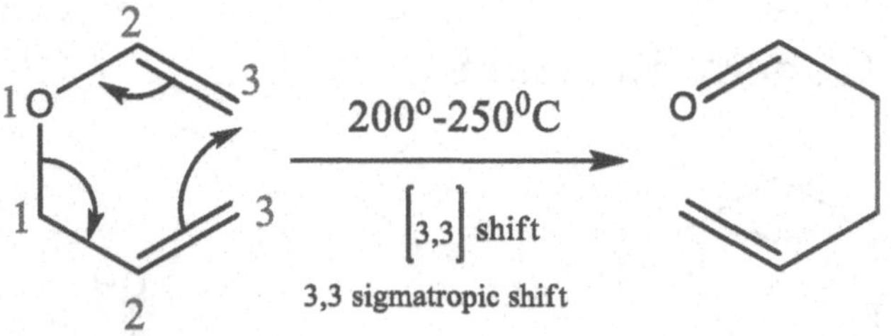

1O 2 3 1 3 2

200°-250°C

[3,3] shift

3,3 sigmatropic shift

Allylic vinyl ether

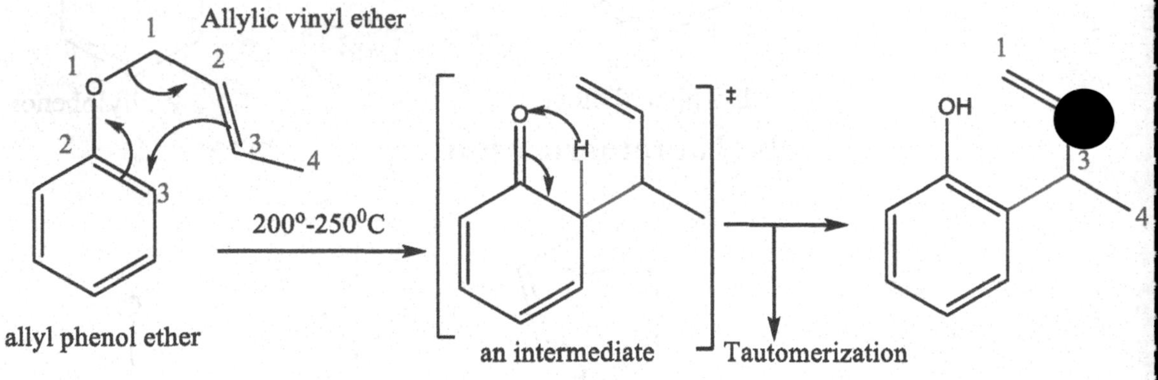

Allylic vinyl ether

200°-250°C

allyl phenol ether

an intermediate

Tautomerization

or

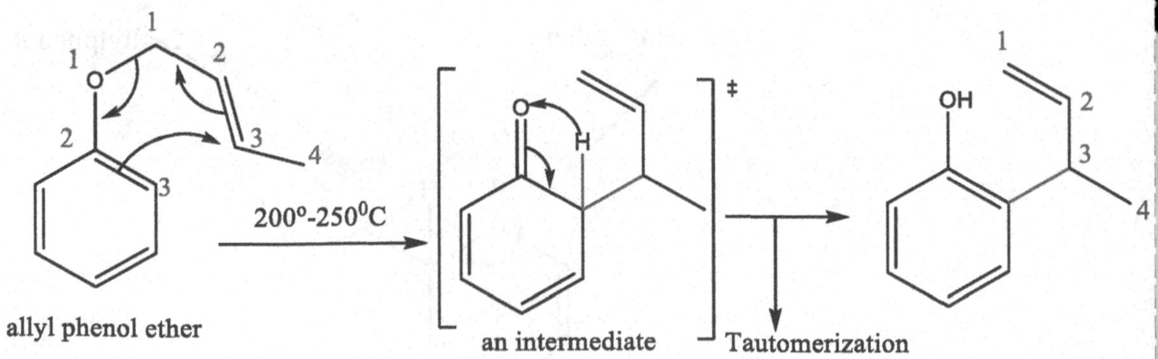

200°-250°C

allyl phenol ether

an intermediate

Tautomerization

3. THE COPE REARRANGEMENT OF 1,5 -DIENES

a. The General Reaction

2 1 3 4 6 5

3,3-Dimethyl-1,5-hexadiene

350°C

a new bond

b. Some examples

3 4 2 5 1 6

350°C

- **Provide a mechanism for:**

OH 2 3 1 4 6 H 5

350°C

O

J. UV-Vis SPECTROSCOPY

1. INTRODUCTION

- **Recall**: Mass Spectrometry: Uses m/z to get molecular size and formula.
- IR spectroscopy: Based on vibrational transition = functional group.
- NMR: Based on nuclear spin flip= C-H framework.
- UV-Vis: Electronic transition = **pi bonds and conjugated pi systems.**

- **Review the EM spectrum; see Fig. on page _______.**

2. BASIC PRINCIPLE

- **UV-Vis** is based on **electronic transitions**. Indeed, absorption of radiation from the UV-Vis region of the electromagnetic spectrum by molecules results in the promotion of **π electrons** from a **lower energy level (ground state)** to a **higher energy level (excited state)**. It is used to show the presence of a **pi bond** or **conjugated pi** system.

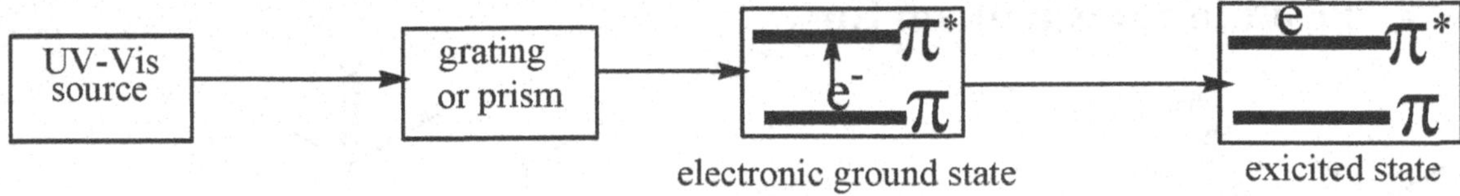

- Read page _______ - _______.
- **Note: σ bonds are insensitive to UV-Vis because the electrons are tightly held in place. Nonconjugated and non pi systems do not absorb in UV-Vis.**

- **The EM ranges used are (roughly):**

  - **UV: 200 nm–400 nm.**
  - **Vis: 400 nm–800 nm.**

- The instrument is called the **UV-Vis spectrophotometer.**

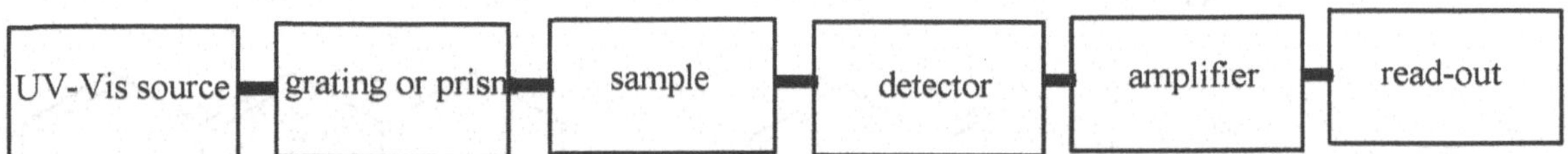

- **See the diagram of a double beam spectrophotometer in Addenda.**

- **The UV-Vis spectrum: Plot of absorbance (A) vs. wavelength (λ) or %transmittance vs. wavelength (λ).**

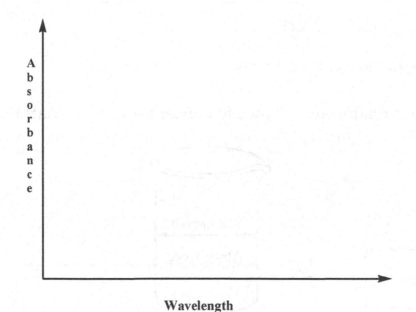

3. THE CONCEPT OF λ_{max}

- λ_{max} is the wavelength at which absorption is **maximum (highest A and lowest** %T).

Ex: See Addenda

- **Note: The greater the conjugation, the higher the λ_{max}.**

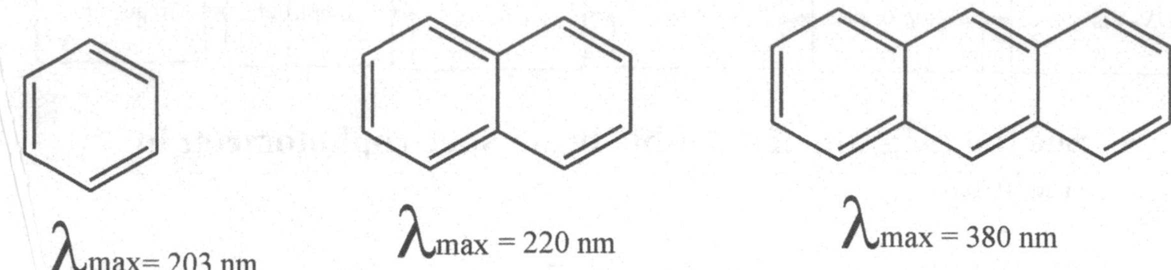

- **Read pages ______ - _______. Read about sunscreens on page _________.**

- **Do problems on page ______.**

4. THE BEER-LAMBERT LAW

- This law relates absorbance to concentration and pathlength.

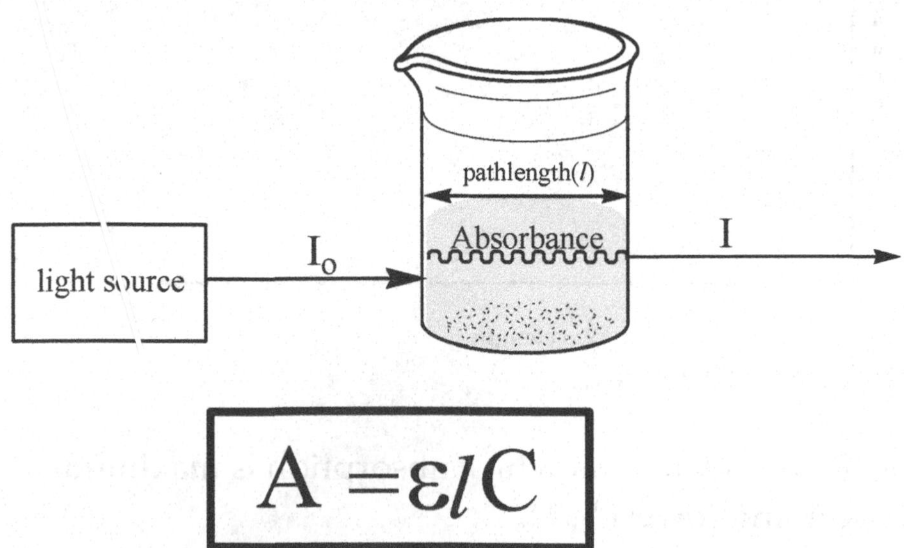

$$A = \varepsilon l C$$

or

$$A = \log \frac{I_o}{I}$$

Where:

- A = **absorbance**
- I_o = **Intensity of light entering the sample cell.**
- I = **Intensity of light not absorbed by the sample**
- C = **Concentration of sample (mol/L)**
- l = **Pathlength**
- ε = **Molar absorptivity**
- **Note: ε is a constant that depends on the sample. The higher the molar absorptivity, the higher the absorbance.**

Ex: The molar absorptivity of acetone at its λ_{max} (195 nm) is 9000 M^{-1} cm^{-1}. Calculate the absorbance of 1.00×10^{-5} M of acetone at this wavelength. Assuming l = 1.20 cm.

K. The MO THEORY: A REVIEW

1. INTRODUCTION

- According to the **MO theory**, electrons in molecules and molecular ions are in the **delocalized** molecular orbitals of each individual molecule (or ion), as electrons in atoms are in atomic orbitals. Molecular orbitals result from the **combinations** (or overlaps) of the atomic orbitals in the molecule. If the atomic orbitals combine or overlap "head to head," a **σ MO** results. However, if the combination or overlap is "sideways," the MO is called a **π MO**.

2. TRAVELING WAVES, STANDING WAVES

a. Traveling Waves

- A **traveling wave** is a wave that moves forward in space.

Ex. Light wave.

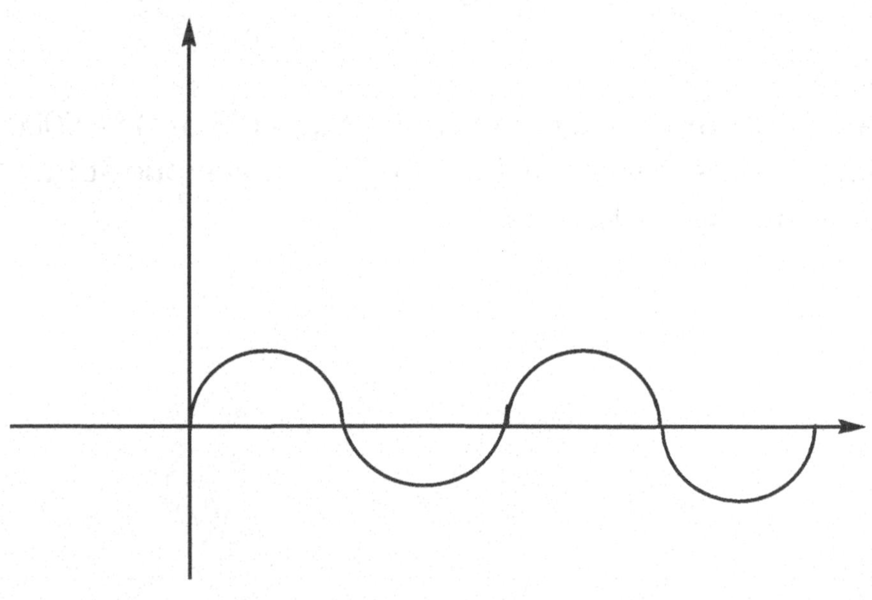

b. Standing Waves

- A **standing wave** is a fixed wave. It is a wave that is confined in a **fixed area** of space.

Ex. The vibrating string of a guitar is a standing wave.

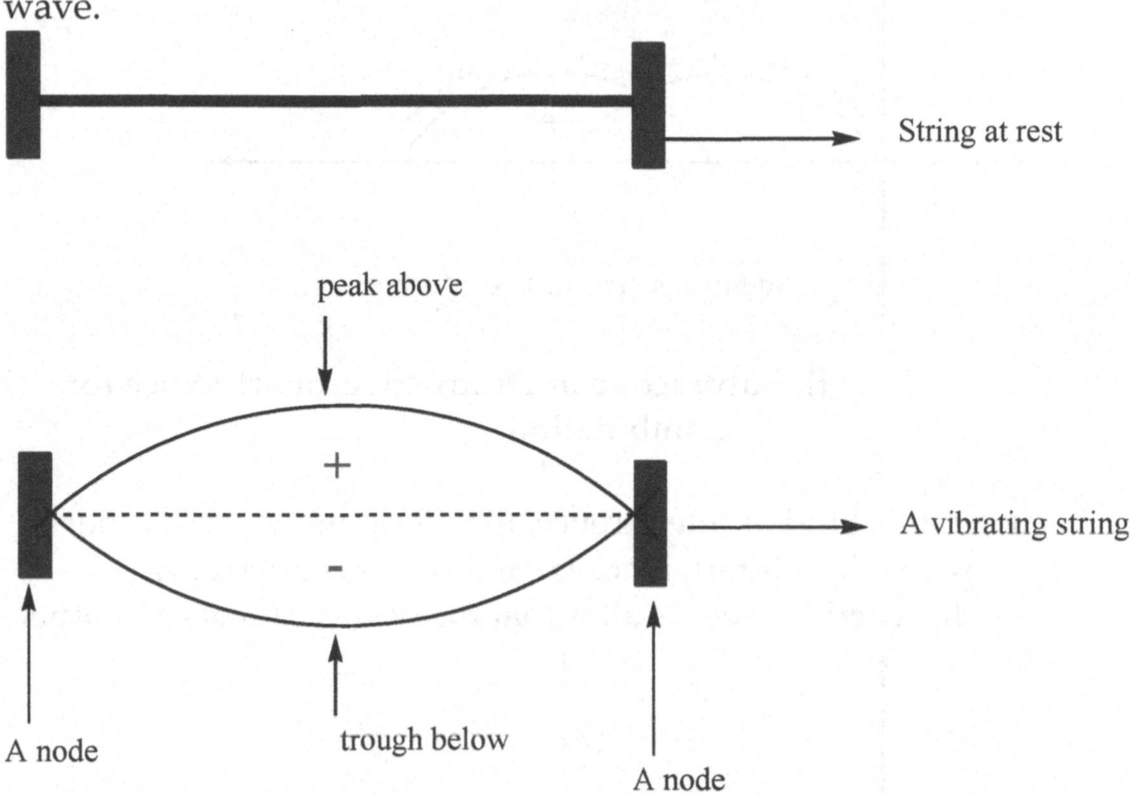

c. Interfering Waves

- In general, waves can interfere **additively** or **subtractively**.

i. Additive or Constructive Interference (Combination)

- When waves are **in phase** (they vibrate the same way), they interfere additively or **constructively**. In this case, the interfering waves **reinforce each other**.
- **See Fig. below:**

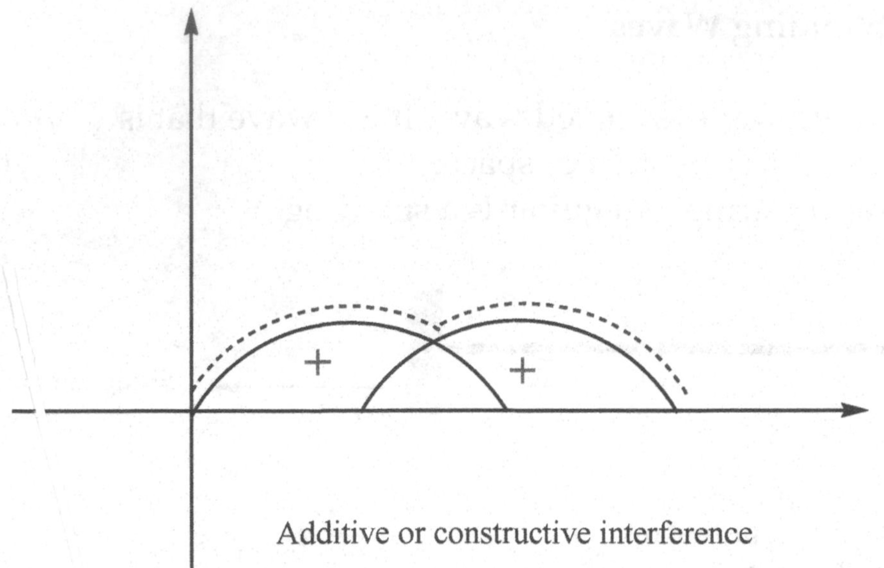

Additive or constructive interference

ii. Subtractive or Destructive Interference (or Combination)

- In this kind of interference, the vibrating waves are **not in phase**. The interference is said to be **subtractive or destructive**. The result is that the waves **cancel each other**.

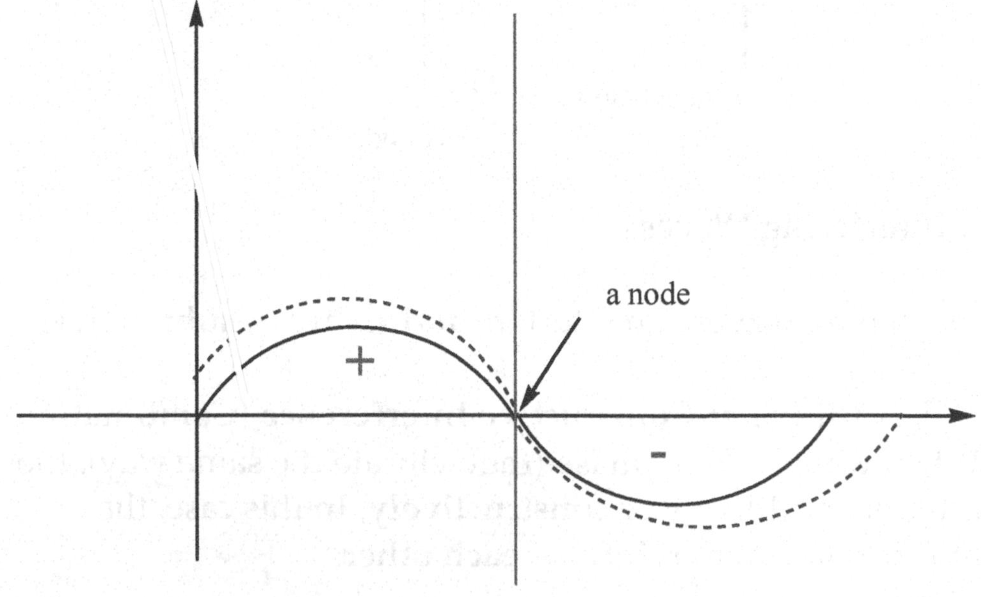

subtractive or destructive interference

3. APPLYING THE STANDING WAVE CONCEPT TO MOS

- **Recall: Atomic orbitals (s, p, d, f) are wave functions.**
- Let's assume that these atomic orbitals are **standing waves.**
- Let's consider the **H_2 molecule:**

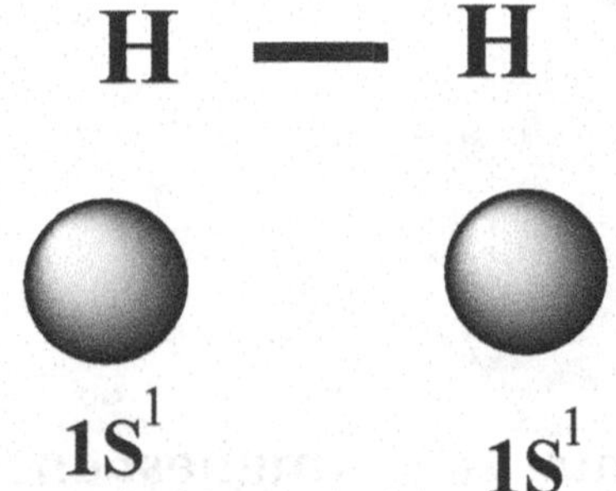

- We can have **constructive** or **destructive** combinations (**interferences**) between the two s orbitals.

4. BONDING AND ANTIBONDING MOLECULAR ORBITALS (MOS)

a. MOs from Two s Orbitals

i. Additive (or Constructive) Combination ofTtwo s Orbitals (In Phase)

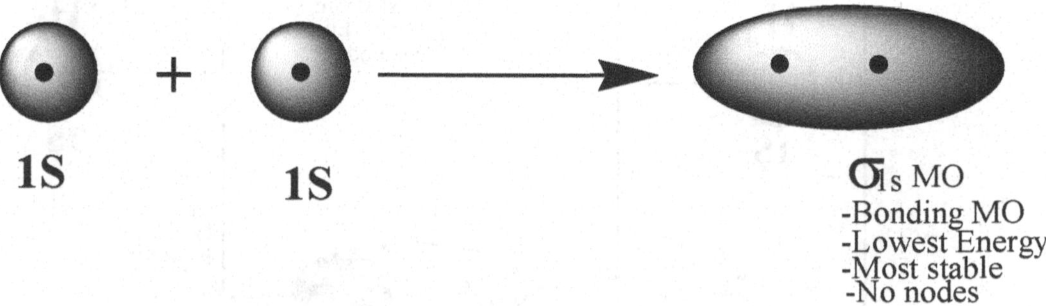

- **→ bonding mo→σ_{1s}→lowest energy→more stable→no node. See Fig. above.**

ii. Subtractive (or Destructive) Combination of Two s Orbitals (Out of Phase)

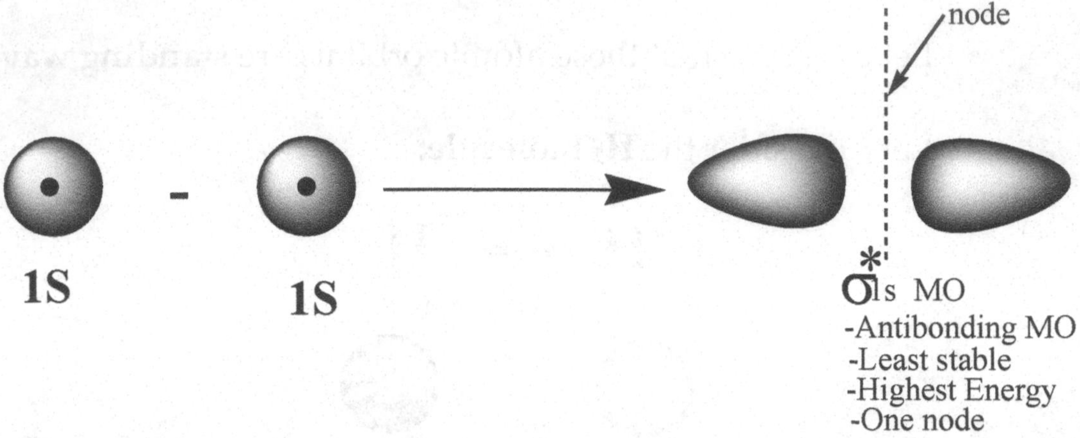

- → antibonding mo→σ^*_{1s}→highest energy→least stable→1 node. See Fig. above.

b. Energy Level Diagram of H_2 Using MOs

Energy

node

σ^*_{1s} MO
-Antibonding MO
-Least stable
-Highest Energy
-One node

H_a
1S

H_b
1S

σ_{1s} MO
-Bonding MO
-Lowest Energy
-Most stable
-No nodes

c. Electron Configuration of the H_2 Molecule

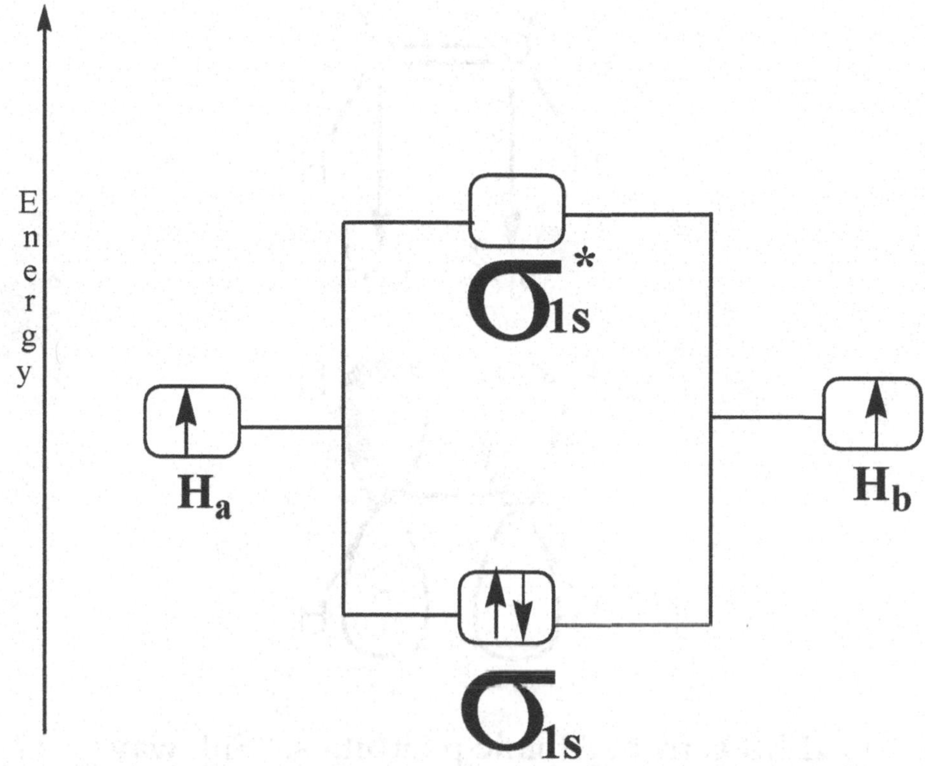

EC of H_2: $(\sigma_{1s})^2(\sigma^*_{1s})^o$

5. THE MO THEORY APPLIED TO THE TWO π ELECTRONS IN ETHENE

a. Introduction

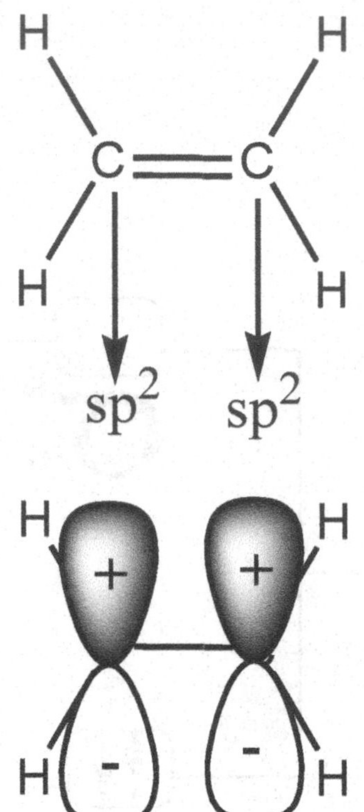

b. Π MOs from Atomic p Orbitals: "Sideways": 2 MOs

i. Constructive or Additive Combination

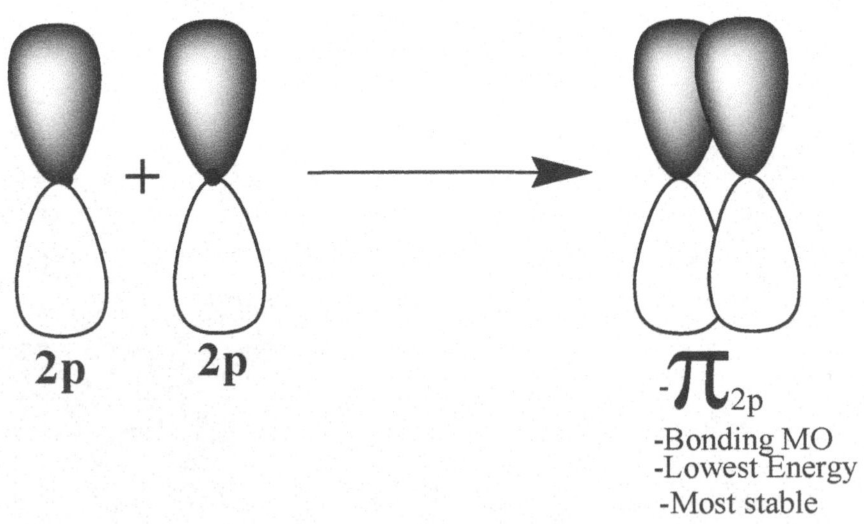

ii. Destructive or Subtractive Combination

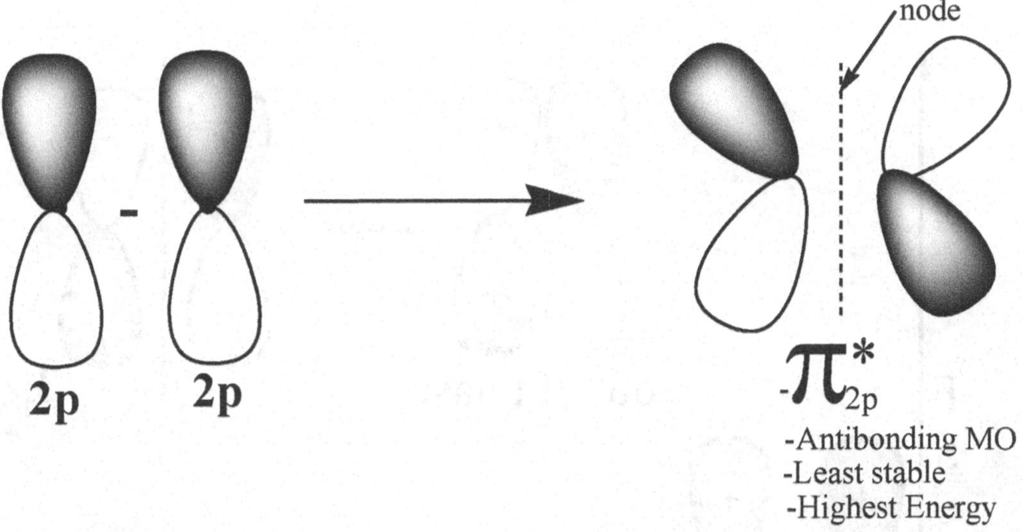

c. Energy Level Diagram of Π MOs from Atomic p Orbitals

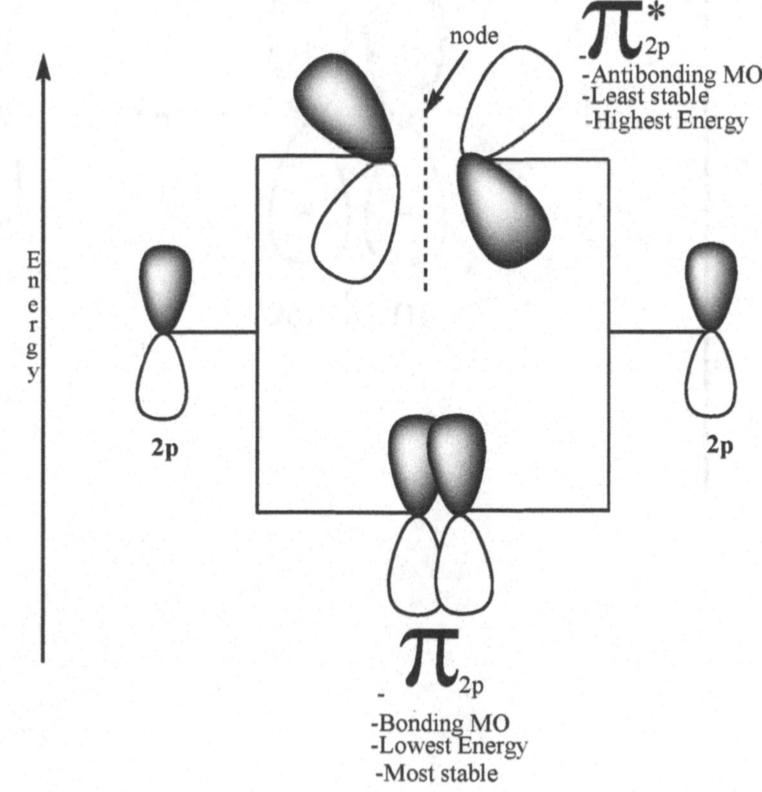

d. Energy Diagram of the pi electrons in Ethene

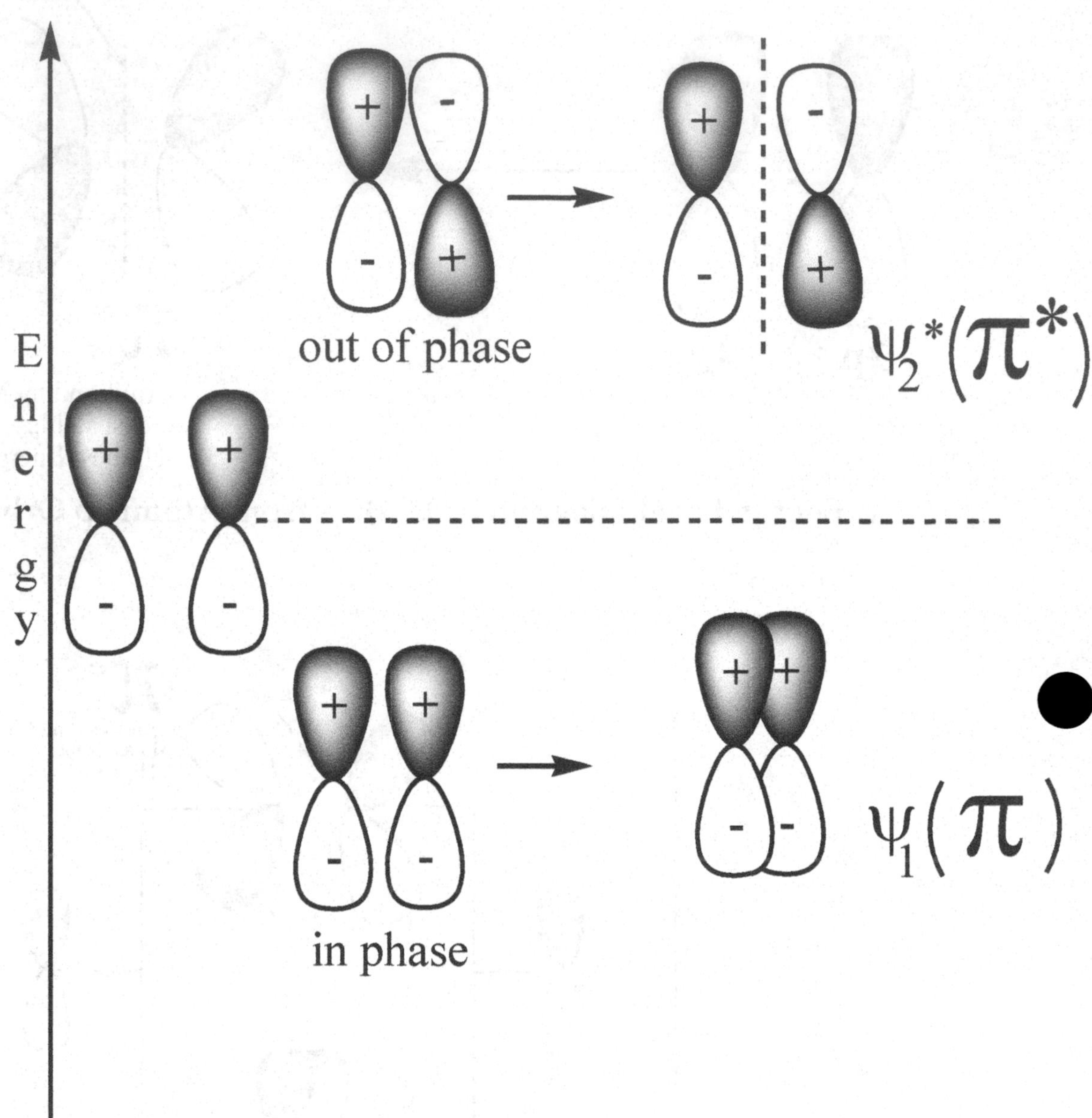

e. MOs of pi electrons in Ethene: A Summary

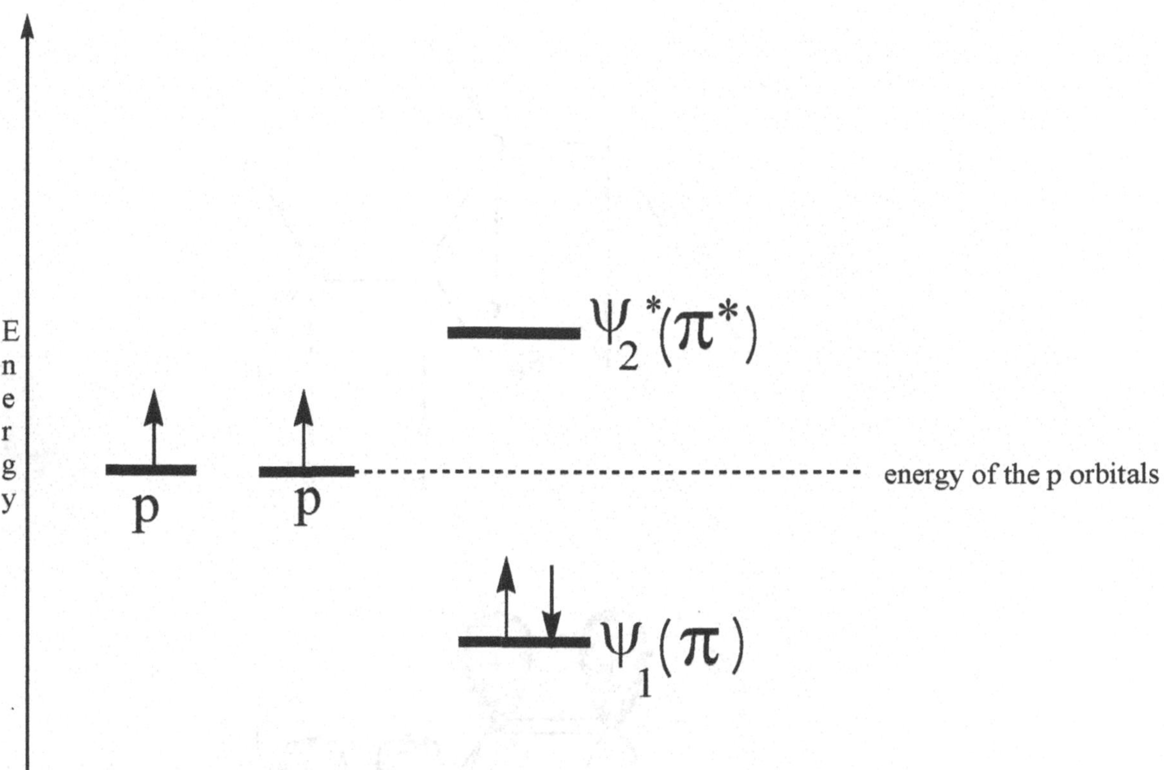

6. THE MO THEORY APPLIED TO THE FOUR π ELECTRONS IN 1,3-BUTADIENE

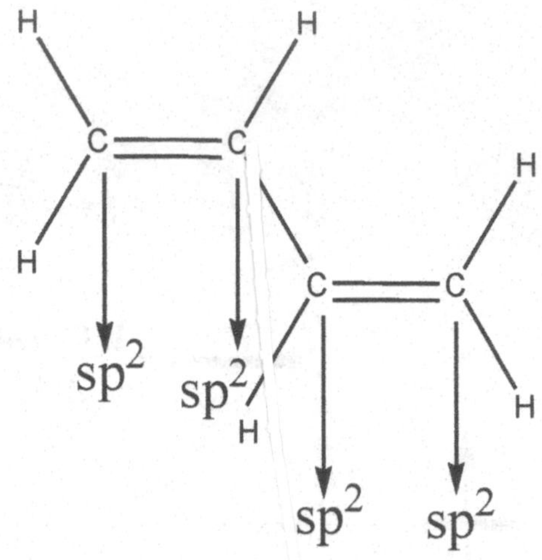

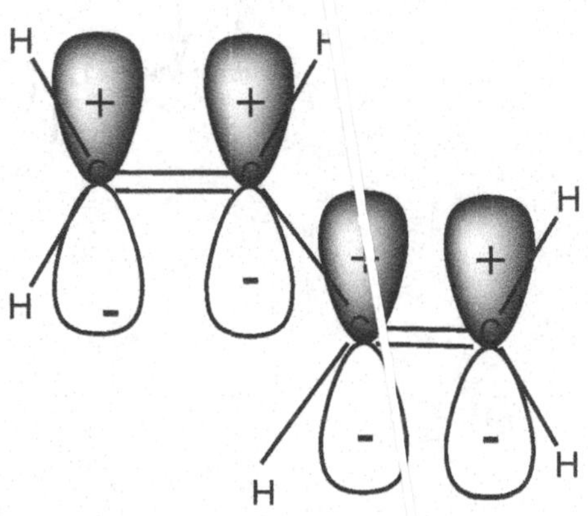

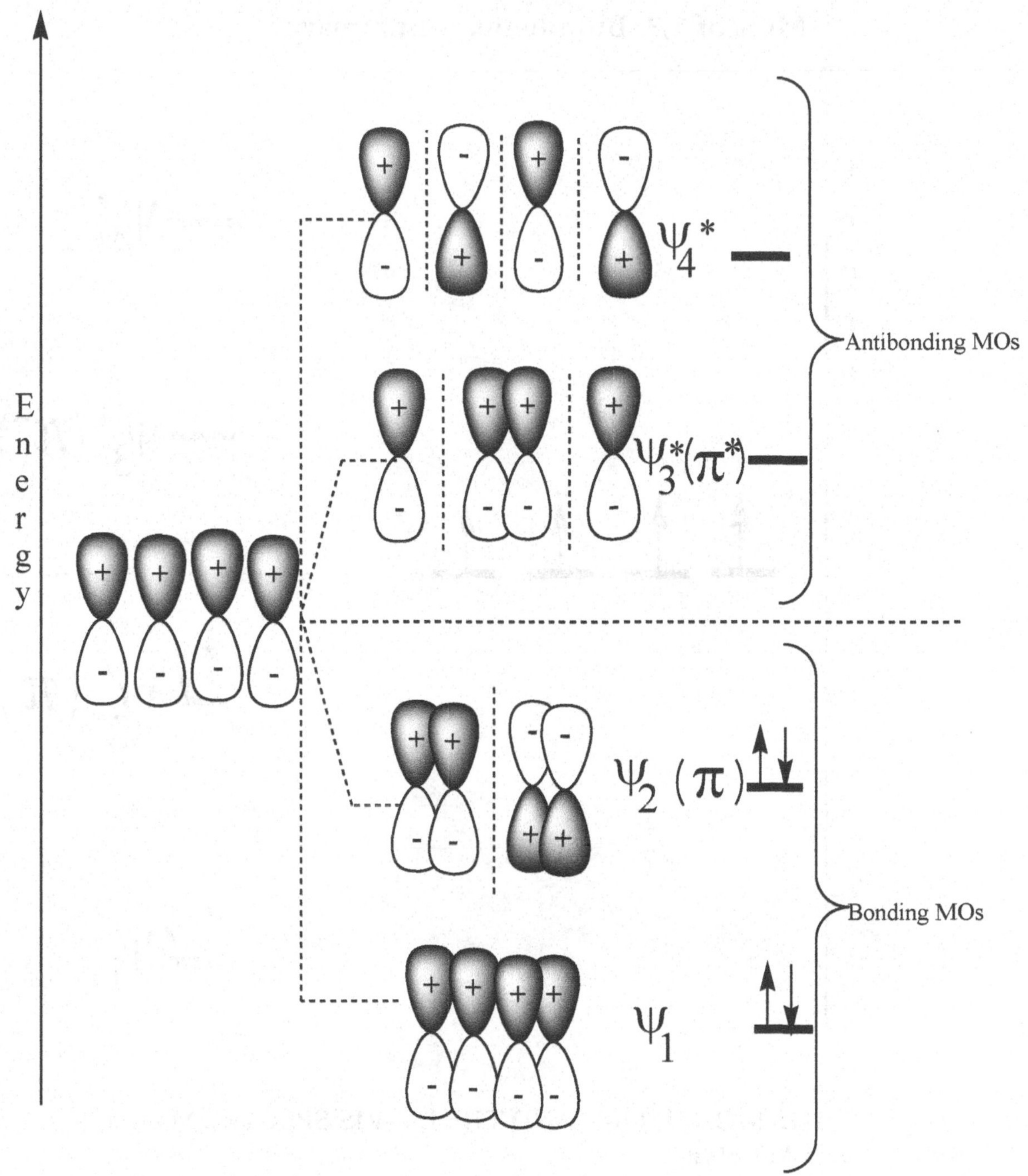
E
n
e
r
g
y
ψ_4^*
Antibonding MOs
$\psi_3^*(\pi^*)$
$\psi_2\ (\pi)$
Bonding MOs
ψ_1

MOs of 1,3-Butadiene: A Summary

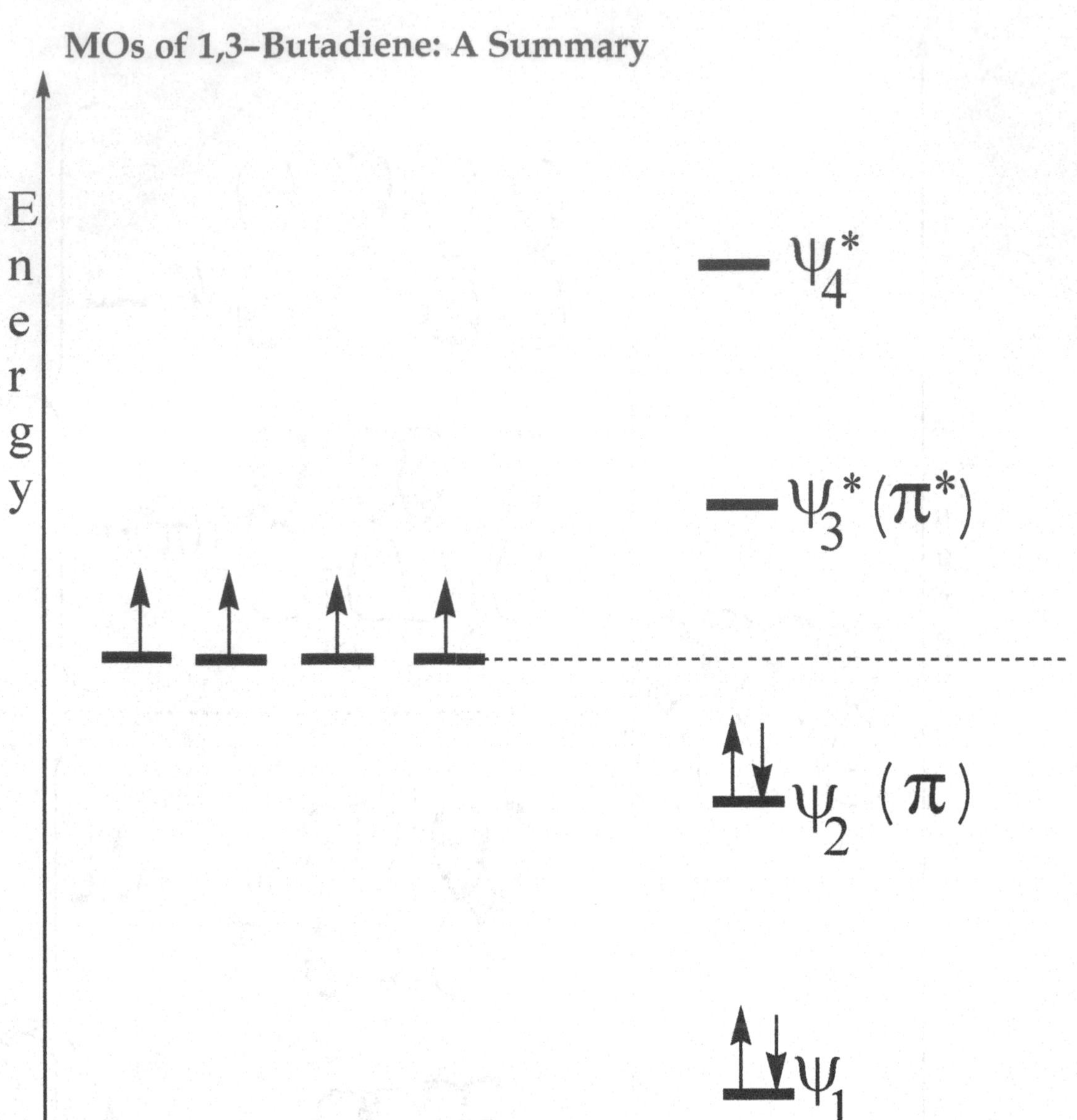

7. THE MO THEORY AND THE UV-VIS SPECTRUM OF 1,3-BUTADIENE

- **Recall**: According to the MO theory, different combinations of the 4 **unhybridized** p orbitals in 1,3-butadiene result in 4 MOs as follows:
  - **Ψ_2 = Highest Occupied MO = HOMO = Π MO.**
  - **Ψ_3^* = Lowest Unoccupied MO = LUMO = Π* MO.**

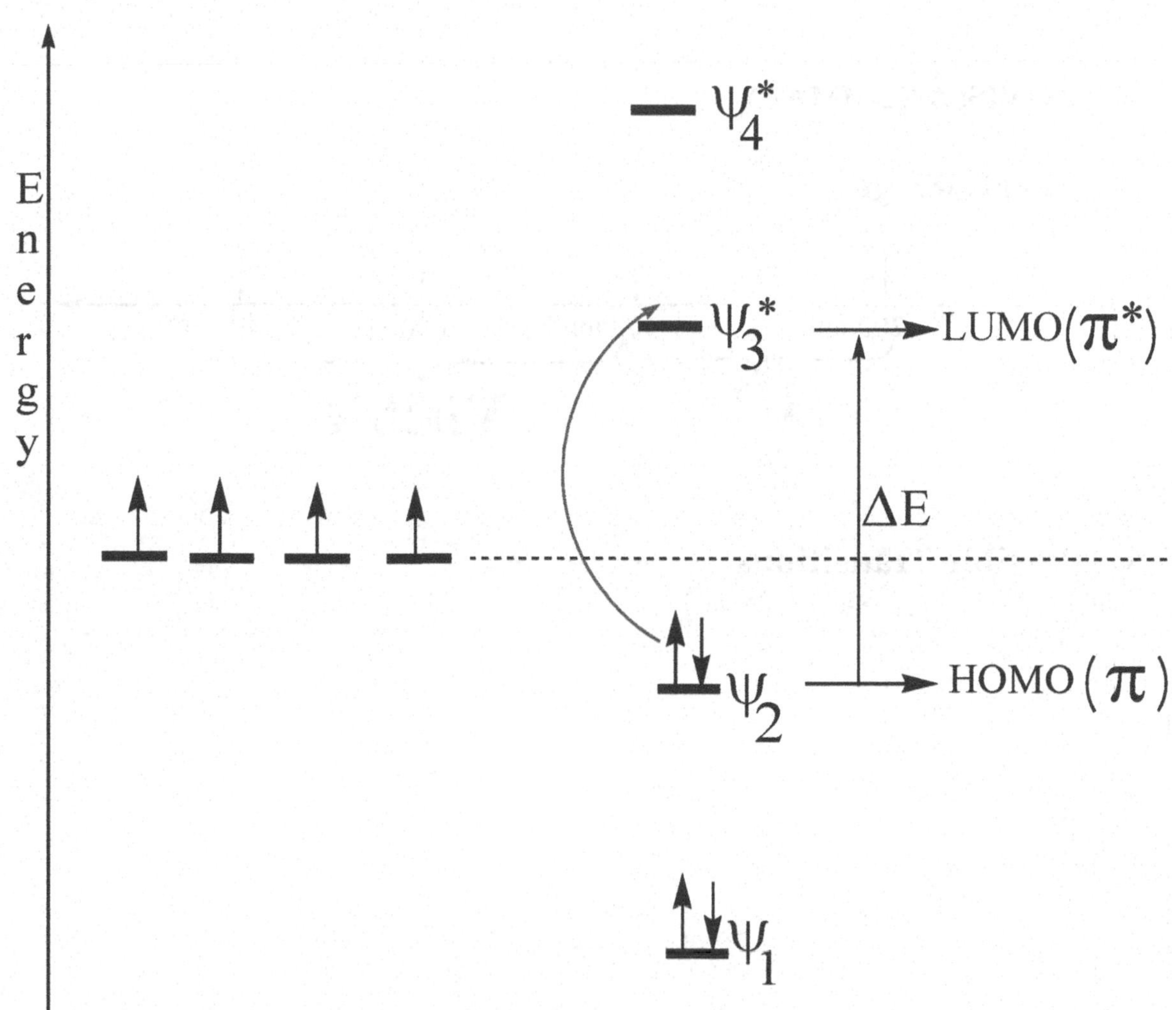

- The energy required to promote 1 electron from the **HOMO** to the **LUMO** has a λ_{max} **of 217 nm. In general,** λ_{max} **depends on the energy gap between the HOMO and LUMO. The MO theory predicts that energy gap decreases with increasing conjugation. Therefore,** λ_{max} **increases with increasing conjugation.**

See page ________.

8. UV-VIS: A SUMMARY

- **UV-Vis Range**

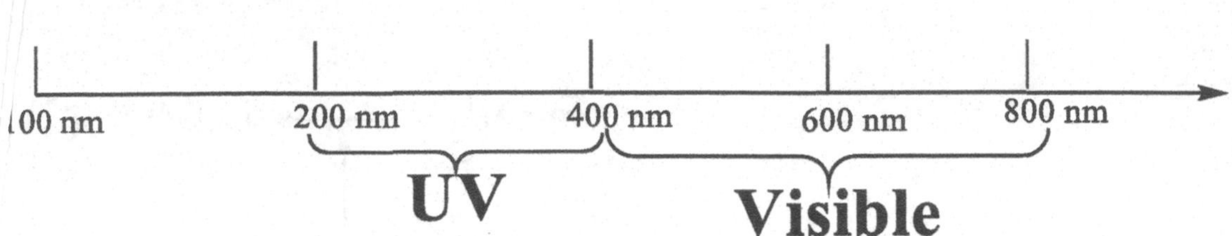

- **Possible Transitions**

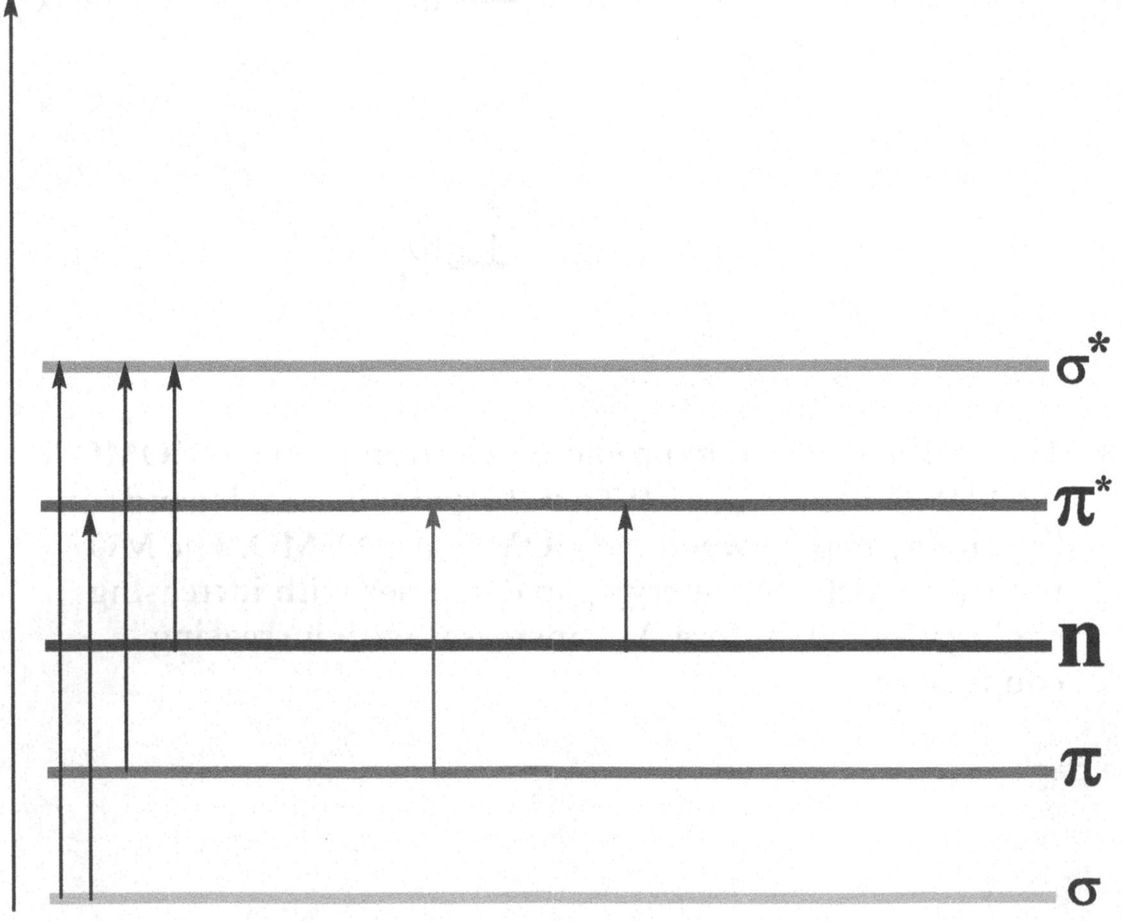

Ex: Acetone has two λ_{max}: one at 190nm and the other at 275 nm. Explain this observation.

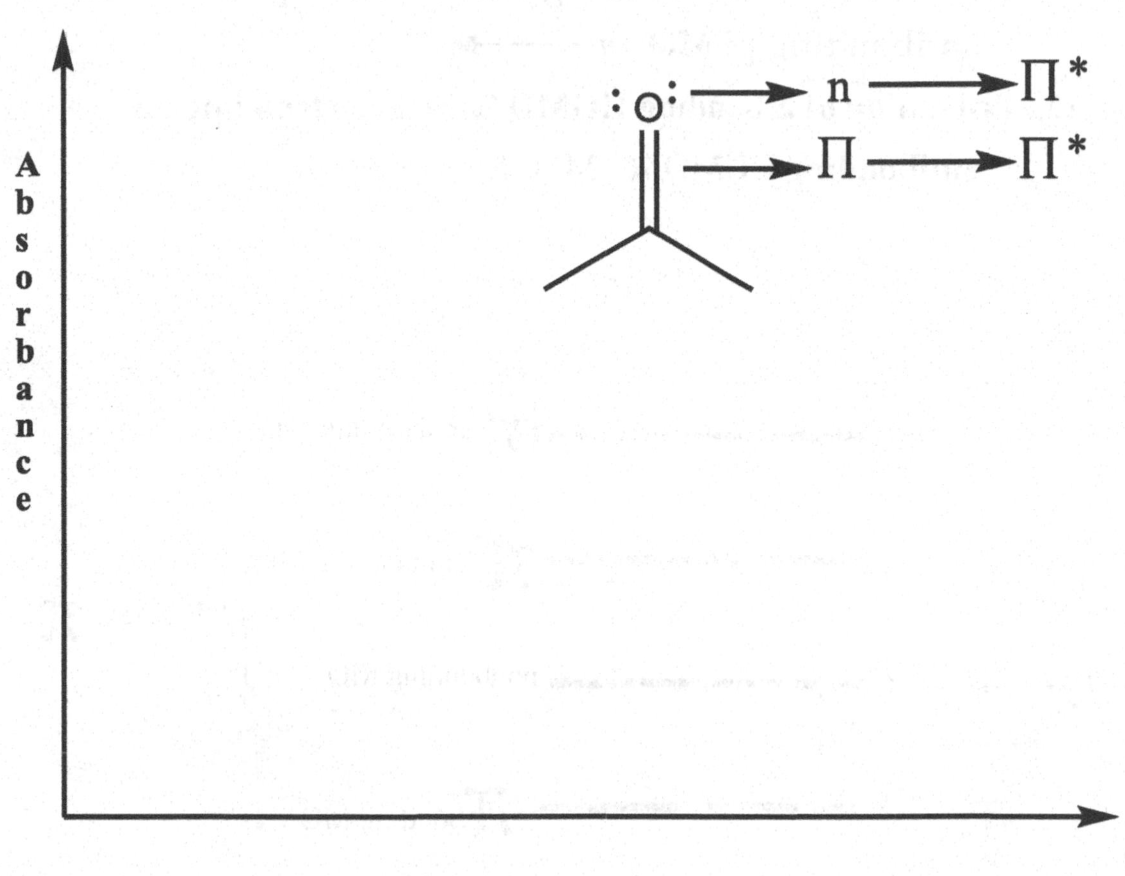

- **Conclusion: UV-Vis** is used to detect pi bonds and conjugated pi bonds in molecules. There are **two types** of electronic transitions that are responsible for UV-Vis transitions:
    - **-Promotion of a nonbonding electron (lone pair) into an antibonding pi MO: $n \longrightarrow \pi^*$.**
    - **-Promotion of a bonding HOMO π MO electron into an antibonding LUMO π^* MO: $\pi \longrightarrow \pi^*$.**

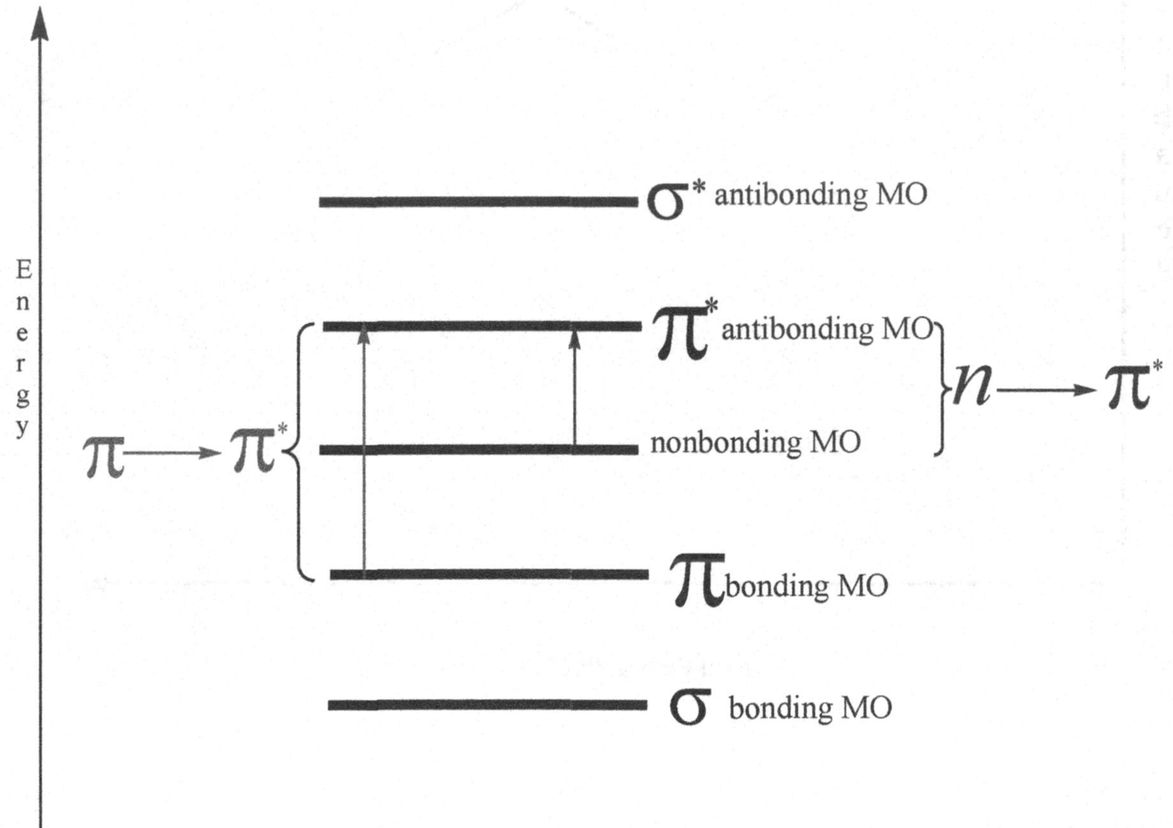

Note: an $n \longrightarrow \pi^*$ has a higher λ_{max} than a $\pi \longrightarrow \pi^*$.

See Key Concepts on pages ______ - _______.

OCHEM II UNIT 7: BENZENE AND AROMATICITY

A. INTRODUCTION

1. DEFINITION: A REVIEW

- **Recall: There are three types of hydrocarbons:**
  - **-Saturated hydrocarbons: Alkanes and cycloalkanes (Chapter ______)**
  - **-Unsaturated hydrocarbons: Alkenes, cycloalkenes, and alkynes (Chapters ______ and ______)**
  - **-Aromatic hydrocarbons or arenes: Benzenelike compounds**
  - **(Chapter______).**

2. AROMATIC HYDROCARBONS

- This chapter is about benzene and derivatives. These compounds are called **aromatic hydrocarbons** because of their characteristic odors (**or aroma or fragrance**). They can be simple or complex in structure. The following examples are listed:

benzene 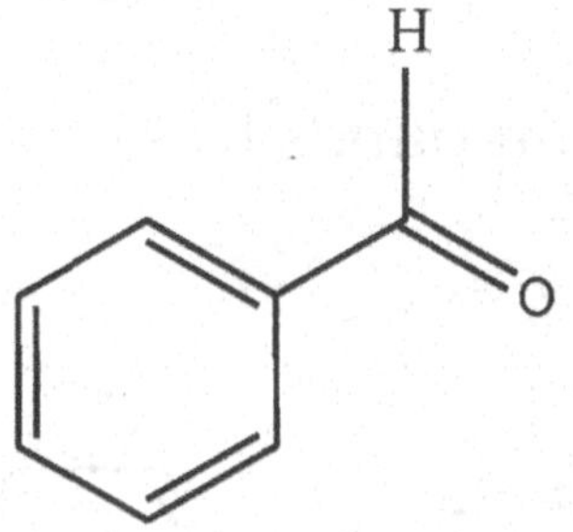 | benzaldehyde 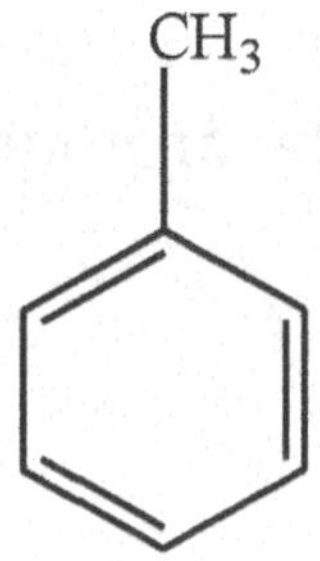 | Toluene 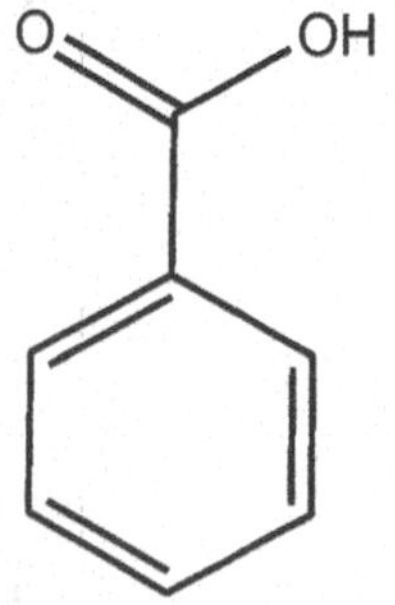 | benzoic acid

valium

estrone

- **Note: Benzene was discovered in 1825 by Michael Faraday. Its ring structure was discovered by August Kékulé. Benzene is carcinogenic.**

B. SOURCES OF AROMATIC HYDROCARBONS

1. INTRODUCTION

- There are **two major sources** of aromatic hydrocarbons:
  - -Coal.
  - -Petroleum (**See page ________: Fossil Fuels).**

2. COAL

$$\text{coal} \xrightarrow[\text{no } O_2]{1000^{\circ}C} \text{mixture of volatile products = coal tar} \xrightarrow[\text{distillation}]{\text{fractional}} \text{aromatic hydrocarbons}$$

- Some of the aromatic compounds from coal are:

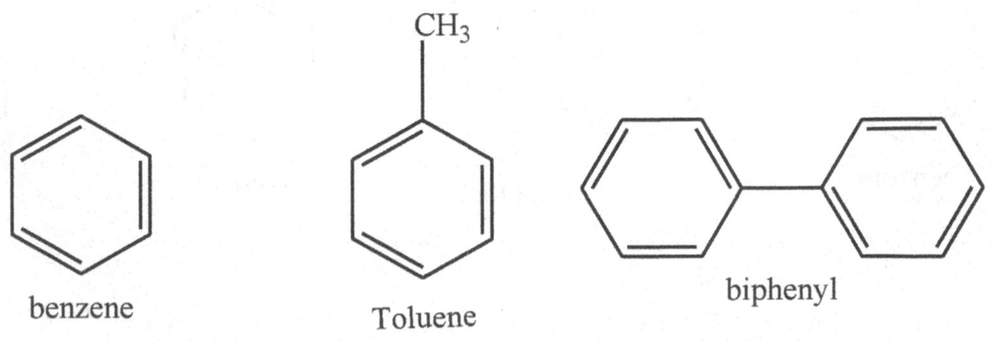

benzene Toluene biphenyl

3. PETROLEUM

- Few aromatic compounds are obtained in **petroleum refineries**. Most of the hydrocarbons thus recovered are alkanes.

- **See Fig. ______, Page ______.**

Ex: Benzene and toluene.

C. NOMENCLATURE OF AROMATIC COMPOUNDS

1. NAMING MONOSUBSTITUTED BENZENES

a. IUPAC Rule

- According to IUPAC, the **parent name is benzene.** Here are some examples:

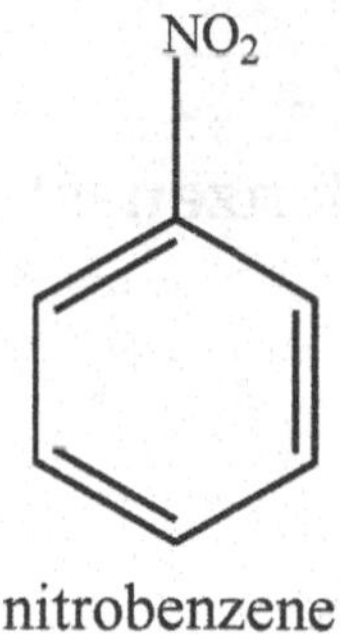

nitrobenzene

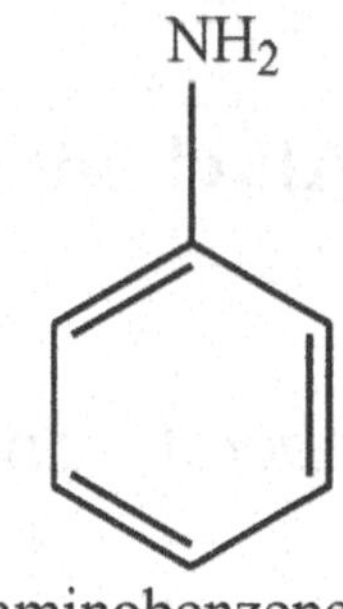

aminobenzene

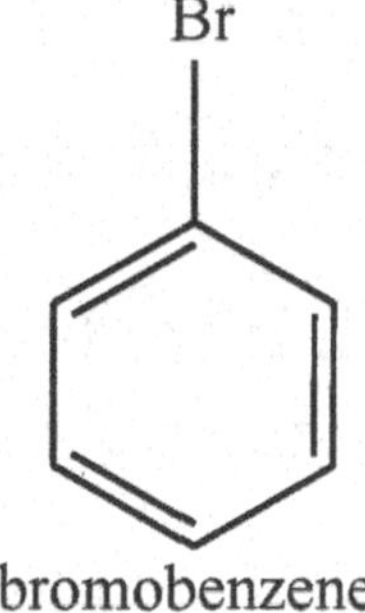

bromobenzene

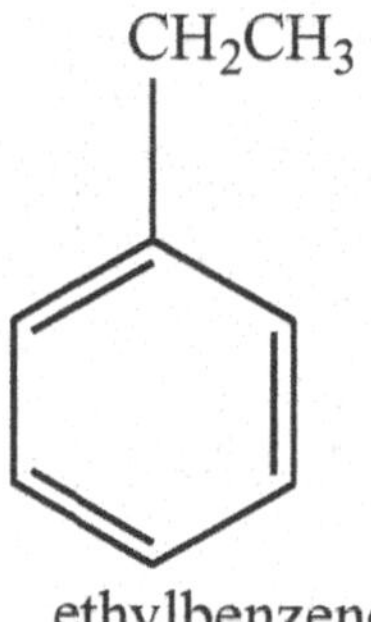

ethylbenzene

b. Common Names

- For the aromatic hydrocarbons, common names are **mostly used**. Please, **become familiar**! Some examples are:

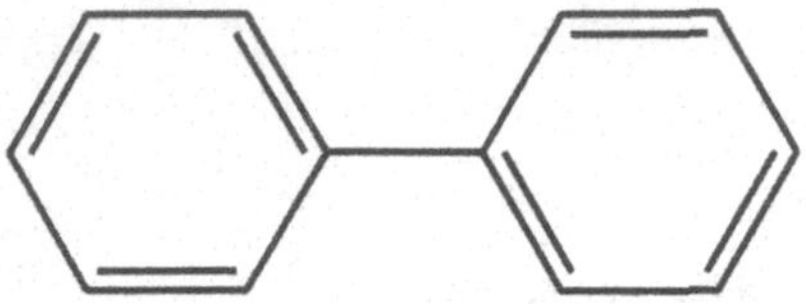

biphenyl

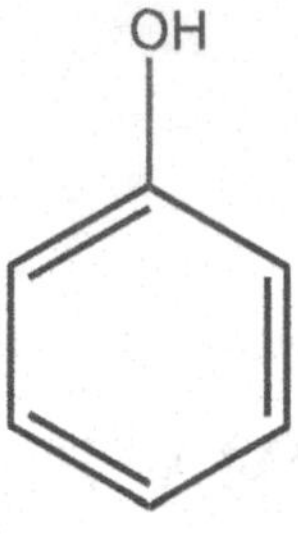

OH

phenol

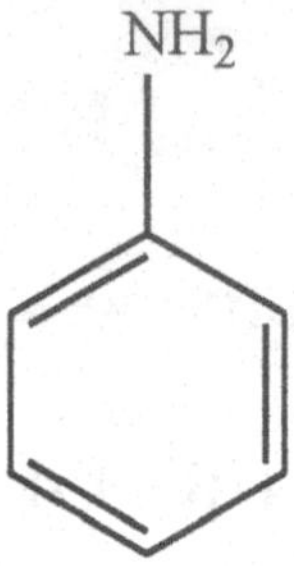

NH_2

aniline

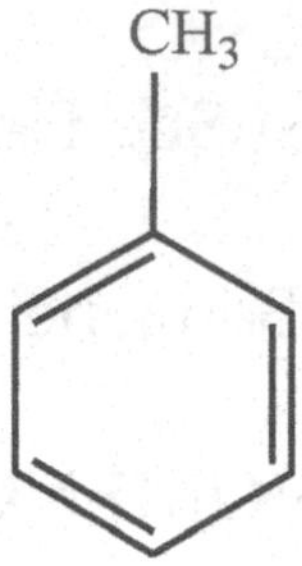

CH_3

Toluene

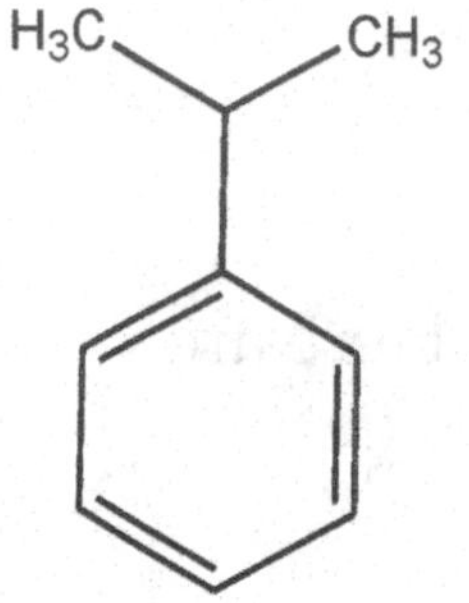

H_3C CH_3

cumene

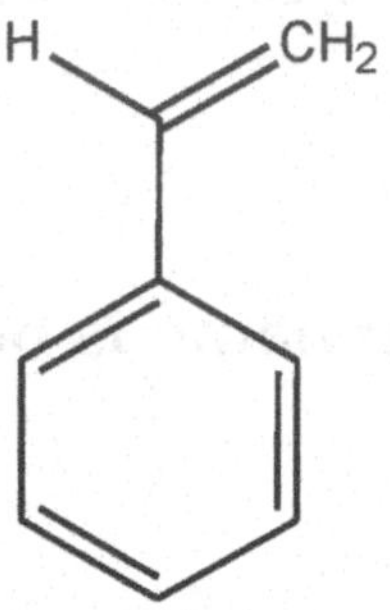

H CH_2

styrene

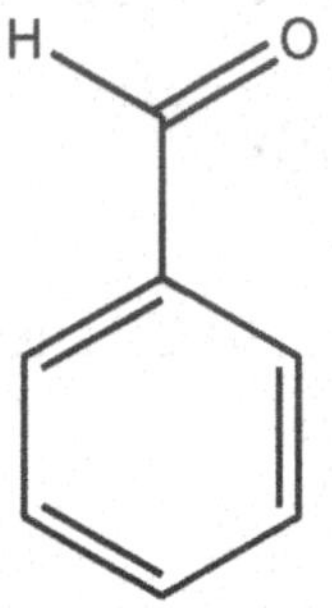

H O

benzaldehyde

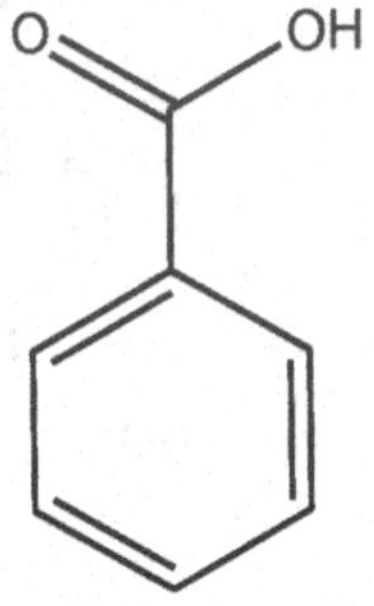

O OH

benzoic acid

c. Naming Arenes (or Alkyl-Substituted Benzenes)

i. Introduction

- Arenes have the following general formula:

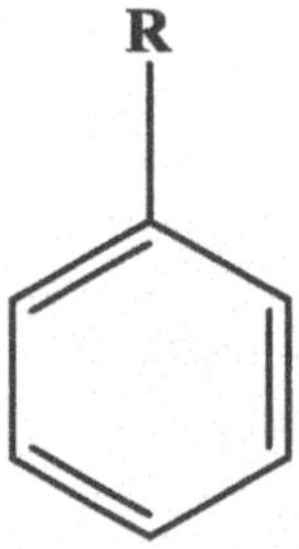

R

Ex:

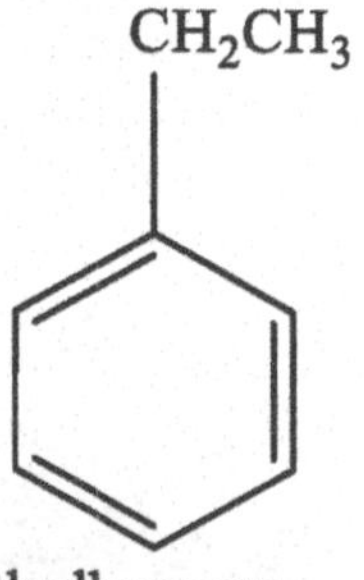

CH_2CH_3

Ethylbenzene

Where **R is an alkyl group.**

ii. Aryl Groups

- There are 2 aryl groups:

  - **Phenyl: C_6H_5-, Ph-, Φ- or:**

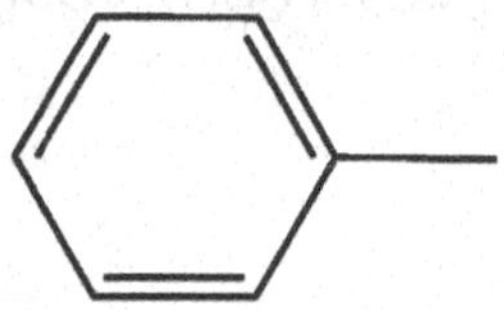

  - **Benzyl: $C_6H_5CH_2$- or:**

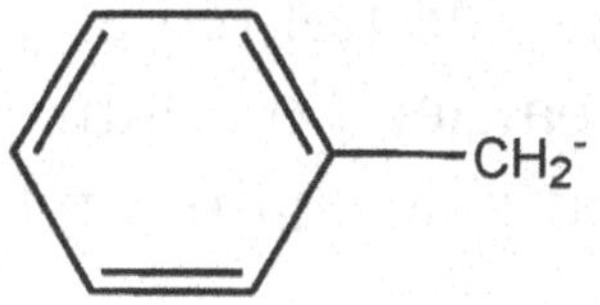

CH₂⁻

iii. Rules:

- **If the number of carbons in the alkyl group, R, is less than 6, R is named as a substituent and the benzene ring is the parent chain.**

Ex:

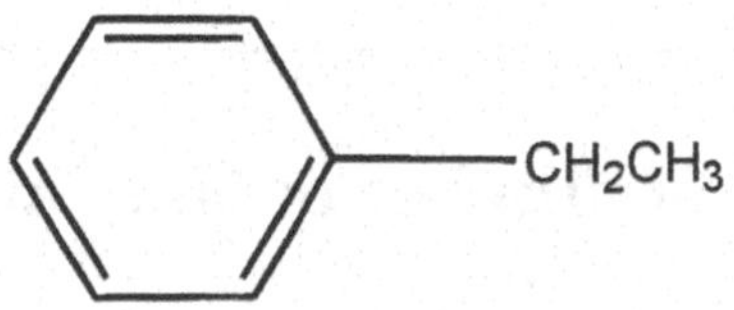

CH_2CH_3

- **If the number of carbons in the alkyl group, R, is greater than 6, R is named as the parent chain, and the benzene ring is a substituent.**

Ex

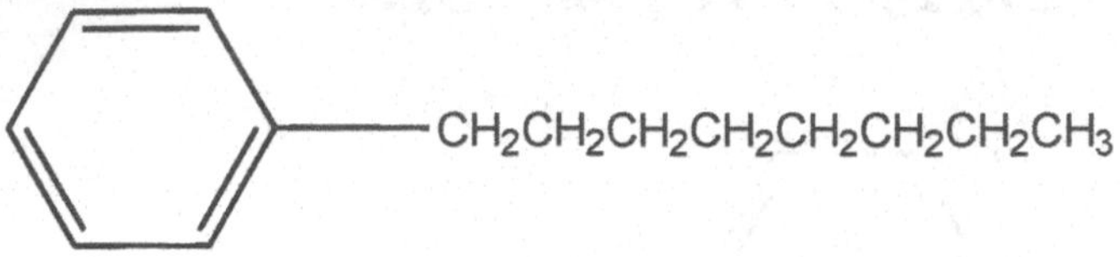

C_6H_5—$CH_2CH_2CH_2CH_2CH_2CH_2CH_2CH_3$

- **Read pages ______ - _______.**

2. NAMING COMPLEX ALKYL BENZENE DERIVATIVES

- Some benzene derivatives do not have "straightforward" structures as described earlier. Sometimes, they have complex structures. Indeed the outside chain can be branched.

Ex:

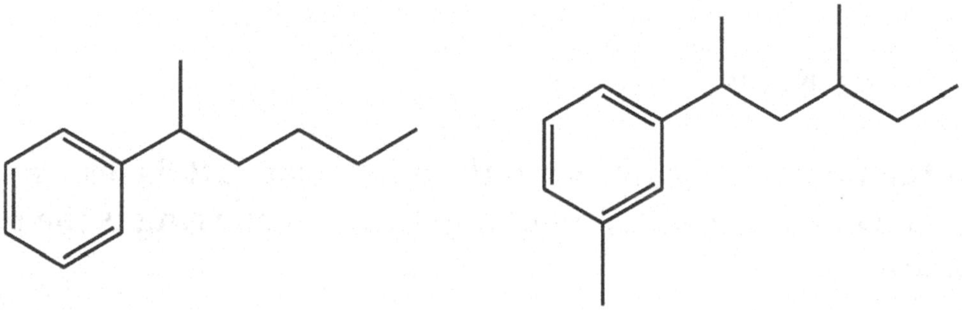

- They are named as:

(name of alkyl group on outside chain) (name of alkyl of outside main chain) ***benzene***

- **Note: Carbon 1 on the benzene ring is the carbon to which the outside main chain is attached.**

Ex:

(1-Ethylpentyl)benzene

3. NAMING DISUBSTITUTED BENZENES

a. Positions on the Benzene Ring

- The second group on a benzene ring can "**orient**" itself in 3 ways: **ortho, meta, or para.** When the **new group** puts itself on the 2 position on the ring, it is said to be **ortho**. If it goes to the 3 position, it is **meta**. Finally, the 4 position is referred as **para**. The following sketch illustrates well the different possible orientations of the second group.

- **Note: The o, p, and m positions are very important in chemical reactions. (See Chapter _______).**

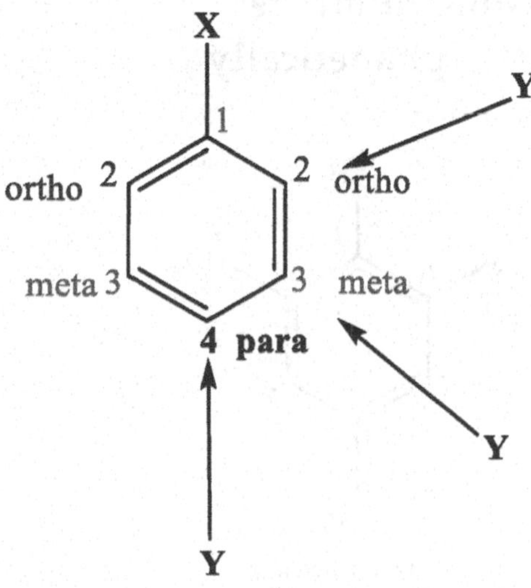

b. IUPAC Names: Some Examples

- **In naming these compounds, you should alphabetize substituents' names. Here are some examples:**

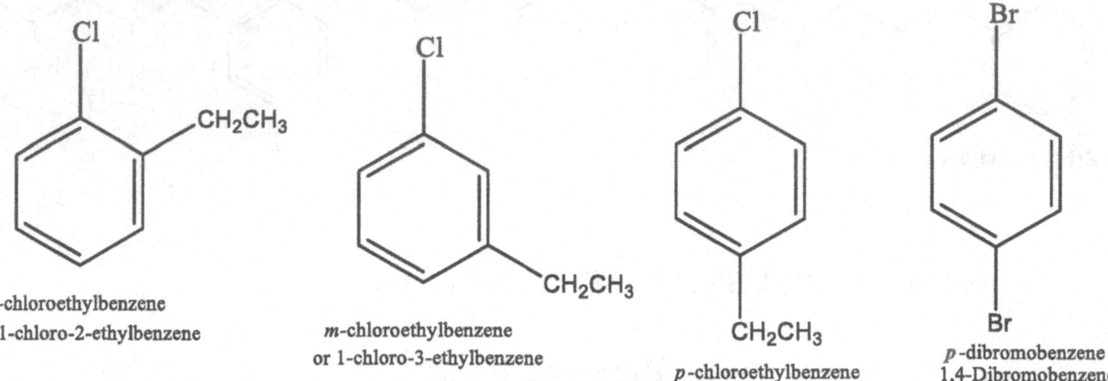

o-chloroethylbenzene or 1-chloro-2-ethylbenzene

m-chloroethylbenzene or 1-chloro-3-ethylbenzene

p-chloroethylbenzene or 4-chloro-3-ethylbenzene

p-dibromobenzene 1,4-Dibromobenzene

- **Note: Common names are usually used.**

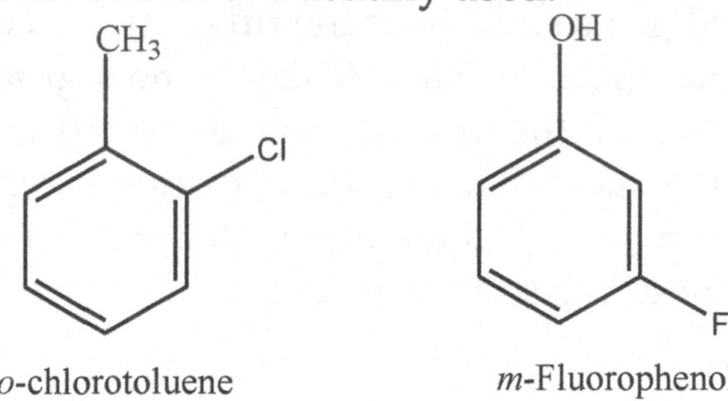

o-chlorotoluene

m-Fluorophenol

4. NAMING POLYSUBSTITUTED BENZENES

- **Number the position of each substituent on the ring**
- **Use the lowest possible numbers**
- **List the substituents alphabetically**

Ex:

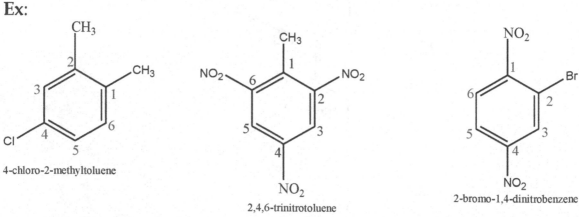

4-chloro-2-methyltoluene

2,4,6-trinitrotoluene

2-bromo-1,4-dinitrobenzene

- **Read pages _____ - ____. Do problems on pages ______ - ____.**

D. SPECTROSCOPY OF AROMATIC COMPOUNDS

1. IR SPECTROSCOPY

- The aryl H absorbs at **3150-3000 cm^{-1}**.

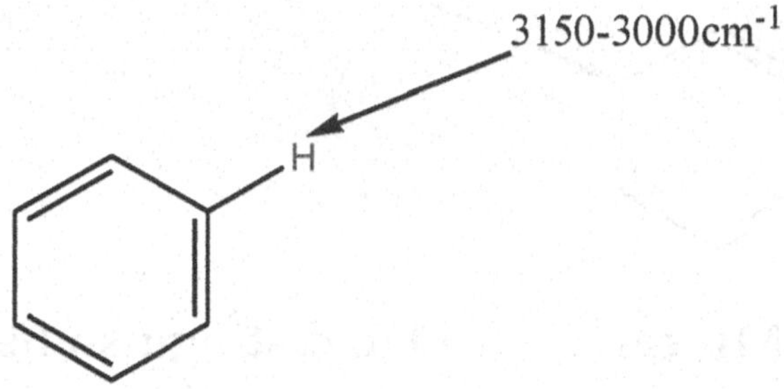

2. NMR SPECTROSCOPY

a. Proton NMR

- The **aryl protons** are highly deshielded because of **the ring current created by the circulating π electrons in a magnetic field.**

Recall: $B_{eff} = B_0 + B_{local}$

6.5-8.0δ

H

- **The benzylic protons** are outside the ring. Therefore, they are not as deshielded as the aryl protons due to a lack of the deshielding effect of a ring current.

1.5-3.0δ

CH_2

b. Carbon-13 NMR

- The aromatic carbons absorb in the same range as normal alkene **sp² carbons: 110-140δ. Deshielded like the sp² Carbon.**

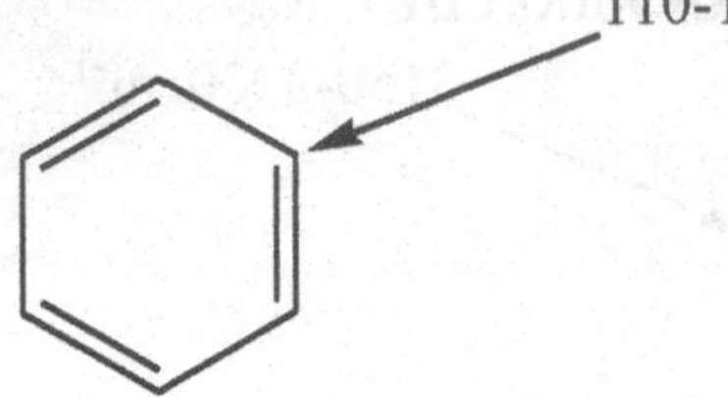

- **Note: C-13 NMR can be used to distinguish between the ortho, para, and meta positions in the identification of a product.**
- **See Fig. ________, page _______.**

3. UV-VIS SPECTROSCOPY

- **λ_{max}** for benzene is **205 nm**. The absorption peak is intense. In general, for aromatic substances, the absorption range is **255-275 nm.**

- **See Table ________, page ________.**
- **Read pages _______ - _______. Do problems on page ______.**

E. INTERESTING AROMATIC COMPOUNDS

- **Read pages _______ - _______.**

- **BTX: A mixture of benzene, toluene, and p-xylene: Used to boost octane ratings.**

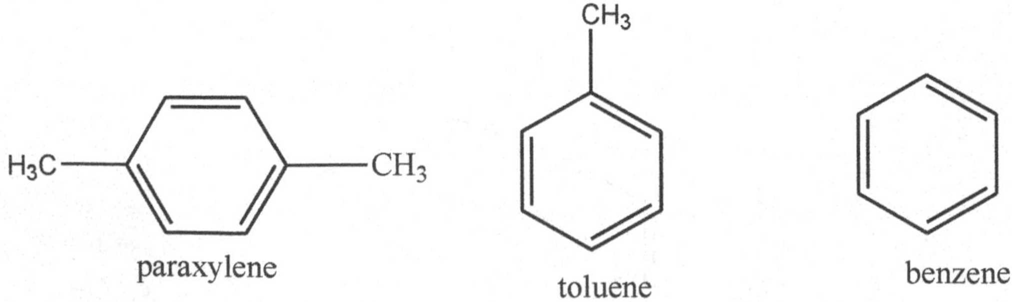

- **Polycyclic Aromatic Hydrocarbons (PAHs) are fused-ring aromatic compounds believed to be carcinogenic. Some examples.**

    - **Naphthalene is found in mothballs.**

    - **Benzo[a]pyrene is produced naturally by the incomplete combustion of organic matter: cigarette smoke, incomplete combustion of gasoline, charcoal burning, etc.**

    - **See the following structures:**

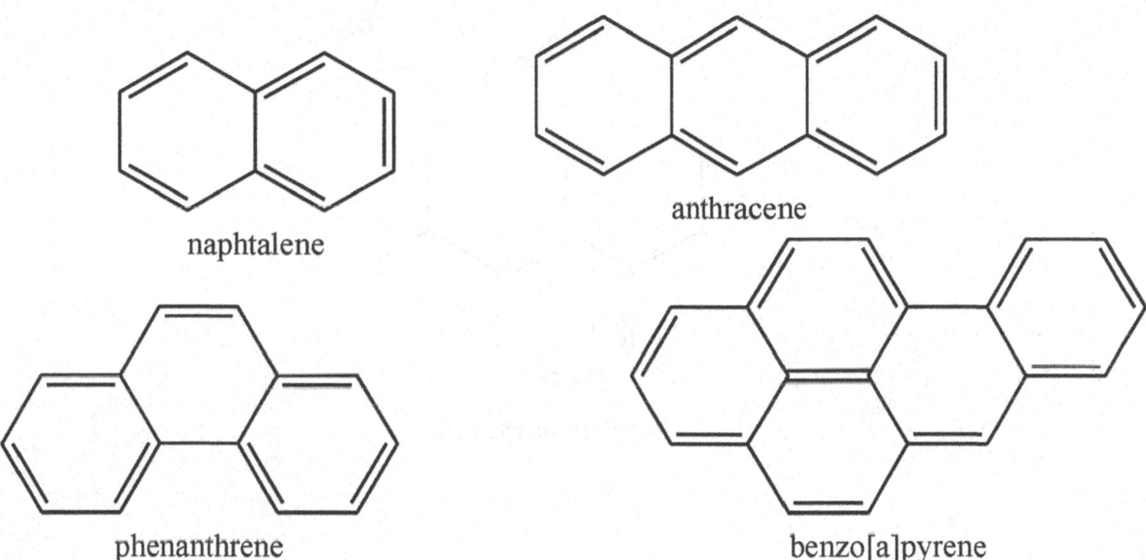

- **Many well-known drugs also contain benzene rings. Here are some examples:**

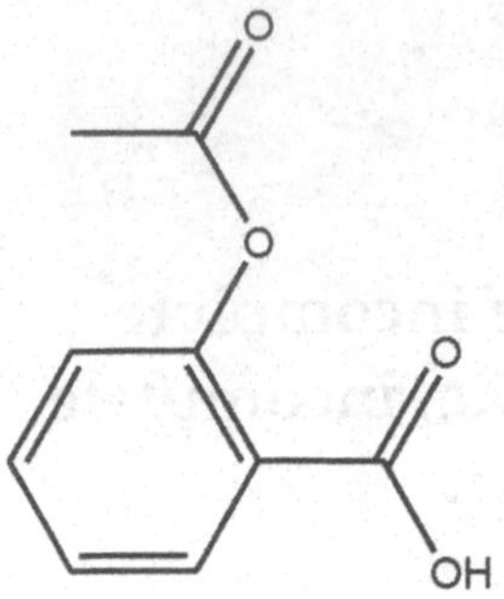

aspirin

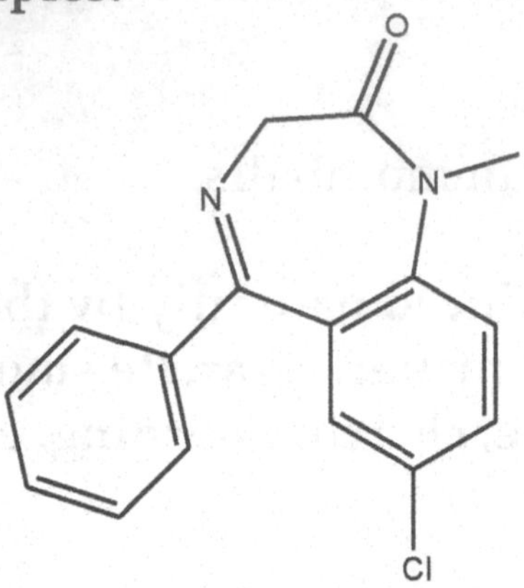

valium

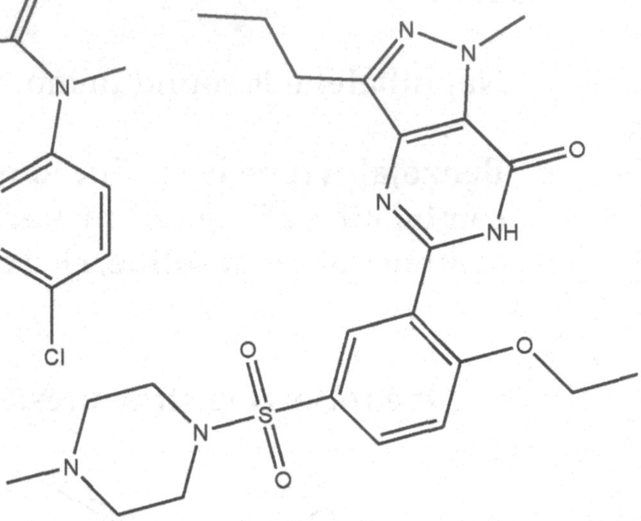

viagra

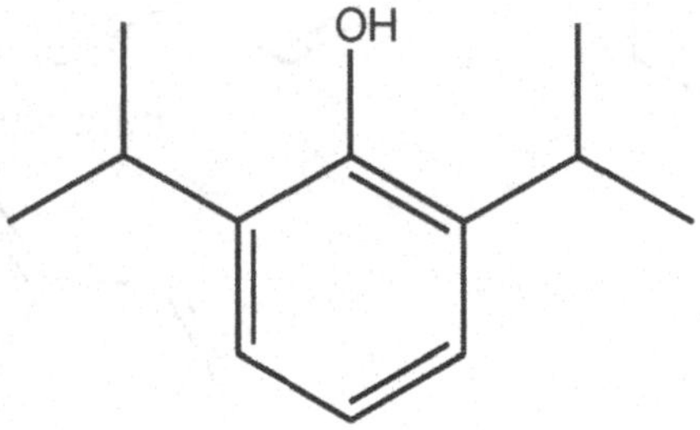

Diprivan or Propofol

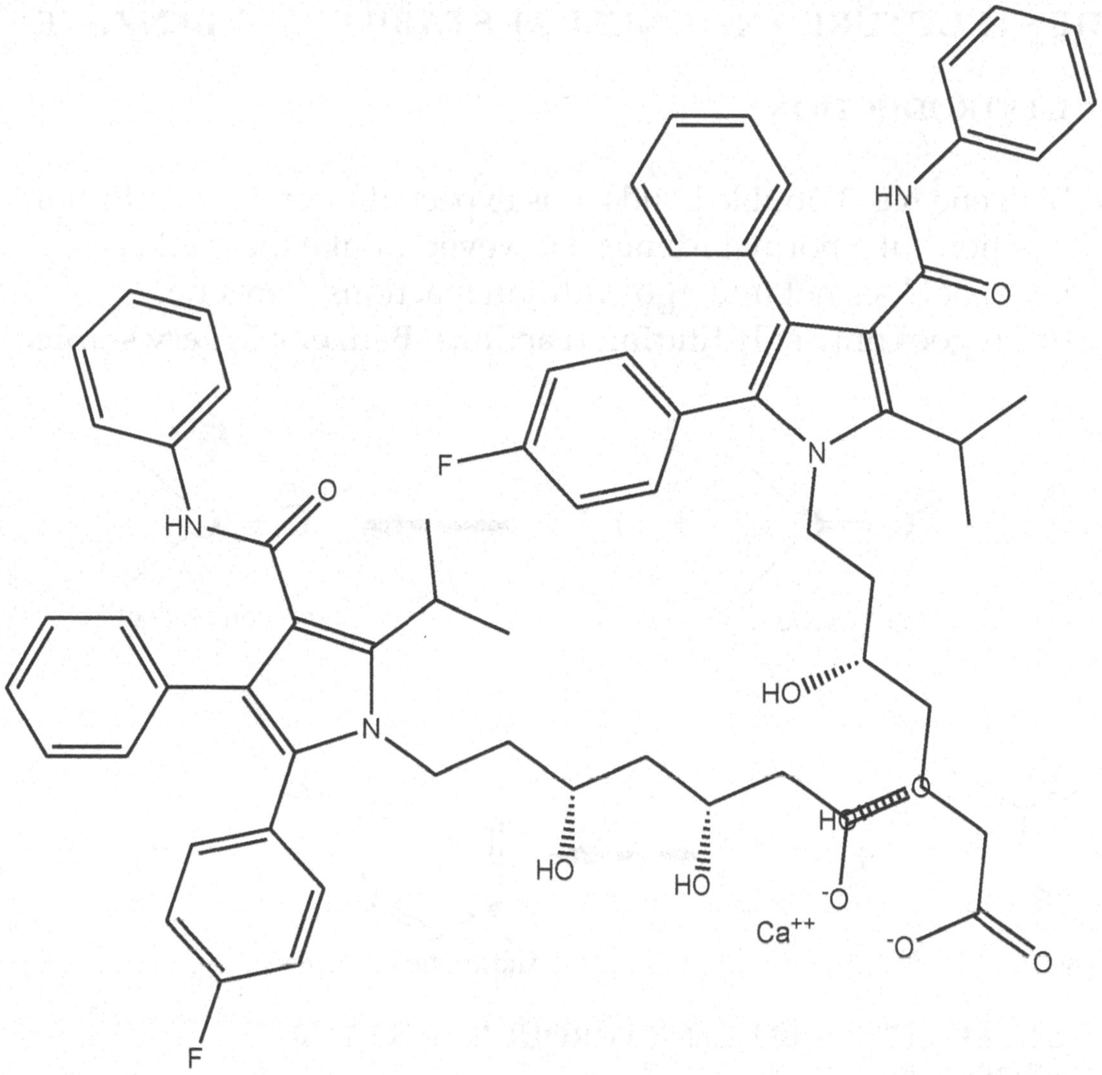

lipitor

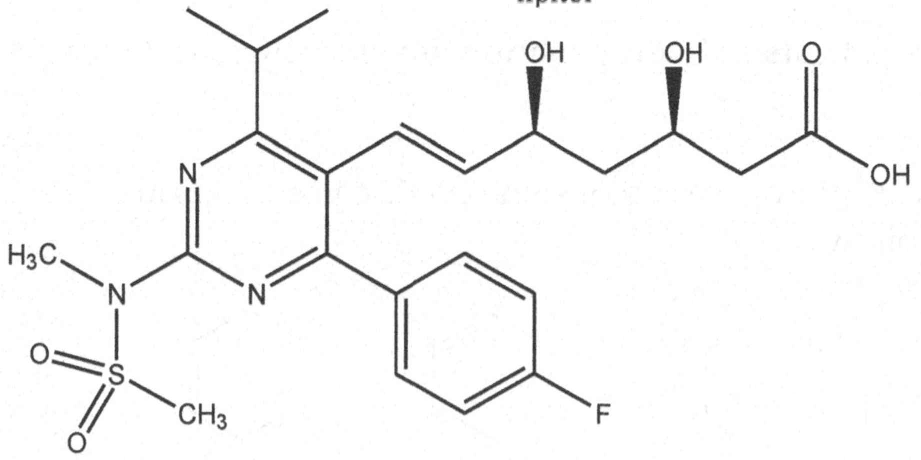

Crestor (Rosuvastatin)

F. THE STRUCTURE AND UNUSUAL STABILITY OF BENZENE

1. INTRODUCTION

- Benzene has 3 double bonds. It is expected to undergo addition reactions like normal **trienes**. However, unlike these alkenes, benzene does not undergo addition reactions. Benzene undergoes only **substitution reactions. Benzene is very stable.**

$C=C + E^+ \longrightarrow \overset{\oplus}{C}-C(E)$

normal alkene　　　　addition product

benzene (C–H) $+ E^+ \longrightarrow$ (C–E) $+ H^+$

benzene　　　　substitution product

2. STABILITY OF BENZENE THROUGH HEATS OF HYDROGENATION

- **Recall: High heats of hydrogenation, low stability, and vice versa.**

- The heats of hydrogenation of some cyclic compounds are listed as follows:

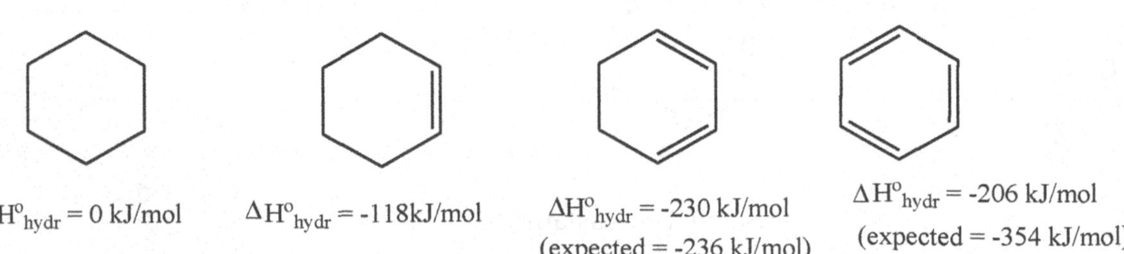

- **Conclusion: Benzene is more stable than expected.**

3. BOND LENGTHS IN BENZENE

bond type	bond length (pm)
Alkane C-C	154
Alkene C=C	134
Benzene C-C	139

- **Conclusion: The C-C bonds in benzene are identical and intermediate in length. Therefore, benzene is not an alkene.**

4. THE ACTUAL STRUCTURE OF BENZENE

- All evidence suggests that benzene is **conjugated.**
    - **It has a planar, hexagonal shape with 120° bond angles.**
    - **All 6 carbons are sp^2 hybridized, using p orbitals that are**
    - **$\perp$ to the plane of the molecule.**
    - **The 6 π electrons are delocalized over the entire ring.**
    - **It is believed that benzene exists as a hybrid of 2 resonance structures.**

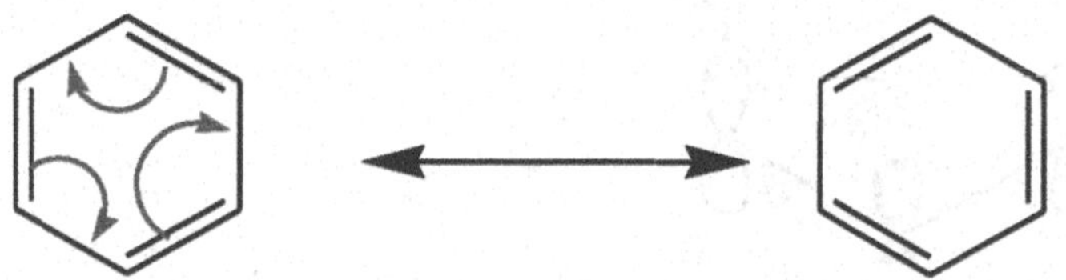

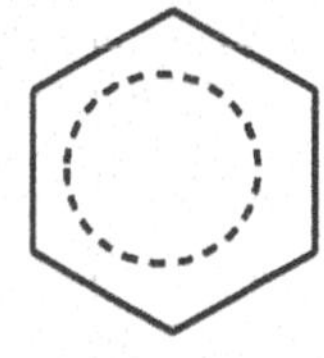

hybrid

- **Read pages ______ - ______.**

- **Do problem on page ______.**

5. THE VB SKETCH OF BENZENE

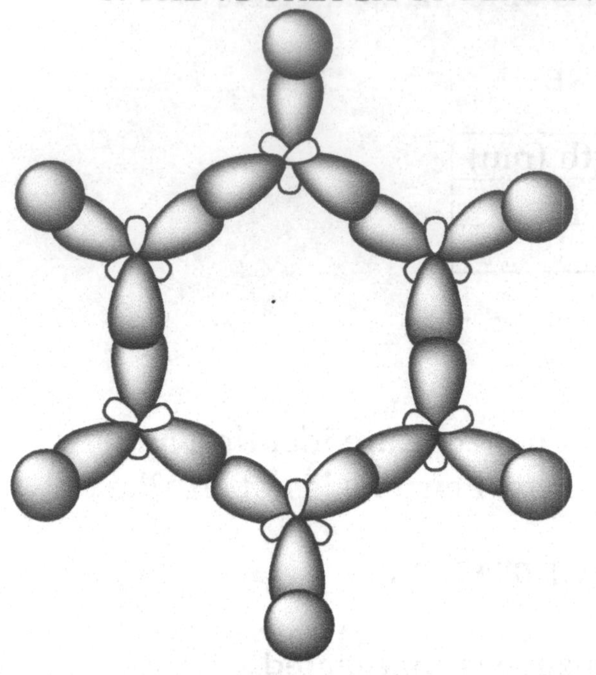

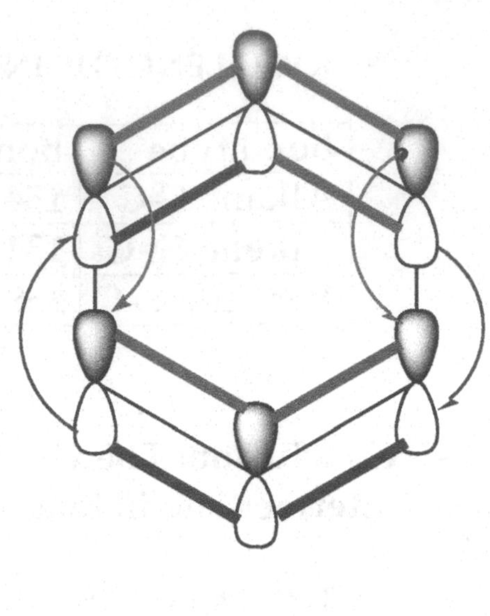

- See page ______.
- See Fig. ________, page ______.

6. THE MOLECULAR ORBITAL VIEW OF BENZENE

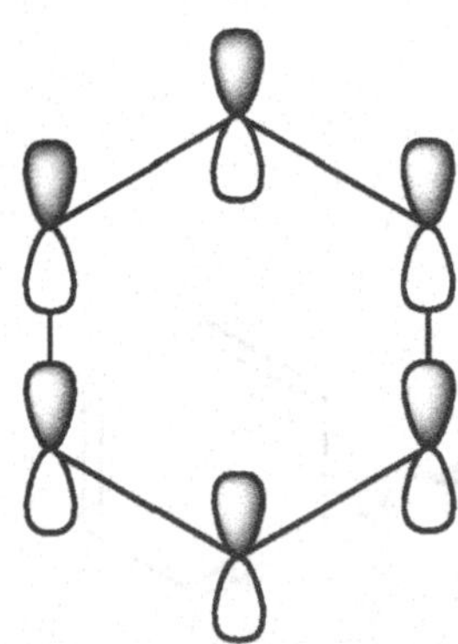

- According to the **MO theory**, the **6 free p orbitals** in benzene **combine** to give **6** molecular orbitals. There are:
    - **3 bonding (low energy) mos:** ψ_{1}, ψ_{2}, and ψ_{3}.
    - **3 antibonding (high energy) mos:** ψ^*_{4}, ψ^*_{5}, and ψ^*_{6}.
    - **The** ψ_{2}, ψ_{3}, and the ψ^*_{4}, ψ^*_{5}, are **degenerate, respectively.**

- The 6 π electrons are in the 3 bonding mos and are delocalized over the entire molecule. This is why benzene is very stable.

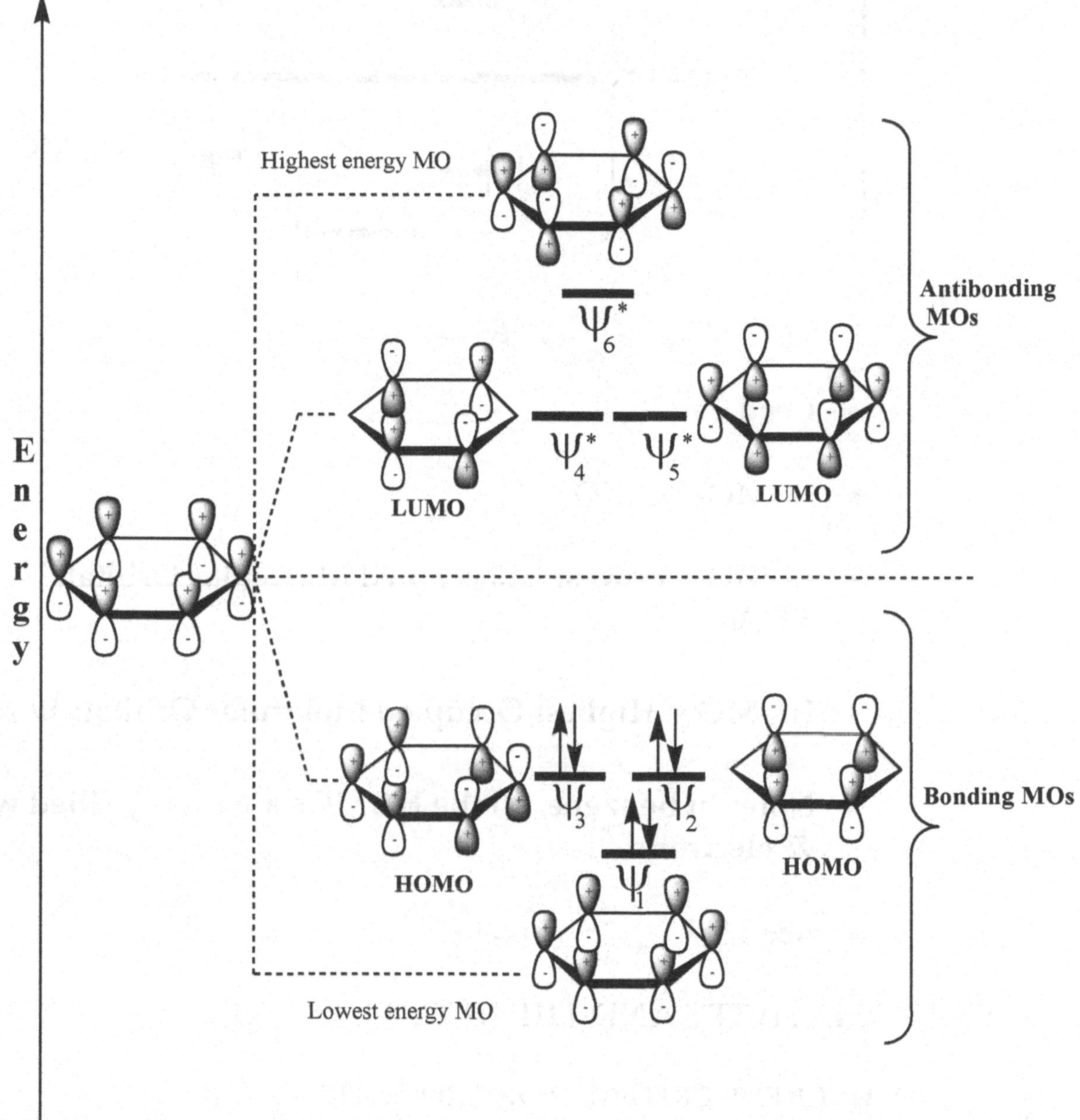

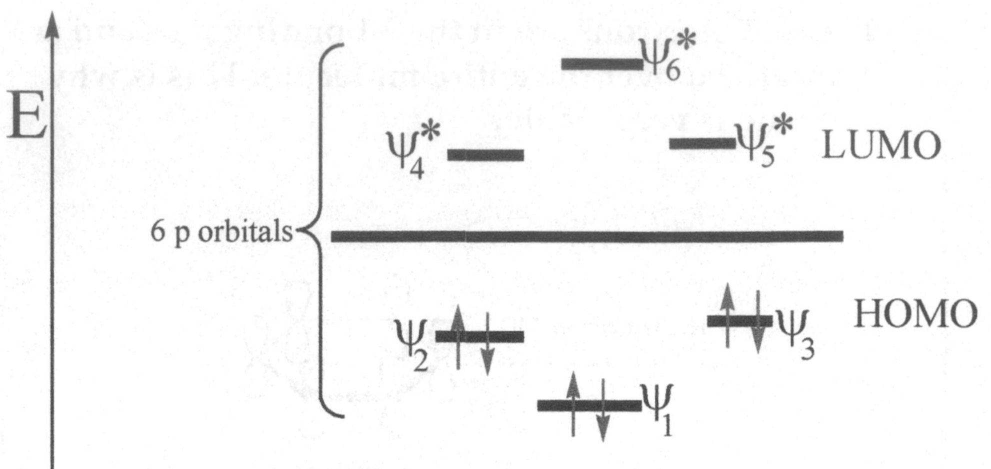

- See Fig. _______, page ______.
- Read pages _______ - ______.
- **LUMO-HOMO**
- **LUMO = Lowest Unoccupied Molecular Orbital:** $\psi^*_4, \psi^*_5,$
- **HOMO = Highest Occupied Molecular Orbital:** $\psi_2, \psi_3,$
- **Note: In benzene, all the HOMOs are totally filled with π electrons.**
- See Fig. ________, page _______.

G. AROMATICITY AND THE HÜCKEL 4n + 2 RULE

1. HÜCKEL'S CRITERIA FOR AROMATICITY

- According to Erich Hückel (1896-1980), a molecule is **aromatic** if:
    - **It is fully conjugated.**
    - **It is planar.**
    - **It is monocyclic.**
    - **It contains a total of $4n + 2$ π electrons (n = 0, 1, 2, 3, 4, 5,....)**

- **Note: The 4n + 2 (or 4n) is called the Hückel number (HN).**

- From the latter statement, a compound can be **aromatic** if it has **2, 6, 10, 14, 18,... π electrons.** These numbers are sometimes referred as **"magic numbers."**

Ex:

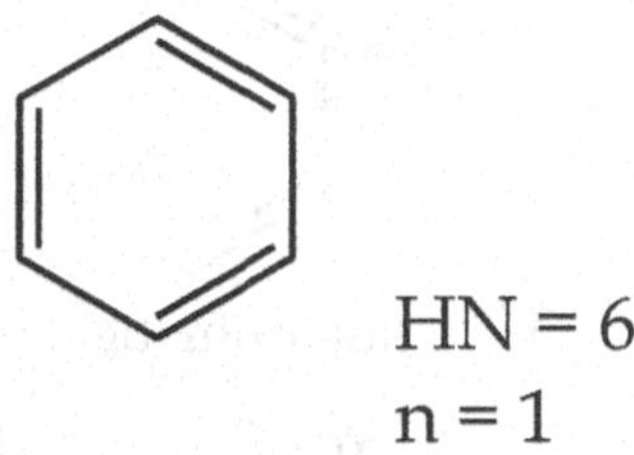

- In general, **an aromatic molecule** is a molecule that is **cyclic, planar, fully conjugated with 4n + 2 π electrons (n = 0, 1, 2, 3, 4, 5,....).**

Ex:

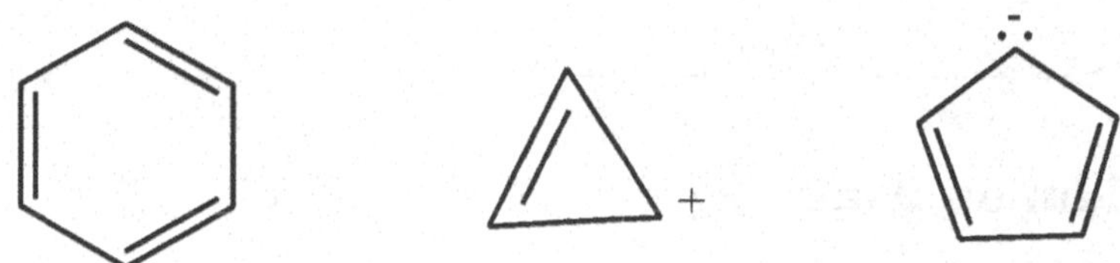

- **An antiaromatic molecule** is **cyclic, planar, fully conjugated with 4n π electrons (n = 0, 1, 2, 3, 4, 5,....).**

Ex:

- A molecule that is neither aromatic nor antiaromatic is said to be **nonaromatic**. In other words, a **nonaromatic** compound is neither cyclic, nor planar, nor fully conjugated.

Ex:

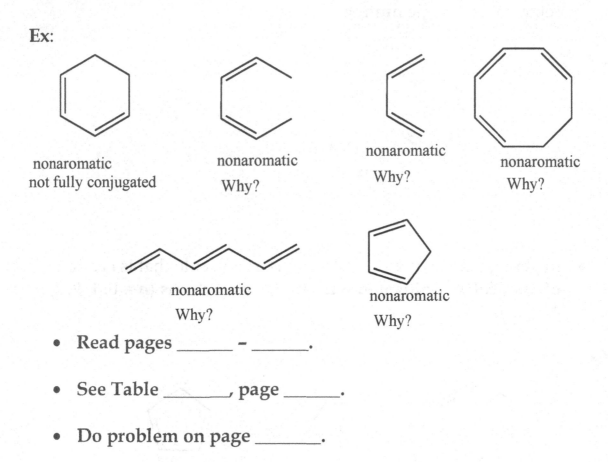

nonaromatic
not fully conjugated

nonaromatic
Why?

nonaromatic
Why?

nonaromatic
Why?

nonaromatic
Why?

nonaromatic
Why?

- **Read pages ______ - ______.**
- **See Table _______, page ______.**
- **Do problem on page _______.**

2. APPLYING HÜCKEL'S RULE TO SOME COMPOUNDS

a. Cyclobutadiene: Antiaromatic

- Has **4 π localized electrons. HN = 4 (n=1).**

b. Benzene: Aromatic

- The HN = 6 (n =1).

c. Cyclooctatetraene: Nonaromatic

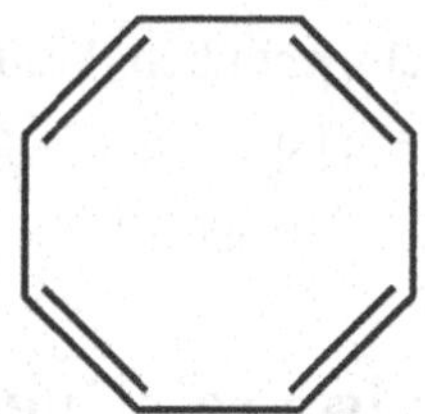

- HN = 8 (n = 2).
- Nonaromatic.
- Not stable.
- Not planar; tub-shaped=not aromatic.

d. 14-Annulene: Aromatic

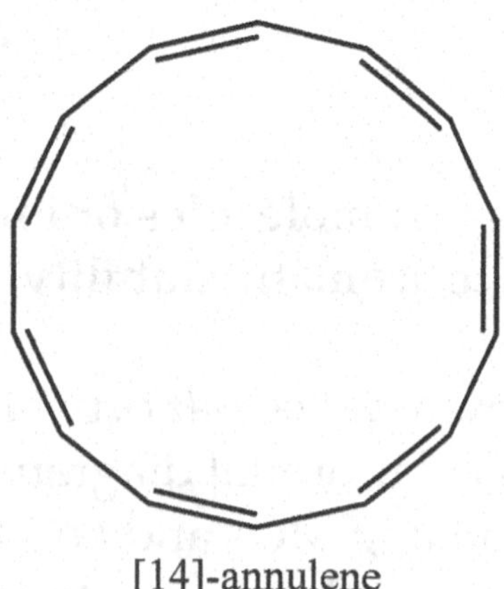

[14]-annulene

- The HN # is 14. Planar aromatic.

3. APPLYING HÜCKEL'S RULE TO SOME CYCLIC IONS

a. Introduction

- Hückel theory can also be applied to ions.

b. Ions of Cyclopentadiene

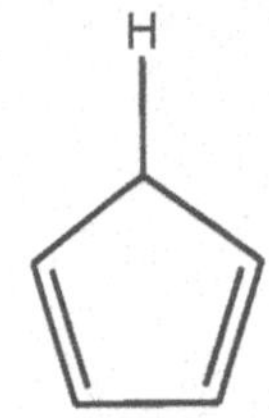

cyclopentadiene
HN = 4:nonaromatic
(unstable)

cyclopentadienyl cation
HN = 4:antiaromatic
(unstable)

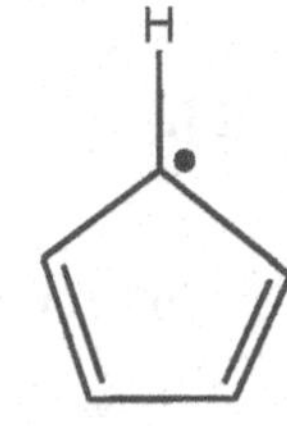

cyclopentadienyl radical
HN = 5:nonaromatic
(unstable)

cyclopentadienyl anion
HN = 6: aromatic
(stable)

- **Note: Cyclopentadiene is a relatively stronger acid than comparable hydrocarbons because its conjugate base, the cyclopentadienyl anion is aromatic and is resonance stabilized (very stable).**

- **Read pages ______ - ______. See the tropylium cation on page ______.**
- **Do problems on pages _______ - ______.**

H. WHY IS AROMATICITY ASSOCIATED WITH 4n + 2?

1. INTRODUCTION

- **Recall: Only conjugated molecules or ions having HN of 2, 6, 10, 14, 18, 22... lead to aromatic stability. Why?**

- The answer comes from molecular orbital calculations. For instance, in the molecular orbital diagram of benzene, there are equal numbers of bonding MOs and antibonding MOs. It happens, that in all **aromatic species, the bonding MOs are filled with all available pi electrons. For each one, the number** of electrons required to fill the bonding MOs is 4n + 2. This is the basis of the 4n + 2 rule.

- See Fig. ________, page ________: MO patterns for fully conjugated cyclic systems.

- Conclusion: -Only compounds having fully filled bonding MOs are aromatic. Partially filled MO compounds are not aromatic and not stable.

2. BENZENE: AN EXAMPLE

- See Fig. _______, page _______; Fig. _______, page _______.

Filled Bonding MOs of Some Aromatic Compounds

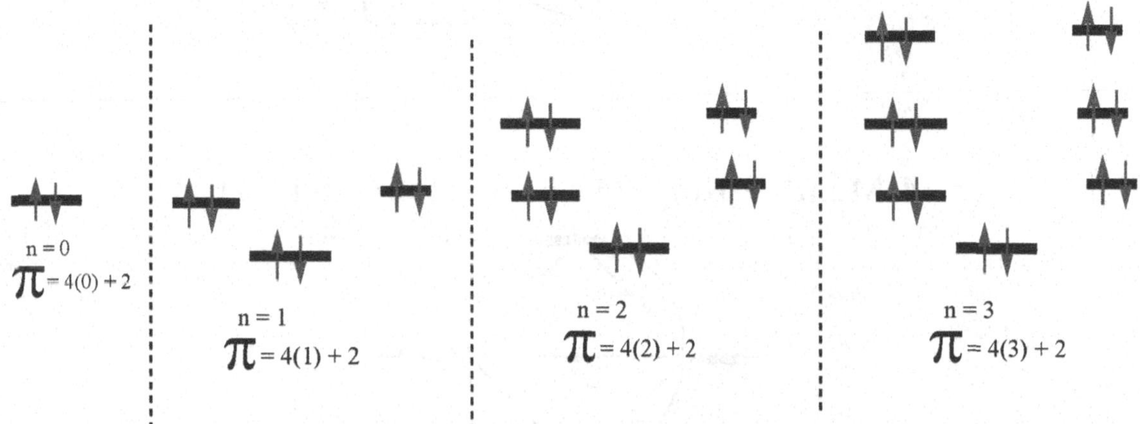

3. GENERAL WAY OF DETERMINING MOS OF MONOCYCLIC CONJUGATED RINGS USING INSCRIBED POLYGONS OR FROST CIRCLES

a. Introduction: Using a circle as a template: Frost's circles

- Above the horizontal line crossing the center of the circle: antibonding MOs.
- Below the horizontal line crossing the center of the circle: bonding MOs.
- At the "horizontal" diameter of the circle: nonbonding MOs.

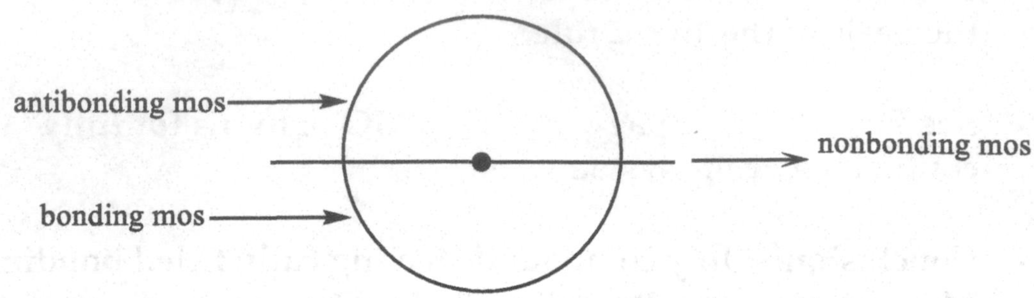

- Read pages _______ - ______.

b. **3-Member Rings (3 carbon atoms in ring)**

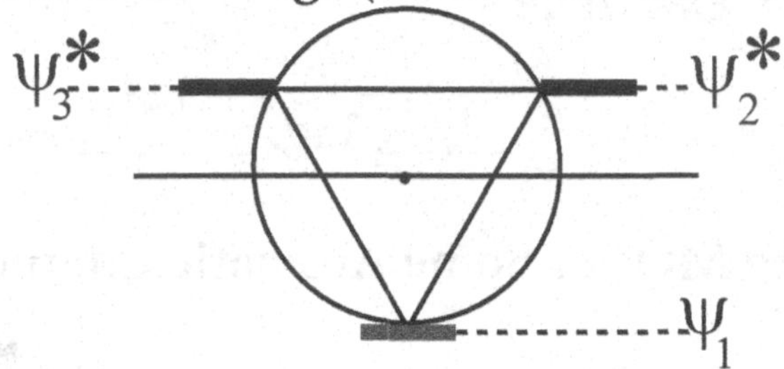

c. **4-Member Rings (4 carbon atoms in ring)**

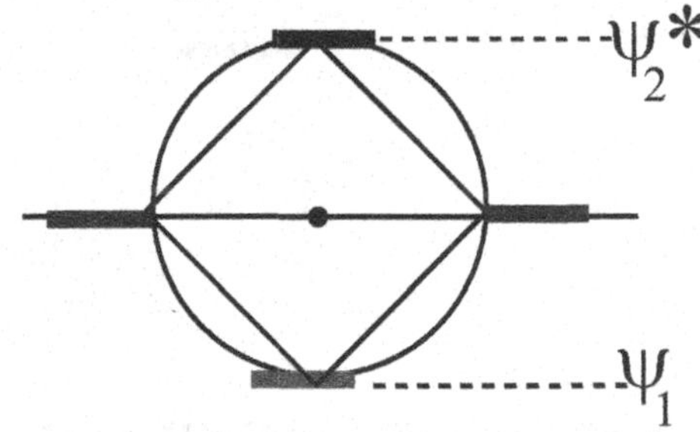

d. **5-Member Rings (5 carbon atoms in ring)**

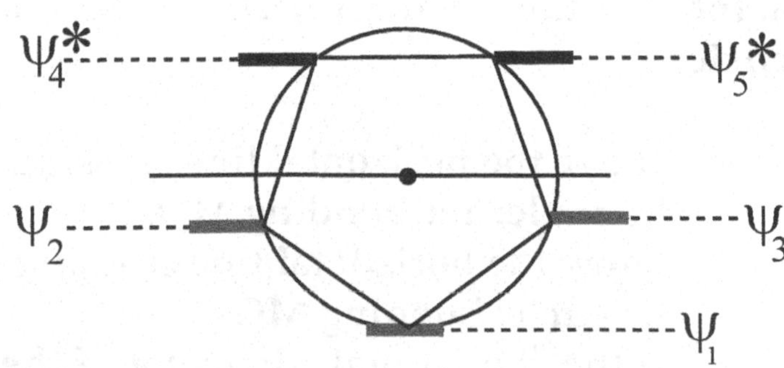

e. **6-Member Rings (6 carbon atoms in ring)**

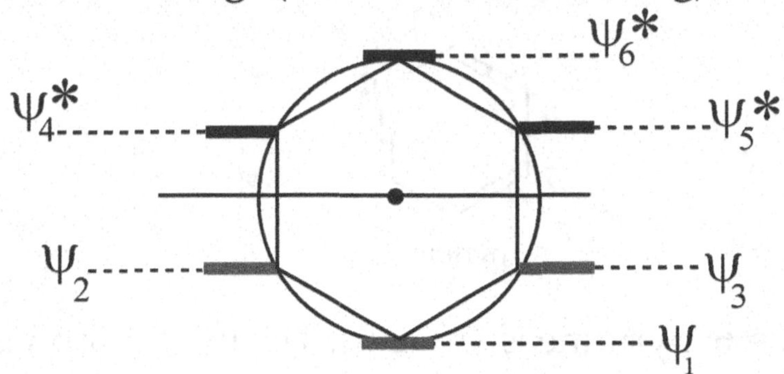

- **Do problems on page _______.**

- **Activity: Cyclobutadiene is antiaromatic. Using the inscribed circle method, explain this fact.**

I. AROMATIC HETEROCYCLIC COMPOUNDS

1. INTRODUCTION

- A **heterocyclic compound or ion** contains an atom (**heteroatom =O, N, S, P)** other than C in ring.

Ex:

N H N O

pyridine pyrrole furan

- **Read pages _______ - ______.**

2. PYRIDINE

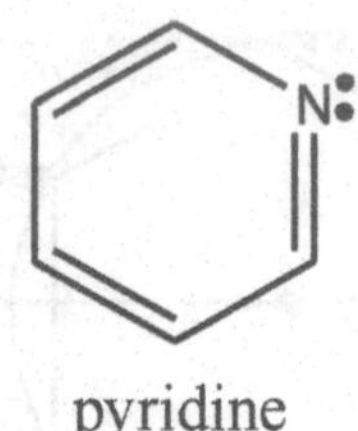

pyridine

- **HN = 6**; pyridine is aromatic. The lone pair on the N is **not a part** of the conjugation system. It is in an sp^2 hybrid orbital that is ⊥ to the **p orbitals of the π bond or vertical plane.**

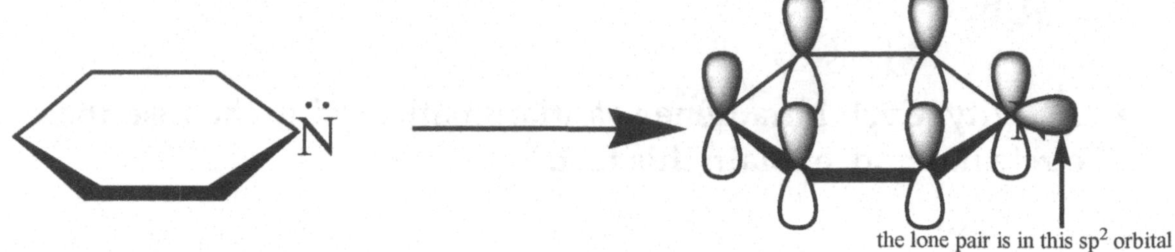

- **See page ______.**

3. PYRROLE

pyrrole

- **The HN is 6**. Pyrrole is aromatic. In this case the lone pair on the N is a part of the conjugation system. In other words, N contributes 2 π electrons.

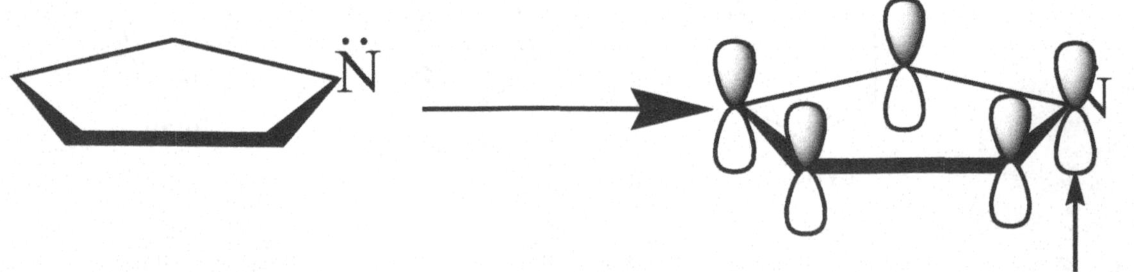

- **See page ________. Do problems on pages ________ - _______.**

J. POLYAROMATIC COMPOUNDS REVISITED

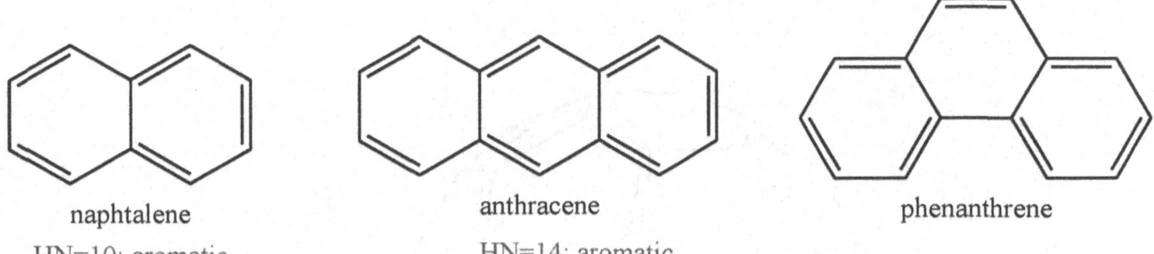

K. AROMATICITY: A SUMMARY

- **Finally**, one can conclude that a molecule is aromatic if:
    - **It is fully conjugated (can go full circle through resonance).**
    - **It is planar.**
    - **All carbons are sp^2 hybridized.**
    - **It is monocyclic or polycyclic.**
    - **It contains a total of 4n+2 π electrons (n = 0, 1, 2, 3, 4, 5, ..).**
    - **All its bonding MOs are full with the available π electrons.**

Ex:

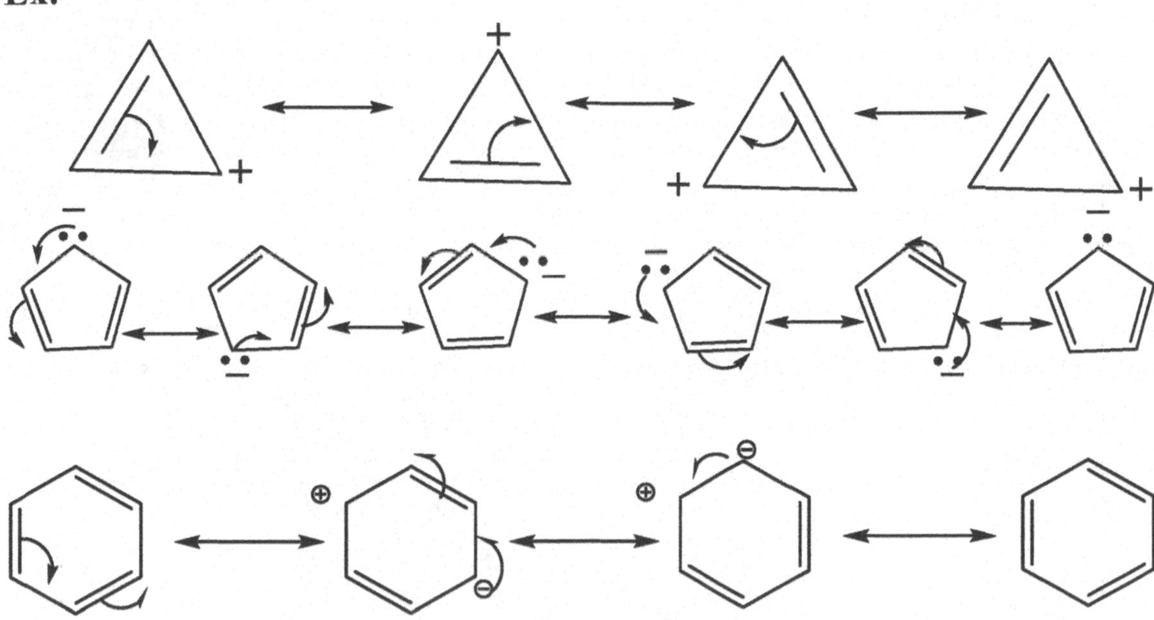

- **Read about graphite, diamond, and the buckyball on page ________.**

- **See Key Concepts on page _______.**

OCHEM II UNIT 8: THE CHEMISTRY OF BENZENE AND DERIVATIVES: ELECTROPHILIC AROMATIC SUBSTITUTION REACTIONS

A. INTRODUCTION

1. REACTIVITY OF BENZENE

- From the last chapter, we learned that the benzene ring is very stable. As a result, it is **unreactive** with many reagents. For instance, benzene does not undergo **addition reactions** as do the other **unsaturated** hydrocarbons: alkenes and alkynes.

Ex:

C=C + Br_2 → Br–C–C–Br

–C≡C– + Br_2 → Br–C=C–Br

(benzene) + Br_2 → No Addition Reaction

2. REACTIONS OF BENZENE

- Although benzene does not undergo addition reactions, it does undergo **substitution reactions.** In other words, it undergoes reactions in which an aryl H atom is replaced by an electrophile as follows:

$$C_6H_5\text{-}H + E^{\oplus} \longrightarrow C_6H_5\text{-}E + H^{\oplus}$$

- **Note: This reaction is called an Electrophilic Aromatic Substitution (or EAS).**

- The reactivity of benzene stems from the fact that the **6 π** electrons in benzene are sterically accessible, exposed and available to an incoming electrophile. Electrophiles (electron acceptors or Lewis acids) that stabilize the benzene ring react with benzene. In these reactions, benzene is the **nucleophile**.

- There are 5 common **EAS** of benzene:

reaction name	E^+	aryl H is replaced by
Nitration	NO_2^+	$-NO_2$
Sulfonation	SO_3H^+	$-SO_3H$
halogenation	X^+	A halogen X (Cl, Br, I)
Friedel-Crafts Alkylation	R^+	An alkyl group -R
Friedel-Crafts Acylation	RCO^+	RCO-

B. THE REACTIONS OF BENZENE

1. GENERAL REACTION

$$C_6H_5\text{–}H + E^{\oplus} \longrightarrow C_6H_5\text{–}E + H^{\oplus}$$

Nucleophile

- **Mechanism: The reaction proceeds in three steps: Formation of a carbocation intermediate followed by the loss of an aryl H, thus restoring the aromatic ring. The first step is the rate determining step. See pages _______ - _______.**
    - **Step 1: Make the E^+**
    - **Step 2: Formation of a Carbocation**
    - **Step 2: Reestablishment of the aromatic ring:**

2. THERMODYNAMICS OF ELECTROPHILIC AROMATIC SUBSTITUTION

- The intermediate carbocation is **less** stable than benzene. As a result, the first step of the reaction is endergonic and has a higher activation energy. However, the 2nd step is exergonic since the substituted product is more stable than the carbocation intermediate. The overall reaction, however, is exergonic.

- **See Fig. ________, page ________.**

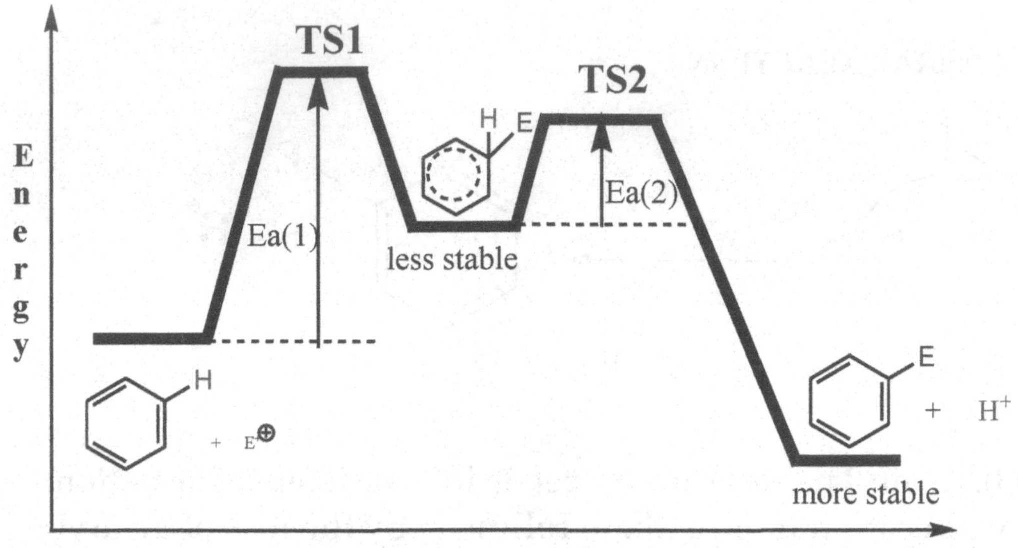

C. EXAMPLES OF ELECTROPHILIC AROMATIC SUBSTITUTION REACTIONS

1. HALOGENATION

a. Introduction

- An **aryl H** is replaced by a halogen atom X (X = Cl_2, Br_2, I_2). Br_2 and Cl_2 react readily. I_2 reacts slowly. F_2 is too reactive to use.

- **The general reaction is:**

H $\xrightarrow[FeX_3]{X_2}$ X + HX

b. Bromination

- The reaction is:

$$C_6H_5\text{-}H \xrightarrow[FeBr_3]{Br_2} C_6H_5\text{-}Br + HBr$$

- Mechanism: 3 steps. See page ________.

$$Br{-}Br + FeBr_3 \longrightarrow Br{-}\overset{+}{Br}{-}\overset{-}{Fe}Br_3$$

$$C_6H_6 + Br{-}\overset{+}{Br}{-}\overset{-}{Fe}Br_3 \longrightarrow [C_6H_6Br]^+ \text{ (three resonance structures)} + FeBr_4^-$$

$$[C_6H_6Br]^+ + Br{-}\overset{-}{Fe}Br_3 \longrightarrow C_6H_5Br + HBr + FeBr_3$$

c. Chlorination

- The reaction is:

$$C_6H_5\text{-}H \xrightarrow[FeCl_3]{Cl_2} C_6H_5\text{-}Cl + HCl$$

- Note: The mechanism is similar to that of bromination.

d. Iodination

- I_2 alone is not reactive. It requires an oxidizing agent, a mixture of H_2O_2 and $CuCl_2$.

- The reaction is:

$$C_6H_5\text{-}H \xrightarrow[CuCl_2/H_2O_2]{I_2} C_6H_5\text{-}I + HI$$

2. AROMATIC NITRATION

  - An aryl H is replaced by NO_2.

  - Recall: NO_2 = nitro.
    NH_2 = amino

- The reaction is:

$$C_6H_5\text{-}H \xrightarrow[H_2SO_4]{HNO_3} C_6H_5\text{-}NO_2 + H_2O$$

- Mechanism: 3 steps

- See page _______.

$$H\text{—}\ddot{O}\text{—}NO_2 + H\text{—}OSO_3H \longrightarrow H\text{—}\overset{+}{O}(H)\text{—}NO_2 + HSO_4^- \longrightarrow H_2O + NO_2^+$$

$$C_6H_6 + NO_2^+ \longrightarrow [C_6H_6NO_2]^+ \text{ (three resonance structures)}$$

$$[C_6H_6NO_2]^+ + HSO_4^- \longrightarrow C_6H_5NO_2 + H_2SO_4$$

3. AROMATIC SULFONATION

- **An aryl H is replaced by SO_3H.**

$$C_6H_5\text{-}H \xrightarrow[H_2SO_4]{SO_3} C_6H_5\text{-}SO_3H$$

benzenesulfonic acid

- **Mechanism: 3 steps**

- **See page ______.**

$$SO_3 + H\text{-}OSO_3H \longrightarrow {}^{+}SO_3H + HSO_4^-$$

$$C_6H_6 + {}^{+}SO_3H \longrightarrow [C_6H_6SO_3H]^{+}$$

$$[C_6H_6SO_3H]^{+} + HSO_4^- \longrightarrow C_6H_5SO_3H + H_2SO_4$$

- **Note: Aromatic sulfonic acids are very useful in the preparation of sulfa drugs and dyes.**

Ex: Sulfanilamide = an antibiotic.

$$H_2N\text{-}C_6H_4\text{-}SO_2\text{-}NH_2$$

sulfanilamide

- **p-methylbenzenesulfonic acid is used to make p-cresol in the so-called alkali fusion reaction.**

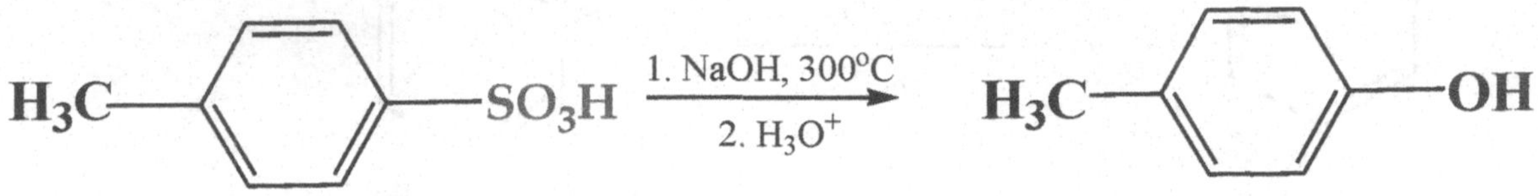

$$H_3C-C_6H_4-SO_3H \xrightarrow[2.\ H_3O^+]{1.\ NaOH,\ 300^\circ C} H_3C-C_6H_4-OH$$

p-cresol

- **Do all problems on pages _______ - ______.**

4. ALKYLATION OF AROMATIC RINGS: THE FRIEDEL-CRAFTS REACTION

- **An aryl H is replaced with an alkyl group.**

- **The general reaction is:**

$$C_6H_5-H \xrightarrow[AlCl_3]{RCl} C_6H_5-R + HCl$$

An alkyl benzene

- **An example:**

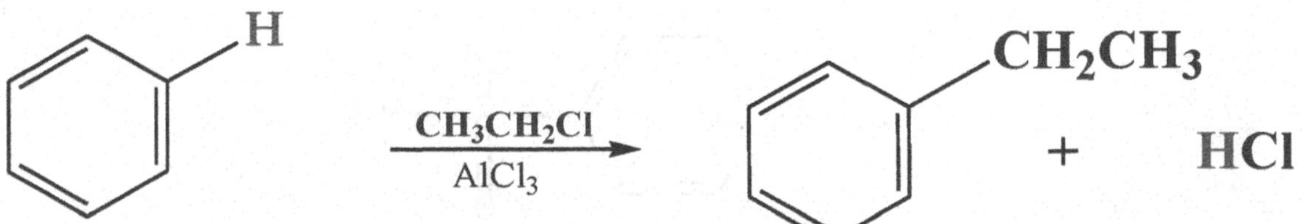

$$C_6H_5-H \xrightarrow[AlCl_3]{CH_3CH_2Cl} C_6H_5-CH_2CH_3 + HCl$$

ethylbenzene

- Mechanism: See pages _______ - ______.
- For CH_3Cl and 1° alkyl halides, have 3 steps; the electrophile is R-Cl – $AlCl_3$.

$$RCH_2{-}Cl + AlCl_3 \longrightarrow RCH_2{-}\overset{+}{Cl}{-}\overset{-}{Al}Cl_3$$

$$C_6H_6 + RCH_2{-}\overset{+}{Cl}{-}\overset{-}{Al}Cl_3 \longrightarrow [C_6H_6(H)(CH_2R)]^+ + AlCl_4^-$$

$$[C_6H_6(H)(CH_2R)]^+ + Cl{-}\overset{-}{Al}Cl_3 \longrightarrow C_6H_5CH_2R + HCl + AlCl_3$$

- When 2° and 3° alkyl halides are used, there are 4 steps. The electrophile is R^+.

$$R_3C{-}Cl + AlCl_3 \longrightarrow R_3C{-}\overset{+}{Cl}{-}\overset{-}{Al}Cl_3 \longrightarrow R_3C^+ + AlCl_4^-$$

$$C_6H_6 + R_3C^+ \longrightarrow [C_6H_6(H)(CR_3)]^+ + AlCl_4^-$$

$$[C_6H_6(H)(CR_3)]^+ + Cl{-}\overset{-}{Al}Cl_3 \longrightarrow C_6H_5CR_3 + HCl + AlCl_3$$

- Read pages _______ - ______.

- **Some facts about the Friedel-Crafts reaction:**
  - **Other alkyl halides can be also used in this reaction: R-F, R-Br, R-I.**
  - **Aryl halides and vinylic halides do not undergo FC reactions since the carbocation intermediates from these halides are relatively unstable. The reaction has a high activation energy.**

Benzene (H) + aryl halide (X) $\xrightarrow{AlCl_3}$ NO REACTION

aryl halide

Benzene (H) + CH=CHX $\xrightarrow{AlCl_3}$ NO REACTION

vinyl halide

  - **Benzene rings with strongly electron withdrawing meta deactivating groups ($-NO_2$, -CN, $-SO_3H$, etc.) cannot undergo FC reactions because these groups "deactivate" the benzene ring, making it less reactive. The same effect is observed with meta $-NH_2$ groups. In general:**

Y, H, Z (meta substituted benzene) + R-X $\xrightarrow{AlCl_3}$ NO REACTION

meta substituted benzenes

Z=$-\overset{+}{N}R_3$, $-NO_2$, -CN, $-SO_3H$, -CHO, $-COCH_3$, $-CO_2H$, $-NH_2$,-NHR, $-NR_2$

Ex:

NO_2 / H / $COCH_3$ (substituted benzene) + CH_3CH_2Cl $\xrightarrow{AlCl_3}$ NO REACTION

- **Another problem with FC reactions is polyalkylation: the following is an example:**

Benzene (H) + $(CH_3)_3CCl$ $\xrightarrow{AlCl_3}$ $C_6H_5C(CH_3)_3$ + $C(CH_3)_3$–C_6H_4–$C(CH_3)_3$

- **We can also have rearrangement due to a hydride or methyl shift that leads to a mixture of products: the following is an example:**

Benzene (H) + $CH_3CH_2CH_2CH_2Cl$ $\xrightarrow{AlCl_3}$ $C_6H_5CH(CH_3)CH_2CH_3$ + $C_6H_5CH_2CH_2CH_2CH_3$

- **Note: Functional groups other than alkyl halides (alkenes and alcohols) can be also used to generate the R^+ electrophile.**

- **Using alkenes as electrophiles.**

$$C_6H_5\text{-}H + CH_2{=}CHR \xrightarrow{H_2SO_4} C_6H_5\text{-}CH(R)CH_3$$

Ex:

$$C_6H_5\text{-}H + CH_2{=}CHCH_2CH_3 \xrightarrow{H_2SO_4} C_6H_5\text{-}CH(CH_2CH_3)CH_3$$

- **Using alcohols in acid as electrophiles.**

$$C_6H_5\text{-}H + R\text{-}OH \xrightarrow{H_2SO_4} C_6H_5\text{-}R$$

Ex:

$$C_6H_5\text{-}H + CH_3CH_2\text{-}OH \xrightarrow{H_2SO_4} C_6H_5\text{-}CH_2CH_3$$

- **Intramolecular FC reactions can also occur.**

Read page ________.

5. ACYLATION OF AROMATIC RINGS: A FRIEDEL-CRAFTS REACTION: REPLACEMENT OF H WITH RCO^+

a. Introduction to the Acyl Group and Acetyl Halides

$$\overset{O}{\overset{\|}{RC}}+ \quad \text{or} \quad RCO^+$$

b. The general reaction is:

$$C_6H_5\text{-}H \xrightarrow[AlCl_3]{RCCl\ (C{=}O)} C_6H_5\text{-}CR\ (C{=}O) + HCl$$

Ex:

$$C_6H_5\text{-}H \xrightarrow[AlCl_3]{CH_3CCl\ (C{=}O)} C_6H_5\text{-}CCH_3\ (C{=}O) + HCl$$

- Read pages _______ - _______.
- See examples on page _____.
- Do problems on pages _____ - _____.

- **Mehanism: The reaction proceeds through an acyl cation intermediate.**

- **See page ______ - ______.**

- **Note: No polysubstitution observed.**

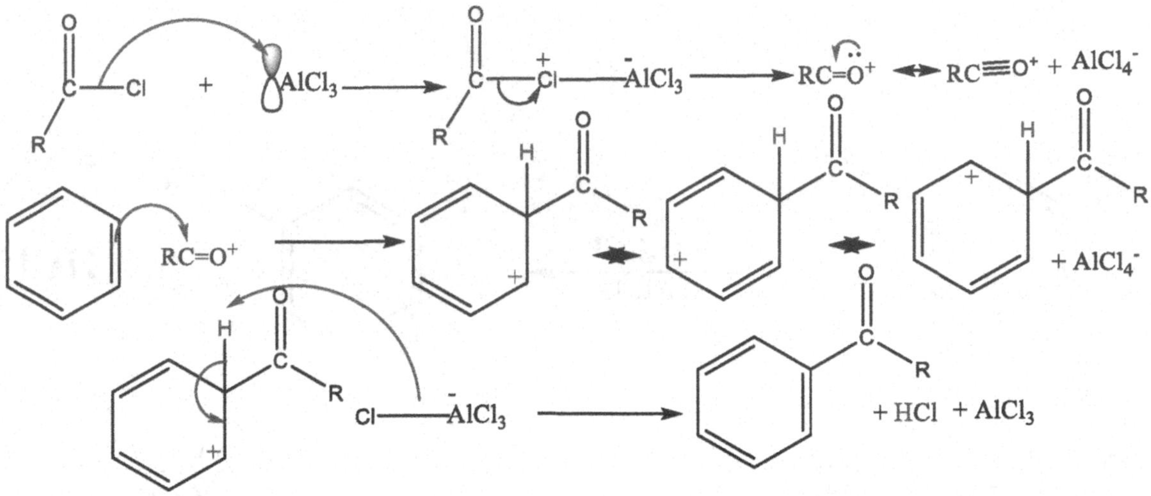

- **EAS of Benzene: A Summary**

reaction name	E+	aryl H is replaced by	catalyst
Nitration	NO_2^+	$-NO_2$	HNO_3/H_2SO_4
Sulfonation	SO_3H^+	$-SO_3H$	SO_3/H_2SO_4
halogenation	X^+	A halogen X (Cl, Br, I)	X_2/FeX_3
Friedel-Crafts Alkylation	R^+	An alkyl group -R	$RCl/AlCl_3$
Friedel-Crafts Acylation	RCO^+	RCO-	$RCOCl/AlCl_3$

or

Reagents	Product substituent
E^+	E
HNO_3 / H_2SO_4	NO_3
SO_3 / H_2SO_4	SO_3H
Cl_2 / $FeCl_3$	Cl
Br_2 / $FeBr_3$	Br
I_2 / $CuCl_2/H_2O_2$	I
RCl / $AlCl_3$	R
RCOCl / $AlCl_3$	COR
ROH / H_2SO_4	R
$RCH{=}CH_2$ / H_2SO_4	CH(R)CH$_3$

D. SUBSTITUENT EFFECT IN SUBSTITUTED AROMATIC RINGS

1. INTRODUCTION

- **Questions: How does the presence of a group already on the benzene ring affect the replacement of a second aryl H? How does the second group orient itself (ortho, para, or meta)?**

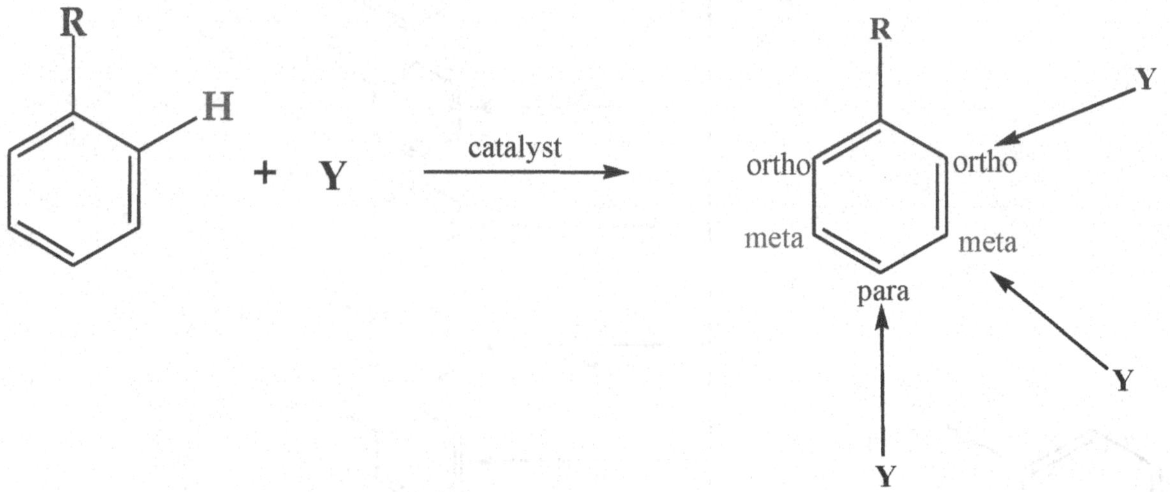

R
H
+ Y
catalyst
R
Y
ortho
ortho
meta
meta
para
Y
Y

- **R** does affect the reaction. It can either make the ring less reactive (**R** = **deactivator** = **EWG**) or more reactive (**R** = **activator** = **EDG**).

Ex: The reaction

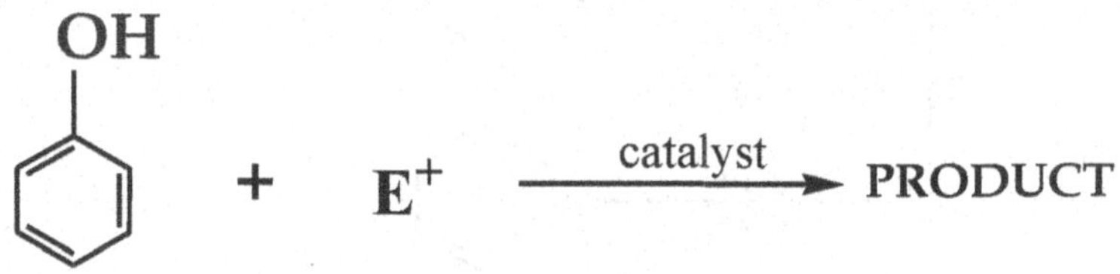

OH
+ E^+ catalyst PRODUCT

is **1,000** times faster than the reaction

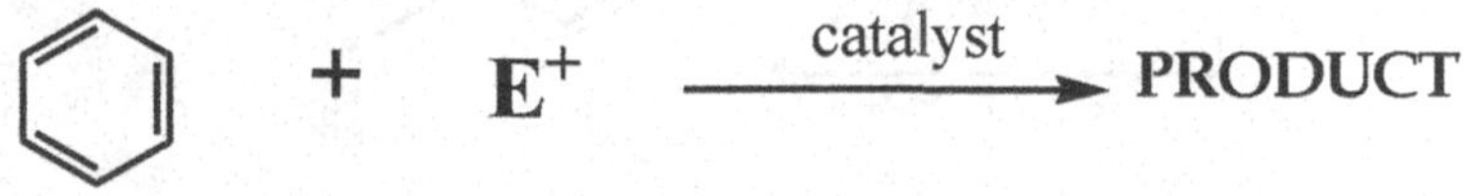

+ E^+ catalyst PRODUCT

On the other hand, the reaction

$$\text{C}_6\text{H}_5\text{NO}_2 + E^+ \xrightarrow{\text{catalyst}} \text{PRODUCT}$$

is **10 million** times slower than the reaction

$$\text{C}_6\text{H}_6 + E^+ \xrightarrow{\text{catalyst}} \text{PRODUCT}$$

- **Conclusion: -OH makes the benzene ring more reactive ==>-OH = activator. On the other hand, NO_2 makes the ring less reactive than benzene; -NO_2 is a deactivator.**

2. GROUP ORIENTATION

a. Introduction

- The **orientation** of a **second substituent** on the benzene ring is dictated by the **group already in place.** The first group can either direct **ortho (o) - para (p) or meta (m).**

Ex:

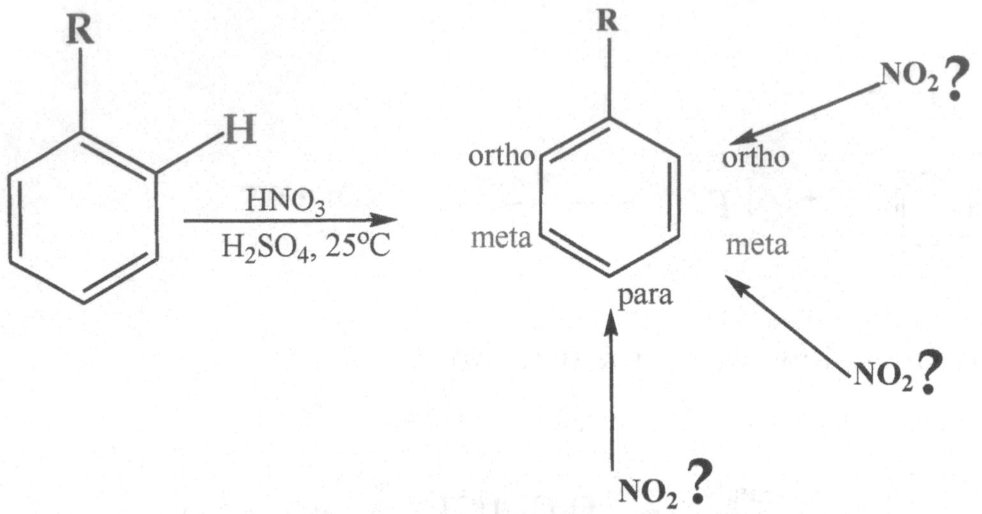

R, H; $\xrightarrow[H_2SO_4,\ 25°C]{HNO_3}$ R; ortho, ortho, meta, meta, para; NO_2? NO_2? NO_2?

b. Types of Substituents

On the basis of group orientation, there are **3 types of substituents.**

- **o-p directors and activators**
- **o-p directors and deactivators**
- **m directors and deactivators**

- **Read pages _______ - ______.**

3. O-P DIRECTING GROUPS AND ACTIVATORS

- They make the **monosubstituted benzene ring more reactive than benzene.** Overall, they "**instruct**" the incoming group to go **ortho and para.**

Ex:

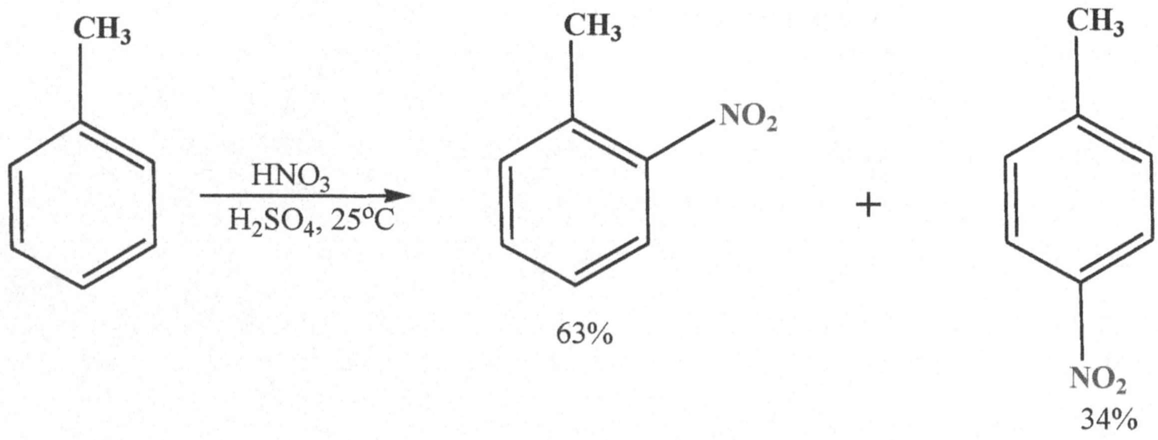

CH_3 $\xrightarrow[H_2SO_4,\ 25°C]{HNO_3}$ CH_3, NO_2 (63%) + CH_3, NO_2 (34%)

- **Conclusion: The $-CH_3$ group is o-p directing.**
- **Hints: o-p directors and activators have only single bonds on the neutral atom of the group that is directly attached to the ring, except phenyl:**
- **The general structure for these groups is: -R (alkyl group) or -Z:**
- **The increasing order of reactivity is:**

$$-R < -NHCOR < -OR < -OH < -NH_2 \approx -NHR \approx \ddot{N}R_2.$$

4. O-P DIRECTING GROUPS AND DEACTIVATORS

- They make the **monosubstituted ring less reactive than benzene itself**. Overall, they also "**instruct**" the incoming group to go **ortho and para.**

Ex:

Cl (C_6H_5Cl) $\xrightarrow[H_2SO_4,\ 25°C]{HNO_3}$ Cl, NO_2 (ortho) 35% + Cl, NO_2 (para) 65%

- **Hint: o-p directors and deactivators are all halogens: -F, -Cl, -Br, -I.**

5. META DIRECTING GROUPS AND DEACTIVATORS

- They make the **monosubstituted ring less reactive than benzene.** Overall, they "**instruct**" the incoming group to go **meta.**

Ex:

$$\overset{+}{N}(CH_3)_3\text{-benzene} \xrightarrow[H_2SO_4,\ 25^{\circ}C]{HNO_3} \overset{+}{N}(CH_3)_3\text{-benzene with } NO_2 \text{ (meta)}$$

- **Hints: meta directors and deactivators have a double bond or a triple bond, or a + or a δ^{+} on the atom of the group that is directly attached to the ring.**
- **The general structure for these groups is: -Y (has + or is δ^{+}).**
- **The decreasing order of reactivity is:**

$$\text{-CHO>-COR>-COOR>-COOH>-CN>–SO}_3\text{H->-NO}_2\text{>-}\overset{+}{N}\text{R}_3$$

or:

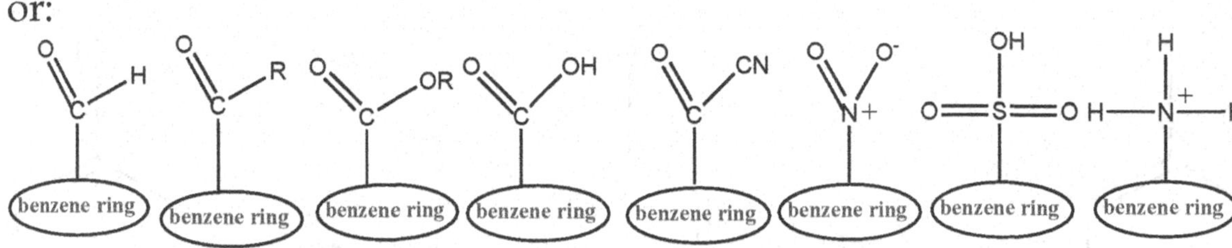

6. ORTHO-PARA AND META DIRECTING: A SUMMARY

- **The overall decreasing order of reactivity is:**
- **o-p directors and activators > o-p directors and deactivators > meta directors and deactivators.**
- **All activating groups direct o. - p.**
- **All halogen deactivating groups direct o.p.**
- **All nonhalogen deactivators direct meta.**
- **Read pages ______ - ______. Do problems on pages _____ - _____.**

E. FACTORS AFFECTING REACTIVITY AND ORIENTATION

1. INTRODUCTION

- There are **two factors** that determine group orientation on the benzene ring:
    - **-Inductive effect.**
    - **-Resonance effect.**

- **Inductive effect** is the **withdrawal** or **donation of electrons by a group thru a σ bond. For instance,** alkyl groups **inductively donate** electrons. On the other hand, -NO_2 **inductively withdraws** electrons.

- **Resonance effect** is the **withdrawal or donation of electrons by a group thru a π bond.** It usually occurs with groups having **lone pairs** or **π** bonds. For instance, -OH has an electron donating resonance effect and -NO_2 has an electron withdrawal resonance effect.

2. INDUCTIVE EFFECT

a. Inductive Electron Withdrawing Groups (→EWG)

- These groups have the general formula:

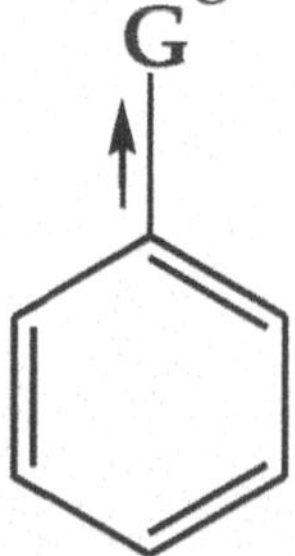

Where G is a group that contains an atom that is **more electronegative than C: N, O, halogen.**

- **The following groups inductively withdraw electrons from the ring.**

-CO-, -C≡N, $-NO_2$, $-NH_2$, -Cl, -OH, -Br, -I, -F
Ex:

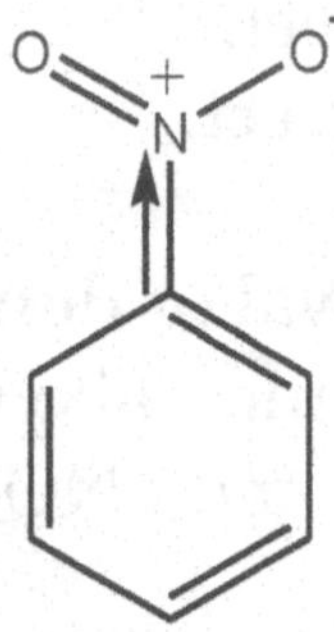

b. Inductive Electron-Donating Groups (EDG→) = Alkyl groups

- The general formula is:

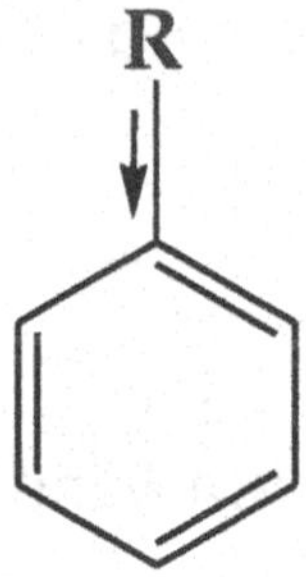

Ex:

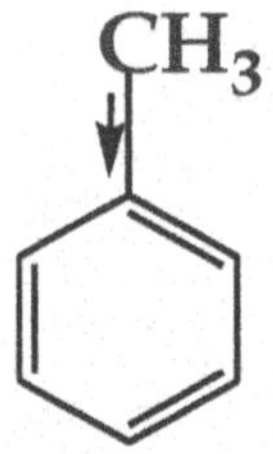

- **Problem: State whether each one of the following groups has an inductive EW effect or an inductive ED effect.**

group	status
$CH_3CH_2CH_2CH_2$-	
Br-	
$CH_3COCH_2CH_2$-	
HO-	
$-NH_2$	
CH_3CH_2O-	
$-CCl_3$	

3. RESONANCE EFFECT

a. Resonance Electron-Withdrawing Groups (R→EWG)

- These are groups that place a **positive (+)** charge on the benzene ring **thru resonance.**

- The general formula is:

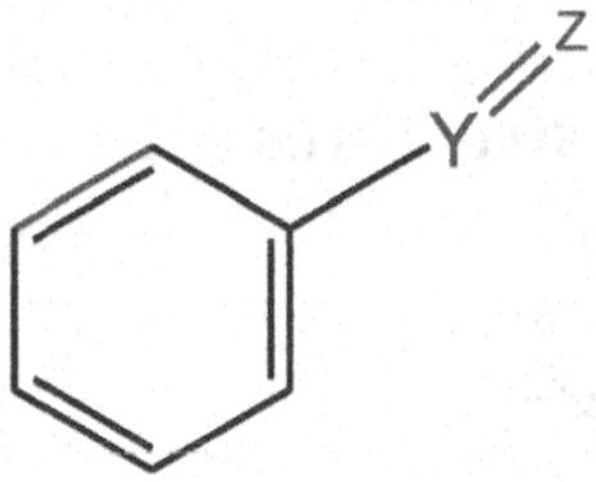

Where **Z** is an atom that is **more electronegative than carbon (except halogens): O, N.**

- **In general, the resonance structures from the EWG are:**

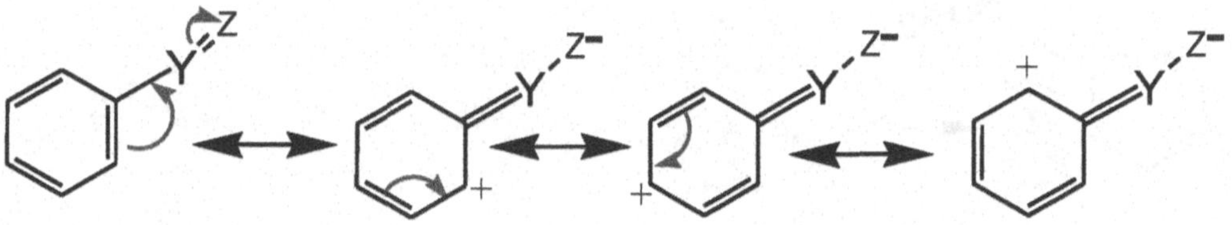

Ex:

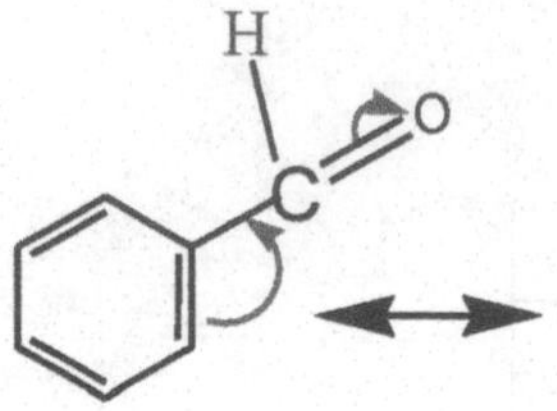

-CO-, -C≡N, $-NO_2$ **have resonance electron withdrawing effect.**

b. Resonance Electron-Donating Groups (R=EDG→)

- These are groups that place a negative (-) charge on the benzene ring thru resonance.
- The general formula is:

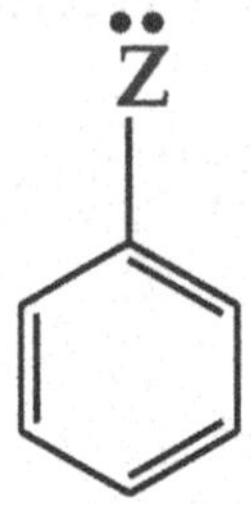

- **The general resonance structures are:**

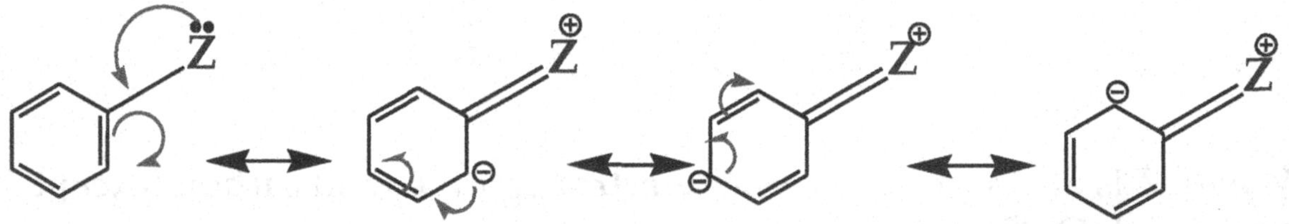

Z = halogens, -OH, $-NH_2$, -OR,$-NR_2$.

Ex:

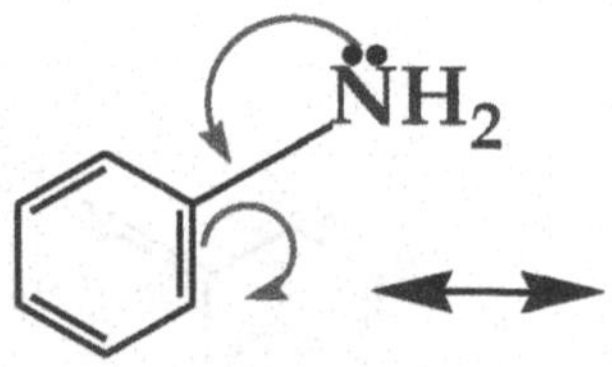

- **Problem:** Assign resonance effect to each one of the following. Please, show your work. **Please draw all resonance structures.**

$C_6H_5\ddot{O}CH_3$ $\quad$ $C_6H_5C(=O)CH_3$ $\quad$ $C_6H_5\ddot{O}H$

4. PUTTING ALL TOGETHER: INDUCTIVE AND RESONANCE EFFECTS, ACTIVATORS AND DEACTIVATORS

a. Introduction

- In an **Electrophilic Aromatic Substitution** reaction, the orientation properties of a given group depend on the balance between its inductive and resonance effects.

- **Recall: EDG = activators: make ring more reactive; EWG = deactivators: make ring less reactive.**

- **See Fig. ________, page _______.**

b. Alkyl Groups ⇒ activators.

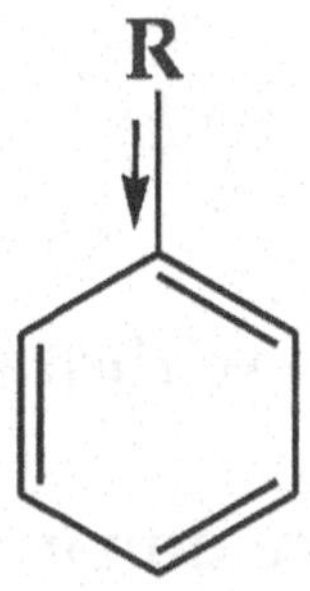

- **Weak electron donating inductive effect.**
- **No resonance effect.**

- **Overall effect: electron donating inductive effect ⇒ make benzene ring more reactive than benzene ⇒ activators.**

c. -Z Groups (Z = O, N) ⇒ activators.

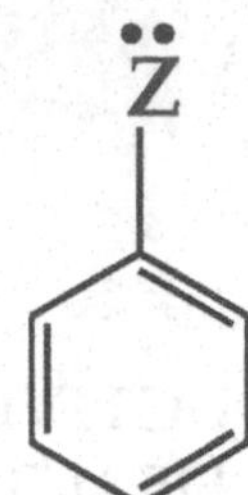

- **Weak electron withdrawing inductive effect.**
- **Strong electron donating resonance effect.**

- **Overall effect: electron donating resonance effect predominates ⇒ make benzene ring more reactive than benzene ⇒ activators.**

d. -X Groups (X = Halogens) ⇒ deactivators.

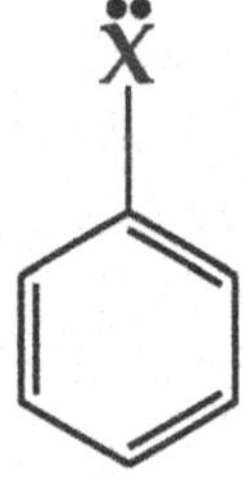

  - **Strong electron withdrawing inductive effect.**
  - **Weak resonance donating effect.**

- **Overall effect: electron withdrawing inductive effect predominates ⇒ make benzene ring less reactive than benzene ⇒ deactivators.**

e. -Y=Z Groups (Z = O, N) ⇒ deactivators.

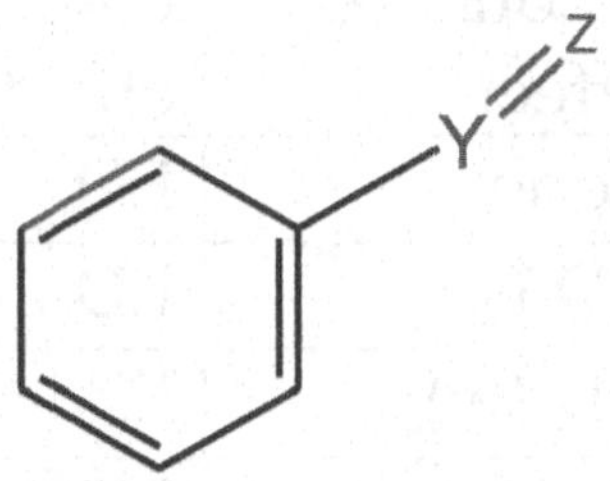

- Strong electron withdrawing inductive effect.
- Strong resonance electron withdrawing effect.

- Overall effect: electron withdrawing ⇒ make benzene ring less reactive than benzene ⇒ deactivators.

f. $-NR_3^+$ Groups (R = Alkyl Group) ⇒ deactivators.

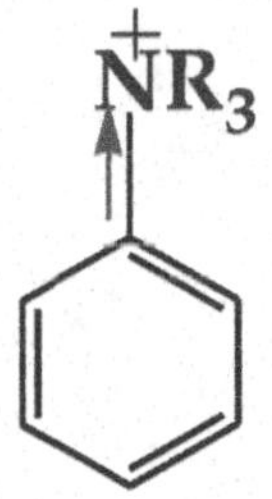

- No resonance electron withdrawing effect.
- Strong inductive electron withdrawing effect.

- Overall effect: electron withdrawing ⇒ make benzene ring less reactive than benzene ⇒ deactivators.

5. ACTIVATORS/DEACTIVATORS: A SUMMARY

group	inductive effect	resonance effect	overall effect	reactivity	orientation
Alkyl	ED (w)	none	ED	**activator**	**o-p**
Z = N, O	EW (w)	ED (s)	ED	**activator**	**o-p**
halogens	EW (s)	ED (w)	EW	**deactivator**	**o-p**
Y (+ or δ^+)	EW (s)	EW (s)	EW	**deactivator**	**meta**
$-NR_3^+$	EW (s)	none	EW	**deactivator**	**meta**

or

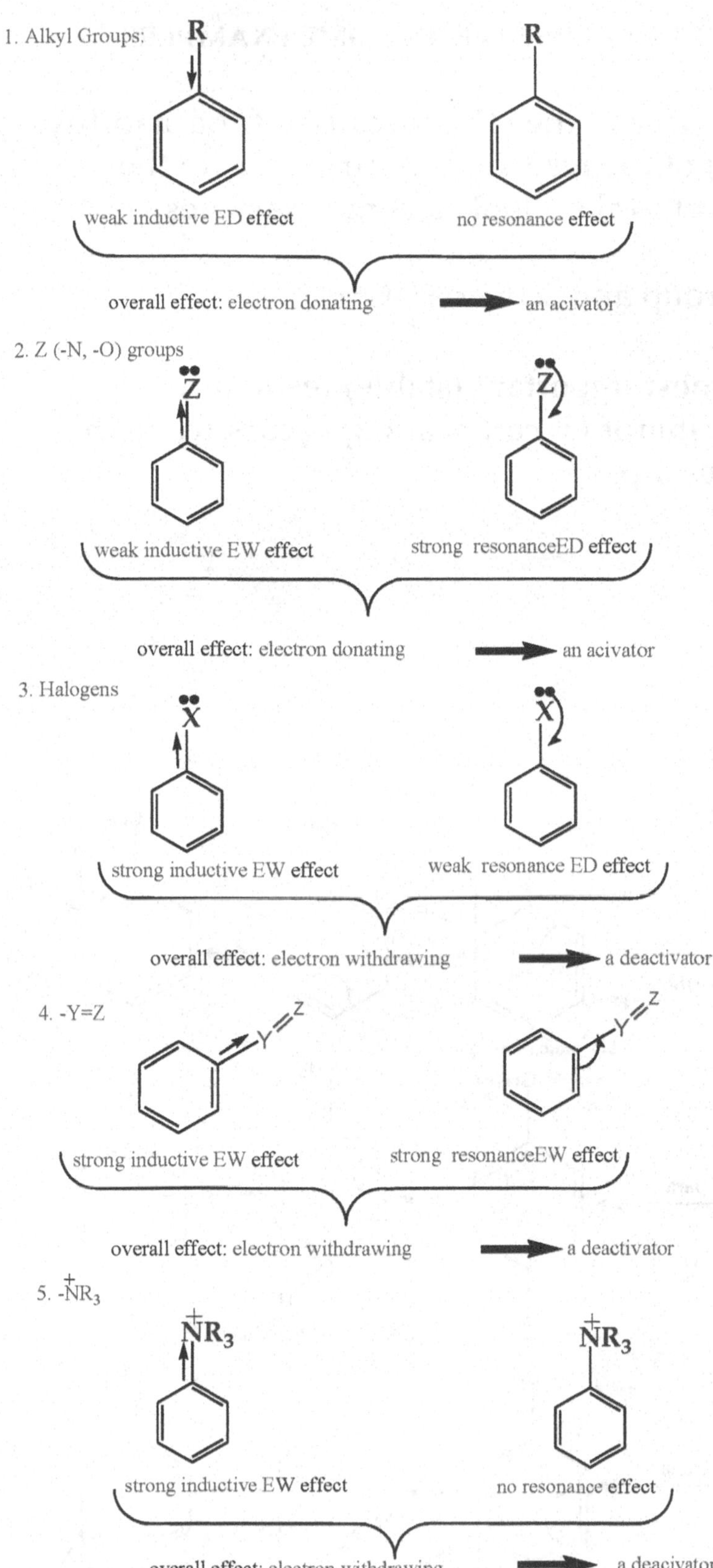
1. Alkyl Groups:
R
R
weak inductive ED effect
no resonance effect
overall effect: electron donating
an acivator
2. Z (-N, -O) groups
Z
Z
weak inductive EW effect
strong resonanceED effect
overall effect: electron donating
an acivator
3. Halogens
X
X
strong inductive EW effect
weak resonance ED effect
overall effect: electron withdrawing
a deactivator
4. -Y=Z
Y
Z
Y
Z
strong inductive EW effect
strong resonanceEW effect
overall effect: electron withdrawing
a deactivator
5. -NR3
NR3
NR3
strong inductive EW effect
no resonance effect
overall effect: electron withdrawing
a deacivator

- See page _______.

6. ORIGIN OF ORIENTATION PATTERNS: SOME EXAMPLES

- **A group directs one way or the other because of the stability (or lack of stability) of the resonance contributors to the resonance hybrid. Let's take a look at some examples.**

Ex. 1: An o-p directing group and activator: $-CH_3$

- **Why? Because the most important (stable) resonance (intermediate) contributor (3° carbocation) occurs when the incoming group goes o-p.**

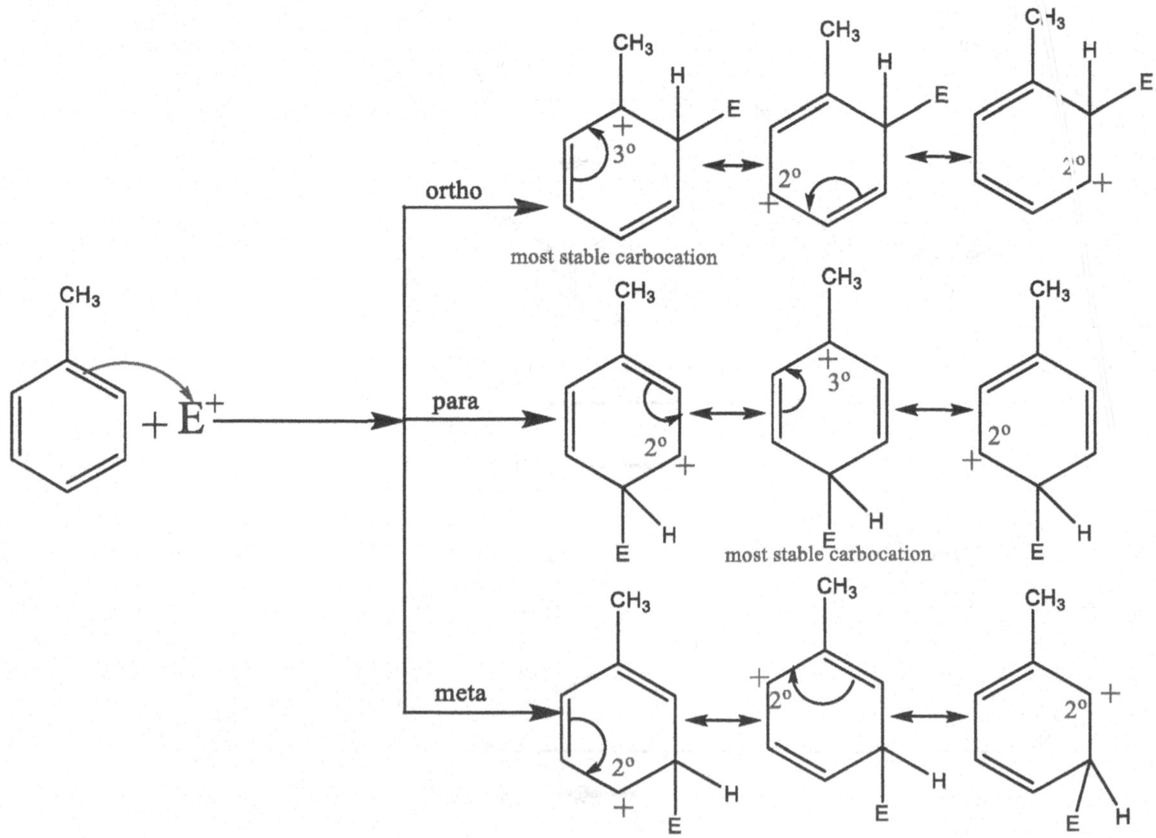

Ex. 2: **An o-p directing group and activator**: -OH

- **Why? Because the most important (stable) resonance (intermediate) contributor (all second row elements have an octet) occurs when the incoming group goes o-p.**

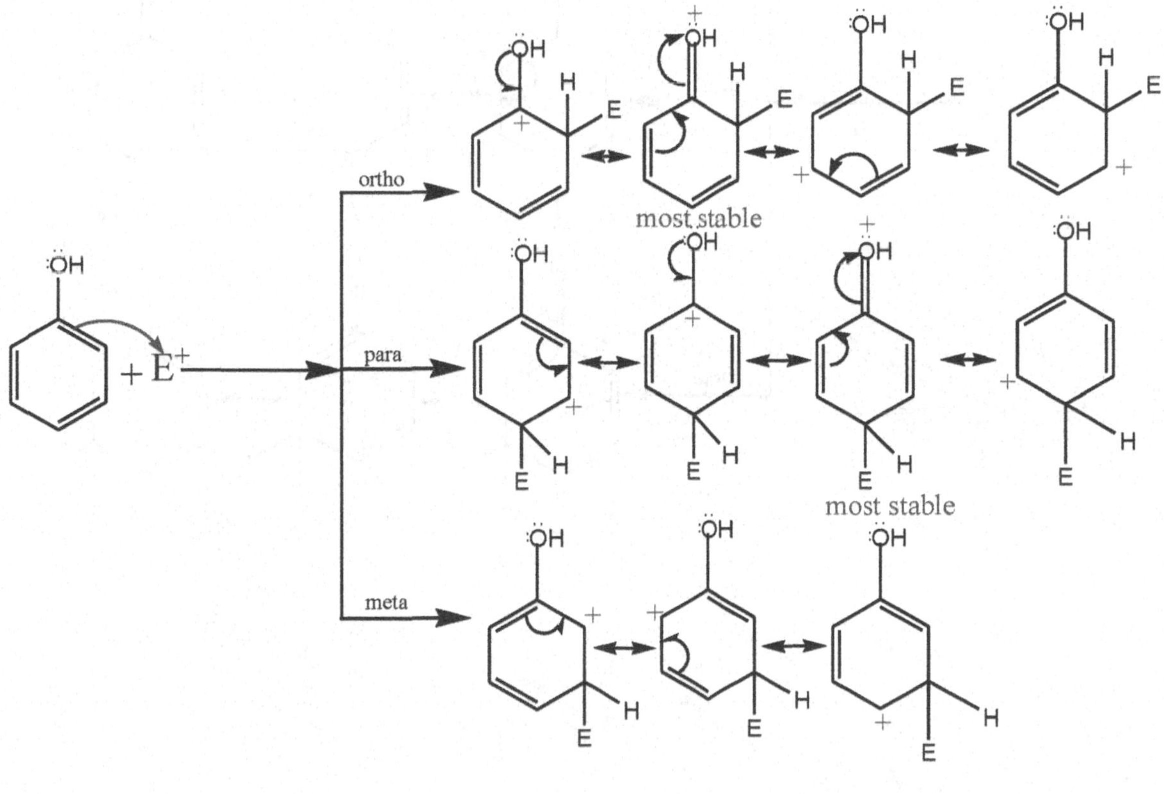

ortho

most stable

para

most stable

meta

Ex. 3: **A halogen o-p directing group and deactivator**: -Br

- **Why? Because the most important (stable) resonance (intermediate) contributor (all second row elements have an octet) occurs when the incoming group goes o-p.**

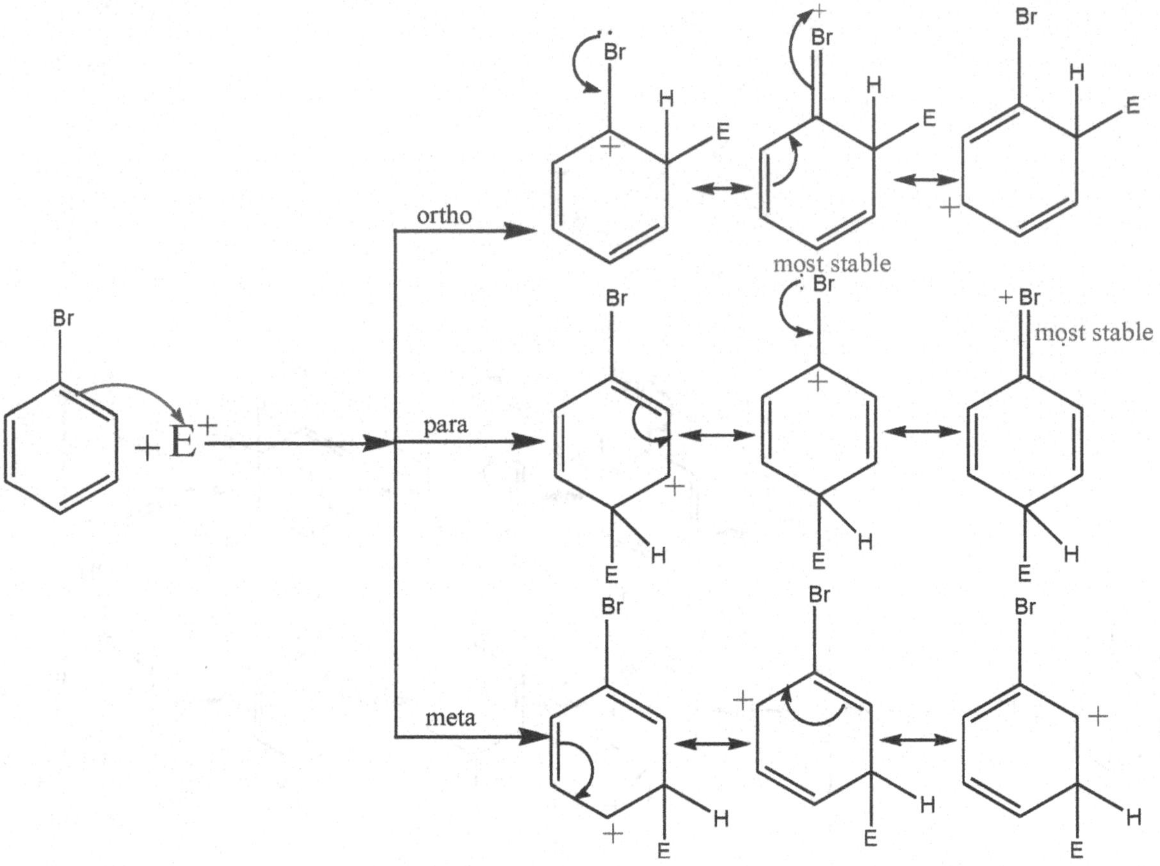

ortho

most stable

para

most stable

meta

Ex. 4: **A meta directing group and deactivator:** -CHO

- **Why? Because the least important (les stable) resonance (intermediate) contributor (a plus next to a delta plus) occurs when the incoming group goes o-p.**

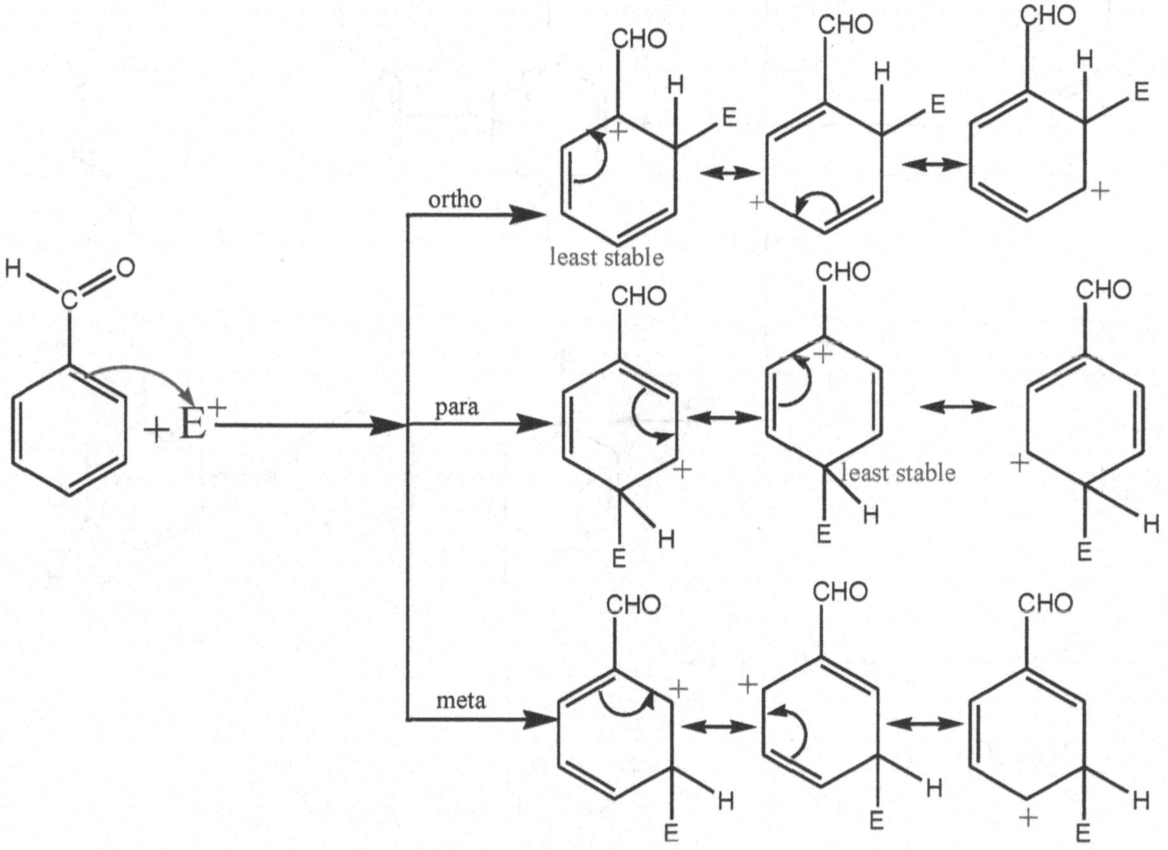

ortho

least stable

para

least stable

meta

Ex. 5: A meta directing group and deactivator: $-NH_3^+$

- **Why? Because the least important (les stable) resonance (intermediate) contributor (a plus next to a plus) occurs when the incoming group goes o-p.**

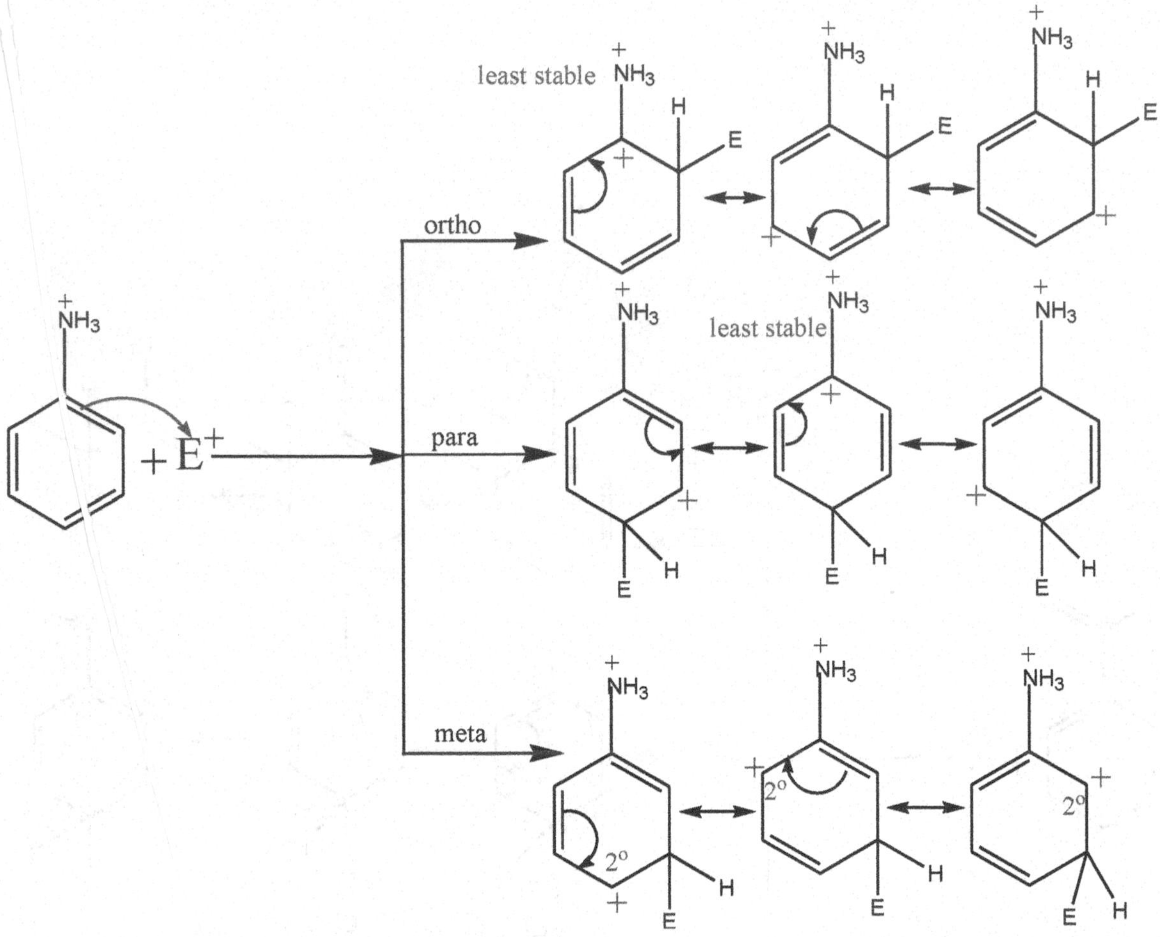

least stable

ortho

para

least stable

meta

2°

2°

2°

- **See examples on pages _______ - _______.**

- **Read pages ______ - ______.**

- **Do all problems pages _____ - ______.**

- **See Fig. _______, page ______: for a summary.**

7. SOME EXAMPLES OF EAS REACTIONS

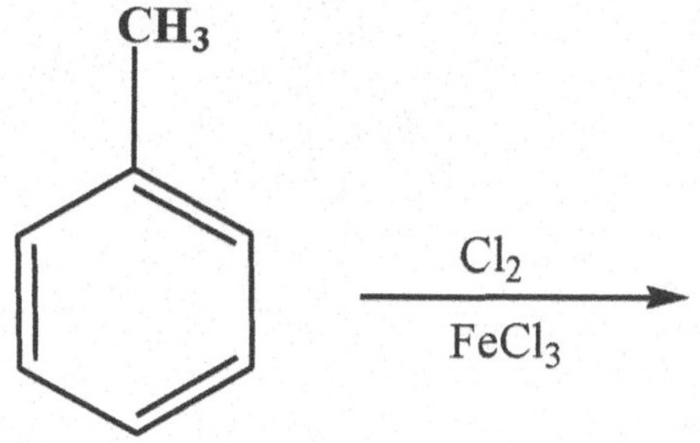

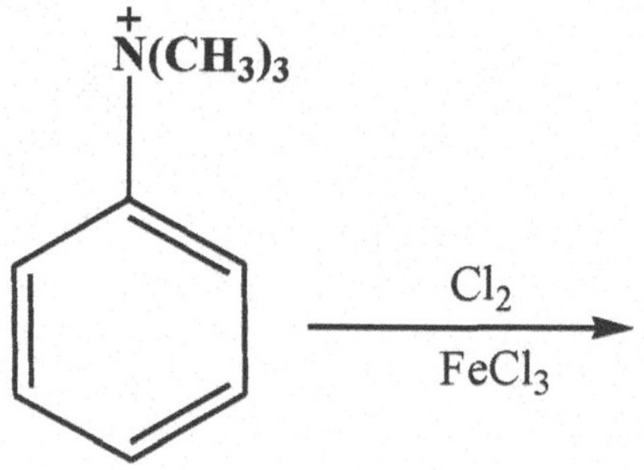

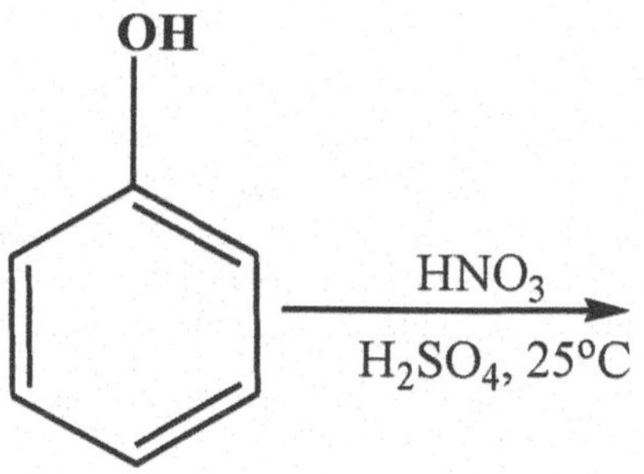

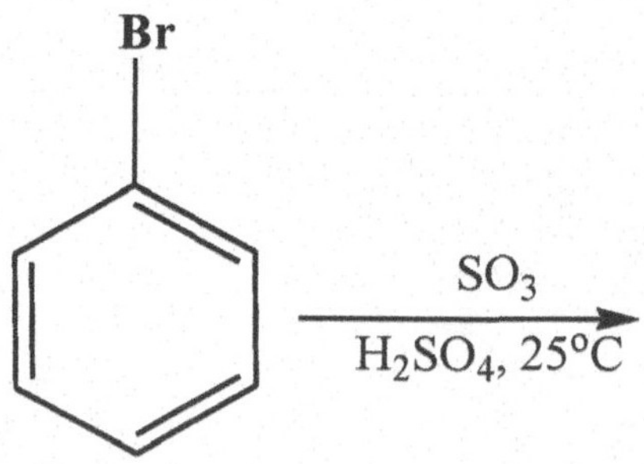

CH_3

$\xrightarrow[FeCl_3]{Cl_2}$

$\overset{+}{N}(CH_3)_3$

$\xrightarrow[FeCl_3]{Cl_2}$

OH

$\xrightarrow[H_2SO_4,\ 25°C]{HNO_3}$

Br

$\xrightarrow[H_2SO_4,\ 25°C]{SO_3}$

$$\text{C}_6\text{H}_5\text{C(=O)CH}_3 \xrightarrow[\text{H}_2\text{SO}_4,\ 25°\text{C}]{\text{SO}_3}$$

$$\text{C}_6\text{H}_5\text{C(=O)OCH}_3 \xrightarrow[\text{H}_2\text{SO}_4,]{\text{HNO}_3}$$

$$\text{C}_6\text{H}_5\text{Cl} \xrightarrow[\text{H}_2\text{SO}_4,\ 25°\text{C}]{(\text{CH}_3\text{CH}_2)_2\text{CHOH}}$$

$$\text{C}_6\text{H}_5\text{NO}_2 \xrightarrow[\text{H}_2\text{SO}_4,\ 25°\text{C}]{\text{HNO}_3}$$

$$\text{C}_6\text{H}_5\text{NH}_2 + \text{CH}_3\text{CH=CH}_2 \xrightarrow{\text{H}_2\text{SO}_4}$$

F. SOME LIMITATIONS ON EAS WITH SUBSTITUTED BENZENES

1. INTRODUCTION

- As we will see it later on in this Unit, EAS reactions are highly useful in preparing benzene derivatives. However there are two limitations that we should not ignore.

2. HALOGENATION OF MONOSUBSTITUTED BENZENES

a. Introduction

- With strong electron-donating groups (**-OH, -NH_2, and derivatives)**, **polyhalogenation** can occur; **(overactivated benzene ring)**

b. In the Absence of a Catalyst

- When **X_2 (X = halogen)** is used alone without the catalyst **FeX_3, no polyhalogenation is observed. The monosubstitution reaction proceeds normally as follows:**

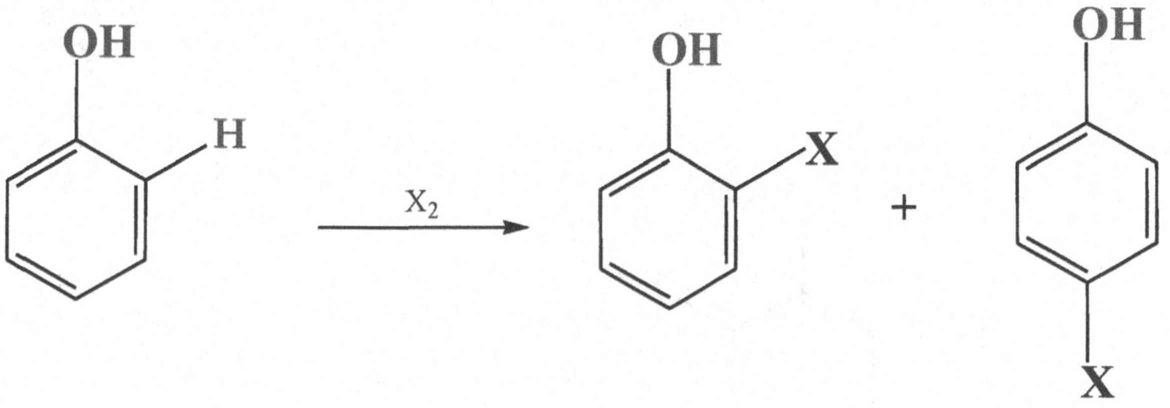

Ex:

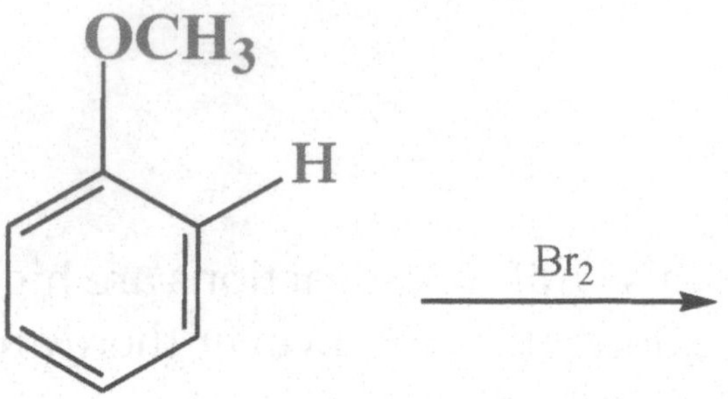

OCH_3, H

$\xrightarrow{Br_2}$

c. In the Presence of a Catalyst FeX_3

- When X_2 **(X = halogen)** is used in the presence of catalyst FeX_3**, polyhalogenation is observed. All ortho and para aryl Hs are replaced. The trisubstitution reaction proceeds as follows:**

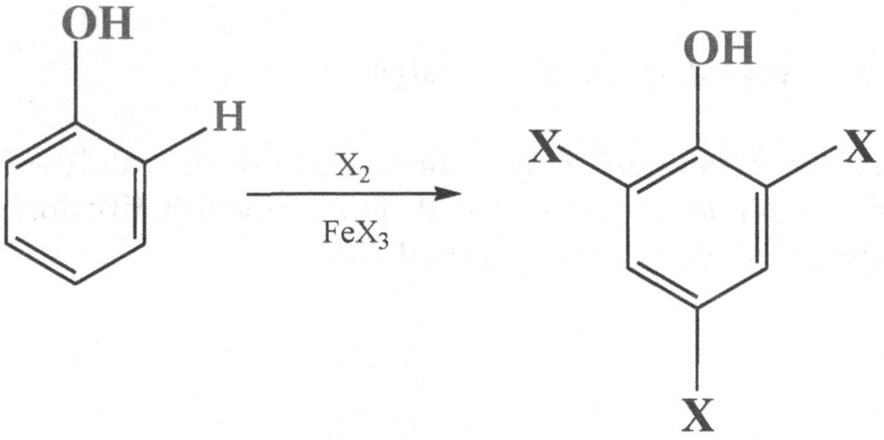

OH, H $\xrightarrow[FeX_3]{X_2}$ OH, X, X, X

Ex:

OCH_3, H

$\xrightarrow[FeCl_3]{Cl_2}$

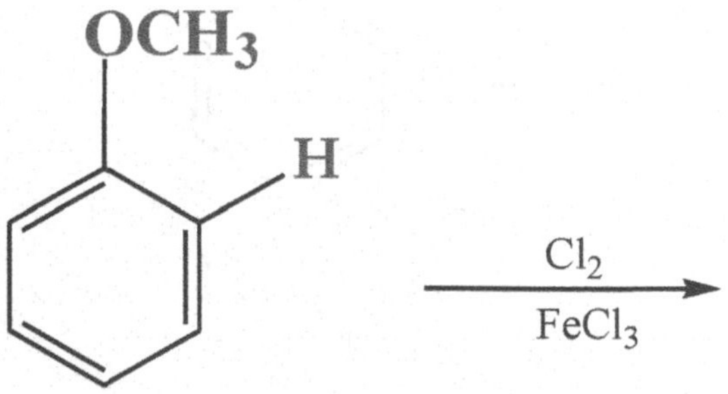

- **Read page________.**

- **Do problems on pages _______ - _______.**

3. FRIDEL-CRAFTS REACTIONS WITH MONOSUBSTITUTED BENZENES

a. Introduction

- In general, Friedel-Crafts reactions are difficult to carry out in the lab. Indeed, they do not occur with benzene **rings having either strong deactivators (-NO_2, SO_3H, -NR_3^+) or strong activators (-NH_2, NHR, NR_2).**

b. Strong Deactivators

- **Strong deactivators (NO_2, etc.)** make the benzene ring **electron poor.** Therefore, monosubstituted benzene rings having these groups are **unreactive**.

NO_2 H $\xrightarrow[AlCl_3]{RCl}$ NO REACTION

c. Strong Activators

- **Strong activating nitrogen containing groups (NH_2-)** (Lewis bases) do not react because these Lewis bases react with the catalyst $AlCl_3$, a **Lewis acid** to give an intermediate **deactivator.**

d. General Reaction for strong activators:

NH_2 + $AlCl_3$ → H, ⊕N, H ⊖$AlCl_3$ $\xrightarrow[RCl]{AlCl_3}$ NR

Lewis base

Lewis acid

EWG = a deactivator

- **Read pages ______ - ______. Do problem on page ______.**

e. **Polyalkylation**

- **Recall: Polysubstitution. Use an excess of benzene in order to avoid it.**
- **Read page _______.**
- **Remember: No polysubstitution occurs in Friedel-Crafts acylation.**

G. ELECTROPHILIC AROMATIC SUBSTITUTION WITH DISUBSTITUTED BENZENES

1. TWO GROUPS WITH THE SAME DIRECTING EFFECTS

- **They reinforce each other.**

Ex: $-CH_3$ and $-NO_2$.

ortho ortho CH_3 meta meta NO_2 $\xrightarrow[FeBr_3]{Br_2}$ CH_3 Br NO_2

2. TWO GROUPS WITH OPPOSING DIRECTING EFFECTS

- **The more powerful activator takes over. Have a mixture of products.**

Ex: -CH_3 and **-OH; both o-p directors, but -OH is a more powerful activator.**

CH_3 ortho ortho ortho ortho OH $\xrightarrow[FeBr_3]{Br_2}$ CH_3 Br OH + CH_3 Br OH

major product

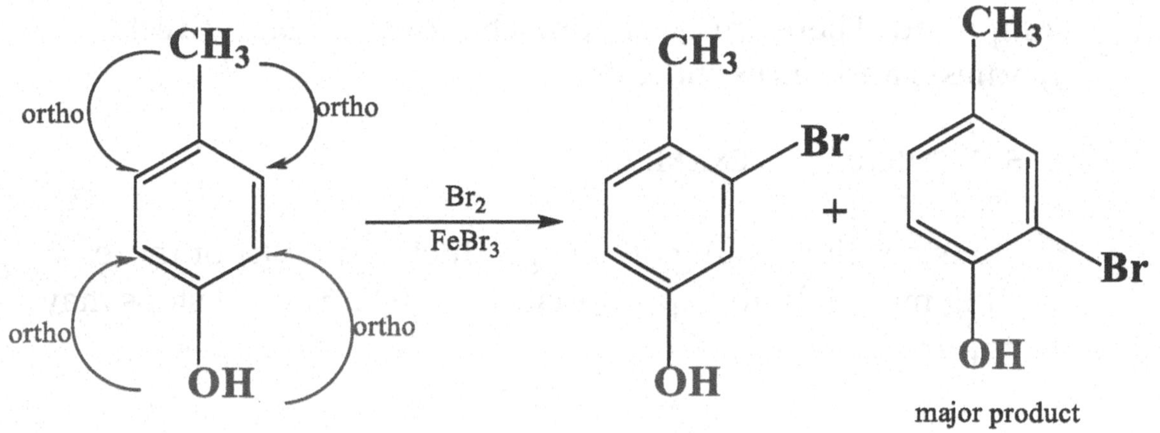

3. META DISUBTITUTED BENZENES

- **In this case, we get a mixture of products.**

Ex:

CH_3 ortho ortho ortho para para ortho Cl crowded $\xrightarrow[FeCl_3]{Cl_2}$ CH_3 Cl Cl + CH_3 Cl Cl + CH_3 Cl Cl

Not observed

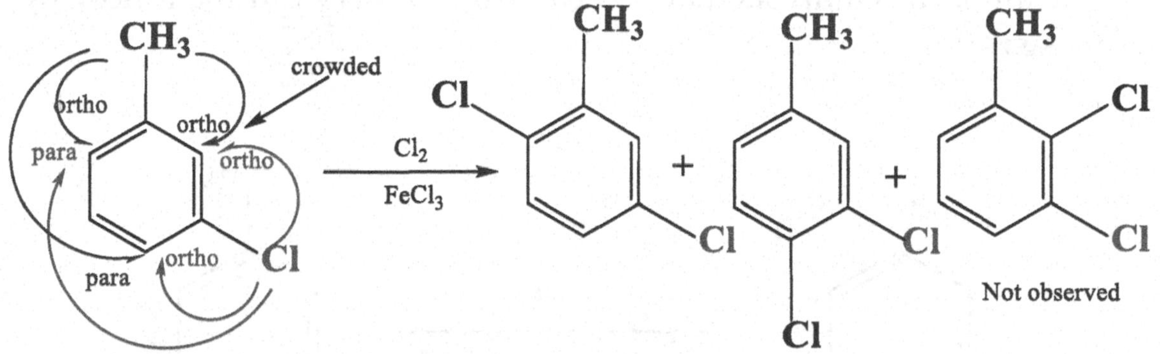

- **Read pages ______ - ______.**

- **See example and do problem on page ______.**

H. USING ELECTROPHILIC AROMATIC SUBSTITUTION TO SYNTHESIZE BENZENE DERIVATIVES

1. INTRODUCTION

- All of the topics that we have covered so far will help us make smart decisions when we want to prepare or **synthesize** a new compound. There are two approaches one can use: **direct synthesis and retrosynthesis.**

2. SYNTHESIS: AN EXAMPLE

- **Synthesis** is the making of new products from one or more starting materials and appropriate catalysts. Several steps may be required.

- In general:

Ex: Use all of the catalysts that you can find to carry out the following synthesis:

?

CH_3

Br

3. RETROSYNTHESIS: AN EXAMPLE

- In **retrosynthesis,** one works **backward** in order to identify all of the different catalysts involved in the reaction.

- In general:

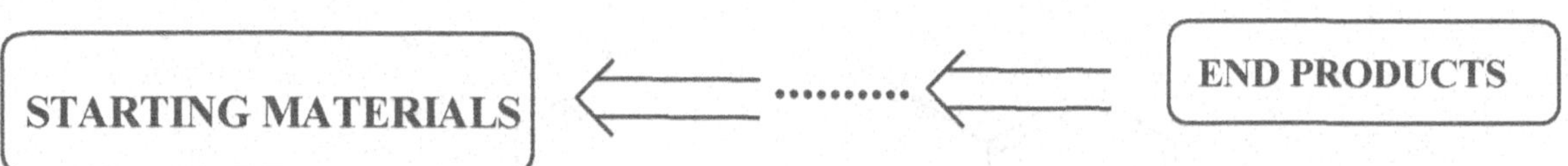

Ex: Devise a stepwise synthesis of the following product using retrosynthesis.

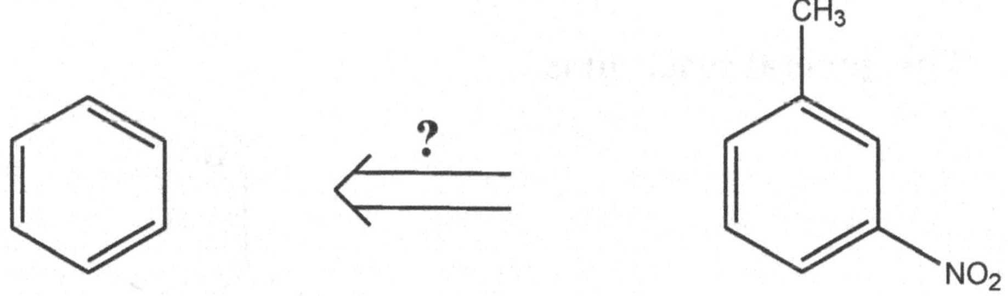

- **Read pages ______ - ______.**

- **Do all examples.**

- **Do Problem _______, page ______.**

I. NUCLEOPHILIC AROMATIC SUBSTITUTION (NAS) REACTIONS

1. INTRODUCTION

- **Disubstituted aryl halides** can act as **electrophiles** and undergo nucleophilic reactions.

- Their general structures are:

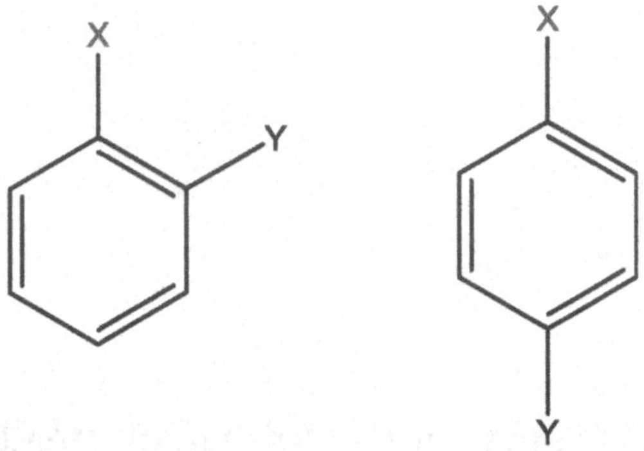

X Y X Y

- Where X = **halogen** and Y = **electron withdrawing group (EWG) in ortho or para position (-NO_2, -CN, -CO, etc.)**

- **The general reaction is:**

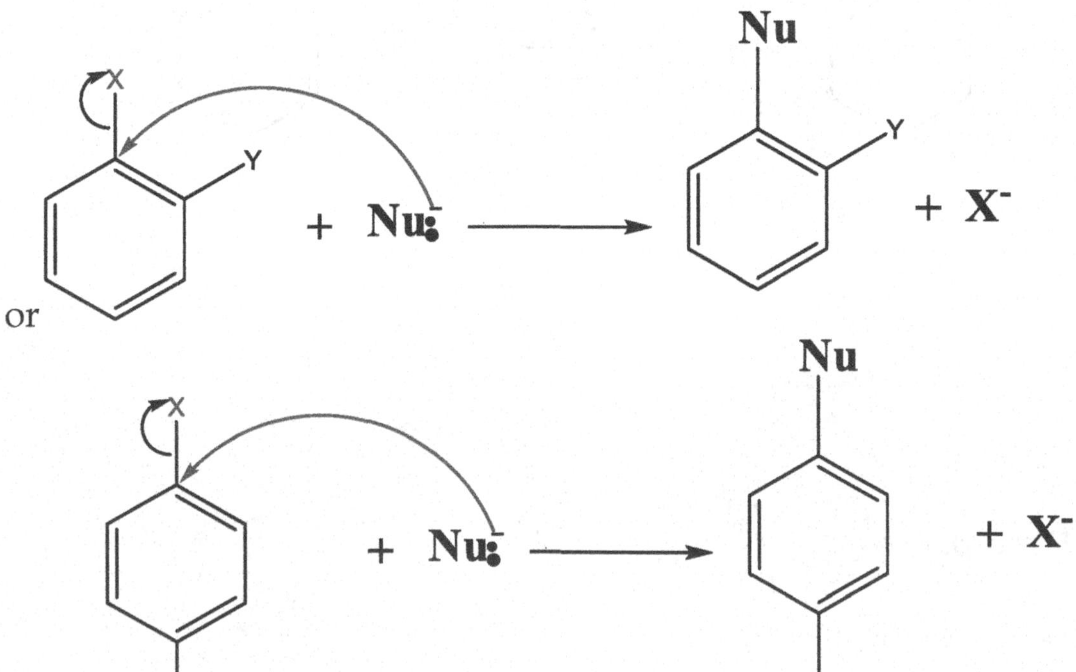

X Y + Nu: ⟶ Nu Y + X^-

or

X Y + Nu: ⟶ Nu Y + X^-

Ex:

$$\text{o-ClC}_6\text{H}_4\text{NO}_2 + OH^- \xrightarrow{130^\circ C} \text{o-HOC}_6\text{H}_4\text{NO}_2 + Cl^-$$

- **Mechanism: The reaction proceeds in two steps: addition of Nu/Elimination of the halogen.**
- **It goes also through a Meisenheimer complex (carbanion).**

Cl, NO2, EWG, + :OH⁻, addition, Meisenheimer complex, elimination, OH, NO2

2. CHARACTERISTICS OF NUCLEOPHILIC AROMATIC SUBSTITUTION REACTIONS

- This reaction is different from SN1 and SN2.
- See Chapter ______.
- **Recall:** Aryl halides inert to both SN1 and SN2 (**steric effect and unstable aryl cation**).
- The reaction proceeds through a **Meisenheimer complex.**
- Only aryl halides with **o-p EWG (-NO_2, ...)** can react. The more the substituent in o-p, the faster the reaction. **Why? Because the Meisenheimer intermediate is stabilized by resonance.**

3. ELECTROPHILIC VS. NUCLEOPHILIC AROMATIC SUBSTITUTION

- EAS:
  - Best for activating groups (EDG).
  - A **carbocation intermediate** is formed.
  - A **H** is replaced on benzene ring.
  - A 2-step reaction: addition/elimination.

- NAS:
  - Best with EWG in o-p.
  - A **Meisenheimer intermediate** is involved.
  - A **halogen** is replaced on benzene ring.
  - A 2-step reaction: addition/elimination.

J. BENZYNE AND PHENOL SYNTHESIS

1. INTRODUCTION

- **Aryl halides with no o-p EWG** can be forced to react at high temperatures and high pressures. For instance, phenol can be synthesized from chlorobenzene as follows:

Cl (chlorobenzene) $\xrightarrow[\text{2. } H_3O^+]{\text{1. NaOH, } H_2O\text{, 340°C, 2500 psi}}$ OH (phenol) + NaCl

2. BENZYNE INTERMEDIATE

- The reaction proceeds in two steps and proceeds through a **benzyne intermediate. We have elimination (of HCl) followed by addition (of H_2O):**

:OH⁻   -HCl / Elimination → [benzyne] $+ Cl^- + H_2O \xrightarrow[\text{Addition}]{H_2O}$ (phenol, OH) + NaCl

2. PHENOL SYNTHESIS VS. NAS

- **In phenol synthesis = elimination/addition.**
- **In NAS, addition/elimination.**
- **In phenol synthesis=benzyne intermediate.**
- **In NAS, have a Meisenheimer intermediate.**

4. STRUCTURE OF BENZYNE

- **In a benzyne, have a triple bond between two carbons of the double bond. As a matter of fact, the second π bond results from the "sideway" overlap of two adjacent, parallel sp^2 orbitals. See sketch.**

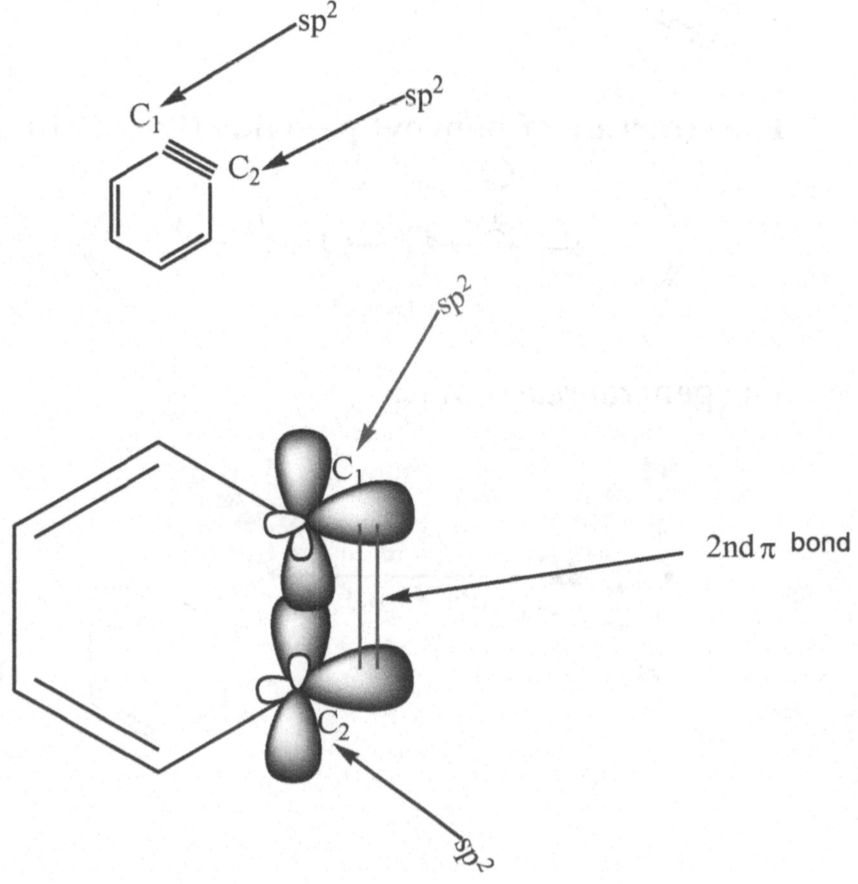

K. OTHER REACTIONS OF BENZENE AND DERIVATIVES

1. BROMINATION OF ALKYLBENZENE SIDE CHAINS WITH A BENZYLIC HYDROGEN

- **The catalyst is: NBS (N-bromosuccinimide) in benzoyl peroxide ($(PhCO_2)_2$) in CCl_4 or Br_2, hv (or Δ)**
- **This is a radical reaction.**
- **One sole product is obtained.**
- **The peroxide initiates the formation of the radical intermediate.**
- **Only alkyl benzenes with benzylic H can react.**

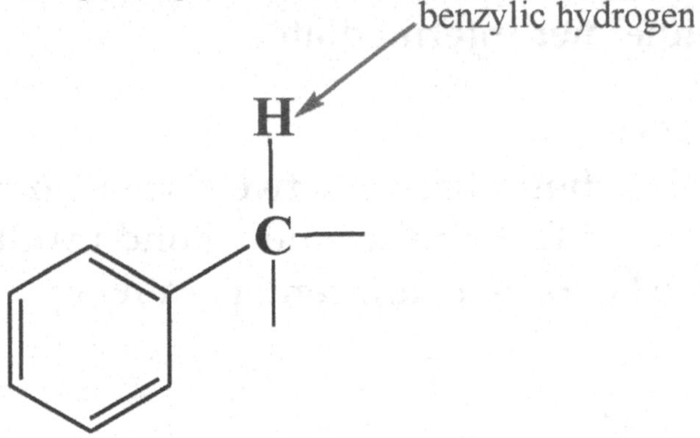

benzylic hydrogen

H

C

- **The structure of benzoyl peroxide ($(PhCO_2)_2$) is:**

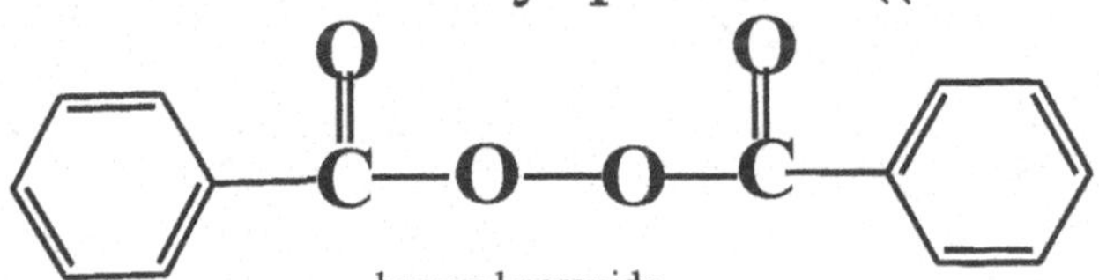

O O

C—O—O—C

benzoyl peroxide

- **The general reaction is:**

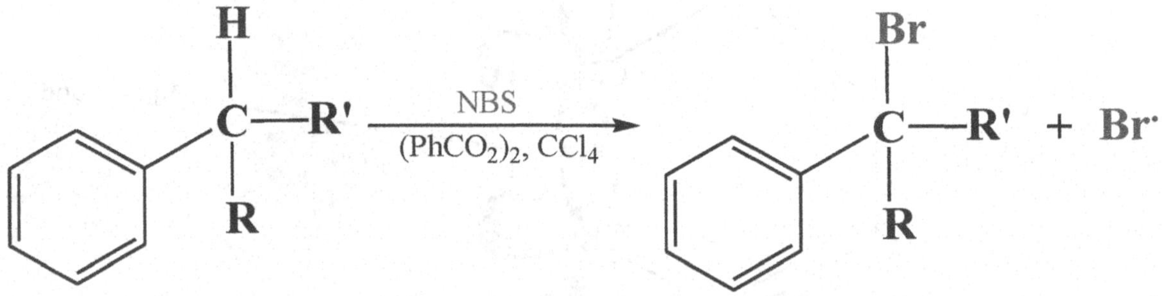

H — C — R' —NBS / $(PhCO_2)_2$, CCl_4→ Br — C — R' + Br·

R R

or

$$\text{C}_6\text{H}_5\text{–CH(R)(R')} \xrightarrow[h\nu \text{ or } \Delta]{Br_2} \text{C}_6\text{H}_5\text{–CBr(R)(R')} + Br^{\cdot}$$

- **Mechanism of the reaction: The reaction proceeds in two steps through a benzylic radical.**
- **See page ______.**

- **Note: The benzylic radical is stabilized by resonance thru the overlap of its p orbital with the π system on the benzene ring.**

Ex:

$$\text{C}_6\text{H}_5\text{–CH(CH}_3\text{)(CH}_2\text{CH}_3\text{)} \xrightarrow[h\nu \text{ or } \Delta]{Br_2} \text{C}_6\text{H}_5\text{–CBr(CH}_3\text{)(CH}_2\text{CH}_3\text{)} + Br^{\cdot}$$

Ex:

$$\text{C}_6\text{H}_5\text{–C(CH}_3\text{)}_2\text{(CH}_2\text{CH}_3\text{)} \xrightarrow[h\nu \text{ or } \Delta]{Br_2}$$

- **Read pages ______ - ______. Do problems on pages ______ - ______.**

- **Note: If the catalyst is $Br_2/FeBr_3$, an Electrophilic Aromatic Substitution reaction occurs.**

- **See page ______.**

2. OXIDATION OF ALKYLBENZENE SIDE CHAINS WITH A BENZYLIC HYDROGEN

a. Introduction

- The benzene ring is inert to **strong oxidizing agents** such as $KMnO_4$ and $K_2Cr_2O_7$. However, alkyl group **side chains with a benzylic hydrogen** react readily with oxidizing agents to give **carboxylic acids.**

b. The General Reaction

H

C—R

R

oxidizing agent

COOH

Ex:

$CH_2CH_2CH_3$

$KMnO_4$

COOH

CH_3

C—CH_3

$KMnO_4$

CH_2CH_3

- **Note: The reaction mechanism is very complex. It involves a benzylic radical intermediate.**

Ex:

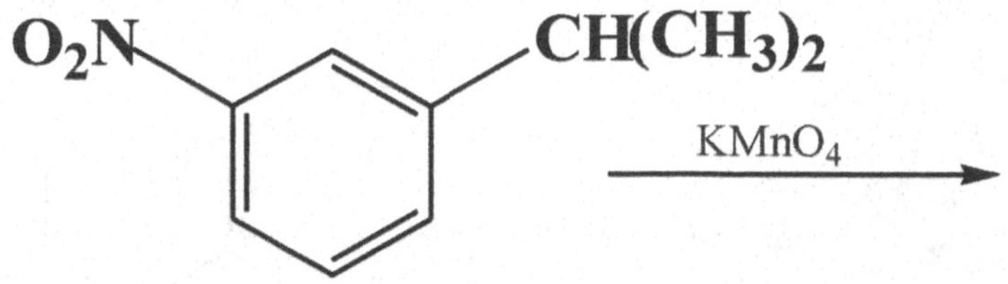

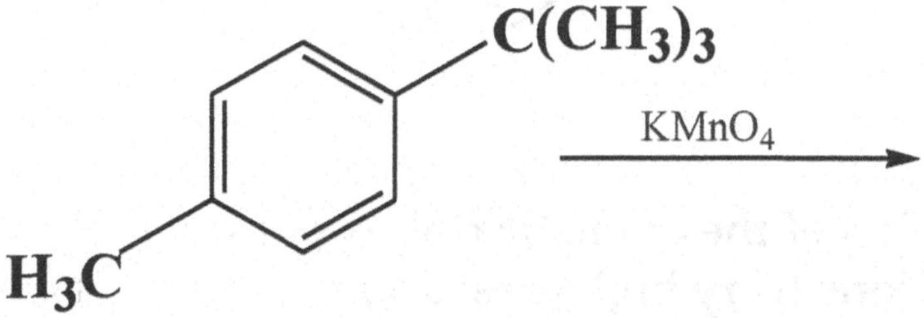

O_2N — $CH(CH_3)_2$ (benzene ring) $\xrightarrow{KMnO_4}$

$C(CH_3)_3$, H_3C (benzene ring) $\xrightarrow{KMnO_4}$

- **Read pages ______ – ______.**

- **Do examples on pages ______ – ______.**

3. REDUCTION OF ALKYLBENZENE WITH UNSATURATED SIDE CHAINS

a. Introduction

- The benzene ring is inert to **reduction under most conditions. However, unsaturated side chains can be reduced.**

b. The General Reaction

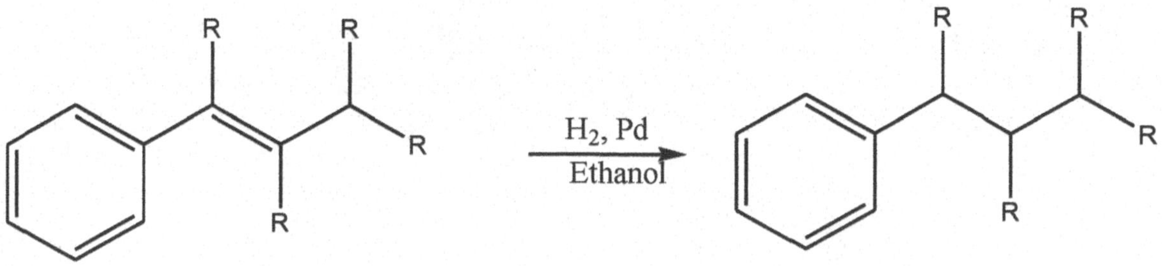

R, R, R, R $\xrightarrow[\text{Ethanol}]{H_2,\ Pd}$ R, R, R, R

Ex:

$$\text{PhC(CH}_3\text{)=C(CH}_3\text{)COCH}_3 \xrightarrow[\text{Ethanol}]{H_2,\ Pd} \text{PhCH(CH}_3\text{)CH(CH}_3\text{)COCH}_3$$

- **Note: The hydrogenation of the aromatic ring is possible under "strong" conditions (very high pressure).**

Ex:

$$C_6H_5CH_3 \xrightarrow[\text{2000 psi, 25°C}]{H_2,\ Pd} C_6H_{11}CH_3$$

Ex:

$$HO{-}C_6H_4{-}C(CH_3)_3 \xrightarrow[\text{2000 psi, 25°C}]{H_2,\ Pd}$$

4. REDUCTION OF ARYL ALKYL KETONES

a. Introduction

- The aromatic ring **activates a neighboring carbonyl** group toward reduction.

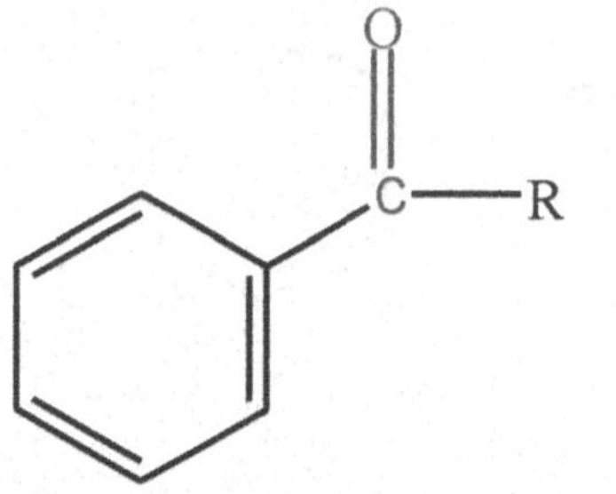

an aryl alkyl ketone

b. The General Reaction

reducing agent

an aryl alkyl ketone

Ex:

H_2, Pd
Ethanol

an aryl alkyl ketone

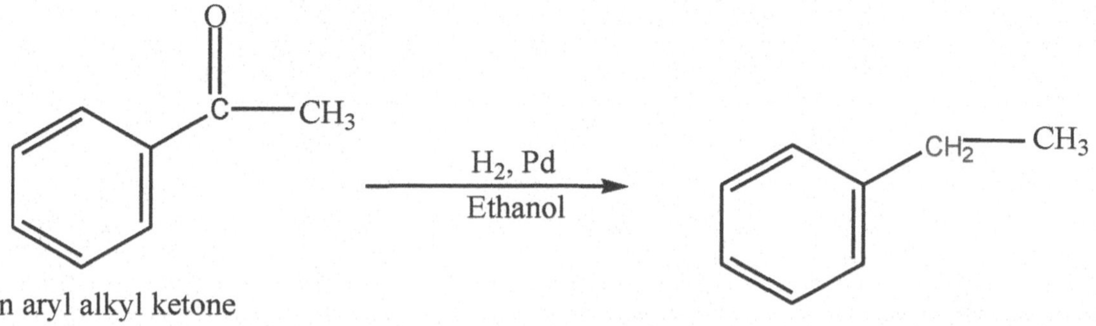

c. The Clemmensen Reduction Reaction

Catalyst: Zn(Hg) + HCl

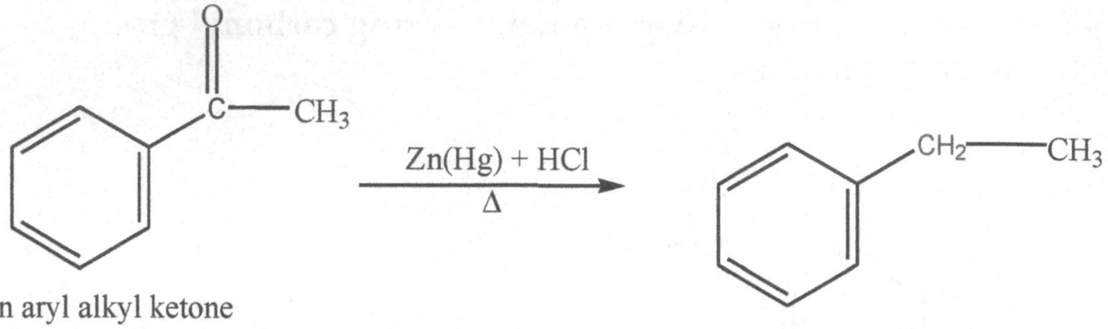

$$C_6H_5\text{-}C(=O)\text{-}CH_3 \xrightarrow[\Delta]{Zn(Hg) + HCl} C_6H_5\text{-}CH_2\text{-}CH_3$$

an aryl alkyl ketone

d. The Wolff-Kishner Reduction Reaction

Catalyst: $NH_2NH_2 + OH^-$

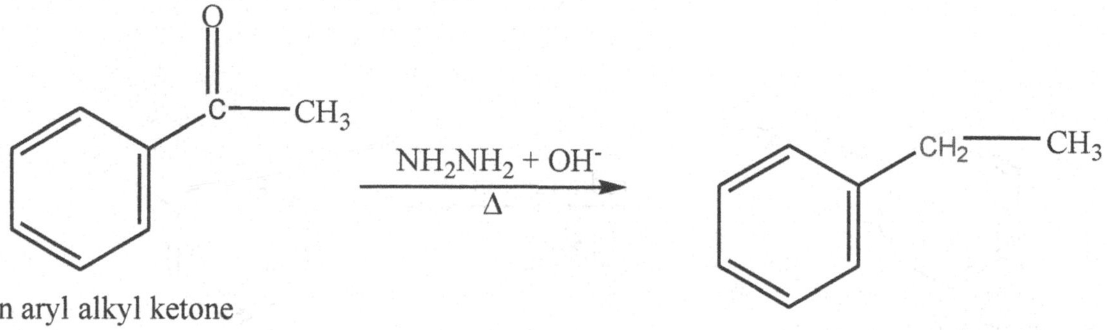

$$C_6H_5\text{-}C(=O)\text{-}CH_3 \xrightarrow[\Delta]{NH_2NH_2 + OH^-} C_6H_5\text{-}CH_2\text{-}CH_3$$

an aryl alkyl ketone

- **Read pages _______ - ______.**
- **Do problems on pages ______ - ______.**

5. REDUCTION OF ARYL NITRO GROUPS TO AMINES

The catalysts are: H_2/Pd, or Fe/HCl, or Sn/HCl.

Ex:

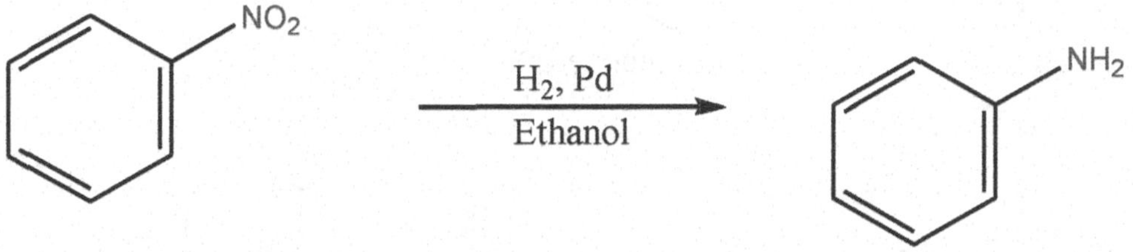

NO_2 — H_2, Pd / Ethanol → NH_2

- **Read pages ______ - ______.**
- **Do problem on page ______.**

L. MULTISTEP SYNTHESIS

- Now, you are ready to carry out multistep synthesis of organic products.

Ex. 1: Devise a way to synthesize p-bromobenzoic acid from benzene.

CO_2H

Br

p-bromobenzoic acid

Ex. 2: Devise a way to synthesize o-aminotoluene from benzene.

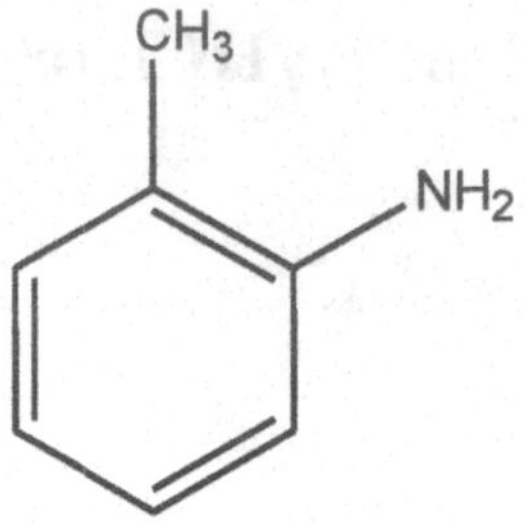

o-aminotoluene

- **Read pages ______ - ______.**
- **Do problems and examples on pages _____ - _____.**

M. A SUMMARY ON EAS REACTIONS: 5

reaction name	E^+	aryl H is replaced by	Catalyst/reagent
Nitration	NO_2^+	$-NO_2$	HNO_3/H_2SO_4
Sulfonation	SO_3H^+	$-SO_3H$	SO_3/H_2SO_4
halogenation	X^+	A halogen X (Cl, Br, I)	X_2/FeX_3
Friedel-Crafts Alkylation	R^+	An alkyl group -R	$RCl/AlCl_3$
Friedel-Crafts Acylation	RCO^+	RCO-	$RCOCl/AlCl_3$

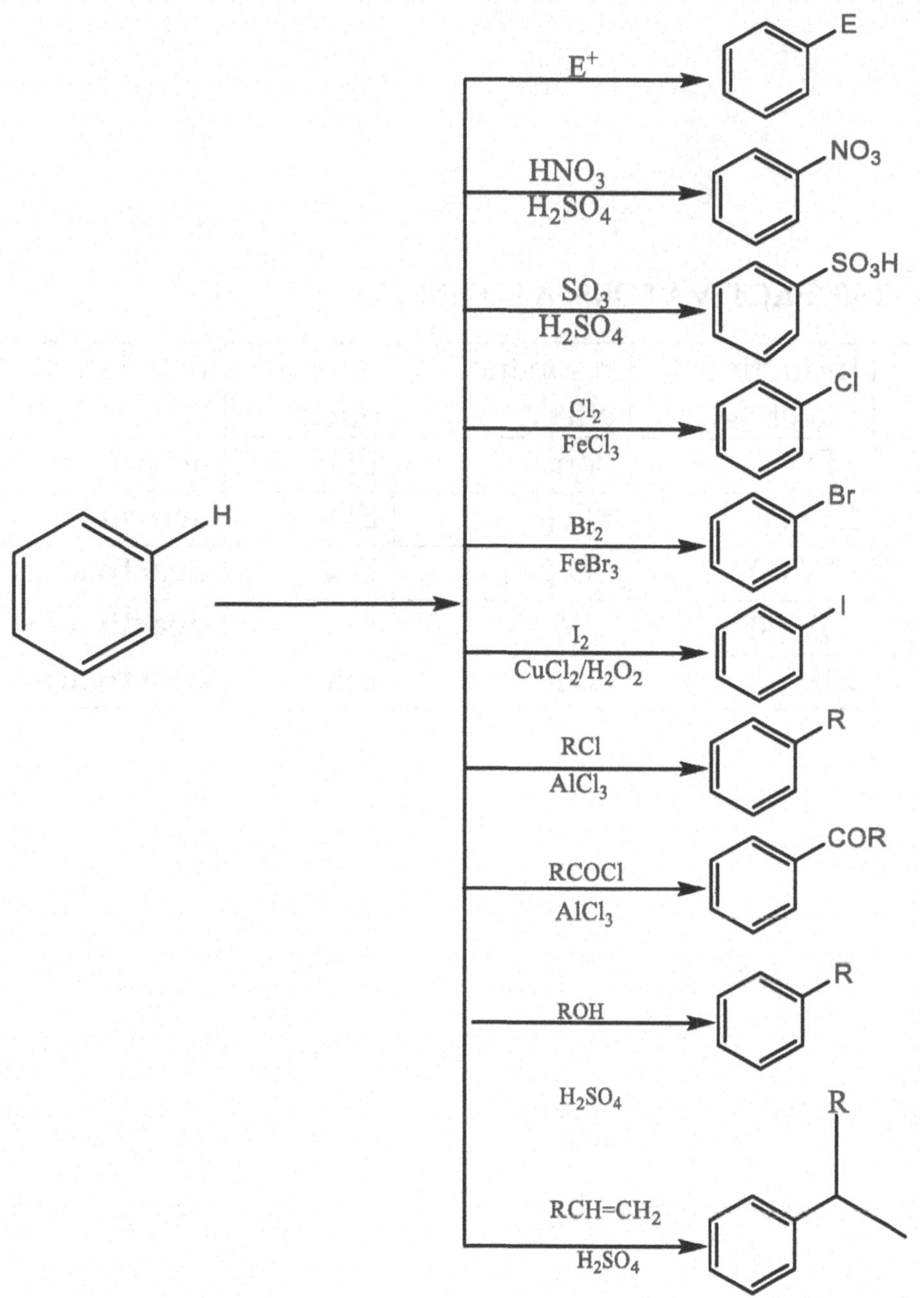
H
E+
E
HNO3
H2SO4
NO3
SO3
H2SO4
SO3H
Cl2
FeCl3
Cl
Br2
FeBr3
Br
I2
CuCl2/H2O2
I
RCl
AlCl3
R
RCOCl
AlCl3
COR
ROH
H2SO4
R
RCH=CH2
H2SO4
R

ACTIVATORS/DEACTIVATORS: A SUMMARY

group	inductive effect	resonance effect	overall effect	reactivity	orientation
Alkyl	ED (w)	none	ED	**activator**	**o-p**
Z = N, O	EW (w)	ED (s)	ED	**activator**	**o-p**
halogens	EW (s)	ED (w)	EW	**deactivator**	**o-p**
Y (+ or δ^+)	EW (s)	EW (s)	EW	**deactivator**	**meta**
$-NR_3^+$	EW (s)	none	EW	**deactivator**	**meta**

ACTIVATORS/DEACTIVATORS: A SUMMARY

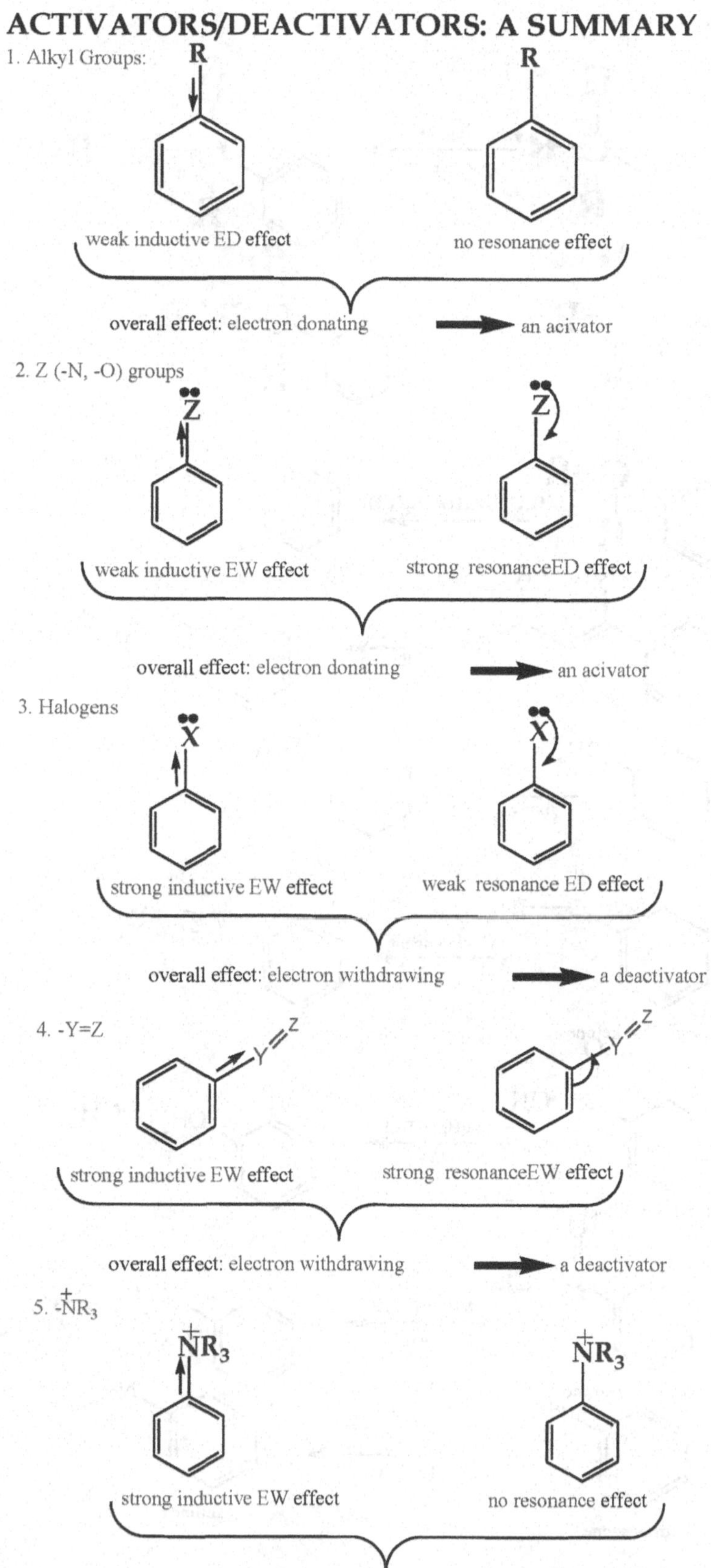
1. Alkyl Groups:
R
R
weak inductive ED effect
no resonance effect
overall effect: electron donating
an acivator
2. Z (-N, -O) groups
Z
Z
weak inductive EW effect
strong resonanceED effect
overall effect: electron donating
an acivator
3. Halogens
X
X
strong inductive EW effect
weak resonance ED effect
overall effect: electron withdrawing
a deactivator
4. -Y=Z
Y
Z
Y
Z
strong inductive EW effect
strong resonanceEW effect
overall effect: electron withdrawing
a deactivator
5. $-\overset{+}{N}R_3$
$\overset{+}{N}R_3$
$\overset{+}{N}R_3$
strong inductive EW effect
no resonance effect
overall effect: electron withdrawing
a deacivator

EAS OTHER REACTIONS

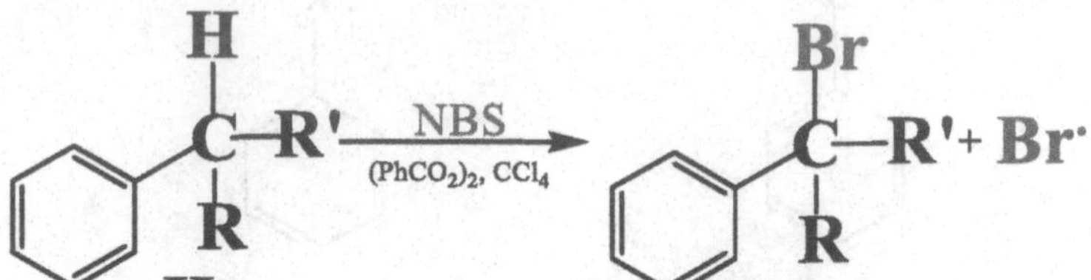

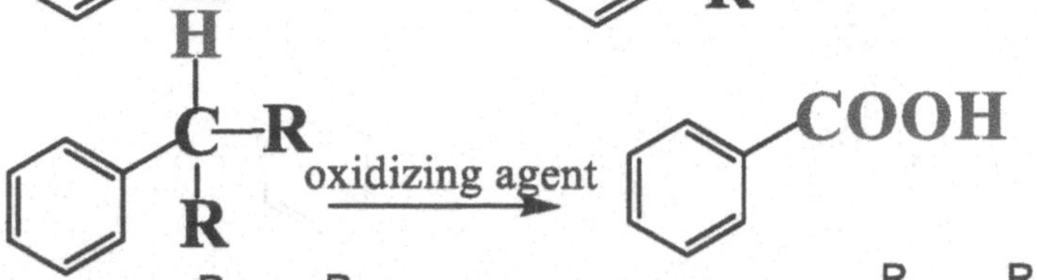

NBS, $(PhCO_2)_2$, CCl_4

Br

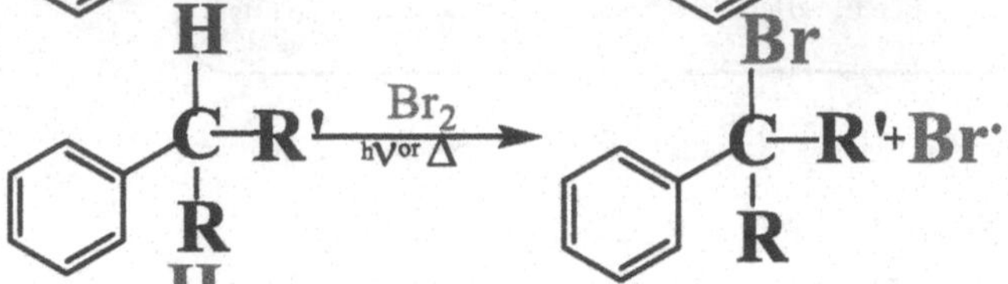

+ Br·

Br_2, hν or Δ

+ Br·

oxidizing agent

COOH

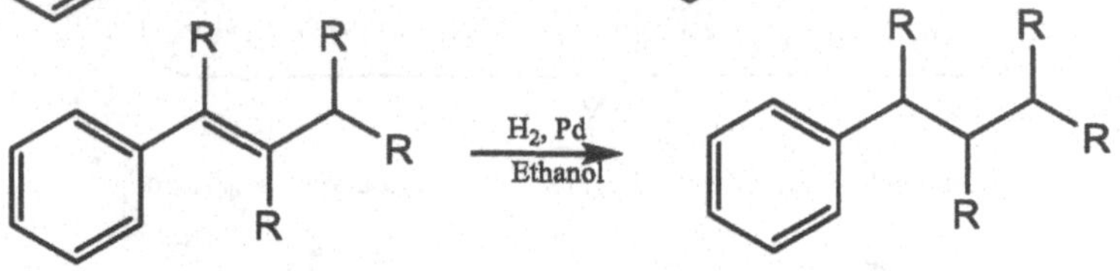

H_2, Pd, Ethanol

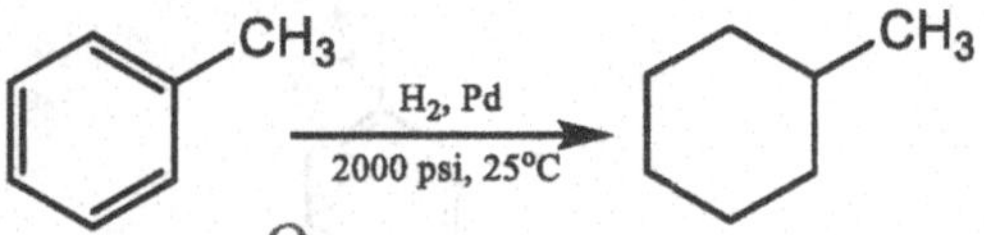

CH_3

H_2, Pd, 2000 psi, 25°C

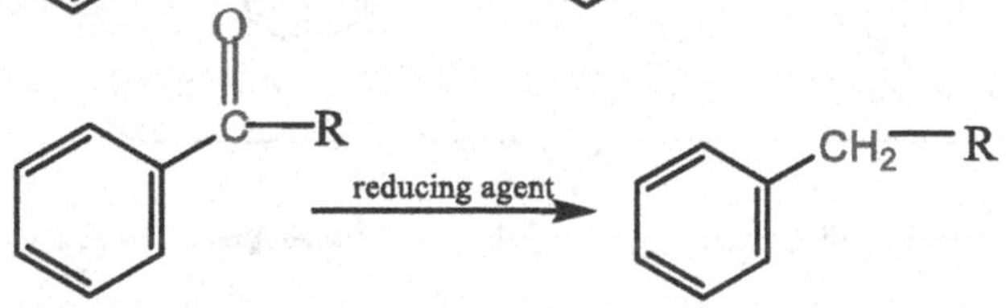

reducing agent

an aryl alkyl ketone

CH_2—R

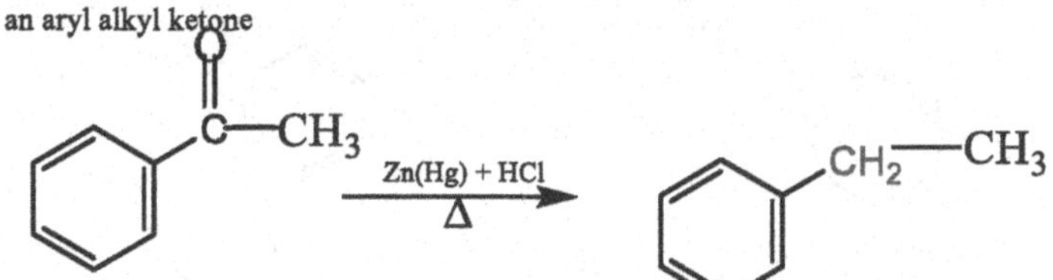

Zn(Hg) + HCl, Δ

an aryl alkyl ketone

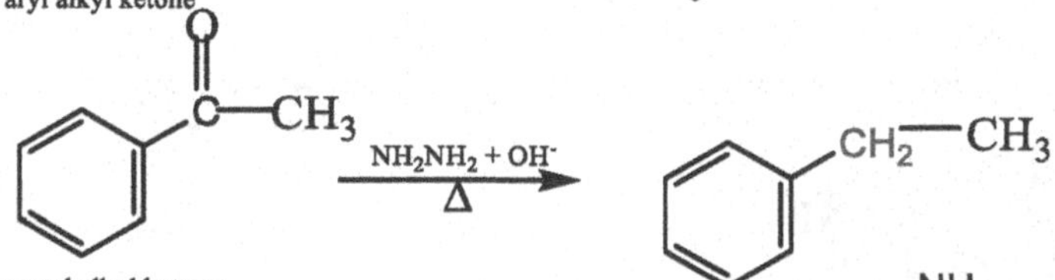

CH_2—CH_3

NH_2NH_2 + OH^-, Δ

an aryl alkyl ketone

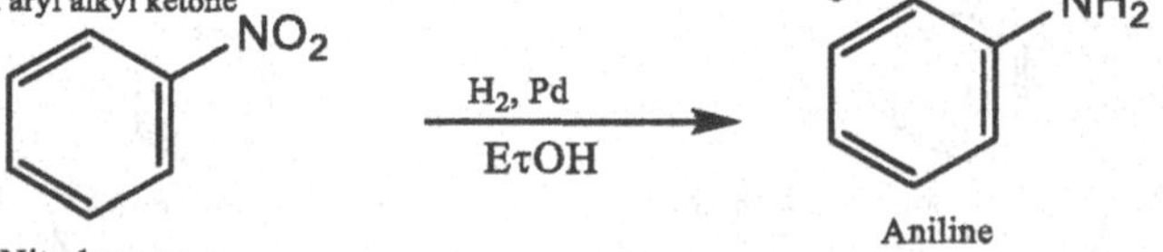

NO_2

H_2, Pd, EtOH

NH_2

Nitrobenzene

Aniline

- **See Key Concepts on pages ______ - _____.**

OCHEM II UNIT 9: CARBOXYLIC ACIDS

A. GENERAL STRUCTURE

1. GENERAL FORMULA

- Carboxylic acids are compounds that contain the **carboxy** group: **-COOH.** The general formula of a carboxylic acid is:

$$R-\overset{\displaystyle O}{\overset{\|}{C}}-OH \quad \text{or} \quad R-COOH \quad \text{or} \quad R-CO_2H$$

Ex:

$$CH_3-\overset{\displaystyle O}{\overset{\|}{C}}-OH \quad \text{or} \quad CH_3-COOH \quad \text{or} \quad CH_3-CO_2H$$

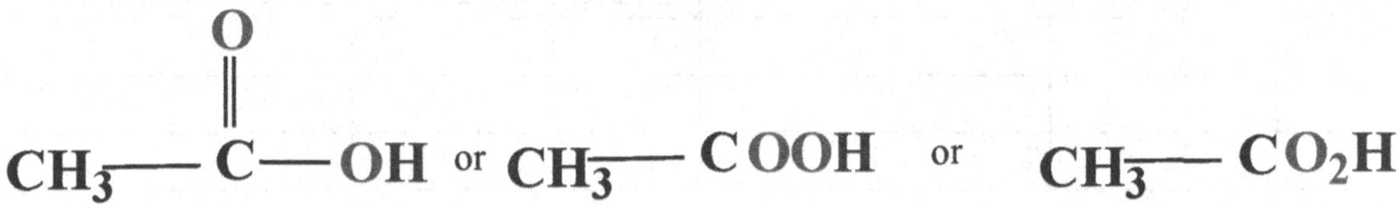

2. HYBRIDIZATION OF THE CARBOXY GROUP CARBON

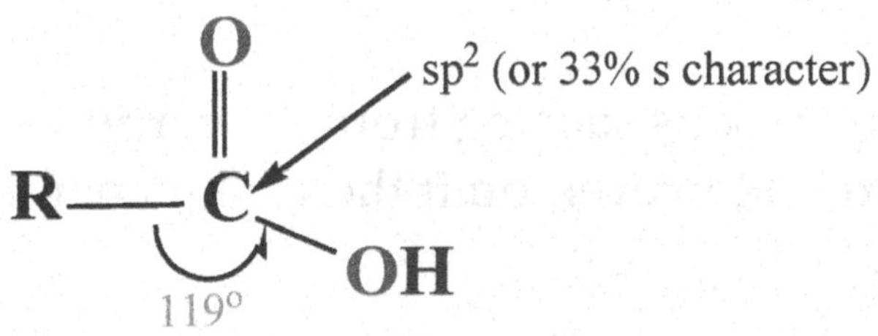

$$R-\overset{\displaystyle O}{\overset{\|}{C}}-OH$$

sp^2 (or 33% s character)

119°

3. POLARITY

- Carboxylic acids are polar compounds. See ESP plot on page _______.

- Read page ______.

B. NOMENCLATURE

1. IUPAC RULES

a. Naming Unbranched Carboxylic Acids

- The names of carboxylic acids end in **-oic acid.**

alkane ~~~ -e + oic acid ~~~ **alkanoic acid**

- Here are some examples:

# of carbons	Alkane	name of alkane	carboxylic acid	name of carboxylic acid
1	CH_4	methane	HCOOH	Methanoic acid
2	CH_3CH_3		CH_3COOH	Ethanoic acid
3				
4				
5				

b. Naming Branched Carboxylic Acids

- Select the **longest continuous carbon chain** (parent chain) that contains the **carboxy** group.

- Number the carbons starting from the **carbon bearing the -COOH group.** However, omit the **1** when naming the compound.

Ex:

$CH_3CH(CH_3)CH_2C(=O)OH$ (carbons numbered 4 3 2 1)

3-methylbutanoic acid

$CH_3CH(CH_3)C(=O)OH$ (carbons numbered 3 2 1)

2-methylpropanoic acid

c. Naming Carboxylic Acids in Which the COOH Group Is Attached to a Ring

- The general rule is:

name of ring + **carboxylic acid**

Ex:

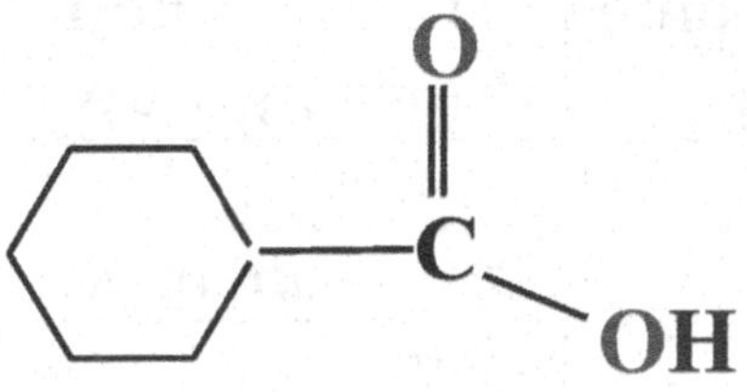

Cyclohexanecarboxylic acid

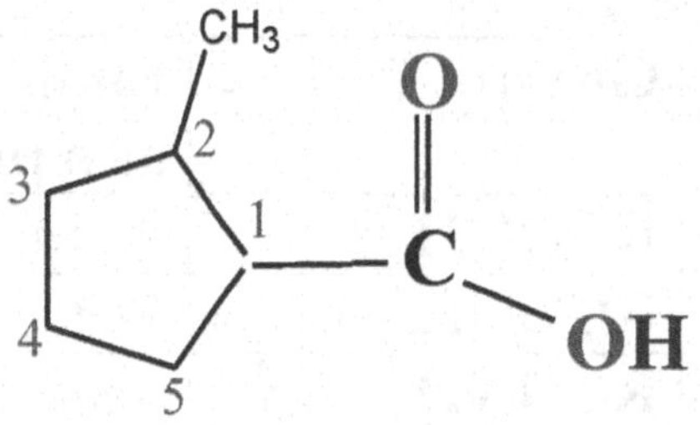

2-methylcyclopentanecarboxylic acid

- Read page ________.
- Do examples on pages _______ - ______.

2. NAMING POLYFUNCTIONAL CARBOXYLIC ACIDS

a. Group Priorities

- When two or more different functional groups are present in a compound, functional group priorities are used. Group priorities are assigned based on the following Table (decreasing order of priority).

Priority order	Group	Ending of name as a priority	Name as a **non Priority group**
Carboxylic acid	**RCOOH**	*-oic acid*	*-carboxy*
Ester	**RCOOR′**	*-oate*	*-alkoxycarbonyl*
Amide	**$RCONH_2$**	*-amide*	*-amido*
Nitrile	**RCN**	*-nitrile*	*-cyano*
Aldehyde	**RCHO**	*-al*	*-oxo(=O) or formyl(-CHO)*
Ketone	**RCOR′**	*-one*	*-oxo*
Alcohol	**ROH**	*-ol*	*-hydroxy*
Amine	**RNH_2**	*-amine*	*-amino*
Alkene	**-C=C-**	*-ene*	*-alkenyl*
Alkyne	**-C≡C-**	*-yne*	*-alkynyl*
Alkane	**-C-C-**	*-ane*	*-alkyl*
Ether	**ROR′**	*-none*	*-alkoxy*
Halide	**R-X**	*-none*	*-halo*

b. Some Examples

oxo

4-*oxopentanoic acid*

3. COMMON NAMES

a. Introduction

- Common names of carboxylic acids are very important since some of them are more used than their IUPAC counterparts.

Ex:

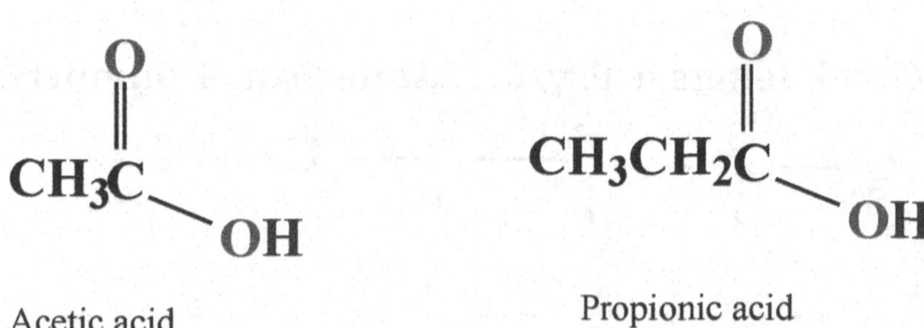

Acetic acid

Propionic acid

b. Roots for Common Names

Number of carbons	Root
1	form-
2	acet-
3	propion-
4	butyr-
5	valer-
6	capro-

c. Common Names for Unbranched Carboxylic Acids

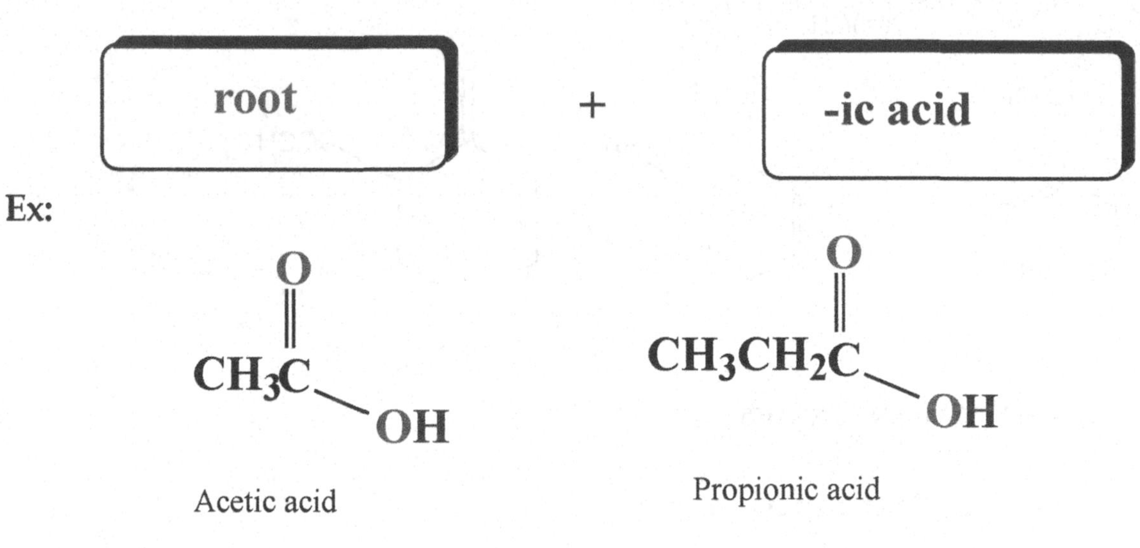

Acetic acid

Propionic acid

- See Table ______, page ______.

d. Common Names for Branched Carboxylic Acids

- Use Greek letters α, β, γ, δ, …,Ω instead of numbers.

$$\overset{5}{\underset{\delta}{C}}—\overset{4}{\underset{\gamma}{C}}—\overset{3}{\underset{\beta}{C}}—\overset{2}{\underset{\alpha}{C}}—\overset{1}{C}OOH$$

OR

Carbon #	2	3	4	5
Greek letter	α	β	γ	δ

Ex:

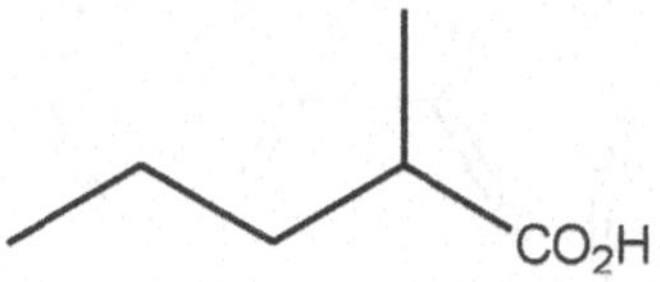

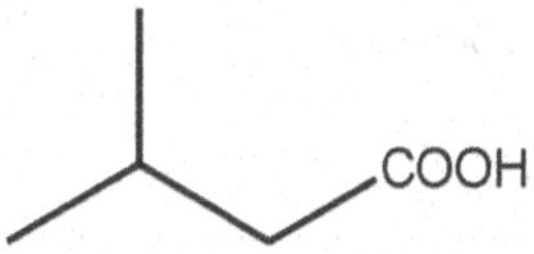

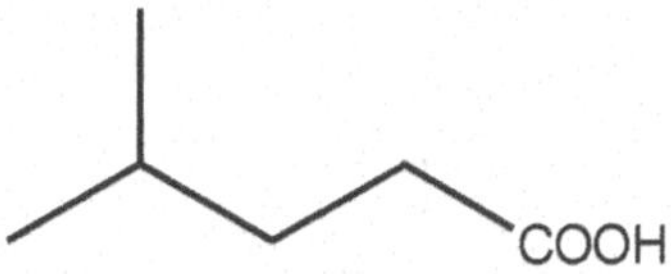

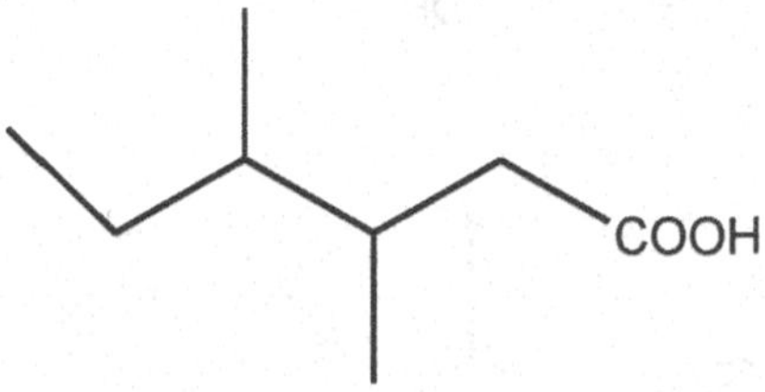

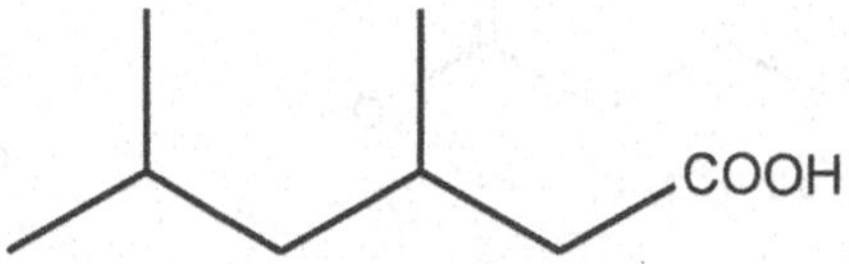

- Read page _______.

- Do Problems _______, ______, page _____ and ______, page _______.

4. NAMING AROMATIC ACIDS

- Aromatic acids are named as benzoic acids.

Ex:

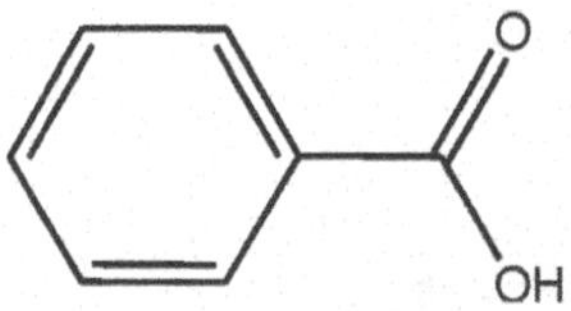

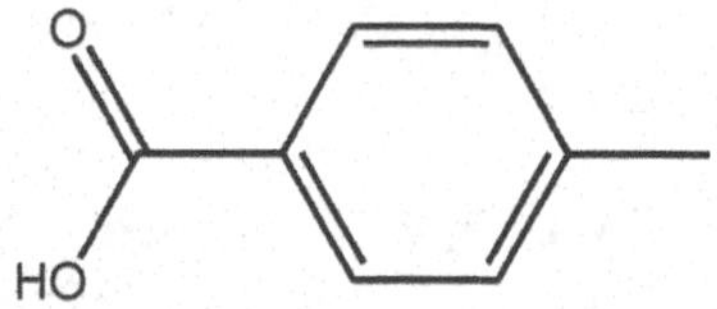

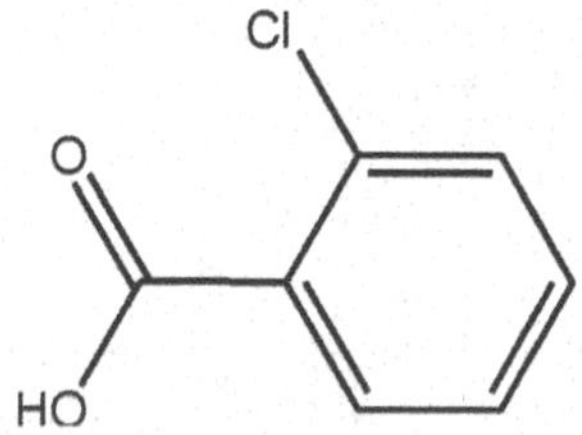

5. NAMING DIACIDS

- Diacids are acids that contain 2 -COOH functional groups.
- IUPAC names end in dioic acid. However, common names are preferred.

- See page _______.

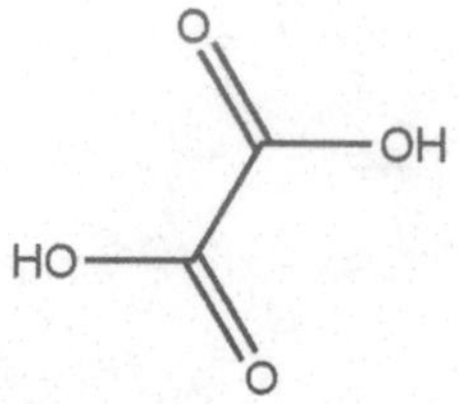

oxalic acid

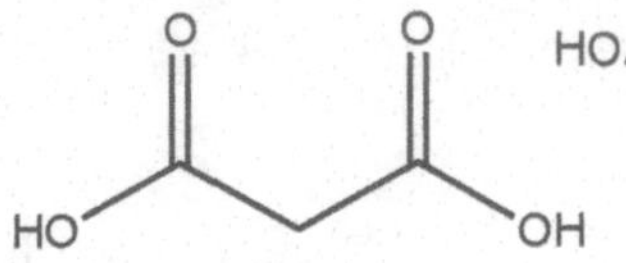

malonic acid

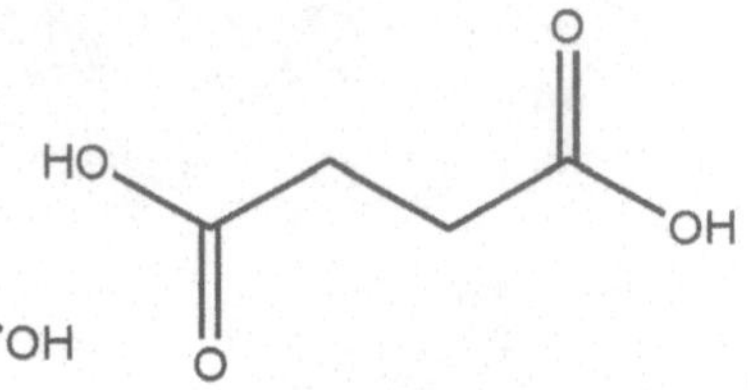

succinic acid

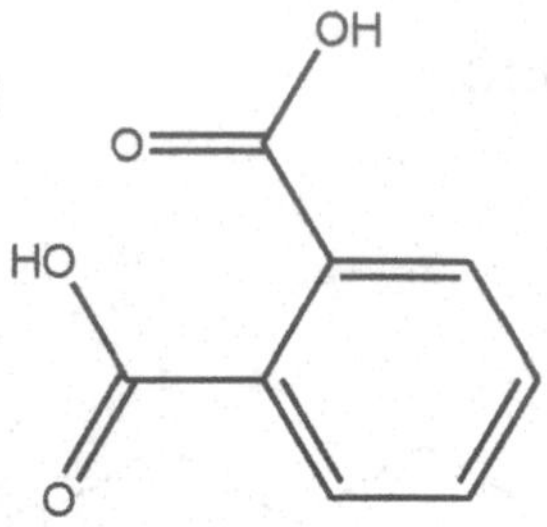

phtalic acid

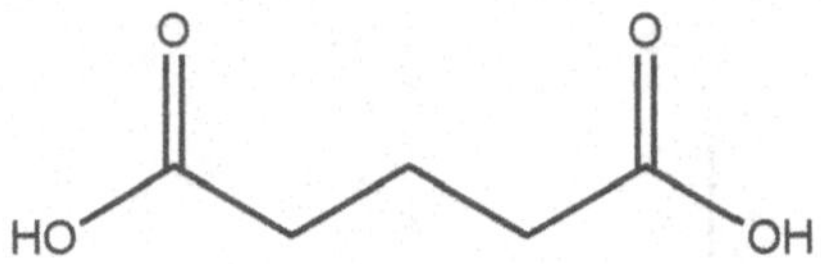

glutaric acid

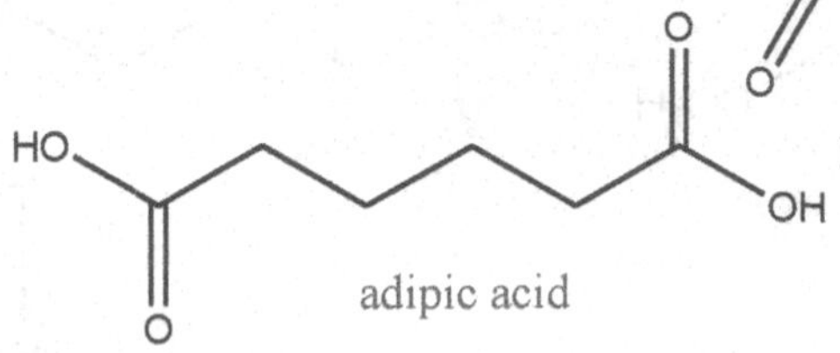

adipic acid

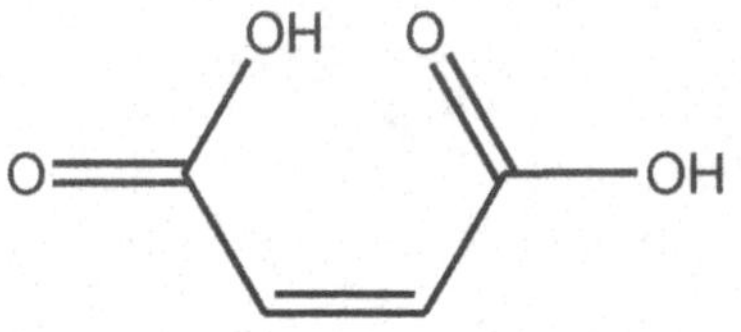

maleic acid

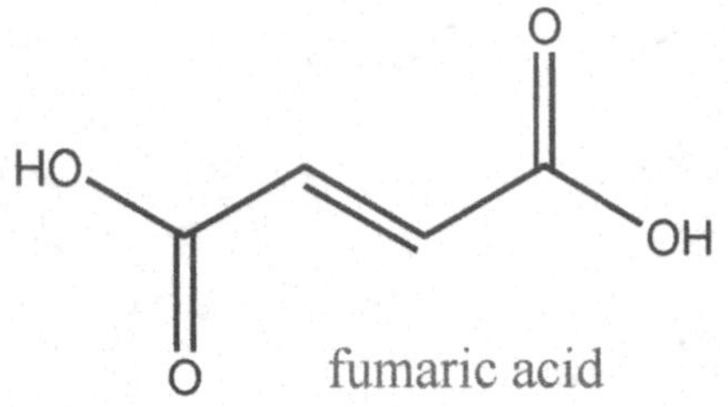

fumaric acid

6. IUPAC NAMES FOR CARBOXYLATE IONS: METAL SALTS

a. IUPAC Names

- **General formula:**

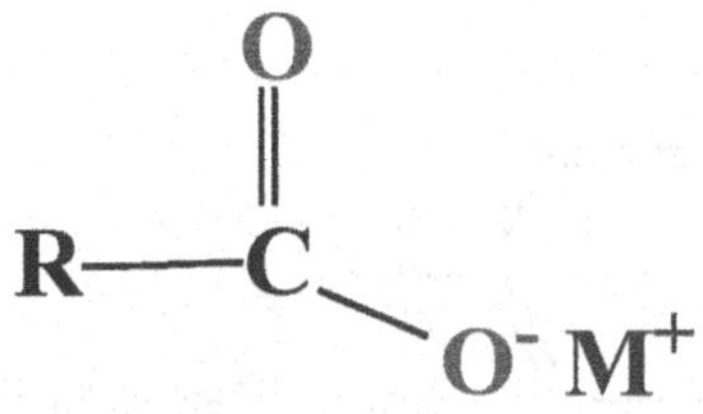

$$R-\overset{\overset{\large O}{\|}}{C}-O^{-}\ M^{+}$$

Name of metal + **alkan- + -oate**

Ex:

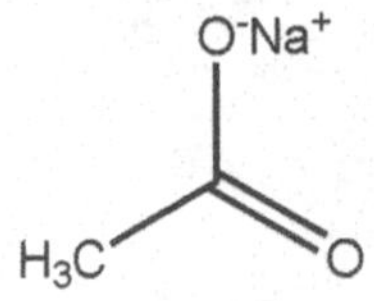

$O^{-}Na^{+}$ ... H_3C ... O

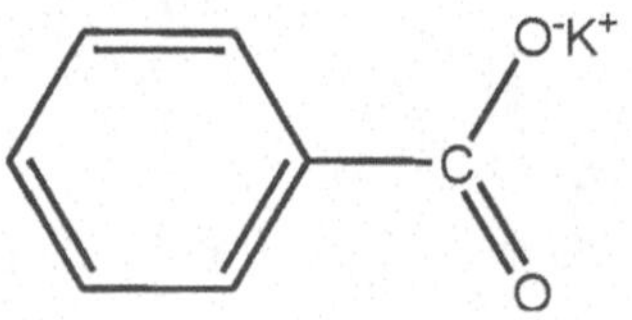

$O^{-}K^{+}$ ... C ... O

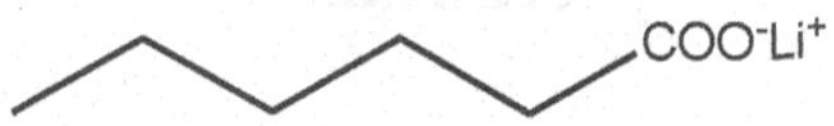

$COO^{-}Li^{+}$

b. Common Names

Name of metal + **root- + -ate**

Ex:

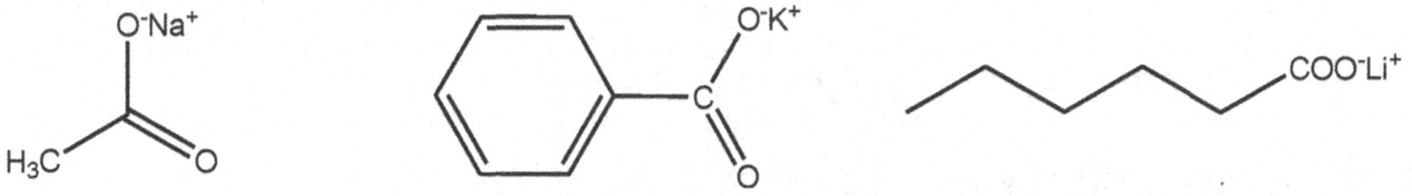

$H_3C{-}C(=O){-}O^-Na^+$ $C_6H_5{-}C(=O){-}O^-K^+$ $CH_3CH_2CH_2CH_2CH_2{-}COO^-Li^+$

- **Read page ________.**
- **Do Problems _______ and ______, page ______.**

C. PHYSICAL PROPERTIES OF CARBOXYLIC ACIDS

1. INTRODUCTION

- Carboxylic acids are compounds that contain **polar C-OH and C=O bonds.** As a result, they are highly **polar**. As a matter of fact, they are the most polar of all functional groups. They are more polar than alcohols, aldehydes, etc. Therefore, they can have VDW forces (LDF), dipole-dipole forces and hydrogen bonds between their molecules. In fact, carboxylic acids exist as **dimers. See Fig. ______, page ______.**

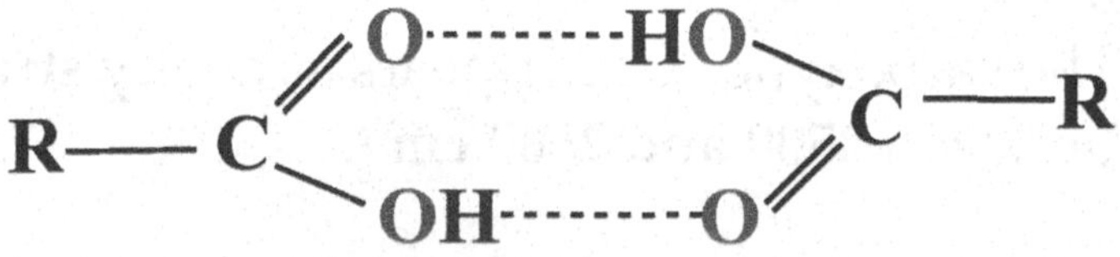

O·········HO
R—C　　　　　C—R
OH·········O

2. BP AND MP

- Since carboxylic acids have very strong intermolecular forces, they have higher BP and MP than other compounds of comparable molar mass.

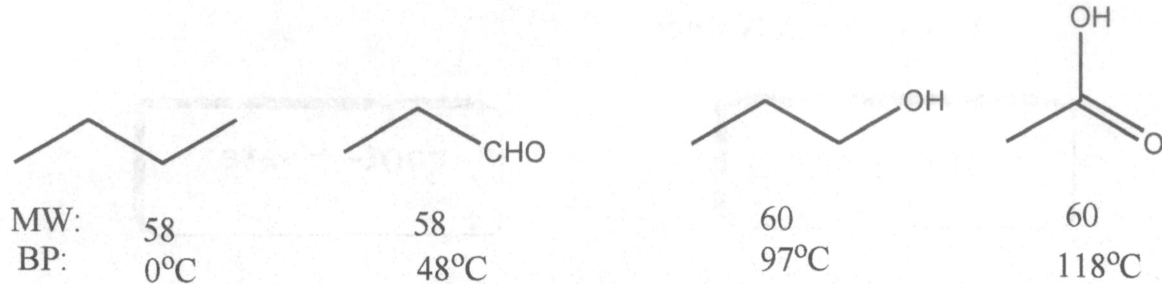

- **See Table _______, page _______.**

3. SOLUBILITY

- All carboxylic acids are soluble in organic solvents.
- Carboxylic acids with a number of carbons ≤ 5 are soluble in water.
- Carboxylic acids with a number of carbons > 5 are insoluble in water.

- **Read page ________.**

- **Do problem _______, page ______.**

D. SPECTROSCOPIC PROPERTIES OF CARBOXYLIC ACIDS

1. IR SPECTROSCOPY

- The carboxylic –C=O appears as a **sharp peak at about 1710 cm^{-1}**.
- The carboxylic –OH appears as a **very strong and broad peak between 3500 and 2500 cm^{-1}**.

- **See Fig. _______, page _______.**

2. NMR SPECTROSCOPY

a. ^{1}H NMR

- The highly deshielded proton of the carboxylic OH group absorbs between 10 and 12 ppm.

- Deshielded hydrogens of the **α carbon** appear between 2-2.5 ppm.

b. ^{13}C NMR

- The C of the C=O is highly deshielded and absorb between 170 and 210 ppm.
- **See Fig. _______, page ______.**
- **Read pages _______ - ______.**
- **Do all problems on page _______.**

E. INTERESTING CARBOXYLIC ACIDS

1. FORMIC ACID

HCOOH = ant sting

2. ACETIC ACID

CH_3COOH = in vinegar; starting material for polymers.

3. BUTANOIC ACID OR BUTYRIC ACID

$CH_3CH_2CH_2COOH$ = byproduct of sweat; **makes butter rancid.**

4. HEXANOIC ACID OR CAPROIC ACID

$CH_3(CH_2)_4COOH$ = gives dirty socks their odor (goat smell).

See the ginkgo story on page _______.

5. OXALIC ACID

HOOCCOOH = **toxic; found in spinach and rhubarb leaves.**

6. LACTIC ACID

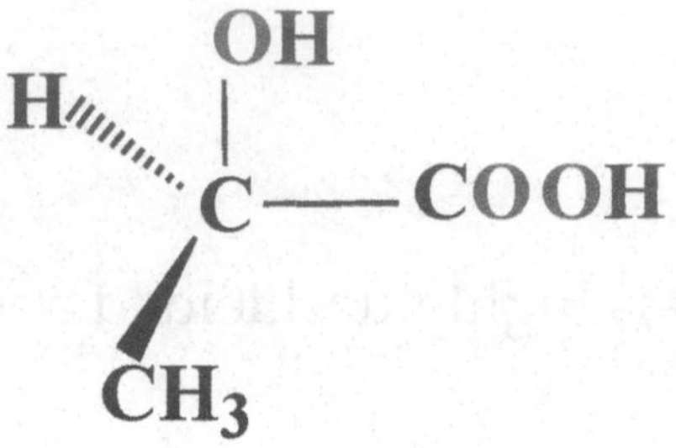

OH
H—C—COOH
CH3

- Lactic acid is a byproduct of the metabolism of glucose. During exercise, its accumulation in muscles causes "**muscle pain.**"

7. SALTS OF CARBOXYLIC ACIDS: CARBOXYLATES

a. Carboxylates as Preservatives

- Sodium benzoate is a **fungal growth inhibitor**; therefore, it is used to preserve can goods.

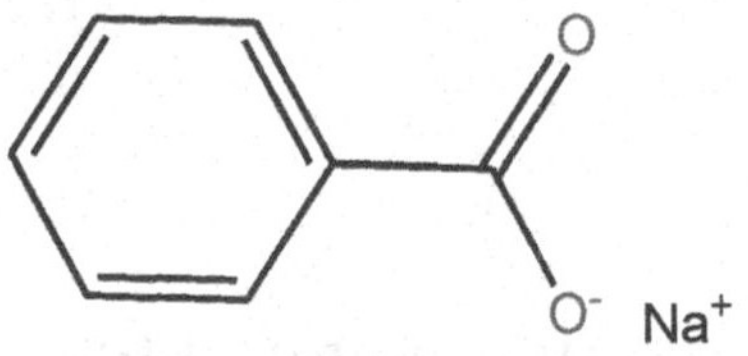

O
O^- Na^+

b. Carboxylates as Soaps

- Soaps are salts of **fatty acids**. Fatty acids are long-chain carboxylic acids.

Ex: $CH_3(CH_2)_{18}COOH$

COOH

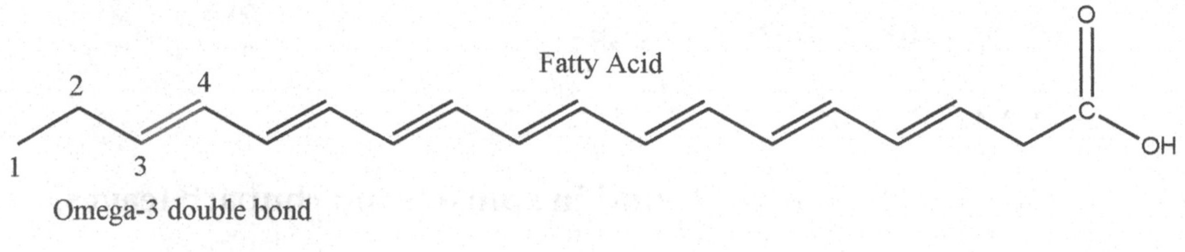

O
C
OH
Fatty Acid
2 4
1 3
Omega-3 double bond

c. Soap Making from Triacylglycerols: Saponification

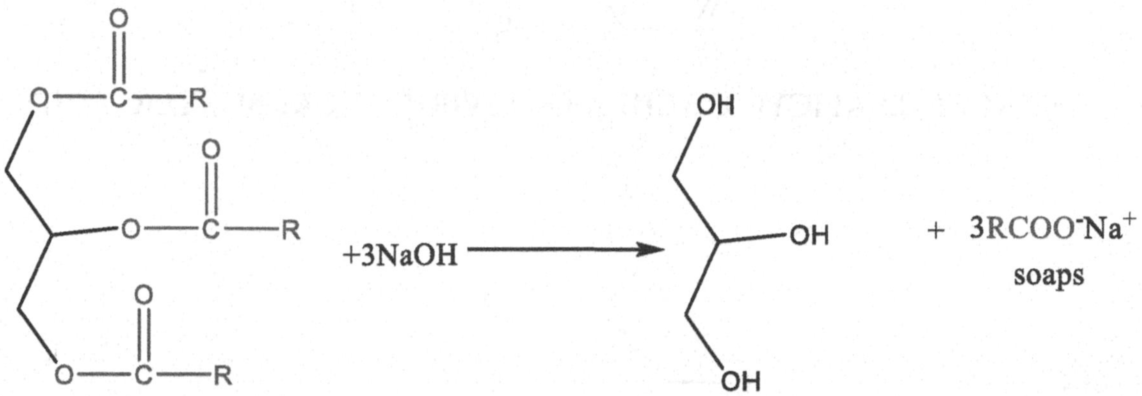

A triacylglycerol (fat or lipid) $+ 3NaOH \longrightarrow$ Glycerol $+ 3RCOO^{-}Na^{+}$ (soaps)

- R = Fatty acid chain.

- **A Word about Detergents**

- **Detergents** are salts of **benzene sulfonic acid** with **nonpolar** long carbon chains at the para position. Like soaps, they also form micelles. However, they are more soluble in aqueous media than soaps. In other words, they **do not** form scum.

- **The general structure of a detergent is:**

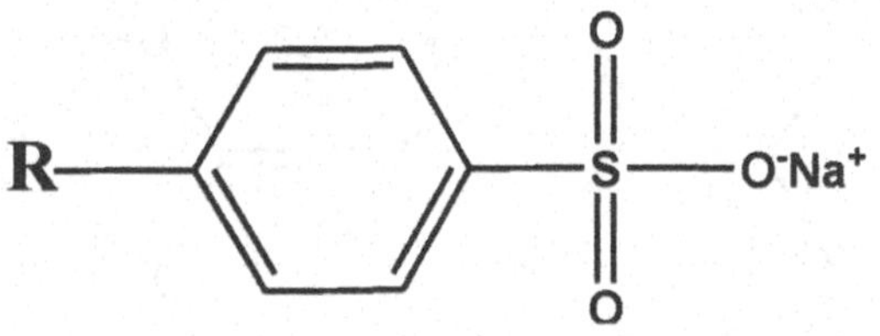

$R-C_6H_4-SO_2-O^{-}Na^{+}$

- **An example:**

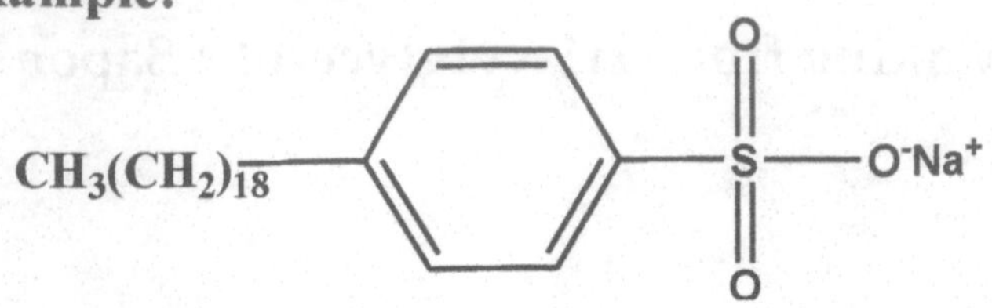

$CH_3(CH_2)_{18}$–C_6H_4–SO_2–O^-Na^+

8. ACETYLSALICYLIC ACID AKA ASPIRIN: A CARBOXYLIC ACID

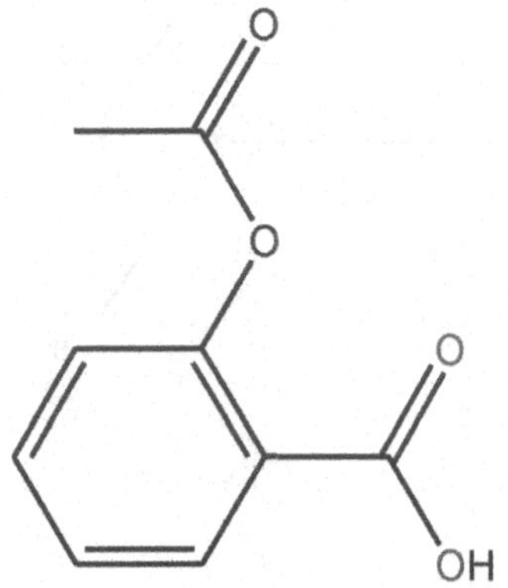

aspirin

- Synthesized in 1899 by **Felix Hoffman**. Aspirin alleviates pain and reduces inflammation. It acts by preventing the synthesis of **prostaglandins** (20-carbon fatty acid with a 5-member ring, responsible for pain and inflammation) from **arachidonic acid**.

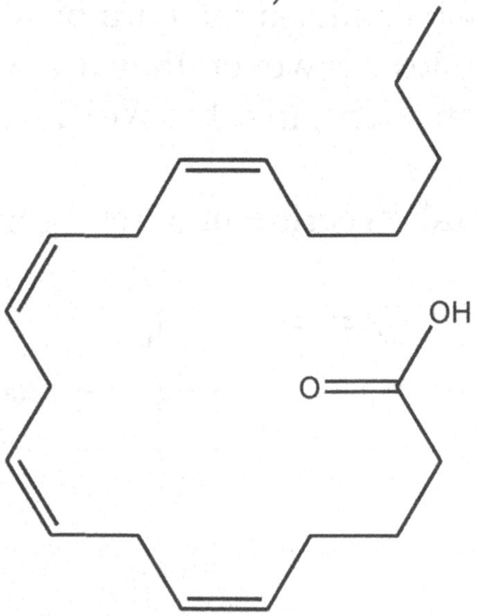

arachidonic acid

- **Read pages _______ - ______.**
- **Do Problem ________, page_______.**

F. SYNTHESIS OF CARBOXYLIC ACIDS: DEJA VUE

1. INTRODUCTION

- We have already covered some ways of preparing carboxylic acids:
  - Oxidation of 1° alcohols.
  - Oxidation of alkyl benzenes with benzylic hydrogens.
  - Oxidative cleavage of alkynes.
  - Oxidative cleavage of alkenes.

2. OXIDATION OF PRIMARY ALCOHOLS

- **The general reaction is (See Unit 1):**

$$RCH_2OH \xrightarrow{[O]} RCOOH$$

$[O] = Na_2Cr_2O_7, K_2Cr_2O_7$, or CrO_3 in H_2O/H_2SO_4.

Ex:

$$CH_3CH_2CH_2OH \xrightarrow{[O]}$$

3. OXIDATION OF ALKYL BENZENE WITH A BENZYLIC HYDROGEN

- **The general reaction is (See Unit 8):**

Ex:

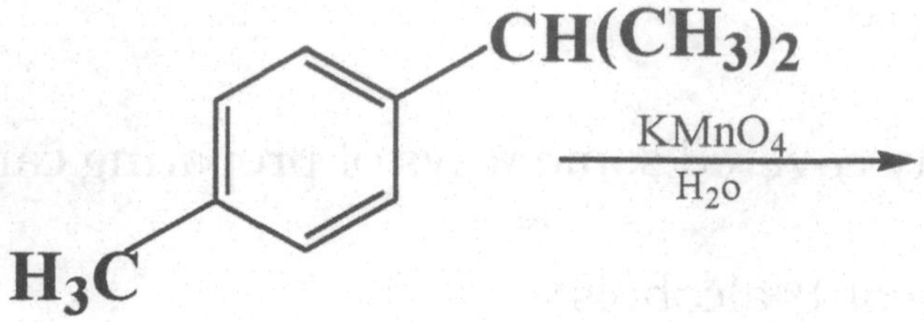

4. **OXIDATIVE CLEAVAGE OF ALKYNES**

- The catalysts: O_3/H_2O or $KMnO_4/H_3O^+$. **Carboxylic acids are produced for internal alkynes. For terminal alkynes, a carboxylic acid and CO_2 are produced. (See Chapters 11 and 12). (see Unit 1)**

$$\mathbf{R-C\equiv C-R'} \xrightarrow[H_3O^+]{KMnO_4}$$

or

$$\mathbf{R-C\equiv C-R'} \xrightarrow[2.\ H_2O]{1.\ O_3}$$

Ex:

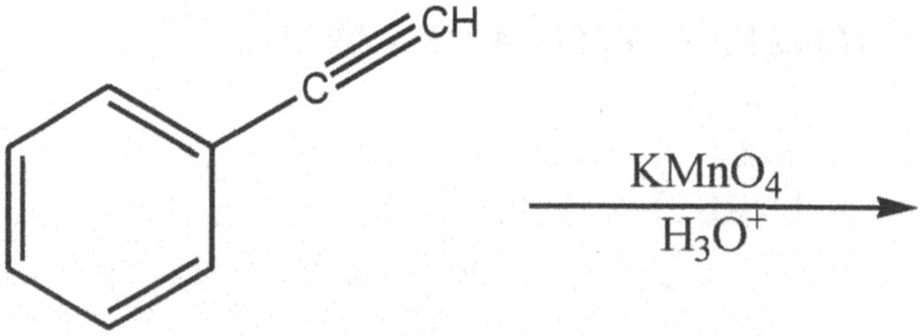

5. OXIDATIVE CLEAVAGE OF ALKENES WITH ACIDIC OR NEUTRAL $KMnO_4$

- The product of the reaction depends on the alkene:
  - If there are **Hs on the carbons** of the double bond, 2 **carboxylic acids** are formed.
  - If there are **2 Hs on a carbon** of the double bond, CO_2 is formed.
  - If there is no H present on the double bond carbons, **2 ketones are formed.**
- **The general reaction is (See Unit 1):**

R(H)C=C(H)R' $\xrightarrow[H_3O^+]{KMnO_4}$

or

R(H)C=CH₂ $\xrightarrow[H_3O^+]{KMnO_4}$

or

R₂C=CR'₂ $\xrightarrow[H_3O^+]{KMnO_4}$

Ex:

C₆H₅–CH=CH₂ $\xrightarrow[H_3O^+]{KMnO_4}$

$$\xrightarrow[H_3O^+]{KMnO_4}$$

6. OXIDATION OF ALDEHYDES USING TOLLEN'S REAGENT

- **The general reaction is:**

Tollen's reagent

$$R-C(=O)H \xrightarrow[NH_4OH]{Ag_2O} R-C(=O)OH + Ag$$

silver mirror

- **Note: This reaction is used as a test for aldehydes.**

Ex:

$$CH_3CH_2-C(=O)H \xrightarrow[NH_4OH]{Ag_2O}$$

- **Read pages ______ - _____.**

- **Do Problems ______, ______, page _____.**

7. SYNTHESIS OF CARBOXYLIC ACIDS: A SUMMARY

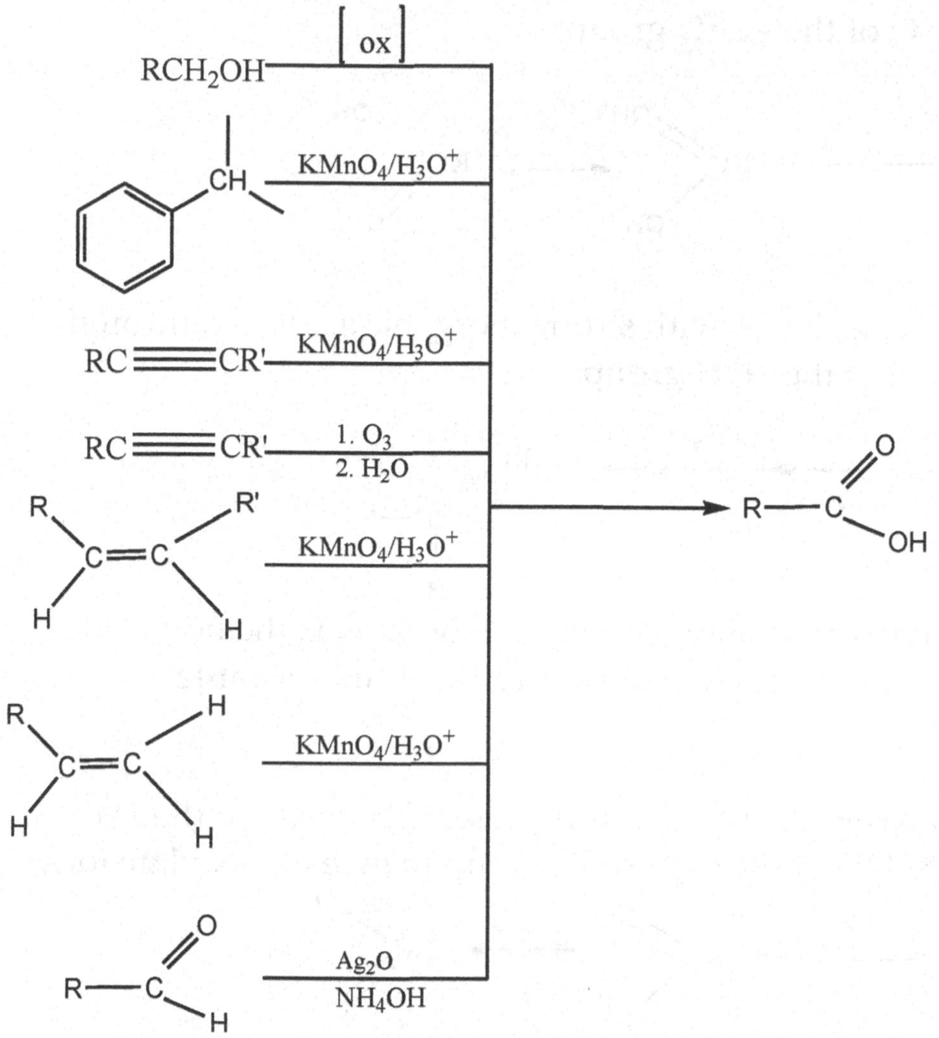

RCH_2OH — [ox] →

Alkylbenzene (CH) — $KMnO_4/H_3O^+$ →

$RC\equiv CR'$ — $KMnO_4/H_3O^+$ →

$RC\equiv CR'$ — 1. O_3 2. H_2O →

R(H)C=C(H)R' — $KMnO_4/H_3O^+$ →

R(H)C=CH$_2$ — $KMnO_4/H_3O^+$ →

R—CHO — Ag_2O, NH_4OH →

→ R—COOH

G. REACTIONS OF CARBOXYLIC ACIDS: AN INTRODUCTION

1. INTRODUCTION

- Carboxylic acids can react in **5 ways**:
  - As **Brønsted-Lowry acids** through the O-H bond.

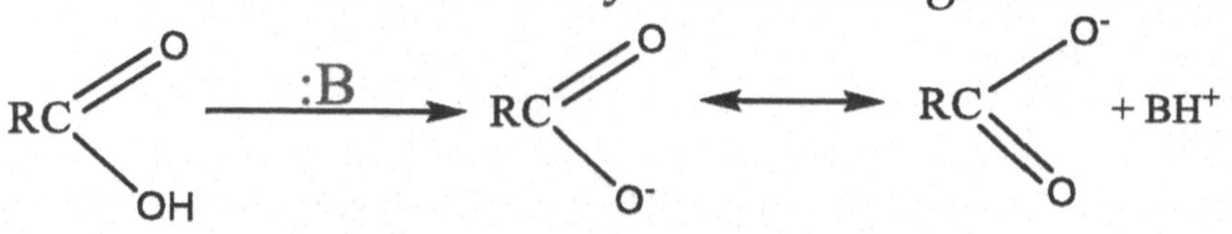

$RCOOH \xrightarrow{:B} RCOO^- \longleftrightarrow RCOO^- + BH^+$

- As **weak bases** with strong inorganic acids HA through the O of the -C=O group.

$$RC(=O)OH \xrightarrow{HA} RC(=OH^+)OH \longleftrightarrow RC(OH)(=OH^+) + A^-$$

- As **weak bases** with strong inorganic acids HA through the O of the **-OH** group.

$$RC(=O)OH \xrightarrow{HA} RC(=O)O^+H_2 + A^-$$

- **Conclusion**: Protonation at the O of the C=O is the best route since the product is **resonance stabilized (more stable product)**.

  - As **weak acids** with strong bases OH^- through the **H** of the **-OH** of the carboxylic group to give carboxylate ions.

$$RC(=O)OH \xrightarrow{OH^-} RC(=O)O^- \longleftrightarrow RC(O^-)(=O) + H_2O$$

  - Can undergo **nucleophilic attack** at the electrophilic C of the carboxylic group.

$$RC(=O)OH \xrightarrow{:Nu} RC(O^-)(Nu)-OH$$

- Read pages ______ - ______.

- Do Problem _______, page ______; Problems ______, and ________, page ______.

2. REACTIONS OF CARBOXYLIC ACIDS: A SUMMARY

$$RCOOH \xrightarrow{:B} RCO_2^- \longleftrightarrow RCO_2^- + BH^+$$

$$RCOOH \xrightarrow{HA} RC(=OH^+)OH \longleftrightarrow RC(OH)(=OH^+) + A^-$$

$$RCOOH \xrightarrow{HA} RC(=O)O^+H_2 + A^-$$

$$RCOOH \xrightarrow{OH^-} RC(=O)O^- \longleftrightarrow RC(O^-)(=O) + H_2O$$

$$RCOOH \xrightarrow{:Nu} RC(O^-)(OH)Nu$$

3. CARBOXYLIC ACIDS AS BRØNSTED- LOWRY ACIDS

- **The general reaction is:**

$$R\!-\!\overset{\delta^+}{C}(=O^{\delta^-})\!-\!\underset{pKa_1}{O\!-\!H} + :B \rightleftharpoons R\!-\!\overset{\delta^+}{C}(=\overset{\delta^-}{O})\!-\!O^- + \underset{pKa_2}{HB^+}$$

- If **$pKa_1 < pKa_2$,** RCOOH is less stable than **HB^+**. The reaction **goes as written.**
- If **$pKa_1 > pKa_2$,** RCOOH is more stable than HB^+. The reaction **will not go as written.**

Ex:

$$CH_3COOH + HCO_3^- \longrightarrow CH_3COO^- + H_2CO_3$$

$pKa_1 = 4.8$ $pKa_2 = 6.3$

$$H_2CO_3 + C_6H_5COO^- \longrightarrow HCO_3^- + C_6H_5COOH$$

$pKa_1 = 6.3$ $pKa_2 = 4.2$

cocaine hydrochloride (Cl⁻) + HCO_3^- $\xrightarrow{\Delta}$ cocaine (crack) + H_2CO_3

$pKa_1 = 8.59$ $pKa_2 = 6.3$

$$R_2NH_2^+ + HCO_3^- \xrightarrow{\Delta} R_2NH + H_2CO_3$$

$pKa_1 = 8.59$ $pKa_2 = 6.3$

- **See Table ______, page _____: common bases used as B.**
- **Note:**
  - **The smaller the pKa, the more reactive the acid.**
  - **The carboxylate ion is resonance stabilized. The more stable the caboxylate ion, the more acidic its conjugate carboxylic acid.**
- **See Fig. _____ on page _____.**
- **See Fig. _____ on page _____.**
- **Do problems on pages _____ - _____.**
- **Read pages _____ - _____.**

4. EWG AND EDG IN ACIDITY OF ALIPHATIC CARBOXYLIC ACIDS

a. **EWG= Halogens**

- They **stabilize the carboxylate ion; they make the acid more acidic.**

Ex:

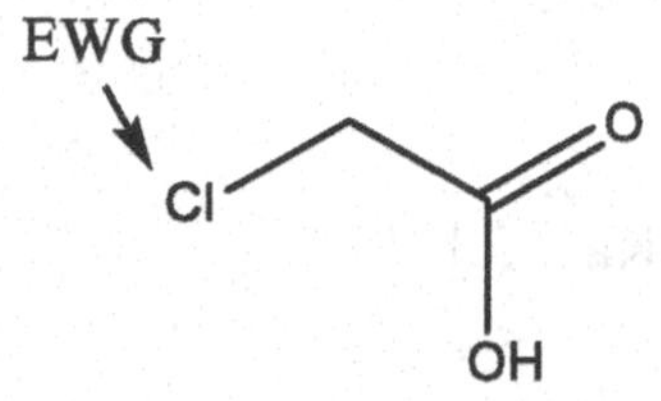

pKa = 2.8

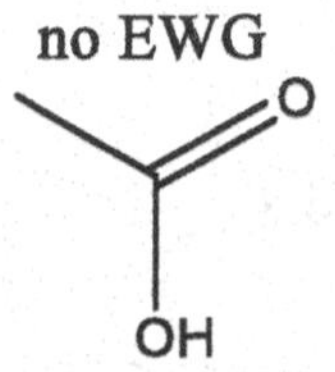

pKa = 4.8

- **Note: The more the EWG on the acid, the more acidic the acid. Furthermore, the closer the EWG to the COOH group, the more acidic the acid.**

Ex:

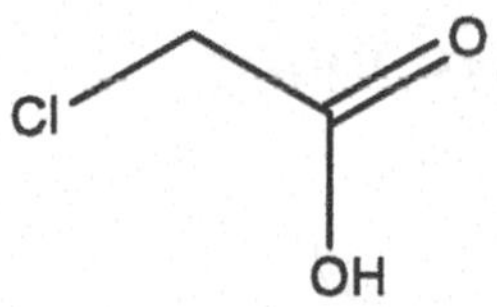

pKa = 2.8

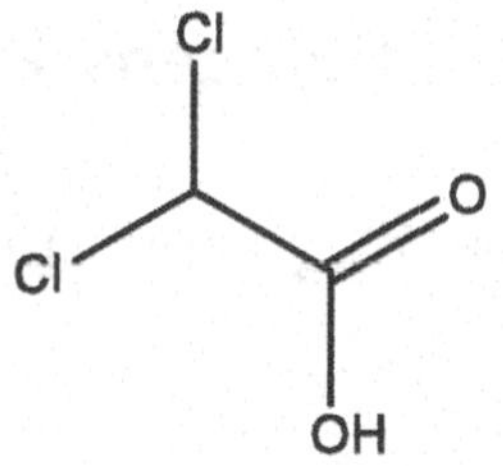

pKa = 1.3

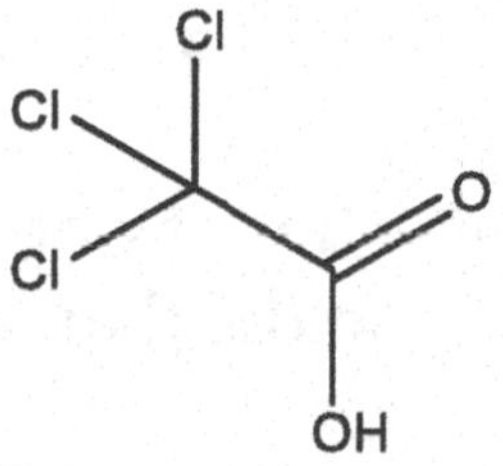

pKa = .90

b. EDG= Alkyl Groups

- They **destabilize the carboxylate ion**; they make the acid **less acidic.**

Ex:

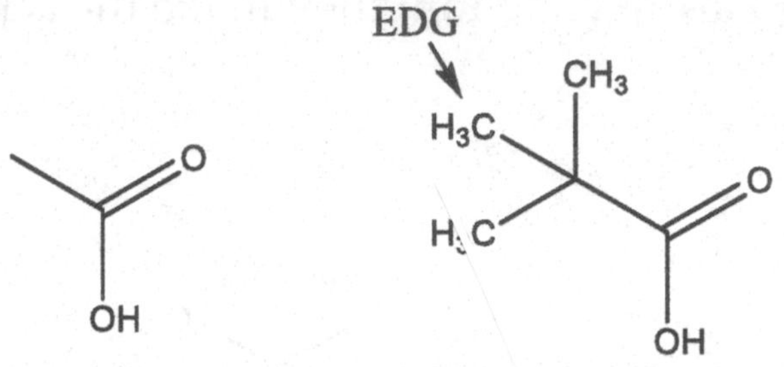

EDG

CH_3

H_3C

O

H_3C

O

OH

OH

pKa = 4.8

pKa = 5.1

Note: The more the EDG on the acid, the less acidic the acid

- **See Fig. on page _______.**

- **Read pages _____ - _____.**

- **Do problems on pages ______ - _____.**

5. EWG AND EDG EFFECTS IN ACIDITY OF SUBSTITUTED BENZOIC ACIDS

a. EWG = Nitrates, Halogens, etc.

- They **stabilize the benzoate ion**; they make the **acid more acidic** than benzoic acid.

Ex:

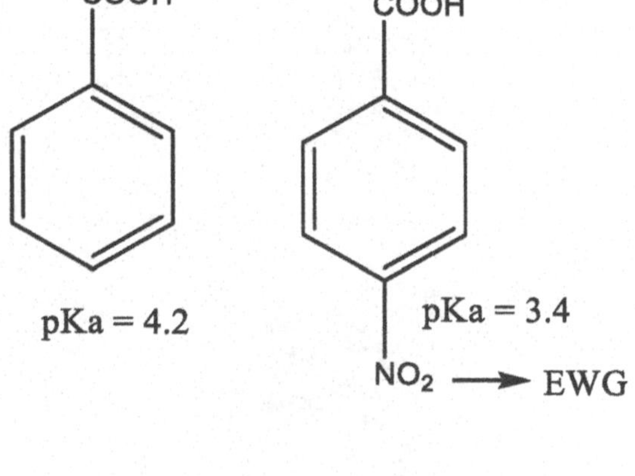

COOH

COOH

pKa = 4.2

pKa = 3.4

NO_2 ⟶ EWG

- **Note: The more the EWG on the acid, the more acidic the acid.**

b. EDG= Alkyl Groups and Others

- They destabilize the benzoate ion; they make the **acid less acidic** than benzoic acid.

Ex:

COOH COOH

pKa = 4.2 pKa = 4.48

OH → EDG

- **Conclusion on Group Effec**

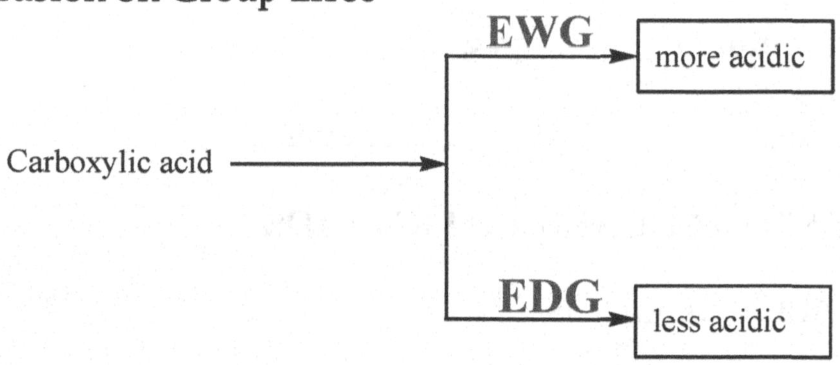

- **See Fig. ________, page ______.**

- **Read pages ______ - ______.**
- **Do problems on pages _______ - _____.**

H. EXTRACTION OF CARBOXYLIC ACIDS FROM ALCOHOLS

Here are the steps:

1. **Convert acid to a water soluble carboxylate salt by adding 10% NaOH or $NaHCO_3$.**
2. **Add to the aqueous solution an organic solvent (CH_2Cl_2, $CH_3CH_2OCH_2CH_3$, hexane, etc.) in a separatory funnel**
3. **The alcohol goes to the organic solvent.**

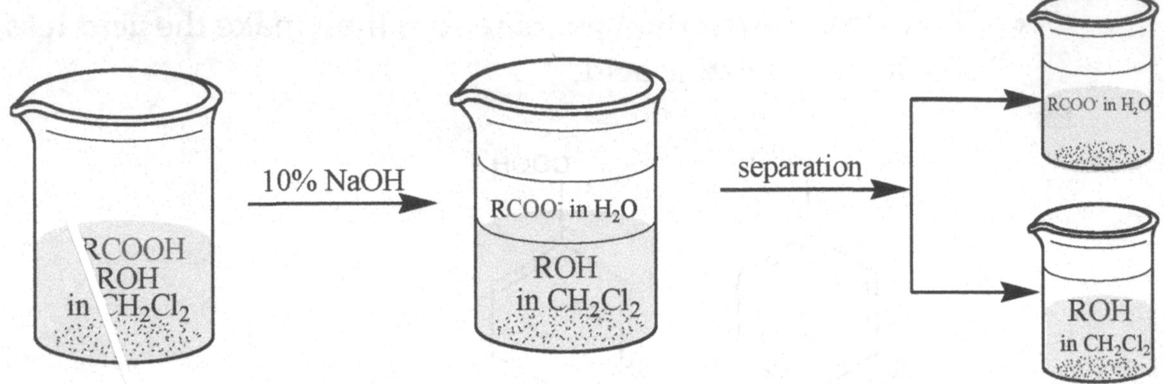

- Read pages ______ - ______.
- Do problem ______, page ______.
- See Figs. ______ and ______, pages ______ - ______.

I. SOME SPECIAL CARBOXYLIC ACIDS

1. SULFONIC ACIDS

General structure:

$$RSO_3H$$

These acids are very strong acids (pKa = -1) because their conjugate bases are **resonance stabilized.**

Ex:

$H_3C-C_6H_4-S(=O)_2-O-H$ = TsOH

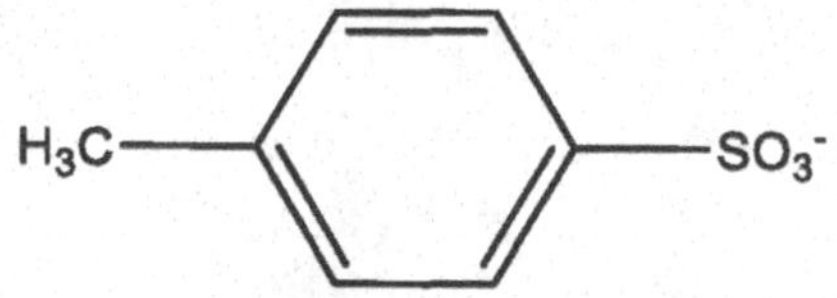

$H_3C-C_6H_4-SO_3^-$

The tosylate ion (TsO^-)

Recall: TsO^- a very good leaving group since it is resonance stabilized.

- Read page ______.
- Do Problem ______, page ______.

2. AMINO ACIDS = BIOMOLECULES

   a. General Structure

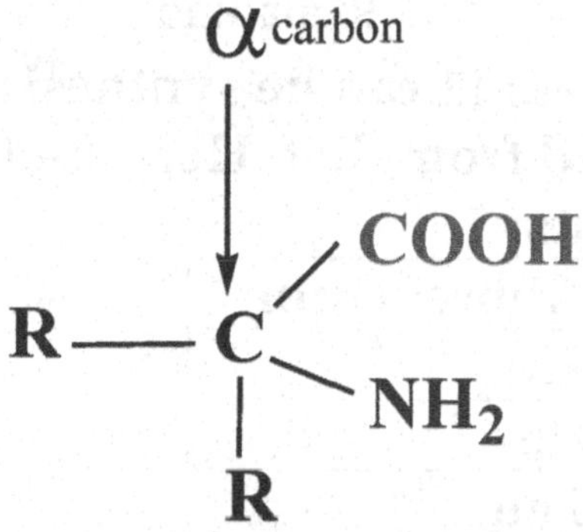

α carbon

R—C(R)(COOH)(NH₂)

Ex: **α-amino acid**

α carbon

H—C(H)(COOH)(NH₂)

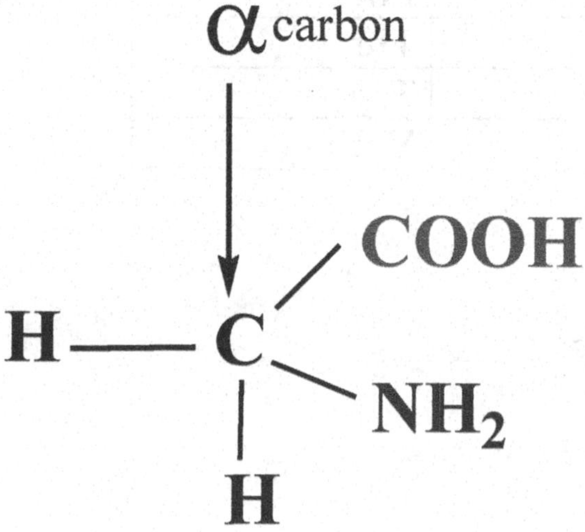

α-amino acid = glycine

Ex: alanine.

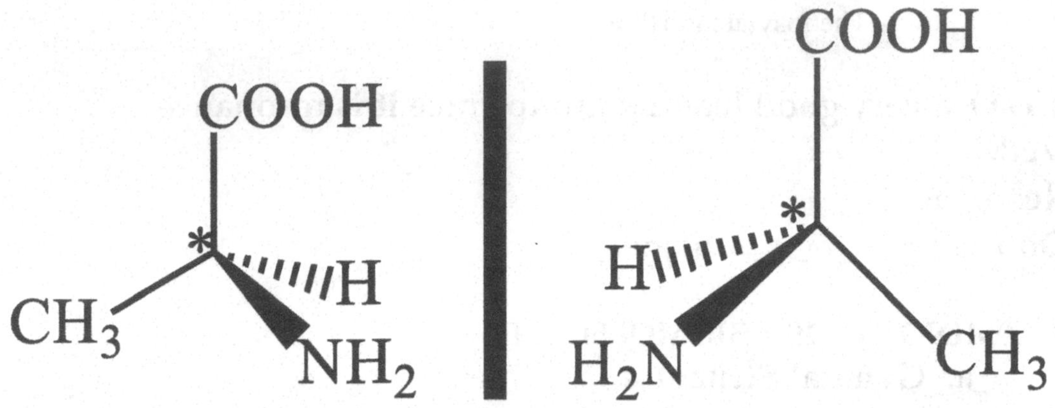

S-alanine R-alanine

- **Note: Human beings need 20 aa; 11 can be synthesized by the body. Essential aa are obtained from diet. Here are the essential amino acids for adults:**

Name	Abbreviation
Histidine	His
Isoleucine	Ile
Leucine	Leu
Lysine	Lys
Methionine	Met
Phenylalanine	Phe
Threonine	Thr
Tryptophan	Trp
Valine	Val

- **See Table _______, page ______.**

- **Do Problem ______, page ______.**

b. Acid-Base Properties of aa

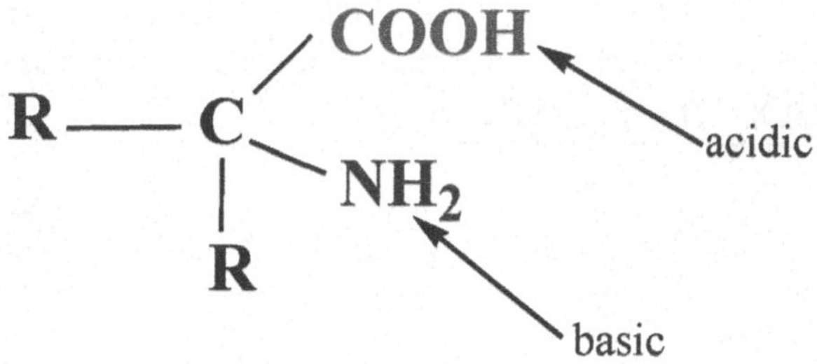

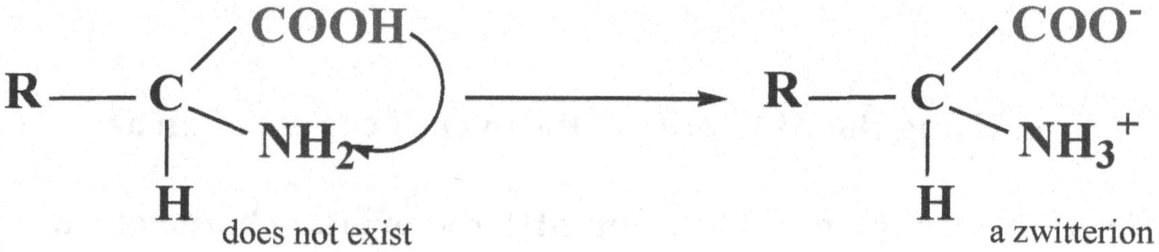

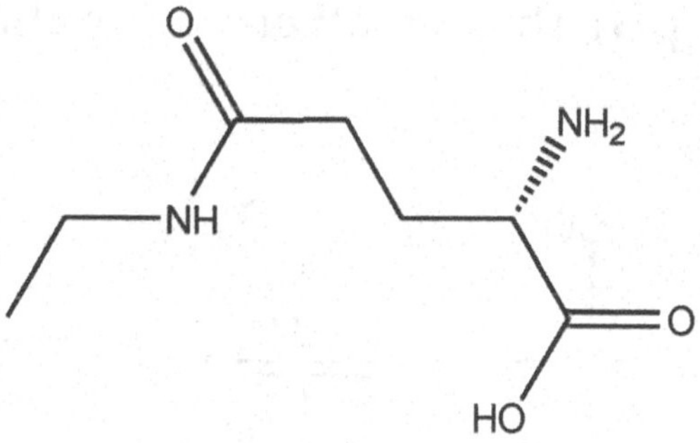

$R-C(R)(COOH)(NH_2)$

COOH — acidic

NH_2 — basic

$R-CH(COOH)(NH_2)$ → $R-CH(COO^-)(NH_3^+)$

does not exist

a zwitterion

L-Theanine

c. Isoelectric Point = pI

- The pI of an aa is the **pH** at which the amino acid exists mainly as a neutral **zwitterion.**

Ex: alanine pI = 6.11.

d. Calculation of Isoelectric Point, pI

- The pI is calculated as follows:

$$pI = \frac{pKa(COOH) + pKa(NH_3^+)}{2}$$

Ex: alanine: pKa(-COOH) = 2.35; pKa($-NH_3^+$) = 9.87. What is pI?

e. Using the pI to predict the overall charge of an aa

- **At pH above the pI (zwitterion pH), the overall charge of the aa is -1.**
- **At pH below the pI (zwitterion pH), the overall charge of the aa is +1.**
- **At pH = pI (zwitterion pH), the overall charge of the aa is 0.**

Ex: Glycine.

$H-C(COOH)(NH_3^+)-H$ $\underset{H^+}{\overset{OH^-}{\rightleftharpoons}}$ $H-C(COO^-)(NH_3^+)-H$ $\underset{H^+}{\overset{OH^-}{\rightleftharpoons}}$ $H-C(COO^-)(NH_2)-H$

pH<pI, overall charge = +1 | pH=pI, overall charge = 0 | pH>pI, overall charge = -1

- **See Fig. ________, page _______.**
- **Do problem on page ________.**
- **Do problems on page _______.**

- **Read page ______.**

f. A word about D and L aa

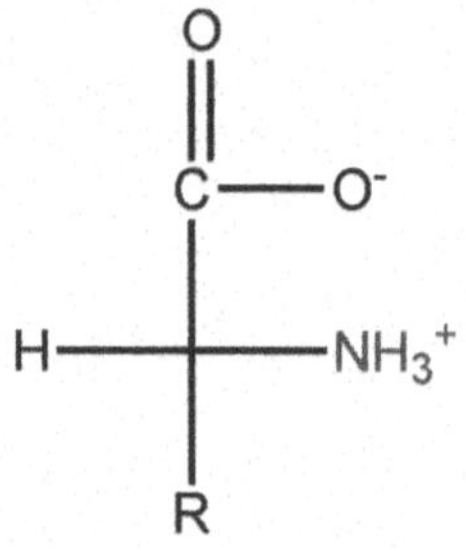

D- amino acid

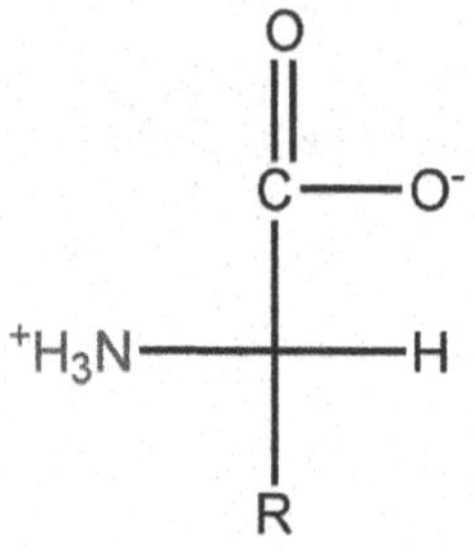

L-amino acid

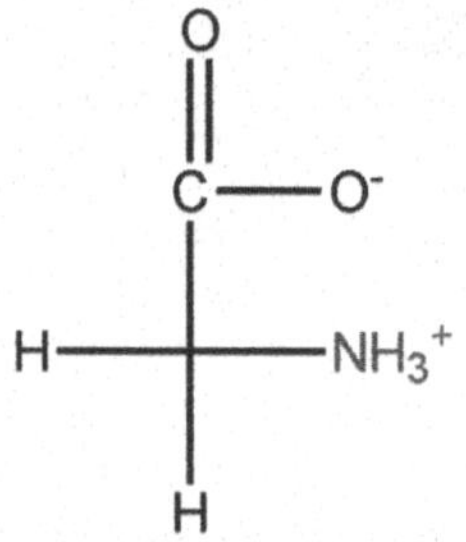

D- glycine

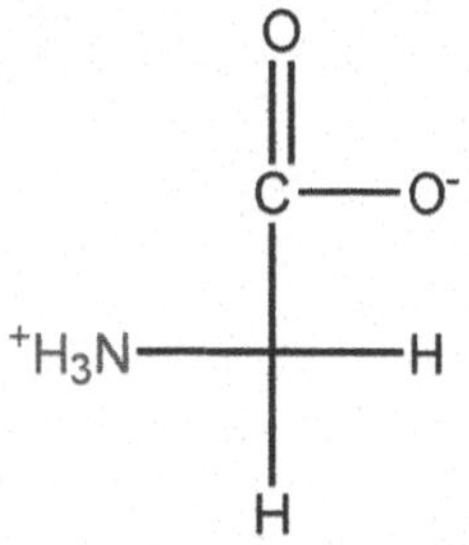

L-glycine

- **See Key Concepts on page ________.**

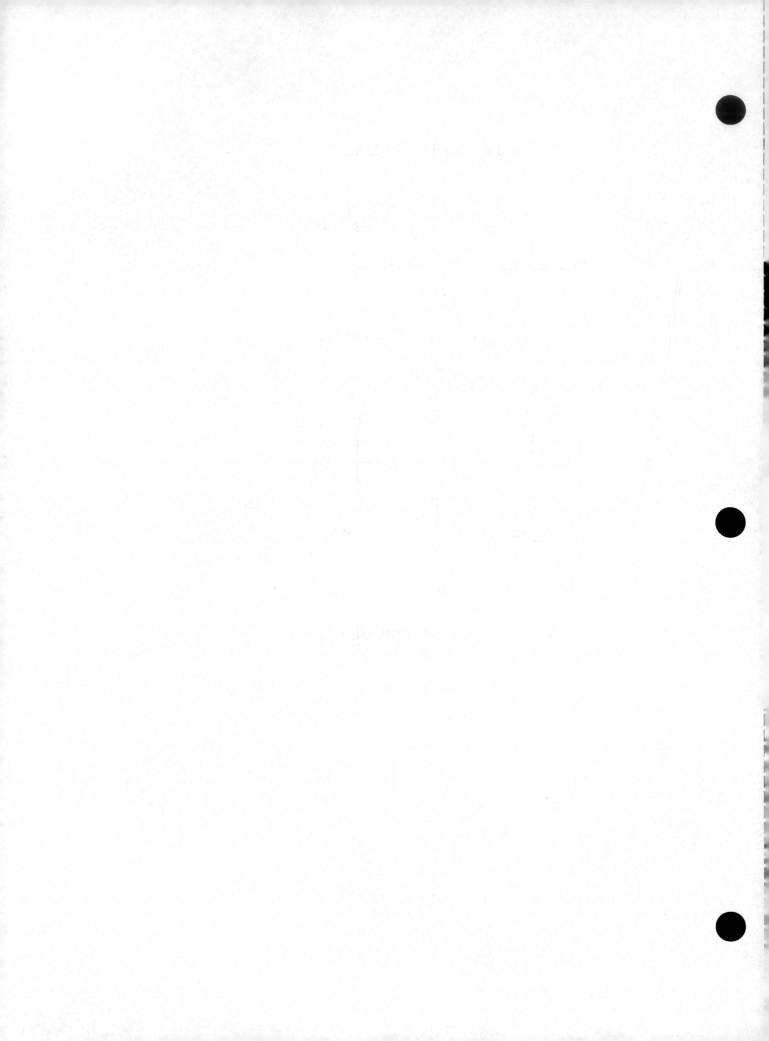

OCHEM II UNIT 10: INTRODUCTION TO THE CHEMISTRY OF THE CARBONYL GROUP

A. INTRODUCTION

- There are **two types** of organic compounds that contain the carbonyl group:

$$-\overset{\overset{\large O}{\|}}{C}-$$

  - Compounds in which we have only H and C bonded to the -CO- group.

Ex: aldehydes and ketones

$$R-\overset{\overset{\large O}{\|}}{C}-R$$

a ketone

$$R-C\begin{smallmatrix} \nearrow\!\!\!\!\!\!= O \\ \searrow H \end{smallmatrix}$$

an aldehyde

Ex:

$$CH_3-\overset{\overset{\large O}{\|}}{C}-CH_2CH_3$$

a ketone

$$CH_3-C\begin{smallmatrix} \nearrow\!\!\!\!\!\!= O \\ \searrow H \end{smallmatrix}$$

an aldehyde

- Compounds in which there is an **electronegative atom** bonded to the -CO- group.

Ex: carboxylic acids and derivatives

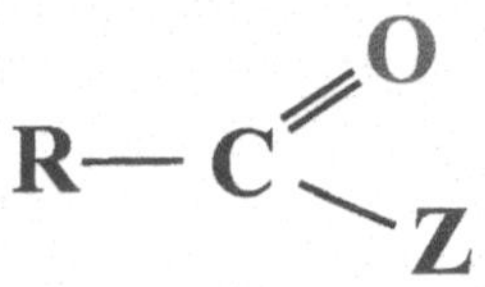

R—C(=O)—Z
a carboxylic acid derivative

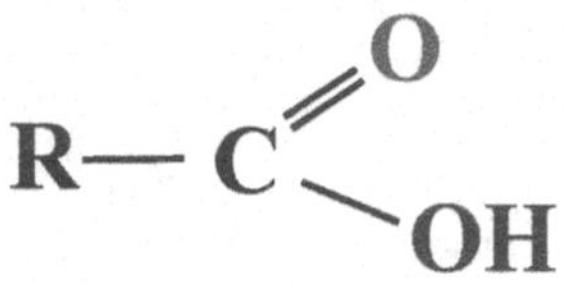

R—C(=O)—OH
a carboxylic acid

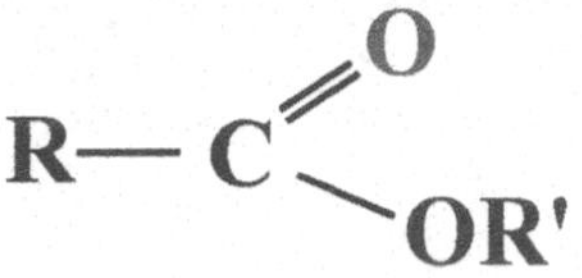

R—C(=O)—OR'
an ester

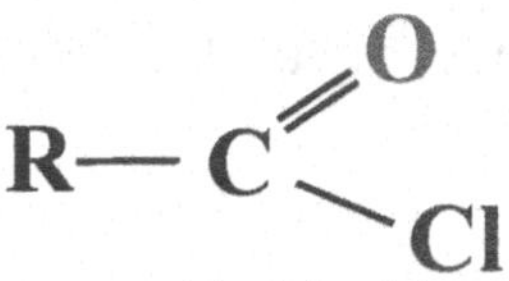

R—C(=O)—Cl
an acid chloride

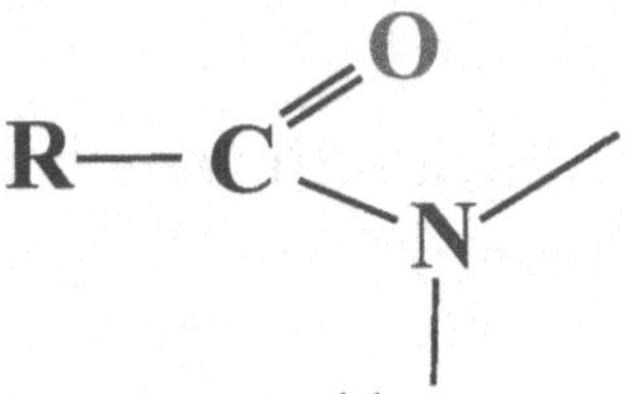

R—C(=O)—N<
an amide

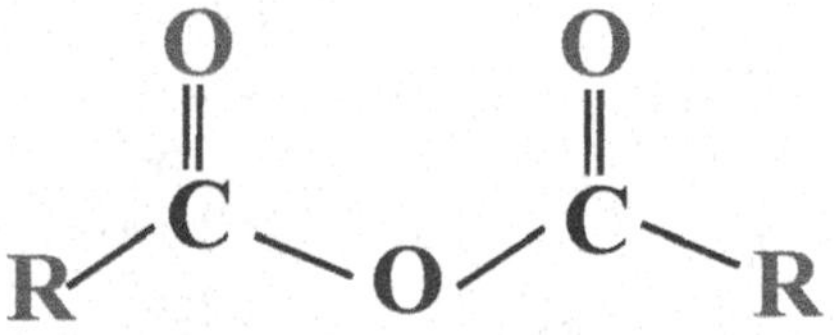

R—C(=O)—O—C(=O)—R
an acid anhydride

B. STRUCTURE OF THE CARBONYL GROUP

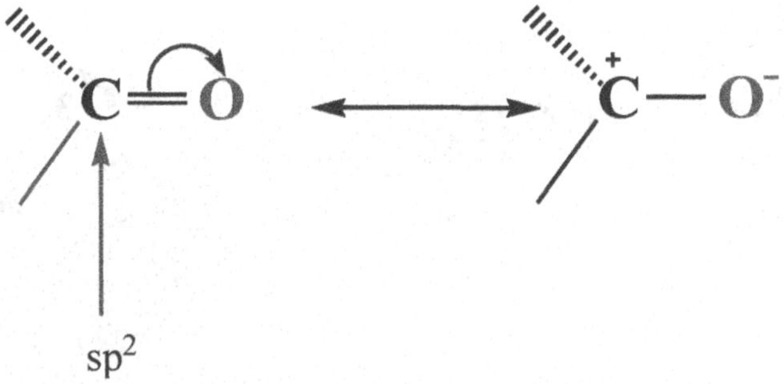

C=O ⟷ $\overset{+}{C}$—O^-

sp^2

- **Read pages** _______ - ______.

C. GENERAL REACTIONS OF THE CARBONYL GROUP

1. INTRODUCTION

- Reactions can occur at the O or at the C.

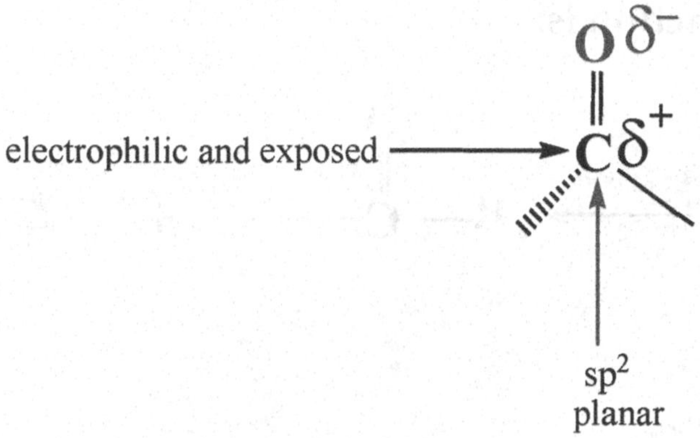

$O^{\delta-}$ (double bond to) $C^{\delta+}$

electrophilic and exposed → $C^{\delta+}$

sp^2
planar

2. ALDEHYDES AND KETONES

- They undergo **nucleophilic addition reactions.**
- **The general reaction is:**

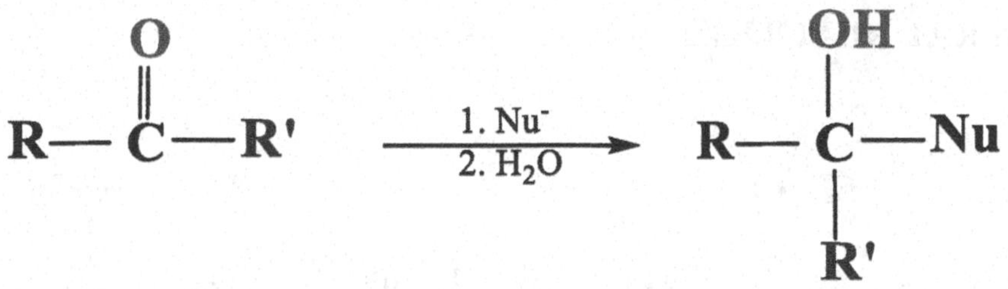

$$R{-}\overset{\overset{\displaystyle O}{\|}}{C}{-}R' \xrightarrow[2.\ H_2O]{1.\ Nu^-} R{-}\underset{\underset{\displaystyle R'}{|}}{\overset{\overset{\displaystyle OH}{|}}{C}}{-}Nu$$

Ex:

$$CH_3CH_2\overset{\overset{\displaystyle O}{\|}}{C}{-}H \xrightarrow[2.\ H_2O]{1.\ HS^-}$$

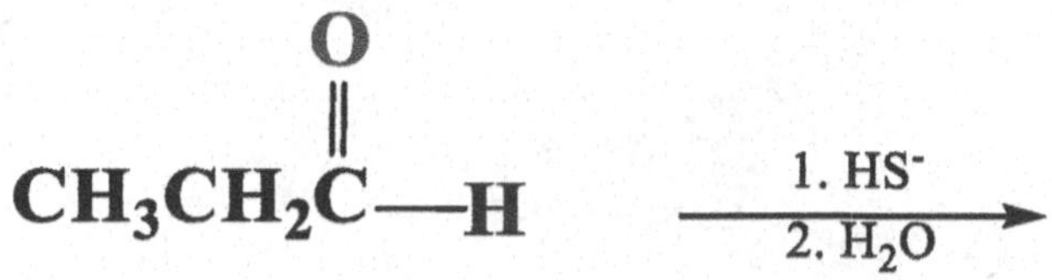

3. COMPOUNDS OF THE FORM RCOZ (CARBOXYLIC ACID, ESTERS, AMIDES, ACID OR ACYL CHLORIDES): Z = LEAVING GROUP

- They undergo **nucleophilic substitution reactions.**
- **The general reaction is:**

$$R-\overset{\overset{\displaystyle O}{\|}}{C}-Z \xrightarrow{:Nu^-} R-\overset{\overset{\displaystyle O}{\|}}{C}-Nu \quad + \quad Z^-$$

Z = OH, Cl, OR, NH_2

Ex:

$$CH_3CH_2\overset{\overset{\displaystyle O}{\|}}{C}-Cl \xrightarrow{OH^-}$$

D. NUCLEOPHILIC ADDITION TO ALDEHYDES AND KETONES

1. GENERAL REACTION

$$R-\overset{\overset{\displaystyle O}{\|}}{C}-R' \xrightarrow[2.\ H_2O]{1.\ Nu^-} R-\underset{\underset{\displaystyle R'}{|}}{\overset{\overset{\displaystyle OH}{|}}{C}}-Nu$$

Ex:

$$CH_3CH_2\overset{\overset{\displaystyle O}{\|}}{C}-CH_3 \xrightarrow[2.\ H_2O]{1.\ CN^-}$$

2. MECHANISM: 2 STEPS

- Read pages _____ - _____.
- See page _____.

E. NUCLEOPHILIC SUBSTITUTION AT THE CARBONYL CARBON FOR RCOZ COMPOUNDS

1. GENERAL REACTION

$$R-\overset{\overset{\large O}{\|}}{C}-Z \xrightarrow{:Nu^-} R-\overset{\overset{\large O}{\|}}{C}-Nu + :Z^-$$

Z = OH, Cl, OR, NH_2

Ex:

$$CH_3CH_2\overset{\overset{\large O}{\|}}{C}-Cl \xrightarrow{SH^-}$$

2. MECHANISM: 2 STEPS

- See page ______.

- Read pages ______ - _____.

- **Note: Reactivity depends on the leaving group Z. The better the leaving group Z, the higher the reactivity of RCOZ. The increasing order of reactivity is:**

$$R-\overset{\overset{\displaystyle O}{\|}}{C}-NH_2 < R-\overset{\overset{\displaystyle O}{\|}}{C}-OH \sim R-\overset{\overset{\displaystyle O}{\|}}{C}-OR' < R-\overset{\overset{\displaystyle O}{\|}}{C}-Cl$$

- **Note: Aldehydes and ketones do not undergo nucleophilic substitution reactions because H and R are very poor leaving groups.**

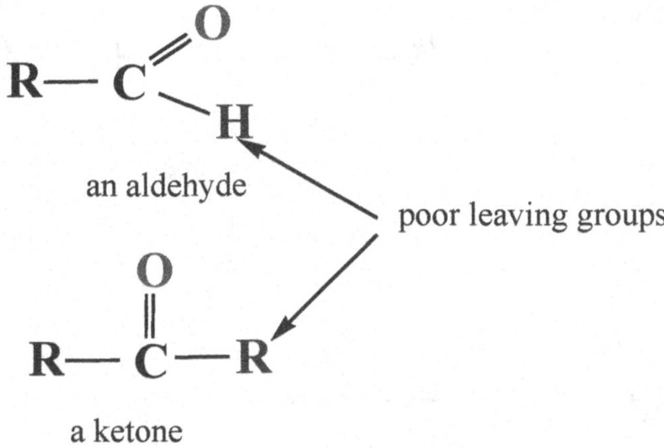

- Do problems on page ______.

F. OXIDATION OF CARBONYL CONTAINING COMPOUNDS

1. OXIDATION REVISITED

- In OChem, an oxidation is a reaction in which there is an **increase** in the number of C-Z bonds (O, halogens) or a **decrease** in the number of C-H bonds.

- See Chapter _____.

Ex:

$$CH_3C(=O)H \xrightarrow{[O]} CH_3C(=O)OH$$

2. OXIDATION OF THE -CO- GROUP

- Carboxylic acids and derivatives are **already** oxidized. Therefore, they **do not undergo oxidation reactions**. However, aldehydes do undergo oxidation reactions to give **carboxylic acids.**

3. OXIDATION OF ALDEHYDES

- The general reaction is :

$$R-\overset{\overset{\displaystyle O}{\|}}{C}-H \xrightarrow{[O]} R-\overset{\overset{\displaystyle O}{\|}}{C}-OH$$

a carboxylic acid

[O] = CrO_3, $Na_2Cr_2O_7$, $K_2Cr_2O_7$, $KMnO_4$, or Ag_2O/NH_4OH

Ex:

$$CH_3C(=O)H \xrightarrow{[O]} CH_3C(=O)OH$$

- Ag_2O/NH_4OH = Tollen's Reagent; a qualitative test for aldehydes = a silver mirror (See Unit 9.)

$$RC(=O)H \xrightarrow[NH_4OH]{Ag_2O} RC(=O)OH + Ag\uparrow$$

silver mirror

- Read pages ______ - ______. Do problems on page ______.

G. REDUCTION REACTIONS OF THE CARBONYL GROUP

1. REDUCTION REVISITED

- In OChem, a reduction reaction is a process in which there is a **decrease** in the number of C-Z bonds (O, halogens) or an **increase** in the number of C-H bonds.

- See Chapter ______.

Ex:

$$CH_3CH_2\overset{O}{\overset{\|}{C}}CH_3 \xrightarrow[Pd]{H_2} CH_3CH_2\overset{OH}{\overset{|}{C}}HCH_3$$

2. REDUCTION OF THE -CO- GROUP

- **All** the carbonyl containing compounds covered in this chapter do undergo reduction reactions. The most common **reducing agents** used are:

    - **$NaBH_4$ and $LiAlH_4$: Sources of H^-, a Nucleophile**

    - **H_2/ Pd-C**

3. REDUCTION OF ALDEHYDES AND KETONES WITH $NaBH_4$ AND $LiAlH_4$

a. Aldehydes

$$R-\overset{\overset{\displaystyle O}{\|}}{C}-H \xrightarrow[\text{OR 1. } LiAlH_4/2.\ H_2O]{NaBH_4/CH_3OH} R-\underset{\underset{\displaystyle H}{|}}{\overset{\overset{\displaystyle OH}{|}}{C}}-H$$

a 1° alcohol

Ex:

- See Mechanism on page ______.

b. Ketones

Ex:

$$CH_3CH_2\overset{\overset{\displaystyle O}{\|}}{C}-CH_3 \xrightarrow[2.\ H_2O]{1.\ LiAlH_4} CH_3CH_2\underset{\underset{\displaystyle H}{|}}{\overset{\overset{\displaystyle OH}{|}}{C}}-CH_3$$

- See Mechanism on page ________.
- Do problems on page ______.

4. HYDROGENATION OF ALDEHYDES AND KETONES

- The general reaction is:

$$R-\overset{\overset{\displaystyle O}{\|}}{C}-R' \xrightarrow[Pd\text{-}C]{H_2} R-\underset{\underset{\displaystyle R'}{|}}{\overset{\overset{\displaystyle OH}{|}}{C}}-H$$

a 1° or 2° alcohol

- **Note: The reduction of an aldehyde yields a primary alcohol. The reduction of a ketone produces a secondary alcohol.**

Ex:

$$CH_3CH_2\overset{\overset{\displaystyle O}{\|}}{C}CH_3 \xrightarrow[Pd]{H_2} CH_3CH_2\underset{\underset{\displaystyle H}{|}}{\overset{\overset{\displaystyle OH}{|}}{C}}CH_3$$

- **Note: The reaction is not selective. See page ______.**

- **Do problem on page _______.**

- **Read pages _____ - _____. See the Synthesis of Ibuprofen and Muscone in Fig. ______ on page _____.**

5. STEREOCHEMISTRY OF THE REDUCTION OF THE CO GROUP

Recall:

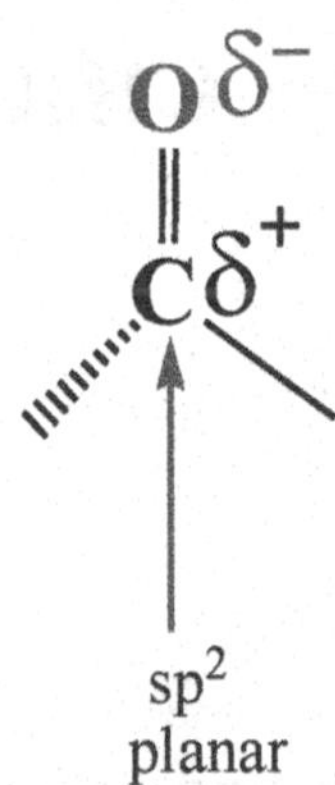

- The H- from the reducing agent can attack the planar sp² carbon on two sides, above and below. As a result, a racemic mixture is obtained during the reaction.

- **Read page _______ and examples.**

- **Do Problem _______, page _______.**

6. ENANTIOSELECTIVITY OR ASYMETRIC REDUCTION

a. Introduction

- If a **chiral reducing agent** is used, then one can get mostly **one enantiomer** as follows:

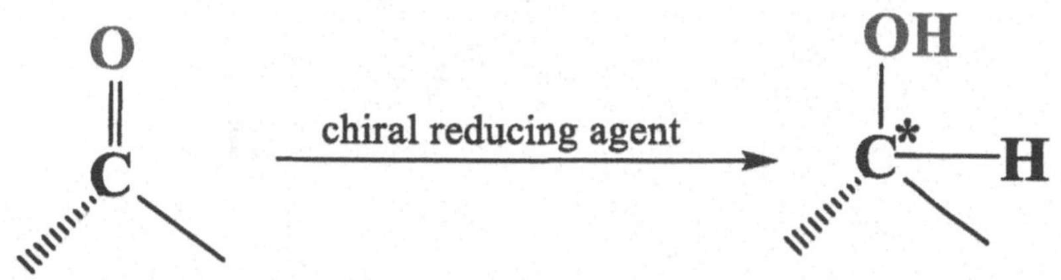

chiral reducing agent

- The chiral reducing agent is **CBS = (R)-CBS or (S)-CBS: Corey-Bakshi-Shibata reagent.**

- See page ________.

- **The general reactions are:**

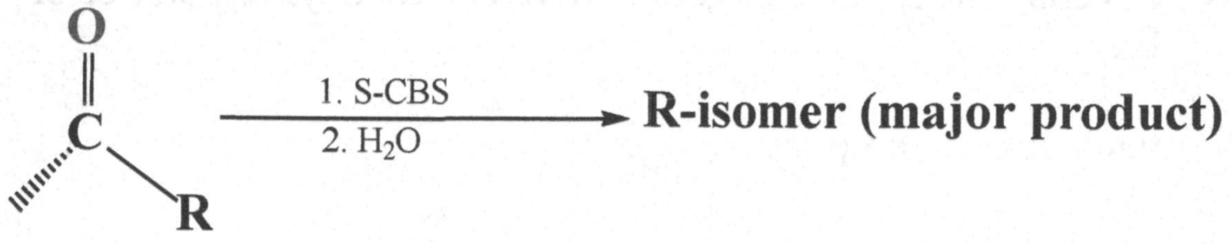

1. S-CBS
2. H_2O

R-isomer (major product)

Ex:

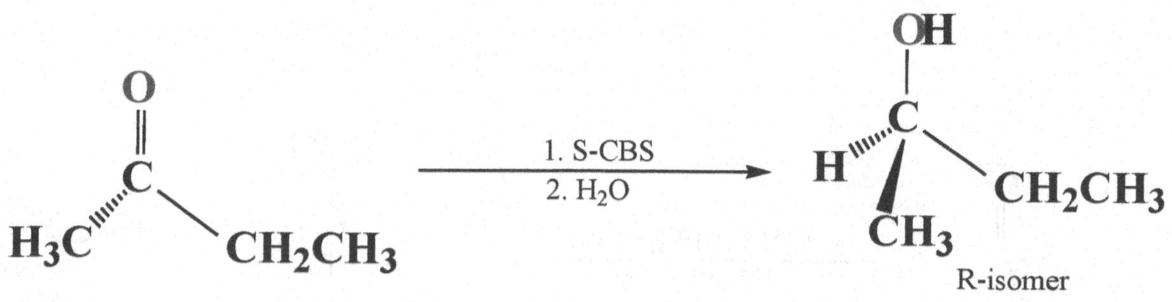

1. S-CBS
2. H_2O

R-isomer

$$\text{R-C(=O)-CH}_3 \xrightarrow[\text{2. } H_2O]{\text{1. R-CBS}} \textbf{S-isomer (major product)}$$

Ex:

$$H_3C\text{-C(=O)-}CH_2CH_3 \xrightarrow[\text{2. } H_2O]{\text{1. R-CBS}} H\text{-C(OH)(}CH_3\text{)(}CH_2CH_3\text{)}$$

S-isomer

- See page ______ for more examples.
- Do Problem ______, page ______.

b. Enantioselective Reduction in the Biological World

- In cells: The reducing agent is NADH a coenzyme = source of H^-.
- See structure on page ______.
- The general reaction is:

$$\text{-C(=O)-R} \xrightarrow[\text{2. } H_2O]{\text{1. NADH/Enzyme}} \text{-C(OH)(H)-} + \textbf{NAD}^+$$

7. REDUCTION OF α, β UNSATURATED ALDEHYDES AND KETONES

a. Structure

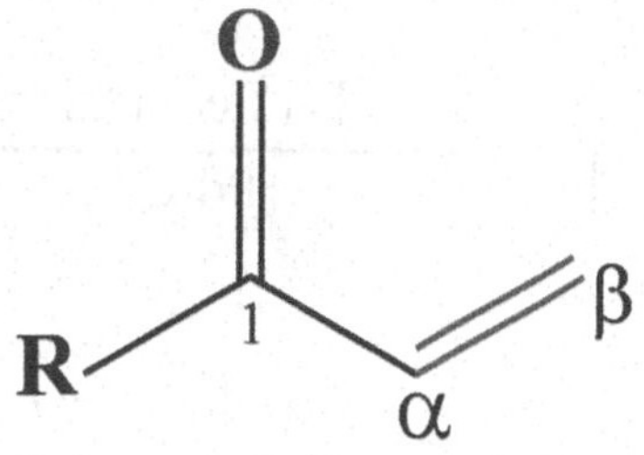

O
R 1 α β

b. Reduction of the CO Only: Use $NaBH_4/CH_3OH$

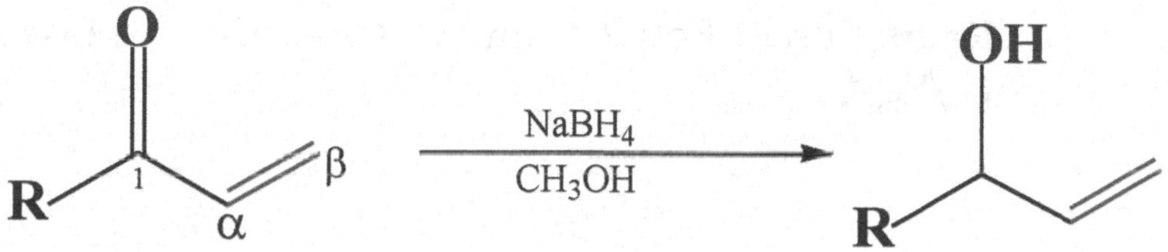

O — R 1 α β → ($NaBH_4$ / CH_3OH) → OH — R

Ex:

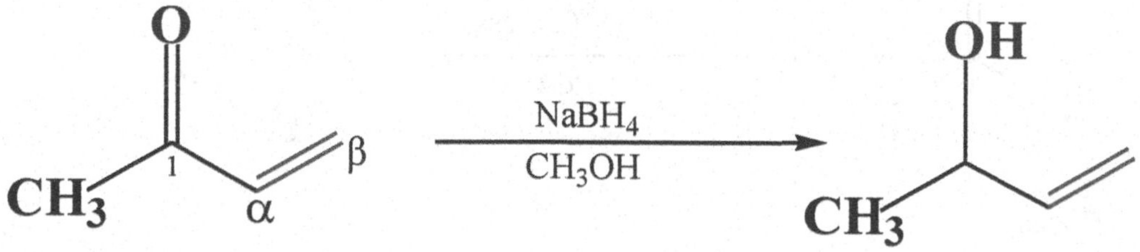

O — CH_3 1 α β → ($NaBH_4$ / CH_3OH) → OH — CH_3

- **Note: The α-β double bond is not reduced.**

c. **Reduction of the C=C Bond Only: Use 1 Equivalent of H_2/Pd-C**

$$R\text{-}C(=O)\text{-}CH_{\alpha}=CH_{2\,\beta} \xrightarrow[\text{Pd-C}]{H_2 \text{ (1 equivalent)}} R\text{-}C(=O)\text{-}CH_2CH_3$$

Ex:

$$CH_3\text{-}C(=O)\text{-}CH_{\alpha}=CH_{2\,\beta} \xrightarrow[\text{Pd-C}]{H_2\text{(1 equivalent)}} CH_3\text{-}C(=O)\text{-}CH_2CH_3$$

d. **Reduction of Both CO and C=C Bonds: Use Excess H_2/Pd-C**

$$R\text{-}C(=O)\text{-}CH_{\alpha}=CH_{2\,\beta} \xrightarrow[\text{Pd-C}]{H_2 \text{ (excess)}} R\text{-}CH(OH)\text{-}CH_2CH_3$$

Ex:

$$CH_3\text{-}C(=O)\text{-}CH_{\alpha}=CH_{2\,\beta} \xrightarrow[\text{Pd-C}]{H_2\text{(excess)}} CH_3\text{-}CH(OH)\text{-}CH_2CH_3$$

H. REDUCTION OF CARBOXYLIC ACIDS AND DERIVATIVES

1. INTRODUCTION

- There are 3 common reducing agents that are used:
  - $LiAlH_4$ = strong reducing agent
  - DIBAL-H = Diisobutylaluminum hydride = milder reducing agent
  - $LiAlH[OC(CH_3)_3]_3$ = Lithium tri-tert-butoxyaluminum hydride = milder reducing agent

- See page ______.

2. REDUCTION OF ACID CHLORIDE

- A stronger reducing agent produces a **primary** alcohol.
- A milder reducing agent produces an **aldehyde.**

A strong reducing agent

$$RCOCl \xrightarrow[2.\ H_2O]{1.\ LiAlH_4} RCH_2OH$$

a 1° alcohol

Ex:

$$CH_3CH_2COCl \xrightarrow[2.\ H_2O]{1.\ LiAlH_4} CH_3CH_2CH_2OH$$

- **Mechanism: 3 steps**

A mild reducing agent

$$\mathrm{R{-}C({=}O){-}Cl} \xrightarrow[\text{2. } H_2O]{\text{1. DIBAL-H}} \mathrm{R{-}C({=}O){-}H}$$

an aldehyde

Ex:

$$\mathrm{CH_3CH_2{-}C({=}O){-}Cl} \xrightarrow[\text{2. } H_2O]{\text{1. DIBAL-H}} \mathrm{CH_3CH_2{-}C({=}O){-}H}$$

- **See page ________ for Mechanism: 2 steps**

3. REDUCTION OF ESTERS

- Same as acid chlorides.
- A stronger reducing agent produces **a primary alcohol and an alcohol.**
- A milder reducing agent produces **an aldehyde and an alcohol.**

A strong reducing agent

$$\mathrm{R{-}C({=}O){-}OR'} \xrightarrow[\text{2. } H_2O]{\text{1. } LiAlH_4} \mathrm{RCH_2OH} + \mathrm{R'OH}$$

a 1° alcohol

Ex:

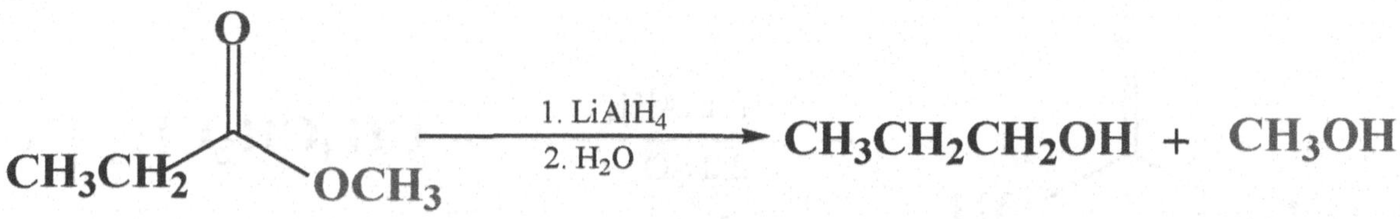

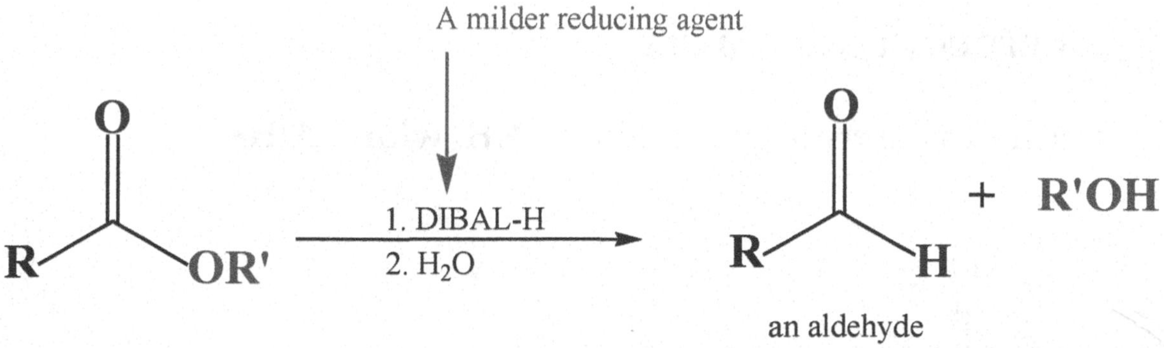

$$CH_3CH_2C(=O)OCH_3 \xrightarrow[2.\ H_2O]{1.\ LiAlH_4} CH_3CH_2CH_2OH + CH_3OH$$

A milder reducing agent

$$RC(=O)OR' \xrightarrow[2.\ H_2O]{1.\ DIBAL\text{-}H} RC(=O)H + R'OH$$

an aldehyde

Ex:

$$CH_3CH_2C(=O)OCH_3 \xrightarrow[2.\ H_2O]{1.\ DIBAL\text{-}H} CH_3CH_2C(=O)H + CH_3OH$$

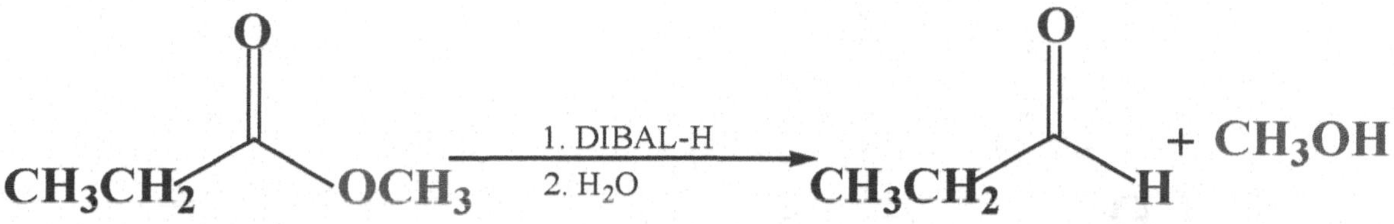

- **See page ________ for Mechanism.**
- **Do problems on page _______.**

4. REDUCTION OF CARBOXYLIC ACIDS

- A **stronger reducing agent** is required. A milder one cannot do the job.

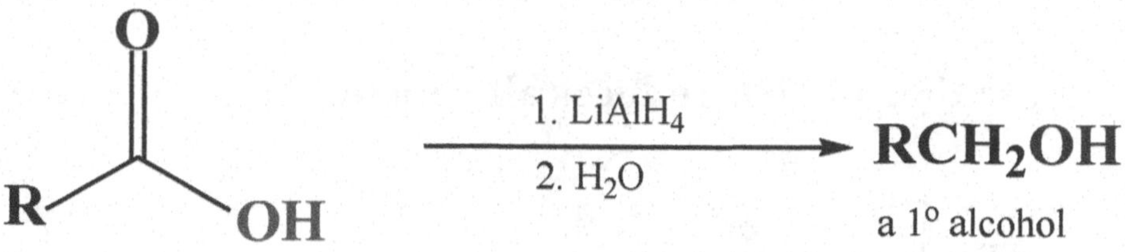

$$RC(=O)OH \xrightarrow[2.\ H_2O]{1.\ LiAlH_4} RCH_2OH$$

a 1° alcohol

Ex:

$$CH_3CH_2C(=O)OH \xrightarrow[2.\ H_2O]{1.\ LiAlH_4} CH_3CH_2CH_2OH$$

5. REDUCTION OF AMIDES

- Amides are reduced to amines, $-NH_2$ with $LiAlH_4$.

$$RC(=O)NH_2 \xrightarrow[2.\ H_2O]{1.\ LiAlH_4} RCH_2NH_2$$

an amine

Ex:

$$CH_3CH_2C(=O)NH_2 \xrightarrow[2.\ H_2O]{1.\ LiAlH_4} CH_3CH_2CH_2NH_2$$

- See mechanism of the reaction on page ____.
- Read page _____.
- Do problems on page _____.
- See Summary of Reducing agents on page ____
- See Table ______, page ______.

- Note that $LiAlH_4$ reduces all functional groups covered here.

- Do Problems page ______.

I. ORGANOMETALLIC REAGENTS

1. INTRODUCTION

- Organometallic reagents are organic compounds that contain a carbon-metal bond. The metal could be **Li, Mg, or Cu, Sn, Si, Tl, Al, Ti, and Hg.** They have a general structure:

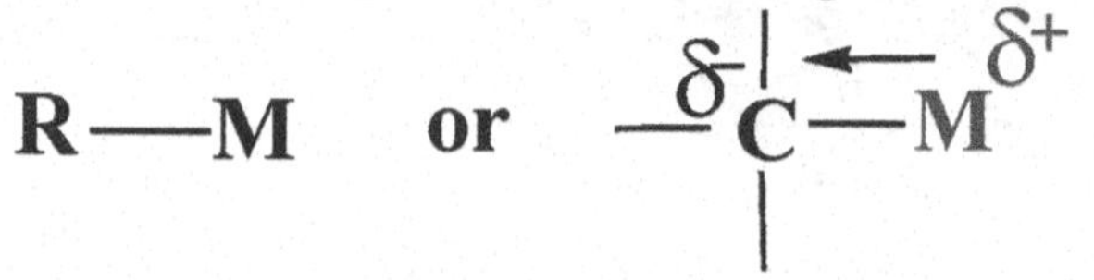

- The three common organometallic reagents are:
  R-Li, R-MgX, and R_2CuLi

- The most reactive are R-MgX and R-Li.
- **RMgX** reagents are called **Grignard reagents (Victor Grignard) = organomagnesium reagents.**
- RLi = organolithium reagents
- R_2CuLi = organocopper or organocuprate reagents.

- **Note: The decreasing order of bond polarity is:**
  C-Li > C-Mg > C-Cu

- **The C-Li is more reactive.**

2. REACTIVITY OF ORGANOMETALLIC REAGENTS

- **An organometallic (OM) reagent reacts like a Lewis base. Therefore, an OM is a nucleophile and reacts thru a carbanion, R^-.**

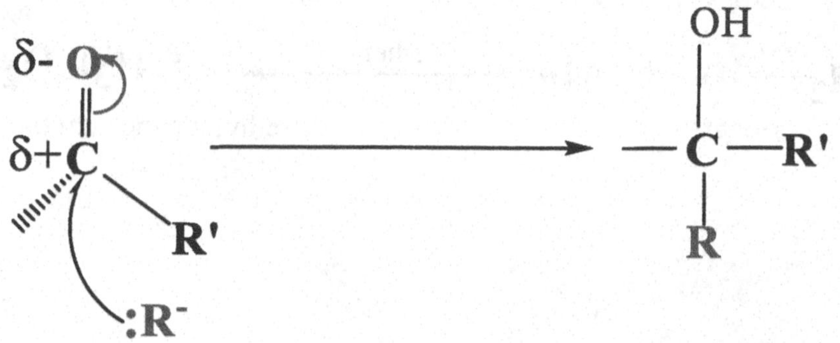

Ex:

$$\overset{\delta-}{O}{=}\overset{\delta+}{C}(CH_3)(H) + CH_3\ddot{C}H_2^- \longrightarrow CH_3{-}C(OH)(H){-}CH_2CH_3$$

3. PREPARATION OF ORGANOMETALLIC REAGENTS

a. Organolithium Reagents

$$R{-}X + 2Li \longrightarrow R{-}Li + LiX$$

X = halogen

an alkyllithium (R—Li)

Ex:

$$CH_3CH_2{-}Br + 2Li \longrightarrow CH_3CH_2{-}Li + LiBr$$

ethyllithium ($CH_3CH_2{-}Li$)

b. Grignard Reagents

$$R{-}X + Mg \xrightarrow{\text{ether}} R{-}MgX$$

X = halogen

an alkyl magnesium halide

Ex:

$$CH_3CH_2{-}Br + Mg \xrightarrow{\text{ether}} CH_3CH_2{-}MgBr$$

X = halogen

ethyl magnesium bromide

c. Coupling Reactions, Organocuprate or Gilman's Reagents

$$R—X + 2Li \longrightarrow R—Li + LiX$$

X = halogen

$$2R—Li + CuI \longrightarrow R_2CuLi + LiI$$

(arrow to Cu) A Gilman's reagent

Ex:

$$CH_3CH_2—Br + 2Li \longrightarrow CH_3CH_2—Li + LiBr$$

$$2CH_3CH_2—Li + CuI \longrightarrow (CH_3CH_2)_2CuLi + LiI$$

(arrow to Cu) Gilman's reagent

4. PREPARATION OF SODIUM REAGENTS: ACETYLIDE ANIONS

a. General Structure

$$RC\equiv C:^- Na^+$$

b. Preparation of Sodium Acetylides

$$RC\equiv CH + Na^+NH_2^- \rightarrow RC\equiv C:^- Na^+ + NH_3$$

Ex:

$$CH_3C\equiv CH + Na^+NH_2^- \rightarrow CH_3C\equiv C:^- Na^+ + NH_3$$

5. PREPARATION OF ORGANOLITHIUM ACETYLIDE ANIONS

a. General Structure

$$RC{\equiv}C{:}^{-}\ Li^{+}$$

b. Preparation of Lithium Acetylides: An Acid-Base Reaction

$$RC{\equiv}CH + CH_3\text{-}Li \rightarrow RC{\equiv}C{:}^{-}\ Li^{+} + CH_3\text{-}H$$

Ex:

$$CH_3C{\equiv}CH + CH_3\text{-}Li \rightarrow CH_3C{\equiv}C{:}^{-}\ Li^{+} + CH_3\text{-}H$$

- Read pages _____ - _____.
- Do problems on page _____.

J. REACTIONS WITH ORGANOMETALLIC REAGENTS

1. REACTIONS IN WHICH THE OM REAGENT ACTS AS A STRONG BASE USING: R^{-}

- The general reaction is:

$$R\text{-}M + R'O\text{-}H \rightarrow \underset{\text{An alkane}}{R\text{-}H} + M^{+}\ {}^{-}OR'$$

Where ROH = H_2O, alcohol, RNH_2, R_2NH, RSH, RCOOH, $RCONH_2$, RCONHR.

R-M = RLi, RMgX, R_2CuLi

Ex:

$$\overset{\ominus}{CH_3}\text{-}\overset{\oplus}{Mg}Br + CH_3OH \longrightarrow CH_4 + CH_3OMgBr$$

- **Note: Avoid moisture when carrying out a Grignard reaction in the lab since the Grignard reagent reacts with water (see the following).**

2. CONVERSION OF AN ALKYL HALIDE TO AN ALKANE: 2 STEPS

$$R\text{-}X + M \rightarrow R\text{-}MX + H_2O \rightarrow RH + M^+\ XOH^-$$

R-M = RLi, RMgX, R_2CuLi.

Ex:

$$\overset{\ominus}{CH_3}\text{-}\overset{\oplus}{Mg}Br + H_2O \longrightarrow CH_4 + MgBrOH$$

- Do problems on page ______.

3. REACTIONS IN WHICH THE OM ACTS AS A NUCLEOPHILE

a. Introduction

- The OM is a **strong nucleophile** that reacts with carbonyl containing compounds such as **aldehydes, ketones, carboxylic acids, and derivatives.**

b. Reaction of OM (RMgX, RLi, HC≡C-Li) with Aldehydes

- **The general reaction is:**

$$RCHO \xrightarrow[2.\ H_2O]{1.\ R'M} R-CH(OH)-R'$$

a 1° or 2° alcohol

Ex:

$$C_6H_{11}CHO \xrightarrow[2.\ H_2O]{1.\ CH_3MgBr} C_6H_{11}-CH(OH)-CH_3$$

- See Mechanism on page ______.

- Note:
  - -With formaldehyde (HCHO), a 1° alcohol is produced.
  - -With other aldehydes, a 2° alcohol is produced.

Formaldehyde (HCHO) + 1. CH_3MgBr; 2. H_2O → H–C(OH)(H)–CH_3

Formaldehyde

a 1° alcohol

Mechanism:

- Read pages ______ - ______.
- Do problem on page ______.

c. Reaction of OM (RMgX, RLi, HC≡C-Li) with Ketones to Give 3° Alcohols

- The general reaction is:

R–CO–R' + 1. R"M; 2. H_2O → R–C(OH)(R")–R'

a 3° alcohol

Ex:

Cyclohexanone + 1. HC≡CMgBr; 2. H_2O → 1-ethynylcyclohexanol (OH, C≡CH)

- **Lab Experiment: Grignard synthesis of triphenylmethanol.**

Br → Mg, Anhydrous ether → MgBr (Grignard reagent) + benzophenone ($C=O$) → H_3O^+ → C—OH (Triphenylmethanol)

- **See Mechanism on page _____.**

- **Do problems ______ and ______, page ______.**

d. Reaction of OM (RMgX, RLi, HC≡C-Li) with Esters to Produce 3° Alcohols

- **The general reaction is:**

$$R-C(=O)-OR' \xrightarrow[2.\ H_2O]{1.\ 2R''M} R-\underset{R''}{\overset{OH}{C}}-R''$$

a 3° alcohol

Ex:

$$\text{cyclopropyl}-C(=O)-OCH_2CH_3 \xrightarrow[2.\ H_2O]{1.\ 2CH_3CH_2\mathbf{MgBr}} \text{cyclopropyl}-\underset{CH_2CH_3}{\overset{OH}{C}}-CH_2CH_3$$

a 3° alcohol

- See Mechanism on page ______.
- Do Problems on pages ______ and ______.

e. **Reaction of OM (RMgX, RLi, HC≡C-Li) with Acid Chlorides to Produce 3° Alcohols**

- **The general reaction is:**

$$RCOCl \xrightarrow[2.\ H_2O]{1.\ 2R''M} R-C(OH)(R'')-R''$$

a 3° alcohol

Ex:

$$\text{cyclobutyl}-COCl \xrightarrow[2.\ H_2O]{1.\ 2CH_3MgBr} \text{cyclobutyl}-C(OH)(CH_3)-CH_3$$

- See Mechanism on page ______.

f. **Reaction of R_2CuLi with Acid Chlorides to Produce Ketones**

- **The general reaction (only a substitution) is:**

$$RCOCl \xrightarrow[2.\ H_2O]{1.\ R'_2\,CuLi} RCOR'$$

a ketone

Ex:

$$\text{cyclobutyl-C(=O)Cl} \xrightarrow[\text{2. } H_2O]{\text{1. } (CH_3)_2CuLi} \text{cyclobutyl-C(=O)-}CH_3$$

- Do problems on page _______.

- Note: Acid chlorides are more reactive because Cl is a good leaving group.

g. Reaction of OM (RMgX, RLi, HC≡C-Li) with Carbon Dioxide to Give Carboxylic acids: Carboxylation

- The general reaction is:

$$\text{R-MgX} \xrightarrow[\text{2. } H_3O^+]{\text{1. } CO_2} \text{R—C(=O)OH} + H_2O$$

a carboxylic acid

- See Mechanism on page _______.

- See Example on page _______.

Ex: How would you synthesize benzoic acid from benzene **using a Grignard?**

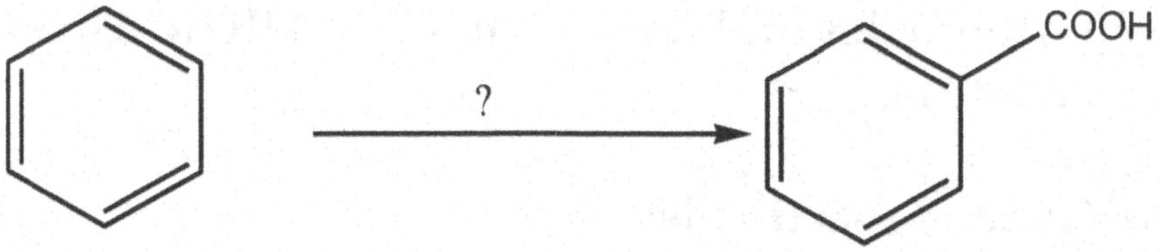

COOH

?

h. Reaction of OM (RMgX, RLi, and R_2CuLi) with Epoxides to Give Alcohols: Ring Opening

- **The general reaction is:**

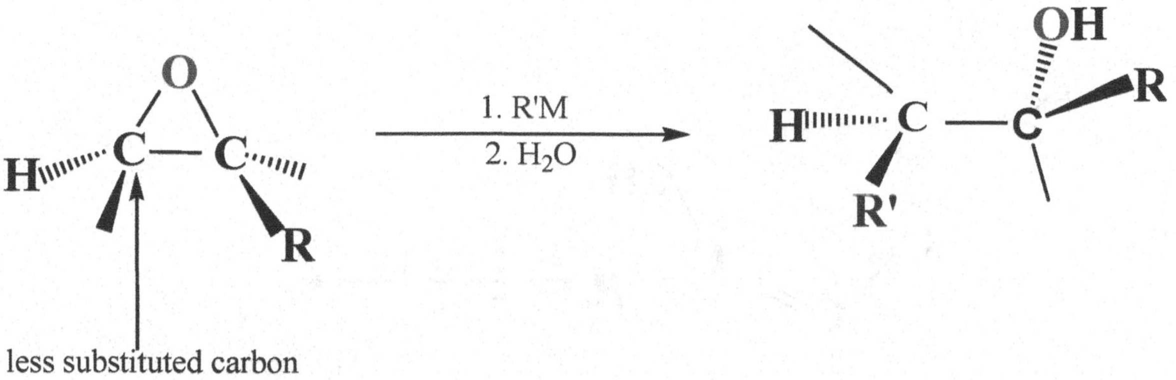

O

H C — C

R

1. R'M

2. H_2O

OH

H C — C R

R'

less substituted carbon

- **Note: The R′ goes to the less substituted carbon.**

Ex:

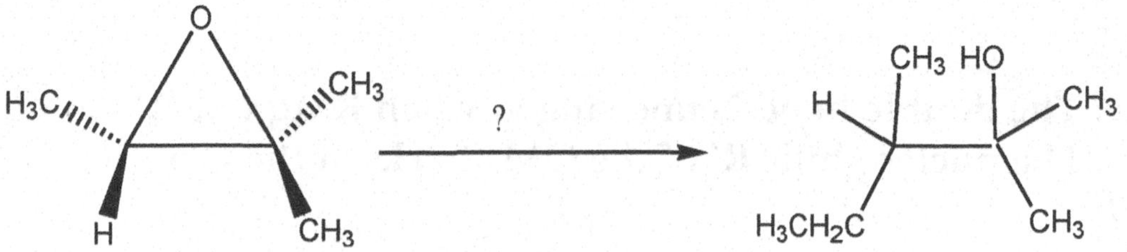

O

H_3C CH_3

H CH_3

?

CH_3 HO

H CH_3

H_3CH_2C CH_3

- **See Mechanism on page _______.**

- **Do Problem ______, page ______.**

i. Reaction of OM (RMgX, RLi) with α, β Unsaturated Aldehydes and Ketones to Give Allylic Alcohols: A 1,2-Addition

- **The General reaction is:**

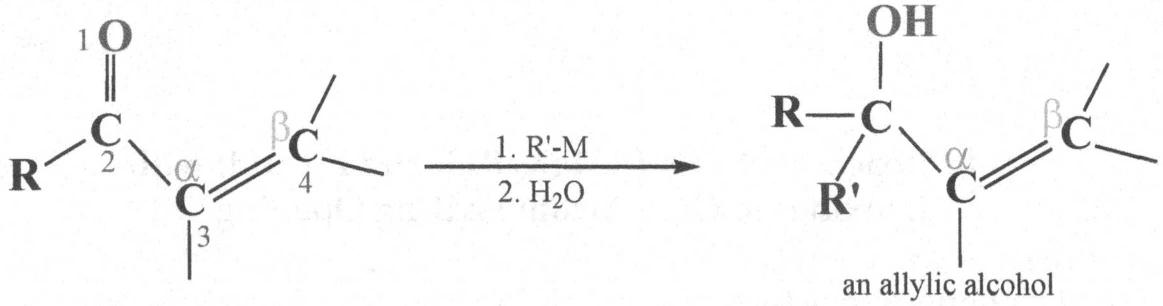

1. R'-M
2. H_2O

an allylic alcohol

Ex:

1. CH_3-Li
2. H_2O

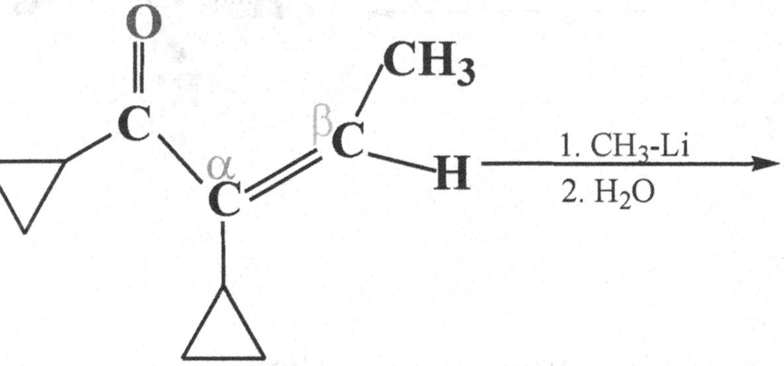

- **Note: The double bond is unchanged when RMgX or RLi is used. The nucleophilic R′ of the OM attacks at the C of the C=O.**

- See Mechanism on pages ______ - ______.

j. **Reaction of R_2CuLi with α, β Unsaturated aldehydes and ketones to Give allylic Ketones: A 1,4-Addition**

- **The General reaction is:**

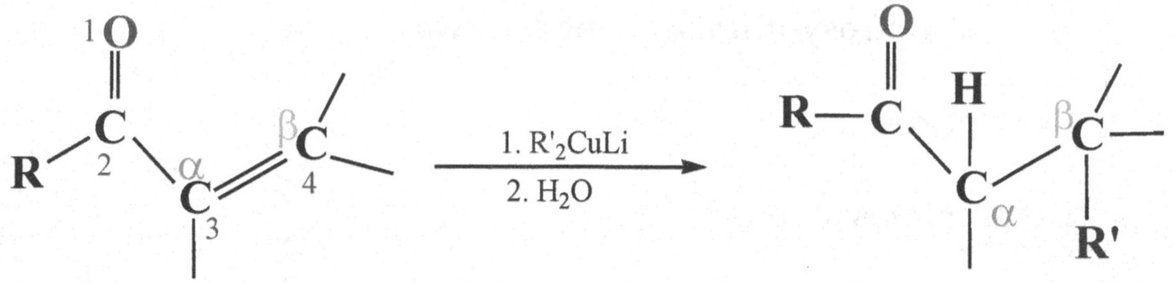

1 O, C 2, R, α C 3, β C 4

1. R'_2CuLi
2. H_2O

O, R—C, H, C α, β C, R'

Ex:

1 O, C 2, α C 3, β C 4, H

1. $(CH_3)_2CuLi$
2. H_2O

1 O, C 2, H, C α 3, β C 4, H, CH_3

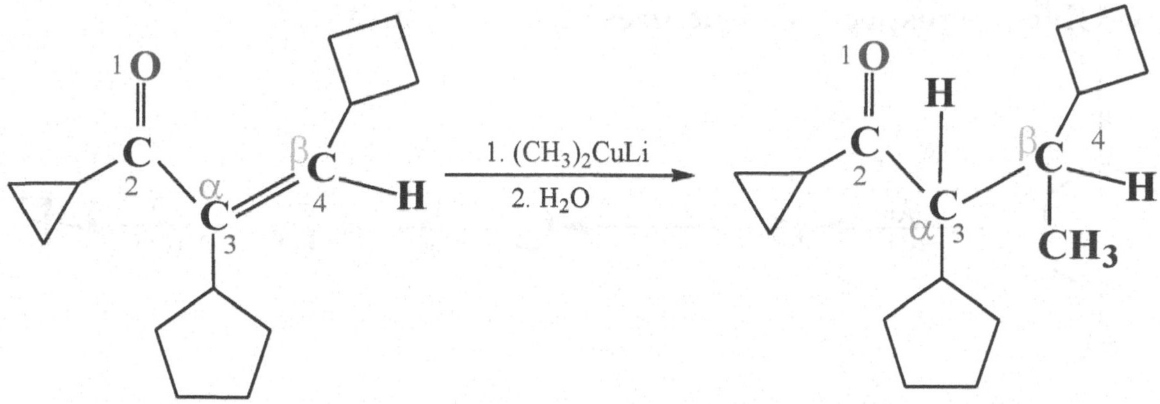

- **Note:**
  - **The C=O is unchanged when R_2CuLi is used.**
  - **The nucleophilic R′ of the OM attacks at the β C of the C=C bond.**

- **See Mechanism on page ______.**

- **Read pages ______ - ______. Do all problems.**

- **See Summary on page _______.**

k. Stereochemistry of Grignard Reactions

- Since the nucleophilic attack can occur from above and below, a **racemic mixture** is produced.

- **See examples and problem on page ________.**

l. Retrosynthesis: Going backward.

- Read pages ______ - ______ and ______ - ______.

- Do problems on page ______.

m. Multistep Synthesis Using Grignard Reagent

A→B→…………… →Z

Ex: Give the respective structures of B, C, D, and E.

A → ($1.\ LiAlH_4$; $2.\ H_2O$) → B → (PBr_3) → C → (Mg; Ether) → D → ($1.\ CO_2$; $2.\ H_3O^+$) → E

A

- Read pages ______ - ______.

K. PROTECTING GROUPS

1. INTRODUCTION

- Organometallic reagents do not react with compounds that contain N-H or O-H as these groups undergo **acid-base** reactions with OM reagents. Before a reaction of an OM reagent with a compound containing these groups (**ROH, RCOOH, RNH_2, R_2NH, $RCONH_2$, RCONHR, and RSH**), the groups are protected with "**blocking**" groups.

Ex:

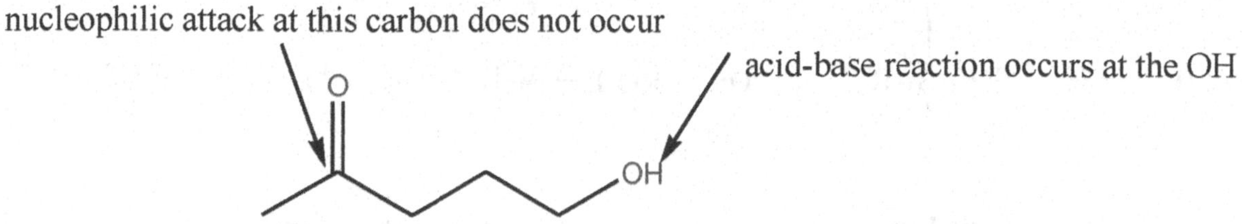

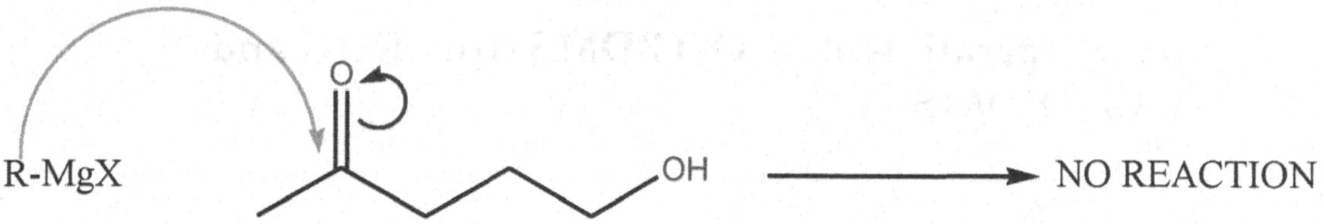

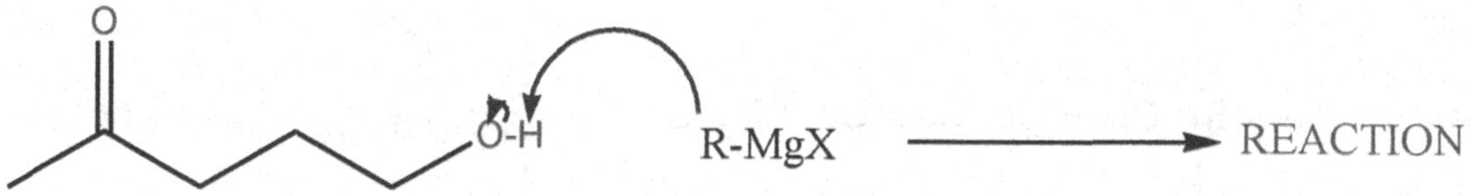

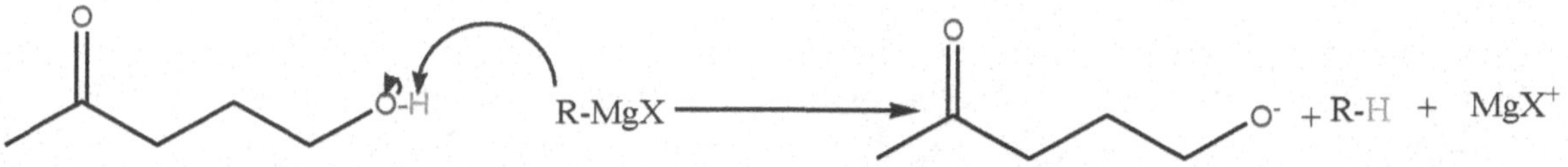

nucleophilic attack at this carbon does not occur

acid-base reaction occurs at the OH

OH

R-MgX OH ⟶ NO REACTION

O-H R-MgX ⟶ REACTION

O-H R-MgX ⟶ O^- + R-H + MgX^+

2. PROTECTION OF OH GROUPS

a. Introduction

- **Silyl ether = tert-butyldimethylsilyl ether (RO-TBDMS).**
- **The general reaction:**

RO-H + PG ⟶ RO-PG

- **Convert first ROH to a Silyl ether, RO-TBDMS.**

$$RO\text{-}H \longrightarrow RO\text{-}\underset{R}{\overset{R}{Si}}\text{-}R \quad \text{OR} \quad RO\text{-}H \longrightarrow RO\text{-}\underset{CH_3}{\overset{CH_3}{Si}}\text{-}C(CH_3)_3 \text{ OR R-O-TBDMS}$$

A silyl ether **A silyl methyl ether**

c. Preparation of R-O-TBDMS from ROH and Cl-TBDMS

Tert-butyldimethylsilyl chloride

Silyl ether

$$RO\text{-}H + Cl\text{-}\underset{CH_3}{\overset{CH_3}{Si}}\text{-}C(CH_3)_3 \xrightarrow{\text{imidazole}} RO\text{-}\underset{CH_3}{\overset{CH_3}{Si}}\text{-}C(CH_3)_3 = \textbf{R-O-TBDMS}$$

or RO-H + Cl-TBDMS ⟶ **R-O-TBDMS**

- By the end of the reaction, the silyl ether is removed with **tetrabutylammonium fluoride (TBAF), ($(CH_3CH_2CH_2CH_2)_4N^+$ F^-).**

- The general **deprotection** reaction is:

$$RO\text{-}Si(CH_3)_2\text{-}C(CH_3)_3 \xrightarrow{(CH_3CH_2CH_2CH_2)_4N^+F^-} ROH + F\text{-}Si(CH_3)_2\text{-}C(CH_3)_3$$

Ex: How would you prepare the following diol?

OH ? OH OH

- **Answer:**

OH 1.Cl-TBDMS 2.CH_3CH_2MgBr 3. H_2O 4.TBAF OH OH

- **See Fig. ______, Page ______.**

- **Read pages _______ - _______.**

- **Do problem on page ______.**

- **See Key Concepts on pages ______ - ______.**

SOME SELECTED REACTIONS FROM UNIT 10

1. Summary on the Reduction Reactions of $LiAlH_4$

Reagents: 1. $LiAlH_4$; 2. H_2O

$$R-C(=O)-H \longrightarrow R-CH(OH)-H$$
a 1° alcohol

$$R-C(=O)-R' \longrightarrow R-CH(OH)-R'$$
a 2° alcohol

$$R-C(=O)-Cl \longrightarrow R-CH(OH)-H$$
a 1° alcohol

$$R-C(=O)-OH \longrightarrow R-CH(OH)-H$$
a 1° alcohol

$$R-C(=O)-OR' \longrightarrow R-CH(OH)-H \quad + \quad R'OH$$
a 1° alcohol

$$R-C(=O)-NH_2 \longrightarrow R-CH_2NH_2$$
a 1° amine

$$R-C\equiv N \longrightarrow R-CH_2NH_2$$
a 1° amine

2. Summary on the Reduction Reactions of RMgX

1. R'MgX
2. H_2O

- $R-C(=O)-H \rightarrow R-C(OH)(R')-H$ a 2° alcohol
- $R-C(=O)-R'' \rightarrow R-C(OH)(R')-R''$ a 3° alcohol
- $R-C(=O)-Cl \rightarrow R-C(OH)(R')-R'$ a 3° alcohol
- $R-C(=O)-OR'' \rightarrow R-C(OH)(R')-R'$ + R''OH a 3° alcohol
- $R-C(=O)-C(CH_3)=C(CH_3)_2 \rightarrow R-C(OH)(R')-C(CH_3)=C(CH_3)_2$ an allylic alcohol
- R—OH → RH an alkane
- 1. CO_2 2. H_3O^+ → $R'-C(=O)-OH$ a carboxylic acid
- $R-C\equiv N \rightarrow R-C(=O)-R'$ a ketone

OCHEM II UNIT 11: ALDEHYDES AND KETONES: NUCLEOPHILIC ADDITION REACTIONS

A. INTRODUCTION

1. GENERAL STRUCTURES

- Aldehydes and ketones are carbonyl containing compounds.

O
||
— C —

R—C(=O)—R
a ketone

R—C(=O)—H
an aldehyde

2. REACTIVITY

- Aldehydes and ketones react with nucleophiles as follows:

$R_2C{=}O$ (O δ−, C δ+) + :Nu⊖ → $R_2C(O^-)Nu$

- **Note: The more the R groups on the carbonyl C, the less reactive the compound. This is due to steric effect. In other words, aldehydes are more reactive than ketones.**

Ex:

H_3C–C(=O)–CH_3 is less reactive than H_3C–C(=O)–H

- **Read page ______.**

- **Do all Problems on pages _____ - ____.**

B. NOMENCLATURE

1. NAMING ALDEHYDES

a. IUPAC NAMES

- Names of aldehydes end in **-al**.
- The **parent chain** is the **longest carbon chain** that contains the **aldehyde functional group.**
- The **-e** in the corresponding alkane name is replaced with suffix **-al**.

Ex:

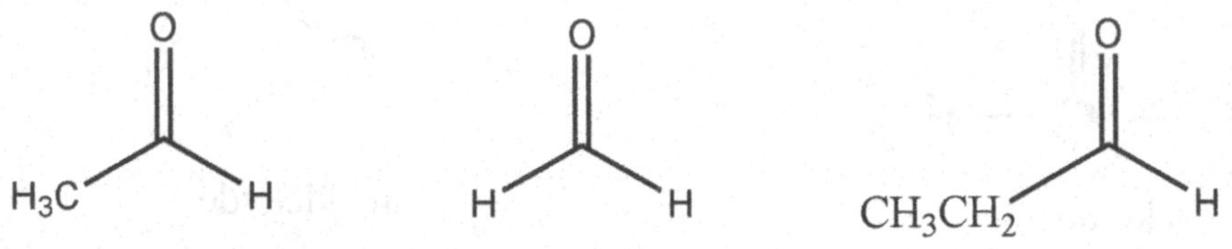

O
H3C H   H H   CH3CH2 H

- If the -CHO group is bonded to a carbon in a ring, the aldehyde is named by using **the name of the cycloalkane + the suffix carbaldehyde.**

Ex:

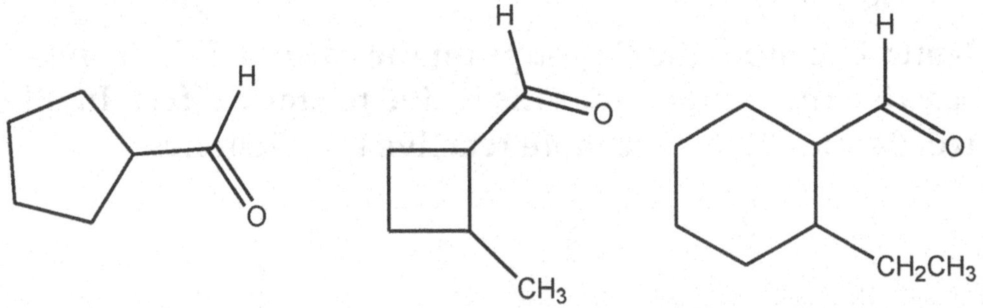

H O   H O CH3   H O CH2CH3

- **Read page ______.**

- **Do Problems ______ and ______, page ______.**

b. Common Names

i. Unbranched Aldehydes

- Nomenclature is similar to that of carboxylic acids (**See Unit 9.**) The following prefixes are used:

number of carbons	root
1	form-
2	acet-
3	propion-
4	butyr-
5	valer-
6	capro-

- They are named as:

root + **aldehyde**

Ex:

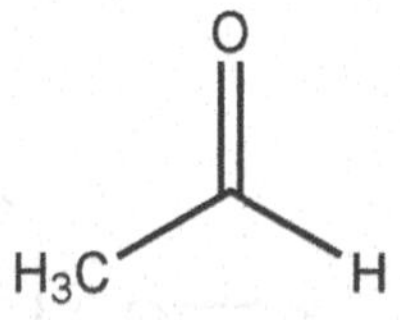

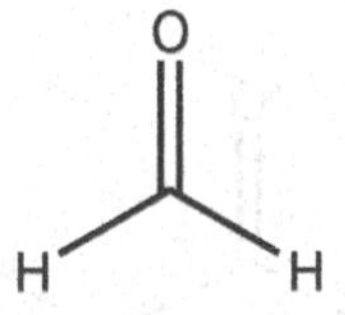

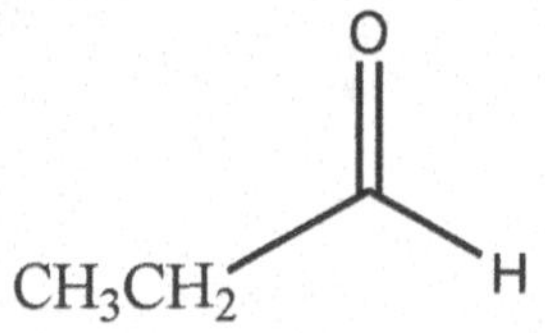

H_3C–C(=O)–H   H–C(=O)–H CH_3CH_2–C(=O)–H

ii. Branched Aldehydes

- Greek letters are used.

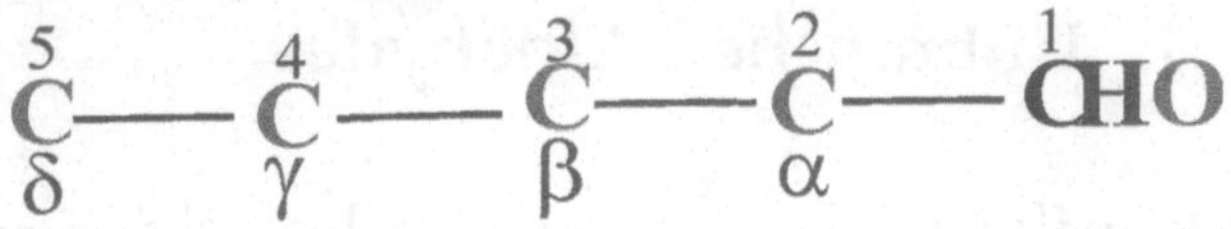

$$\overset{5}{\underset{\delta}{C}}—\overset{4}{\underset{\gamma}{C}}—\overset{3}{\underset{\beta}{C}}—\overset{2}{\underset{\alpha}{C}}—\overset{1}{C}HO$$

Ex:

CHO CHO CHO

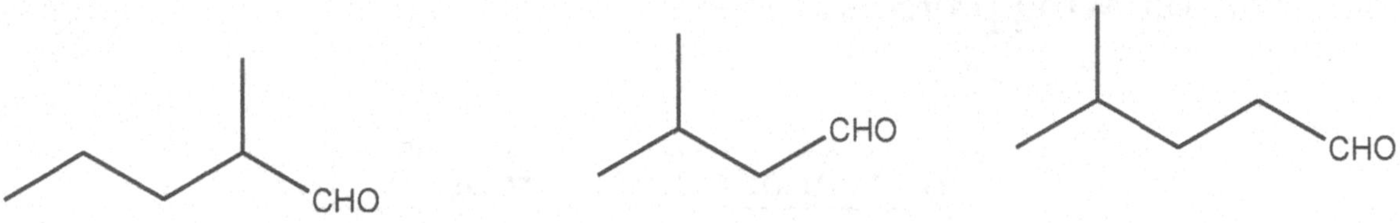

2. NAMING KETONES

a. IUPAC

- Names of ketones end in **-one.**
- **Parent chain = longest carbon chain** that contains the -C=O group.
- Use the **lowest number** to locate the -C=O.
- Replace the **-e** in corresponding alkane name with **-one.**

Ex:

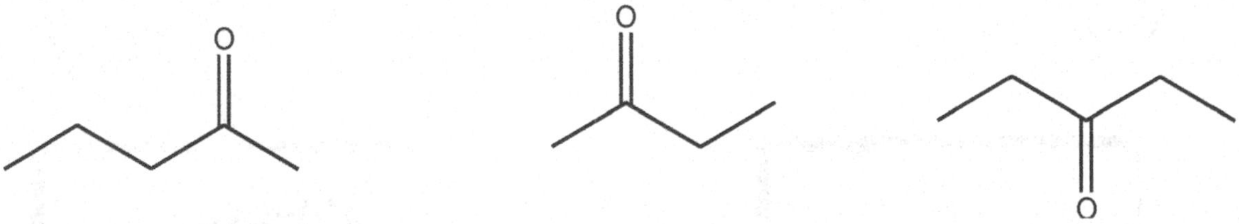

O O O

- If the =O is attached to a carbon in a ring, the ketone is named **by replacing the e in the cycloalkane name with -one.**

Ex:

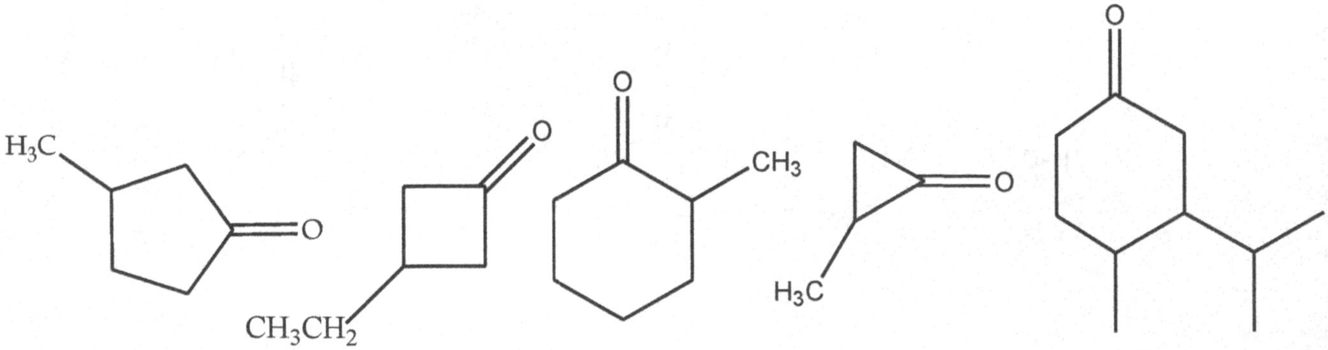

H_3C O O CH_3CH_2 O CH_3 O H_3C O

- **Do Problem _______, page _______.**
- **Read page ______.**

b. Common Names of ketones

- Name the 2 R groups **alphabetically** and add the word **ketone**.

Ex:

c. Acyl Groups:

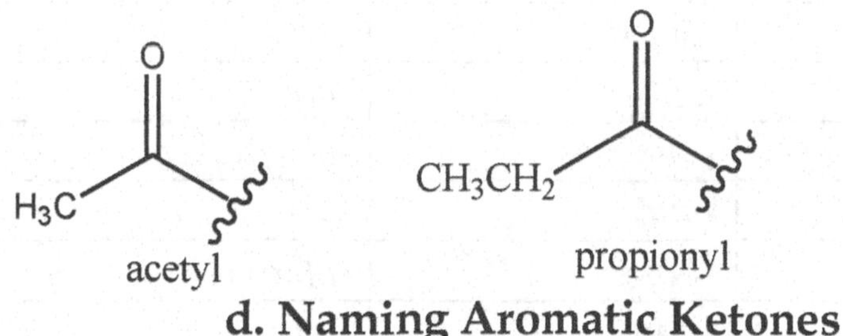

acetyl

propionyl

formyl

d. Naming Aromatic Ketones

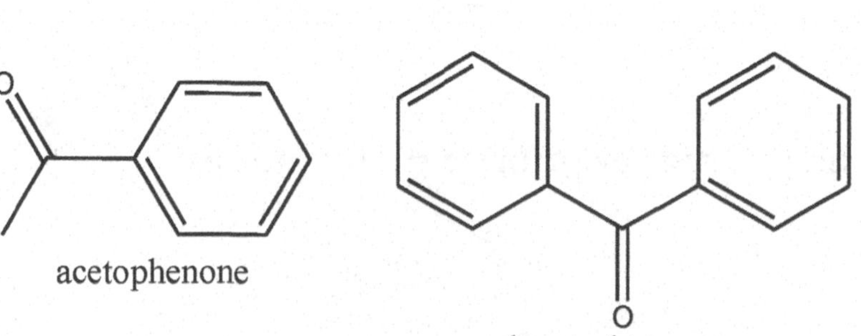

acetophenone

benzophenone

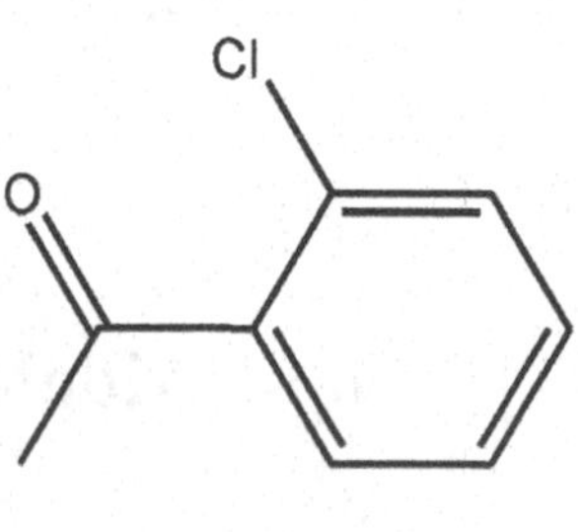

- **Read page _______.**
- **Do Problems on page _____.**

3. NAMING POLYFUNCTIONAL ALDEHYDES AND KETONES

a. Group Priorities: Déjà Vue

- When two or more different functional groups are present in a compound, functional group priorities are used. Group priorities are assigned based on the following Table (decreasing order of priority).**(See Unit 9).**

Priority order	Group	Ending of name as a priority	Name as a non **Priority group**
Carboxylic acid	**RCOOH**	*-oic acid*	*-carboxy*
Ester	**RCOOR′**	*-oate*	*-alkoxycarbonyl*
Amide	**$RCONH_2$**	*-amide*	*-amido*
Nitrile	**RCN**	*-nitrile*	*-cyano*
Aldehyde	**RCHO**	*-al*	*-oxo(=O) or formyl(-CHO)*
Ketone	**RCOR′**	*-one*	*-oxo*
Alcohol	**ROH**	*-ol*	*-hydroxy*
Amine	**RNH_2**	*-amine*	*-amino*
Alkene	**-C=C-**	*-ene*	*-alkenyl*
Alkyne	**-C≡C-**	*-yne*	*-alkynyl*
Alkane	**-C-C-**	*-ane*	*-alkyl*
Ether	**ROR′**	*-none*	*-alkoxy*
Halide	**R-X**	*-none*	*-halo*

- **Note: Aldehydes have a higher priority than ketones.**

b. **Some Examples**

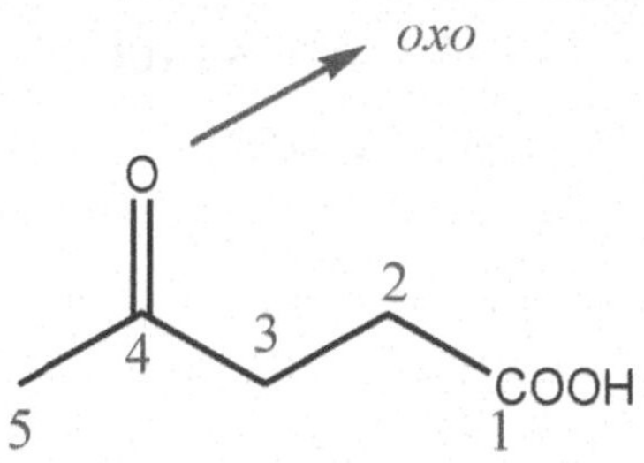

oxo

O, 5, 4, 3, 2, COOH, 1

4-oxopentanoic acid

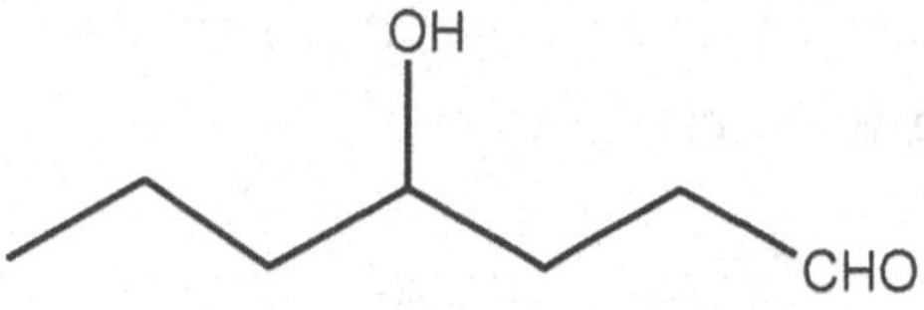

OH, CHO

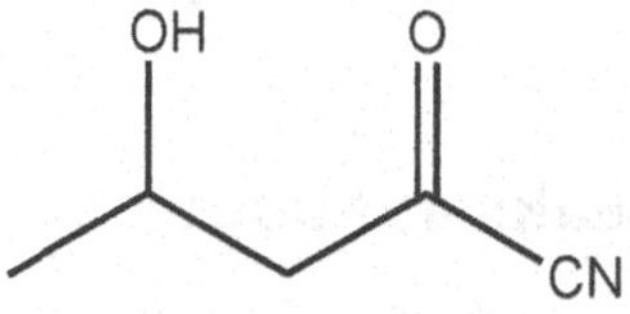

OH, O, CN

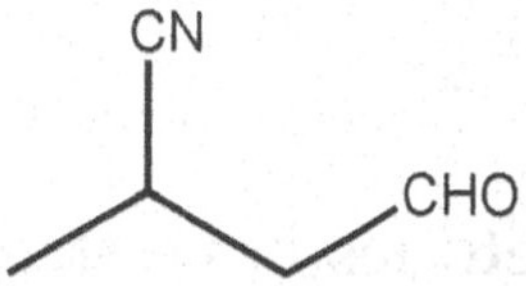

CN, CHO

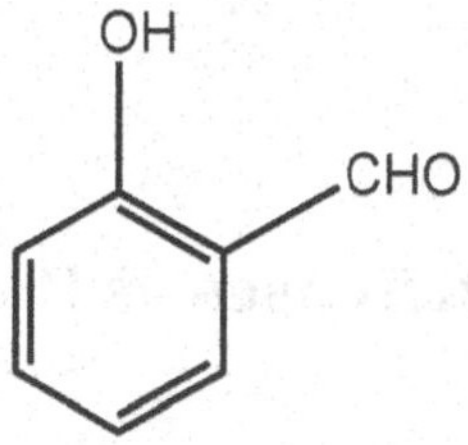

OH, CHO

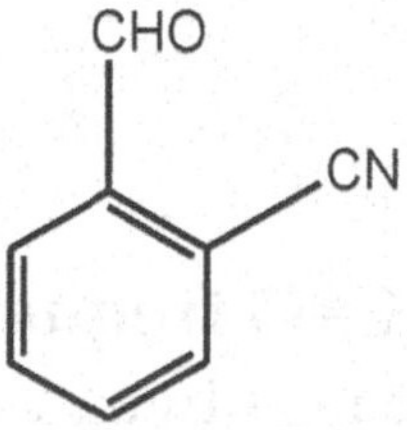

CHO, CN

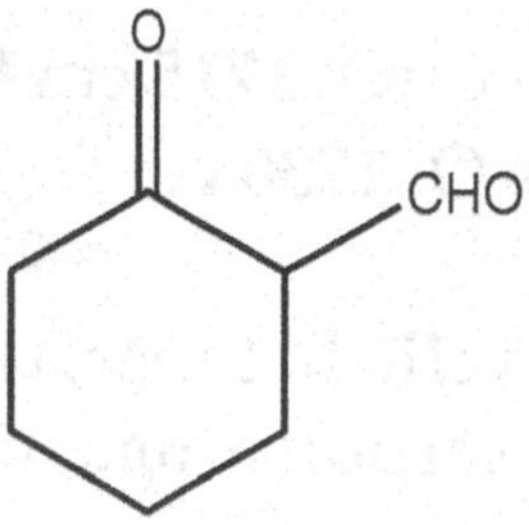

O, CHO

C. PHYSICAL PROPERTIES OF ALDEHYDES AND KETONES

1. INTRODUCTION

- Aldehydes and ketones do not contain OH groups. As a result, there is no possibility of H bonding. They have only VDW forces and dipole-dipole interactions between their molecules. Consequently, they have lower MP + BP than alcohols and carboxylic acids of comparable molar mass.

Ex:

BP = 36°C

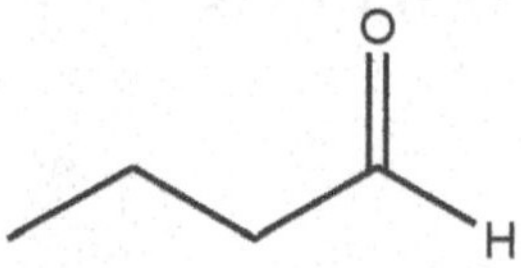

O, H, BP = 76°C

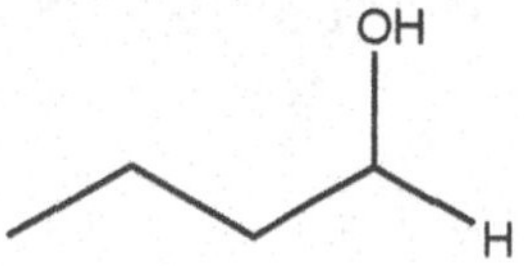

OH, H, BP = 118°C

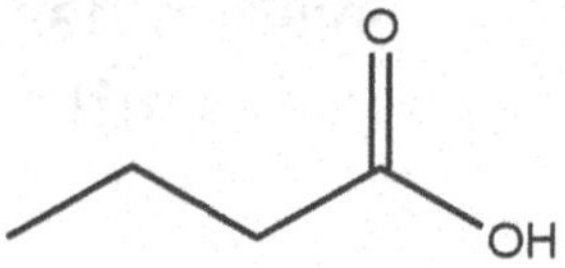

O, OH, BP = 165°C

2. SOLUBILITY

- All sizes of aldehydes and ketones are soluble in organic solvents.

- Aldehydes and ketones with a number of carbons ≤ 5 are soluble in water because of their ability to H bond with water.
- Aldehydes and ketones with a number of carbons > 5 are insoluble in water.

- Read page _______.

D. SPECTROSCOPY

1. IR

- **Aldehydes and ketones have strong, sharp peaks at about 1700 cm^{-1}.**
- **Aldehyde -C=O: 1715 cm^{-1}.**
- **Ketone -C=O: 1730 cm^{-1}.**

- **Note: For cyclic ketones, the -C=O frequency increases as the number of carbon atoms decreases because of ring strain (See Unit 6)**

Ex:

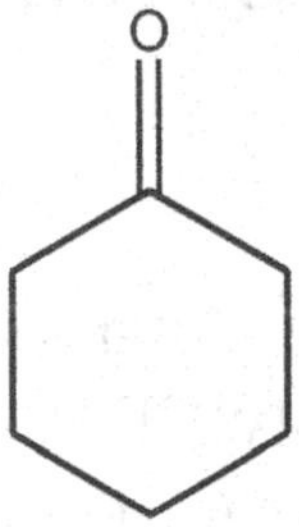

1715 cm^{-1} 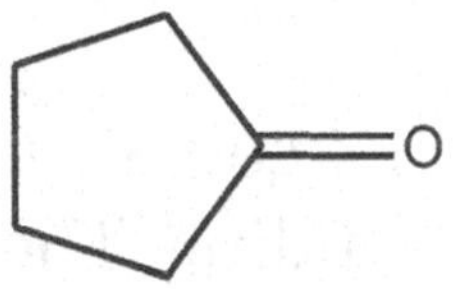 1745 cm^{-1} 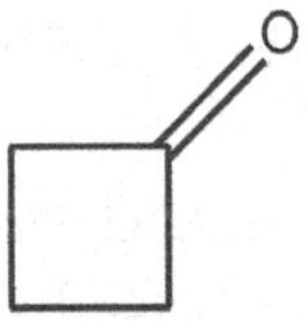 1780 cm^{-1}

- **Note: For conjugated systems, frequency decreases with increasing conjugation because the -C=O in the conjugated system has a partial single bond character due to resonance (See Unit 6).**

Ex:

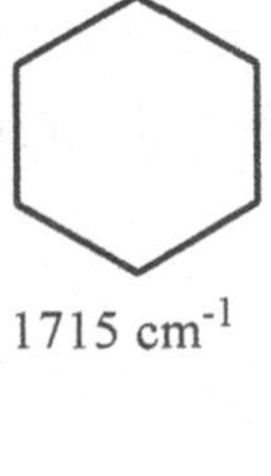

1715 cm^{-1} 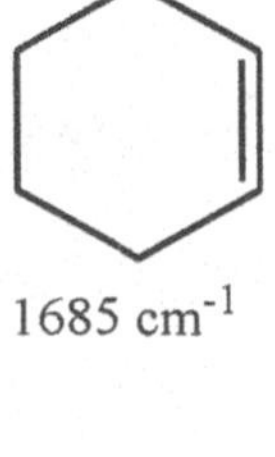 1685 cm^{-1}

No resonance stronger bond

Resonance weaker bond

single bond character

- Read page ______.
- See Fig. ______ on page ______.
- Do Problem ______, page ______ and ______, page ______.

2. NMR

c. ^{1}H NMR

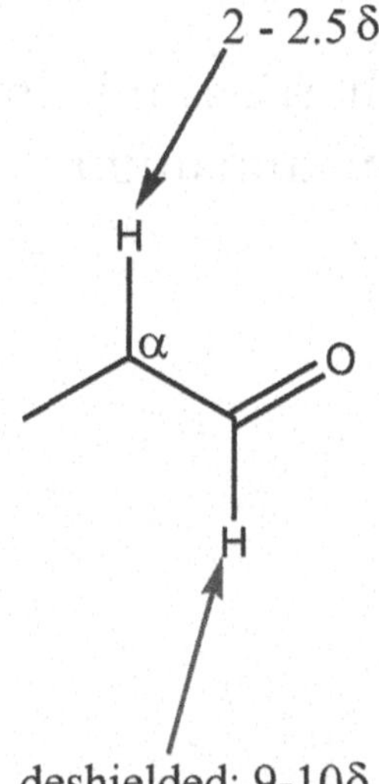

c. ^{13}C NMR

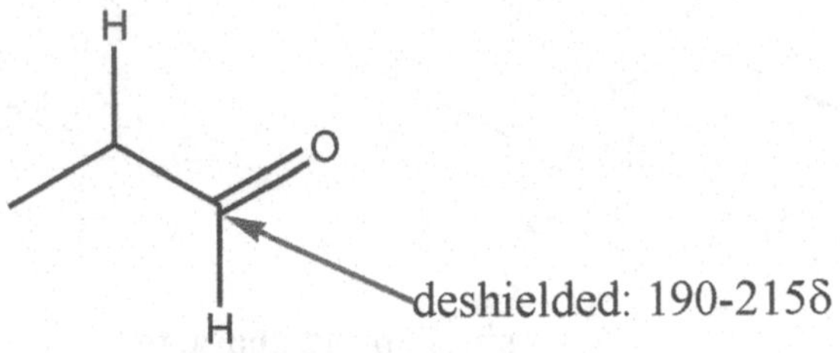

H
O
H
deshielded: 190-215δ

- See Fig. _______, page ______.

- Read page ______.

- Do Problem ______, page _____.

E. INTERESTING ALDEHYDES AND KETONES

- o Formaldehyde (37% = Formalin).
- o Acetone = solvent and starting material.

- Read page _________.

    - o Benzaldehyde from cherries and almonds.
    - o Cinnamaldehyde from cinnamon.
    - o Citral.
    - o Cortisone.
    - o Progesterone.
    - o Testosterone.

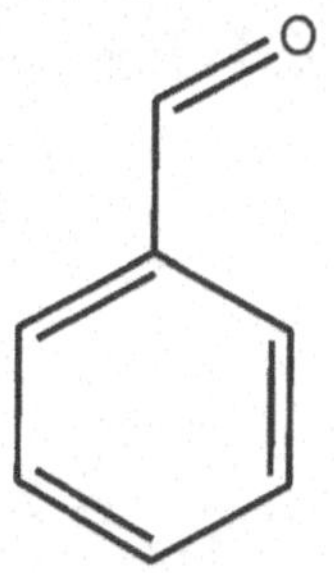

O

benzaldehyde

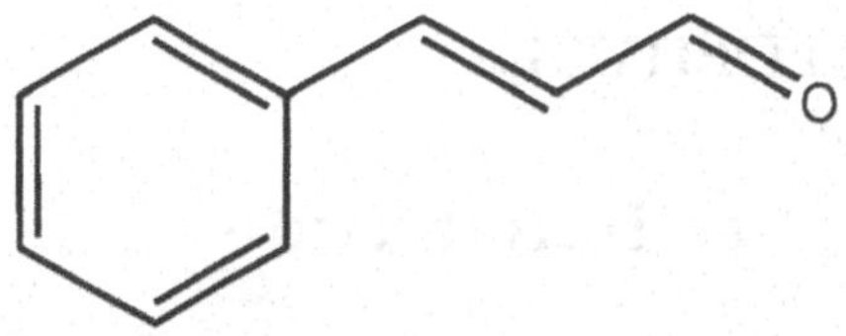

O

cinnamaldehyde

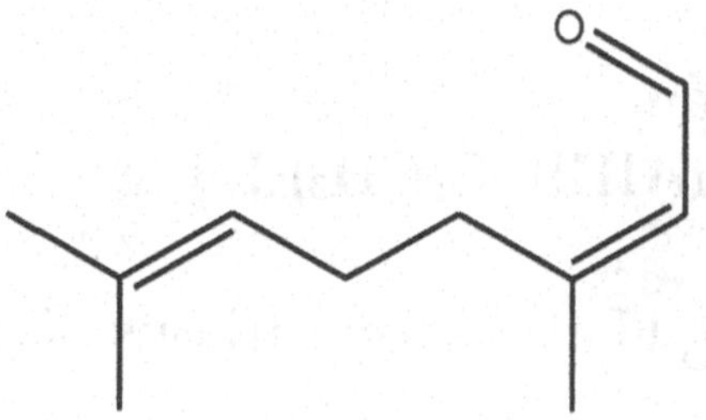

O

citral

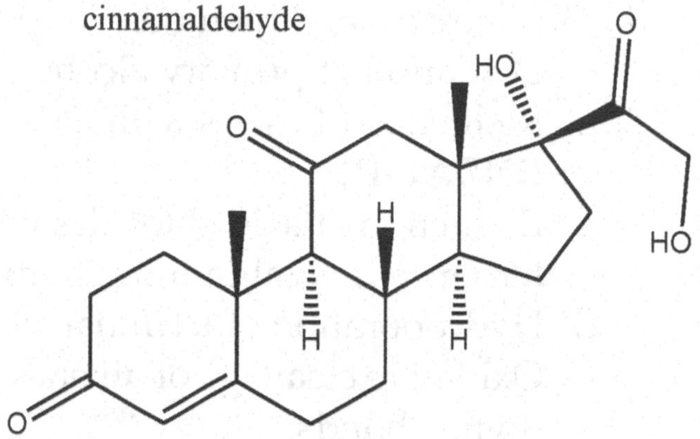

O HO O O H HO H H O

cortisone

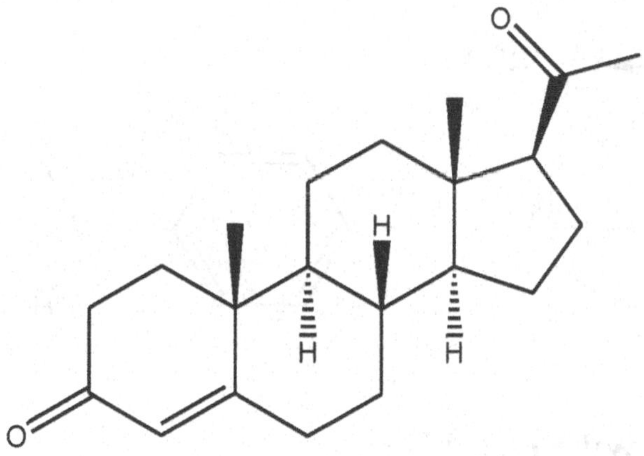

O H H H O

progesterone

OH H H H O

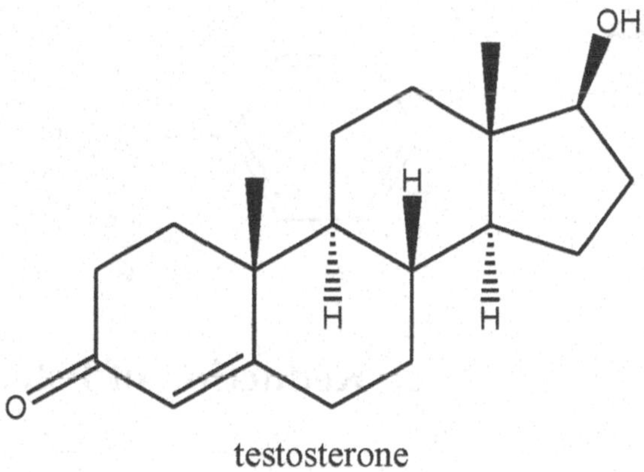

testosterone

F. PREPARATION OF ALDEHYDES AND KETONES

1. ALDEHYDES

a. Introduction

- There are several methods:
  - o Oxidation of primary alcohols with PCC.
  - o Reduction of esters with **Diisobutyl Aluminum Hydride (DIBAL-H).**
  - o Reduction of acid chlorides with Lithium tri-tertiarybutoxyaluminum hydride **($LiAlH[OC(CH_3)_3]_3$.)**
  - o Hydroboration of terminal alkynes.
  - o Oxidative cleavage of alkenes having H atoms on carbons of double bonds.

b. Oxidation of Primary Alcohols with PCC

$$RCH_2OH \xrightarrow{PCC} RC(=O)\text{-}H$$

Ex:

$$C_6H_5\text{-}CH_2OH \xrightarrow{PCC} C_6H_5\text{-}C(=O)H$$

c. Reduction of Esters with DIBAL-H

$$RC(=O)\text{-}O\text{-}R' \xrightarrow[\text{2. } H_2O]{\text{1. DIBAL-H}} RC(=O)\text{-}H + R'OH$$

Ex:

$$C_6H_5\text{-}C(=O)OCH_2CH_3 \xrightarrow[\text{2. } H_2O]{\text{1. DIBAL-H}} C_6H_5\text{-}C(=O)H + CH_3CH_2OH$$

d. Reduction of Acid Chlorides with $LiAlH[OC(CH_3)_3]_3$

$$RC(=O)Cl \xrightarrow[2.\ H_2O]{1.\ LiAlH[OC(CH_3)_3]_3} RC(=O)H$$

Ex:

$$C_6H_5C(=O)Cl \xrightarrow[2.\ H_2O]{1.\ LiAlH[OC(CH_3)_3]_3} C_6H_5C(=O)H$$

e. Hydroboration (Non-Markonikov) of Terminal Alkynes

H → RC, OH → CH

$$RC{\equiv}CH \xrightarrow[2.\ H_2O_2,\ OH^-]{1.\ BH_3} RCH_2C(=O){-}H$$

Ex:

$$CH_3CH_2C{\equiv}CH \xrightarrow[2.\ H_2O_2,\ OH^-]{1.\ BH_3}$$

f. Ozonolysis: Oxidative Cleavage of Alkenes

$$(H)(R)C{=}C(H)(R') \xrightarrow[2.\ Zn,\ H_2O\ or\ CH_3SCH_3]{1.\ O_3} (H)(R)C{=}O + O{=}C(H)(R')$$

Ex:

$$(H)(H_3C)C{=}C(H)(CH_2CH_3) \xrightarrow[2.\ Zn,\ H_2O\ or\ CH_3SCH_3]{1.\ O_3} (H)(H_3C)C{=}O + O{=}C(H)(CH_2CH_3)$$

Ex:

$$C_6H_5\text{-}CH{=}CH_2 \xrightarrow[\text{2. Zn, } H_2O \text{ or } CH_3SCH_3]{\text{1. } O_3} C_6H_5\text{-}CHO + CO_2$$

g. Synthesis of Aldehydes: A summary

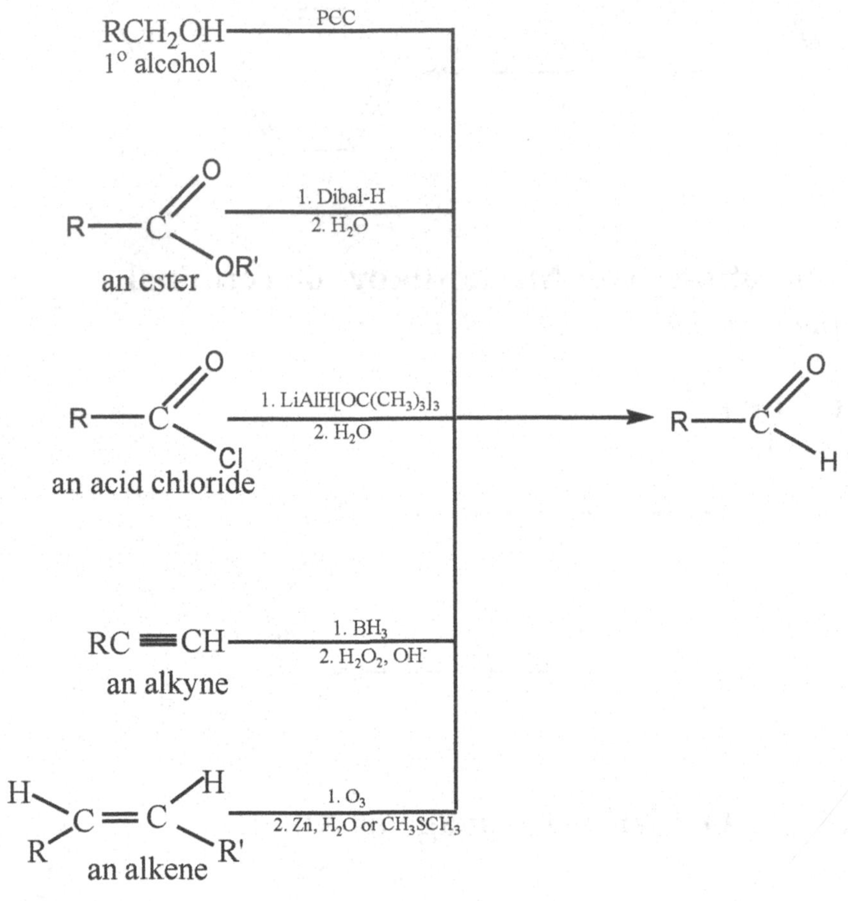

2. SYNTHESIS OF KETONES

a. Introduction

- There are several ways of synthesizing ketones:
  - -Oxidation of 2° alcohols.
  - -Reaction of organocuprates with acid chlorides.
  - -Friedel-Crafts acylation.
  - -Hydration of an alkyne: oxymercuration.
  - -Oxidative cleavage of alkenes: ozonolysis.
  - -Hydroboration of internal alkynes.

b. Oxidation of 2° Alcohols

$$RCH(OH)R' \xrightarrow{[O]} RC(=O)\text{-}R'$$

A secondary alcohol

[O] = CrO_3, $Na_2Cr_2O_7$, $K_2Cr_2O_7$, PCC.

Ex:

$$\text{(cyclopropyl)(cyclobutyl)CH—OH} \xrightarrow[H_3O^+]{K_2Cr_2O_7} \text{(cyclopropyl)(cyclobutyl)C=O}$$

c. Reaction of Organocuprates with Acid Chlorides

$$RC(=O)\text{-}Cl \xrightarrow[2.\ H_2O]{1.\ R'_2CuLi} RC(=O)\text{-}R'$$

Ex:

$$C_6H_5C(=O)Cl \xrightarrow[2.\ H_2O]{1.\ (CH_3)_2CuLi} C_6H_5C(=O)CH_3$$

d. Friedel-Crafts Acylation (See Unit 8): Aromatic Ketones

$$C_6H_6 + RC(=O)\text{-}Cl \xrightarrow{AlCl_3} C_6H_5C(=O)\text{-}R$$

Ex:

$$\text{C}_6\text{H}_6 + CH_3CH_2\overset{O}{\overset{\|}{C}}\text{-Cl} \xrightarrow{AlCl_3} C_6H_5\text{-}\overset{O}{\overset{\|}{C}}\text{-}CH_2CH_3$$

e. Mercuration or Hydration of an Alkyne (Follows Markovnikov)

OH H

$$RC \equiv CH \xrightarrow[HgSO_4]{H_2O,\ H_2SO_4} R\overset{O}{\overset{\|}{C}}\text{-}CH_3$$

Ex:

$$\text{(cyclopropyl)}\text{—}C \equiv CH \xrightarrow[HgSO_4]{H_2O,\ H_2SO_4}$$

f. Ozonolysis: Oxidative Cleavage of Tri and Tetrasubstituted Alkenes

$$R'(R)C{=}C(R'')(R''') \xrightarrow[\text{2. Zn, } H_2O \text{ or } CH_3SCH_3]{\text{1. } O_3} R'(R)C{=}O + O{=}C(R'')(R''')$$

Ex:

$$CH_3CH_2CH_2(\text{cyclopropyl})C{=}C(CH_3)(CH_2CH_3) \xrightarrow[\text{2. Zn, } H_2O \text{ or } CH_3SCH_3]{\text{1. } O_3} CH_3CH_2CH_2(\text{cyclopropyl})C{=}O + O{=}C(CH_3)(CH_2CH_3)$$

g. Non-Markonikov Hydroboration of Internal Alkynes

$$RC \equiv CR' \xrightarrow[\text{2. } H_2O_2,\ OH^-]{\text{1. } BH_3} RCH_2\overset{O}{\overset{\|}{C}}\text{-}R' + R\overset{O}{\overset{\|}{C}}CH_2R'$$

Ex:

$$\text{(cyclopropyl)}\text{—}C \equiv CCH_3 \xrightarrow[\text{2. } H_2O_2,\ OH^-]{\text{1. } BH_3}$$

h. Reaction of Nitriles with R′MgX

$$\underset{\text{a nitrile}}{RC\equiv N} \xrightarrow[\text{2. } H_2O]{\text{1. R'MgX}} R-\overset{\overset{\displaystyle O}{\|}}{C}-R'$$

- **Read page ______.**
- **Do Problems page _______.**

i. Synthesis of Ketones: A Summary

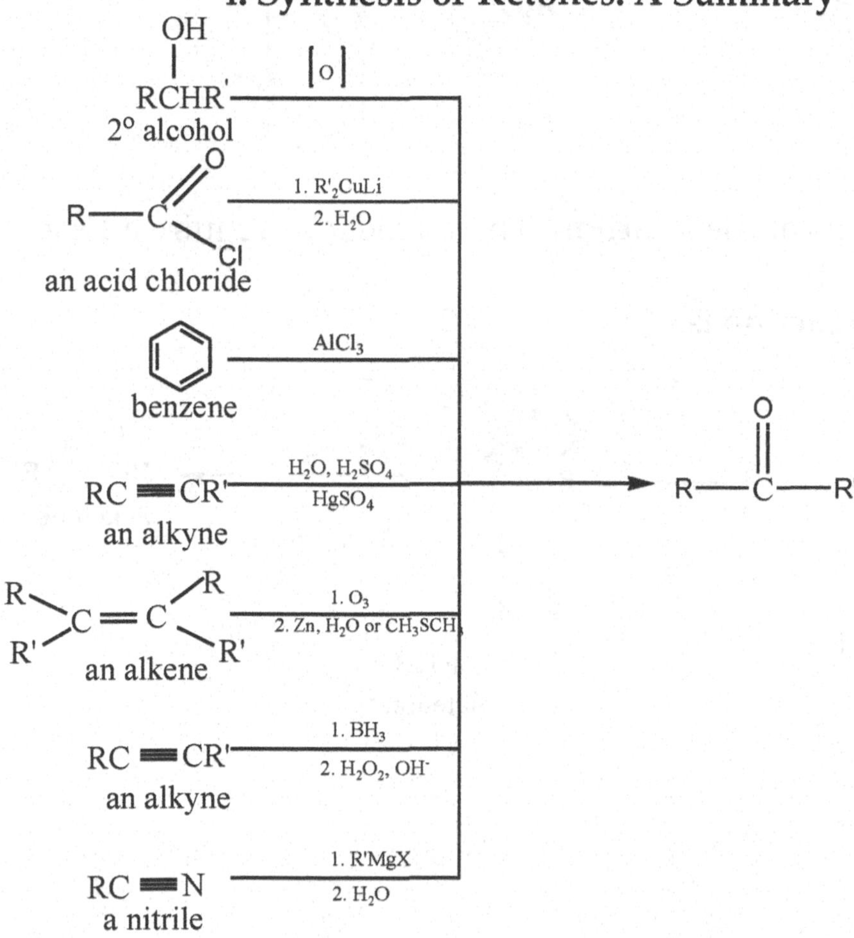

G. REACTIONS OF ALDEHYDES AND KETONES

1. INTRODUCTION

- There **are 2 general kinds** of reactions:
    - **Nucleophilic addition reactions** at the carbonyl carbon. **(See Unit 10.)**
- **The general reaction** is:

$$R-\overset{\overset{\Large O}{\|}}{C}-R' \xrightarrow[2.\ H_2O]{1.\ Nu^-} R-\underset{\underset{\Large R'}{|}}{\overset{\overset{\Large OH}{|}}{C}}-Nu$$

- **Read pages** _______ - ______.
    - **Reactions at the α carbon.** These reactions require a **base (Unit 14)**
- **The general reaction** is:

$$R-\overset{\overset{O}{\|}}{C}-CH_2-H \quad :B \longrightarrow R\overset{\overset{O}{\|}}{C}-\overset{\ominus}{C}H_2 \xrightarrow{E^+} R\overset{\overset{O}{\|}}{C}-CH_2E$$

α carbon; acidic H

$$\updownarrow$$

$$R\overset{\overset{O^-}{|}}{C}=CH_2 + HB^+$$

an enolate ion

A ketone

2. NUCLEOPHILIC ADDITION REACTIONS AT THE CO: A REVIEW

- Read pages ______ - ______.

a. Addition of H^- to Aldehydes and Ketones

- The general reaction of aldehydes is:

$$R-\overset{\overset{\displaystyle O}{\|}}{C}-H \xrightarrow[\text{OR 1. LiAlH}_4\text{/2. H}_2\text{O}]{\text{NaBH}_4\text{/CH}_3\text{OH}} R-\overset{\overset{\displaystyle OH}{|}}{\underset{\underset{\displaystyle H}{|}}{C}}-H$$

a 1° alcohol

Ex:

$$CH_3CH_2\overset{\overset{\displaystyle O}{\|}}{C}-H \xrightarrow{\text{NaBH}_4\text{/CH}_3\text{OH}} CH_3CH_2\overset{\overset{\displaystyle OH}{|}}{\underset{\underset{\displaystyle H}{|}}{C}}-H$$

- The general reaction of ketones is:

$$R-\overset{\overset{\displaystyle O}{\|}}{C}-R' \xrightarrow[\text{OR 1. LiAlH}_4\text{/2. H}_2\text{O}]{\text{NaBH}_4\text{/CH}_3\text{OH}} R-\overset{\overset{\displaystyle OH}{|}}{\underset{\underset{\displaystyle R'}{|}}{C}}-H$$

a 2° alcohol

Ex:

$$CH_3CH_2\overset{\overset{\displaystyle O}{\|}}{C}-CH_3 \xrightarrow[\text{2. H}_2\text{O}]{\text{1. LiAlH}_4} CH_3CH_2\overset{\overset{\displaystyle OH}{|}}{\underset{\underset{\displaystyle H}{|}}{C}}-CH_3$$

b. Addition of R^- to Aldehydes and Ketones

- **The general reaction of aldehydes is:** R′M = R′MgX, R′Li, HC≡C-Li.

1. R'M
2. H_2O

a 1° or 2° alcohol

Ex:

1. CH_3MgBr
2. H_2O

- **The general reaction of ketones is:** R″M = R″MgX, R″Li, HC≡C-Li

1. R"M
2. H_2O

a 3° alcohol

Ex:

1. $HC \equiv CMgBr$
2. H_2O

- Read page ________.
- Do all problems on page _______.

c. Addition of CN^- in Acidic Media: Cyanohydrin Formation

- **The general reaction** with **aldehydes** is:

$$\text{RCHO} \xrightarrow[\text{HCl}]{\text{NaCN}} \text{R–C(OH)(H)–CN}$$

A cyanohydrin

Ex:

$$\text{cyclopropyl–CHO} \xrightarrow[\text{HCl}]{\text{NaCN}} \text{cyclopropyl–C(OH)(H)–CN}$$

A cyanohydrin

- **The general reaction** with **ketones** is

$$\text{RC(O)R'} \xrightarrow[\text{HCl}]{\text{NaCN}} \text{R–C(OH)(R')–CN}$$

A cyanohydrin

Ex:

$$\text{cyclobutyl–C(O)–cyclopentyl} \xrightarrow[\text{HCl}]{\text{NaCN}} \text{cyclobutyl–C(OH)(CN)–cyclopentyl}$$

A cyanohydrin

- See mechanism on page ______.
- Read pages ______ - _____. Do Problems on page _____.
- Note: The hydrolysis of a cyanohydrin or nitrile leads to an
- α-hydroxy carboxylic acid as follows:

$$\mathrm{R-\underset{C\equiv N}{\overset{OH}{C}}-R'} \xrightarrow[\text{(H}^+\text{ or OH}^-)]{H_2O} \mathrm{R-\underset{COOH}{\overset{OH}{C}}-R'}$$

cyanohydrin or nitrile α-hydroxy carboxylic acid

Ex:

$$\mathrm{Ph-\underset{C\equiv N}{\overset{OH}{C}}-C_3H_5} \xrightarrow[\text{(H}^+\text{ or OH}^-)]{H_2O} \mathrm{Ph-\underset{COOH}{\overset{OH}{C}}-C_3H_5}$$

cyanohydrin or nitrile α-hydroxy carboxylic acid

d. The Wittig Reaction: Conversion of a Ketone (or an aldehyde) to an Alkene [Named after German Georg Wittig (1897 - 1987); Nobel Prize in Chemistry in 1979]

- **Wittig's Reagent:**

$$Ph_3P{=}CH_2 \quad = \quad (C_6H_5)_3P{=}CH_2$$

Wittig reagent

- The general Wittig reaction is:

$$R(C{=}O)R' + Ph_3P{=}CH_2 \longrightarrow RR'C{=}CH_2 + Ph_3P{=}O$$

a ketone
(R' = H for an aldehyde)
Wittig reagent
an alkene

Ex:

$$\text{(cyclopropyl)}C(=O)CH_3 + Ph_3P{=}CH_2 \longrightarrow \text{(cyclopropyl)}(CH_3)C{=}CH_2 + Ph_3P{=}O$$

a ketone
Wittig reagent
an alkene

- Synthesis of the Wittig Reagent:

$$Ph_3P + CH_3Br \xrightarrow{BuLi} Ph_3P{=}CH_2 \longleftrightarrow Ph_3\overset{+}{P}\text{-}\overset{-}{C}H_2$$

- See Mechanism of the Reaction on page _____.

- Read pages _____ - _____ about retrosynthesis.

- In summary:

$$RR'C{=}O + Ph_3P{=}CH_2 \longrightarrow RR'C{=}CH_2 + Ph_3P{=}O$$

- Do Problems on pages _____ - _____.

e. The Cannizzaro Reaction of Aldehydes

- **It occurs with aldehydes and ketones with no alpha hydrogens (formaldehyde and benzaldehyde) in NaOH or KOH. The aldehyde is oxidized and reduced at the same time = disproportionation.**

- General Reaction:

2 an aldehyde (no α carbon) — 1. :HO⁻ 2. H_2O → a carboxylic acid (oxidation product) + anl alcohol (reduction product)

Ex:

2 an aldehyde (no α carbon) — 1. :HO⁻ 2. H_2O → formic acid + methyl alcohol

- This reaction occurs when benzaldehyde is treated with NaOH (or KOH) as follows:

Benzaldehyde — 1. :HO⁻ 2. H_2O → benzoic acid (oxidation product) + benzyl alcohol (reduction product)

Mechanism:

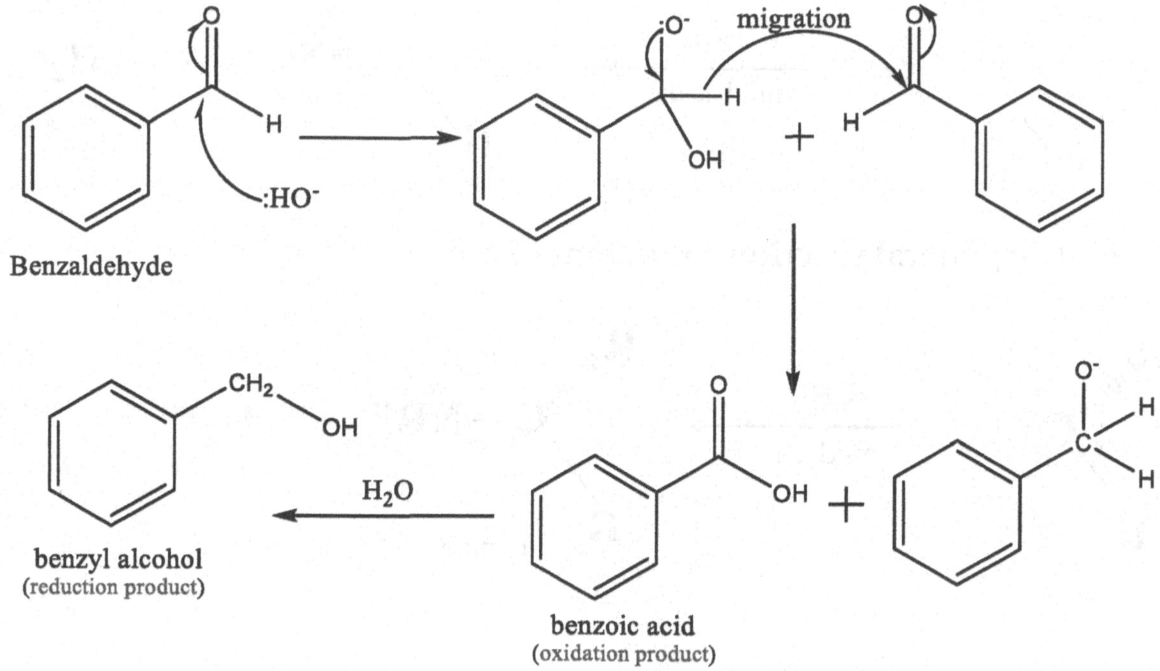

Benzaldehyde → (:HO⁻, migration) → benzoic acid (oxidation product) + ... → (H_2O) → benzyl alcohol (reduction product)

benzyl alcohol
(reduction product)

benzoic acid
(oxidation product)

f. Addition of 1° Amines to Aldehydes and Ketones: Imines

- The reaction is faster at **pH 4-5**. The general reaction of **aldehydes** is:

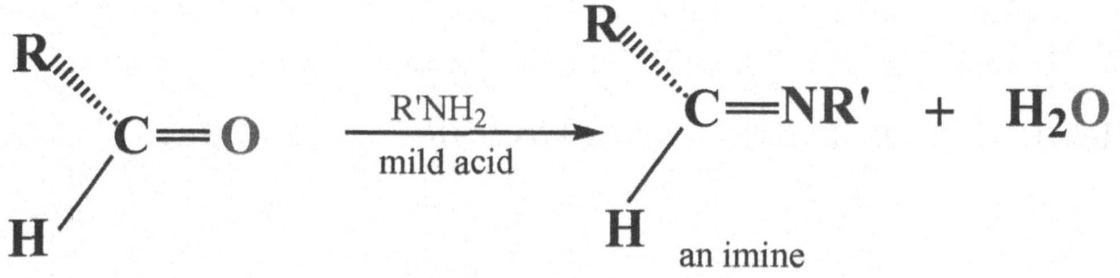

$$\mathrm{RHC{=}O} \xrightarrow[\text{mild acid}]{\mathrm{R'NH_2}} \mathrm{RHC{=}NR'} + \mathrm{H_2O}$$

an imine

Ex:

$$\text{cyclopentyl–CHO} \xrightarrow[\text{mild acid}]{CH_3NH_2} \text{cyclopentyl–CH}{=}NCH_3 + H_2O$$

- The **general reaction** for **ketones** is:

$$R(R')C{=}O \xrightarrow[\text{mild acid}]{R''NH_2} R(R')C{=}NR'' + H_2O$$

an imine

Ex:

$$\text{cyclopropyl–}C(CH_3){=}O \xrightarrow[\text{mild acid}]{CH_3NH_2} \text{cyclopropyl–}C(CH_3){=}NCH_3 + H_2O$$

- **See Mechanism on page _______. Do problems on pages ______ - ______.**

- **Read page ______ about the chemistry of vision.**

g. Addition of 2° Amines to Aldehydes and Ketones: Enamines

- **The general reaction** of the **aldehydes** is:

O=C(H)–C(H)– + R'_2NH (mild acid) → NR'_2–C(H)=C< + H_2O

an enamine

- **Note: Replace O with NR_2' and put a double bond between C_1 and α carbon.**

Ex:

O=C(H)–C(H)(CH_3)–CH_3 + $(CH_3CH_2)_2NH$ (mild acid) → $N(CH_2CH_3)_2$–C(H)=C(CH_3)–CH_3 + H_2O

an enamine

- **The general reaction** of **ketones** is:

O=C(R)–C(H)– + R'_2NH (mild acid) → NR'_2–C(R)=C< + H_2O

an enamine

Ex:

$$CH_3C(=O)CH(CH_3)_2 \xrightarrow[\text{mild acid}]{(CH_3CH_2)_2NH} CH_3C(N(CH_2CH_3)_2)=C(CH_3)_2 + H_2O$$

an enamine

- See Examples on page ______.
- See Mechanism on page ______.
- See Fig. ______, page ______.
- Do problems on page ______.

h. Addition of H_2O to Aldehydes and Ketones in Acidic or Alkaline Media: Hydration

i. General Reaction

- The **general reaction** of the hydration of **aldehydes** is:

$$RCHO \underset{H^+ \text{ or } OH^-}{\overset{H_2O}{\rightleftharpoons}} R{-}C(OH)_2{-}H$$

an aldehyde — A gem-diol (hydrate)

Ex:

$$\text{an aldehyde (cyclobutyl–CHO)} \underset{H^+ \text{ or } OH^-}{\overset{H_2O}{\rightleftharpoons}} \text{cyclobutyl–C(OH)}_2\text{–H}$$

an aldehyde

A gem-diol (hydrate)

- **The general reaction** of the hydration of **ketones** is:

$$R\text{–}C(=O)\text{–}R' \underset{H^+ \text{ or } OH^-}{\overset{H_2O}{\rightleftharpoons}} R\text{–}C(OH)_2\text{–}R'$$

a ketone

A gem-diol (hydrate)

Ex:

$$\text{cyclobutyl–C(=O)–}CH_3 \underset{H^+ \text{ or } OH^-}{\overset{H_2O}{\rightleftharpoons}} \text{cyclobutyl–C(OH)}_2\text{–}CH_3$$

a ketone

A gem-diol (hydrate)

- **See Mechanism in Acidic Media on _____. The acid protonates the carbonyl group first giving a species that is in resonance with a carbocation.**

- **See Mechanism in Alkaline media on page _____. The base converts the H_2O into OH^-, a stronger nucleophile than water. Then the OH^- attacks the nucleophilic center C of the CO .**

- Read pages _____ - _____.

- Do problems on pages _____ - _____.

ii. Thermodynamics of Hydration

- The reaction is **reversible and does reach equilibrium**. More **stable** reactants give **small yields. Less stable** reactants lead to **larger yields.**

- **Note: Since alkyl groups (EDG) stabilize the carbonyl groups through inductive electron donation, the more the alkyl groups (electron rich groups) on the carbon of the carbonyl group, the smaller the yield at equilibrium.**

- **Read pages ____ - ____.**

Ex:

$H_2C{=}O$ ⟶ 99% yield at equilibrium

less stable

$H_3C(H)C{=}O$ ⟶ 58% yield at equilibrium

more stable

$(H_3C)_2C{=}O$ ⟶ .2% yield at equilibrium

more stable

- **Note: EDG stabilize the CO. The presence of EDG in the compound leads to small yields at equilibrium. On the other hand, EWG destabilize the CO. As a result, larger yields are obtained at equilibrium when EWG are present.**

- **Read pages _____ - _____.**

Ex:

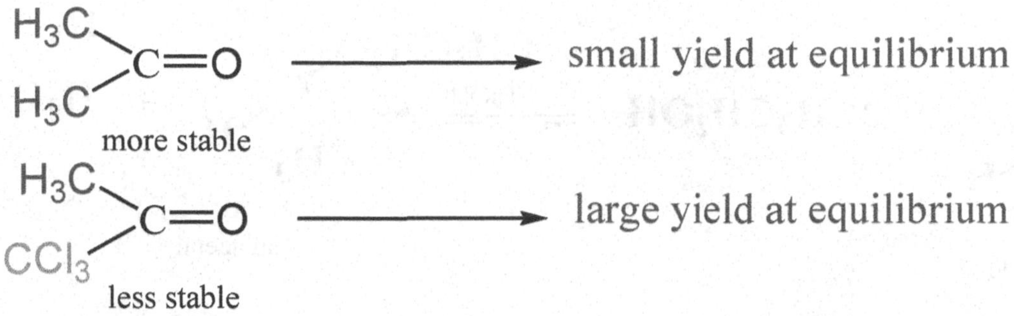

H_3C–C(=O)–CH_3 ⟶ small yield at equilibrium

more stable

H_3C–C(=O)–CCl_3 ⟶ large yield at equilibrium

less stable

- **Do Problem _______, page ______.**

iii. The Kinetics of the Reaction

- **See Mechanisms on page ______. An acid or a base is needed to speed up the reaction.**

- **See Problem ______, page ______.**

- **Note: The speed of the reaction does not affect the yield of the reaction since it is reversible.**

i. Addition of Alcohols to Aldehydes and Ketones: Acetal Products

i. General Reaction

- The reaction is catalyzed by acids such as **p-toluenesulfonic acid** (TsOH). The general reaction for **aldehydes** is:

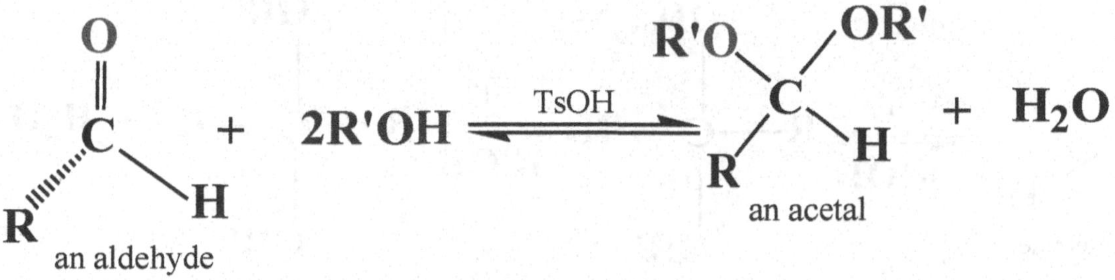

$$\underset{\text{an aldehyde}}{R\text{–}C(=O)\text{–}H} + 2R'OH \underset{}{\overset{TsOH}{\rightleftharpoons}} \underset{\text{an acetal}}{(R'O)_2C(R)H} + H_2O$$

Ex:

$$CH_3CHO + 2CH_3CH_2OH \underset{}{\overset{TsOH}{\rightleftharpoons}} CH_3CH(OCH_2CH_3)_2 + H_2O$$

an aldehyde — an acetal

- **The general reaction** for **ketones is:**

$$RC(=O)R' + 2R''OH \overset{TsOH}{\rightleftharpoons} R''O\text{–}C(OR'')(R)(R') + H_2O$$

a ketone — a ketal

Ex:

$$CH_3C(=O)\text{–cyclopropyl} + 2CH_3CH_2OH \overset{TsOH}{\rightleftharpoons} (CH_3CH_2O)_2C(CH_3)\text{–cyclopropyl} + H_2O$$

a ketone — a ketal

- **See Mechanism on pages ______ - ______. A hemiacetal is first formed.**

$$RC(=O)H \underset{R'OH}{\overset{H^+}{\rightleftharpoons}} R\text{—}C(OH)(OR')\text{—}H \underset{R'OH}{\overset{H^+}{\rightleftharpoons}} R\text{—}C(OR')(OR')\text{—}H + H_2O$$

an aldehyde — a hemiacetal — an acetal

Ex:

$$H_2C{=}O \underset{CH_3OH}{\overset{H^+}{\rightleftharpoons}} H{-}C(OH)(OCH_3){-}H \underset{CH_3OH}{\overset{H^+}{\rightleftharpoons}} H{-}C(OCH_3)_2{-}H + H_2O$$

an aldehyde — a hemiacetal — an acetal

- Acetals of ethylene glycol: cyclic acetals.

$$RCHO + HOCH_2CH_2OH \overset{H^+}{\rightleftharpoons} \text{cyclic acetal} + H_2O$$

an aldehyde — Ethylene glycol — an acetal

Ex:

$$\text{cyclopentanone} + HOCH_2CH_2OH \overset{H^+}{\rightleftharpoons}$$

- Read pages _____ - _____ about acetals (page _____) and hemiacetals (page ______.)

- Do Problems on pages _____ - _____.

- Note: Since the reaction is reversible, removing the water byproduct increases the yield according to Le Chatelier's principle. Therefore, a Dean-Stark trap is used to remove the water. See Fig. ______, page _____. Water Collected in benzene.

- Read pages _____ - _____.

ii. Hydrolysis of Acetals

- Since the reaction is reversible, one can get the aldehyde or the ketone back through **hydrolysis**. Just use an **excess of water** to drive the reaction to **the left in acidic media Le Chatelier's principle).**

- **The general reaction is:**

$$\text{R''O(OR'')C(R)(R')} + H_2O \underset{}{\overset{H^+}{\rightleftharpoons}} \text{R–C(=O)–R'} + 2\text{R''OH}$$

a ketal; H_2O: large excess; a ketone

Ex:

$$\text{CH}_3\text{O(OCH}_3\text{)C(CH}_3\text{CH}_2\text{)(cyclopropyl)} + H_2O \overset{H^+}{\rightleftharpoons} \text{CH}_3\text{CH}_2\text{–C(=O)–cyclopropyl} + 2CH_3OH$$

a ketal; H_2O: large excess

- See Mechanism on page _____.

- Read pages _____ - _____.

- Do Problems on page _____.

iii. Protection of Aldehydes and Ketones through Acetal Formation

- An "**interfering**" CO group can be protected through acetal formation. Here is an example:
- **Question: How can you reduce the ester group without reducing the keto group on the following keto-ester? Use ethylene glycol: $HOCH_2CH_2OH$.**

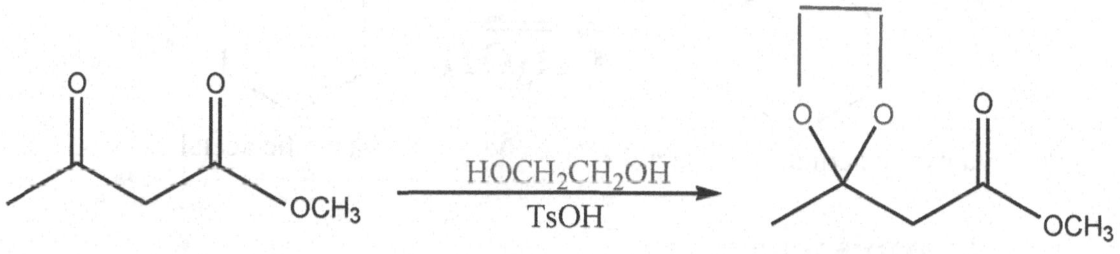

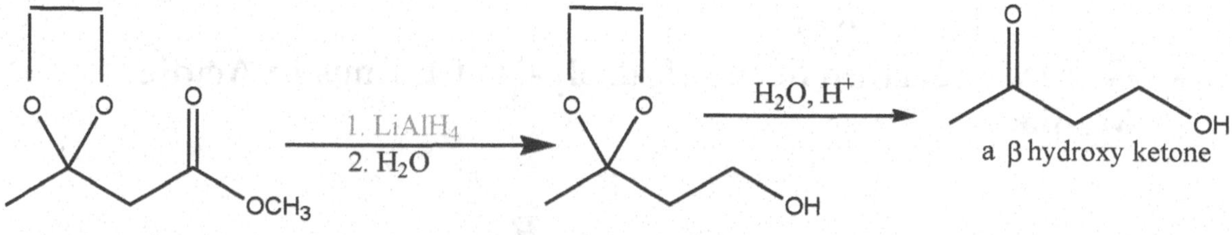

O O OCH3 → ? → O OH + CH_3OH

O O OCH3 → $HOCH_2CH_2OH$ / TsOH → O O O OCH3

an acetal

O O O OCH3 → 1. $LiAlH_4$ 2. H_2O → O O OH → H_2O, H^+ → O OH

an acetal

an acetal

a β hydroxy ketone

- **Read page ______.**
- **Do Problem ______, page ______.**

iv. Cyclic Hemiacetals: Rings

- **Acyclic** hemiacetals are unstable. As a result, they are difficult to isolate. On the other hand, **cyclic** hemiacetals, **lactols**, are stable and can therefore be isolated. **Hydroxy aldehydes** exist mostly as cyclic hemiacetals. The reversible conversion occurs through an **intermolecular cyclization in acidic media**. Here is an example:

- **See Mechanism on page ______.**

- **Do Problem _______, page _____.**

- **Note: A cyclic hemiacetal or lactol can always be converted to an acetal. An example:**

OH, O — $\xrightleftharpoons[CH_3OH]{H^+}$ — OCH_3, O

a cyclic hemiacetal = lactol

a cyclic acetal

- **See Mechanism on page _______.**

- **Do problems on page ______ - _____.**

i. Hydrolysis of Imines and Enamines in Acidic Media: Reverse of G. -2. f

- **General Reaction of the Hydrolysis of Imines in Acidic Media:**

$$R(R')C{=}NR'' \xrightarrow{H_2O/H^+} R(R')C{=}O + R''NH_2$$

an imine

a ketone (or aldehyde if R' = H)

a primary amine

Ex:

$$\text{(cyclopropyl)(phenyl)C=NCH}_3 \xrightarrow{H_2O/H^+} \text{(cyclopropyl)(phenyl)C=O} + CH_3NH_2$$

- **General Reaction of the Hydrolysis of Enamines in Acidic Media:**

$$\text{R–C(NR'}_2\text{)=C<} \xrightarrow{H_2O/H^+} \text{R–C(=O)–CH<} + R'_2NH$$

an enamine — a ketone

Ex:

$$\text{(cyclopropyl)C(N(CH}_2\text{CH}_3)_2\text{)=C(cyclopentyl)–CH}_3 \xrightarrow{H_2O/H^+} \text{(cyclopropyl)C(=O)–CH(cyclopentyl)–CH}_3 + (CH_3CH_2)_2NH$$

j. Cyclic Hemiacetals and Carbohydrates: Internal Cyclization

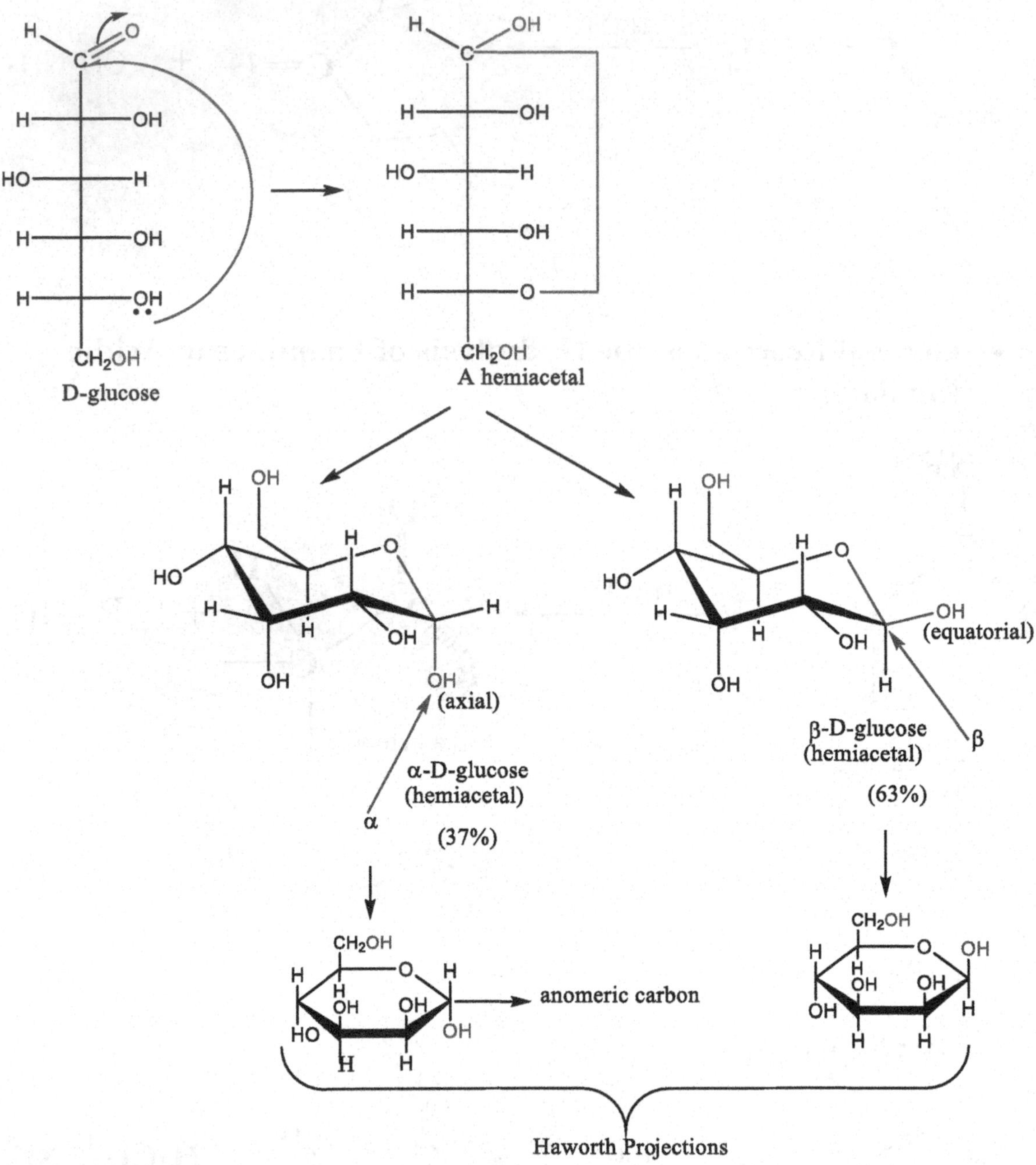

- Read about **hemiacetals** in carbohydrates on page ______.

- **Do problems on pages ______ - _____.**

- **See the structure of lactose on page _____.**

Ex:

$$\text{(cyclopropyl)(phenyl)C=NCH}_3 \xrightarrow{H_2O/H^+} \text{(cyclopropyl)(phenyl)C=O} + CH_3NH_2$$

- **General Reaction of the Hydrolysis of Enamines in Acidic Media:**

$$\underset{\text{an enamine}}{R-C(NR'_2)=C<} \xrightarrow{H_2O/H^+} \underset{\text{a ketone}}{R-C(=O)-CH<} + R'_2NH$$

Ex:

$$\text{cyclopropyl}-C(N(CH_2CH_3)_2)=C(CH_3)(\text{cyclopentyl}) \xrightarrow{H_2O/H^+} \text{cyclopropyl}-C(=O)-CH(CH_3)(\text{cyclopentyl}) + (CH_3CH_2)_2NH$$

j. Cyclic Hemiacetals and Carbohydrates: Internal Cyclization

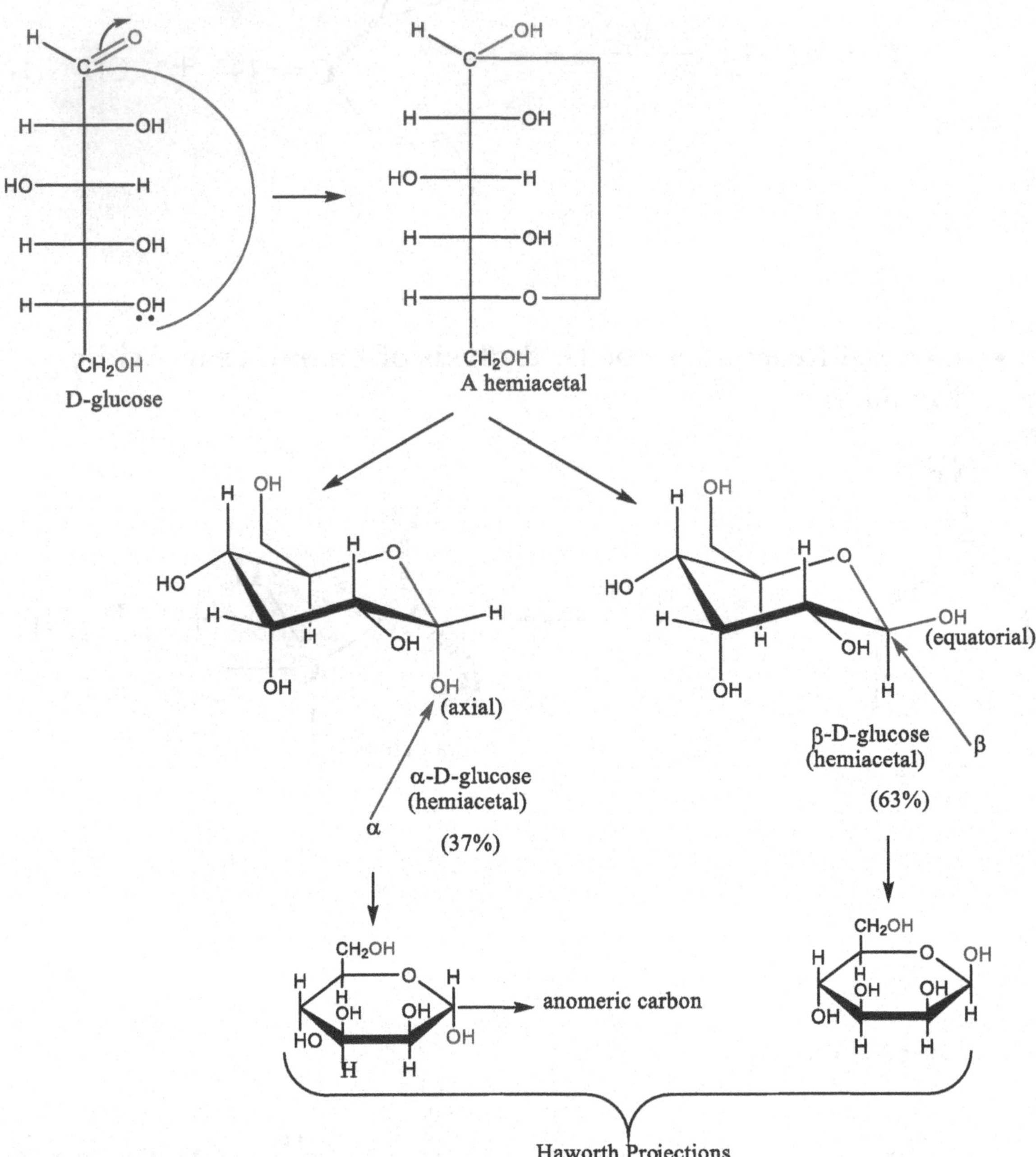

- Read about **hemiacetals** in carbohydrates on page ______.
- **Do problems on pages ______ - ______.**
- **See the structure of lactose on page ______.**

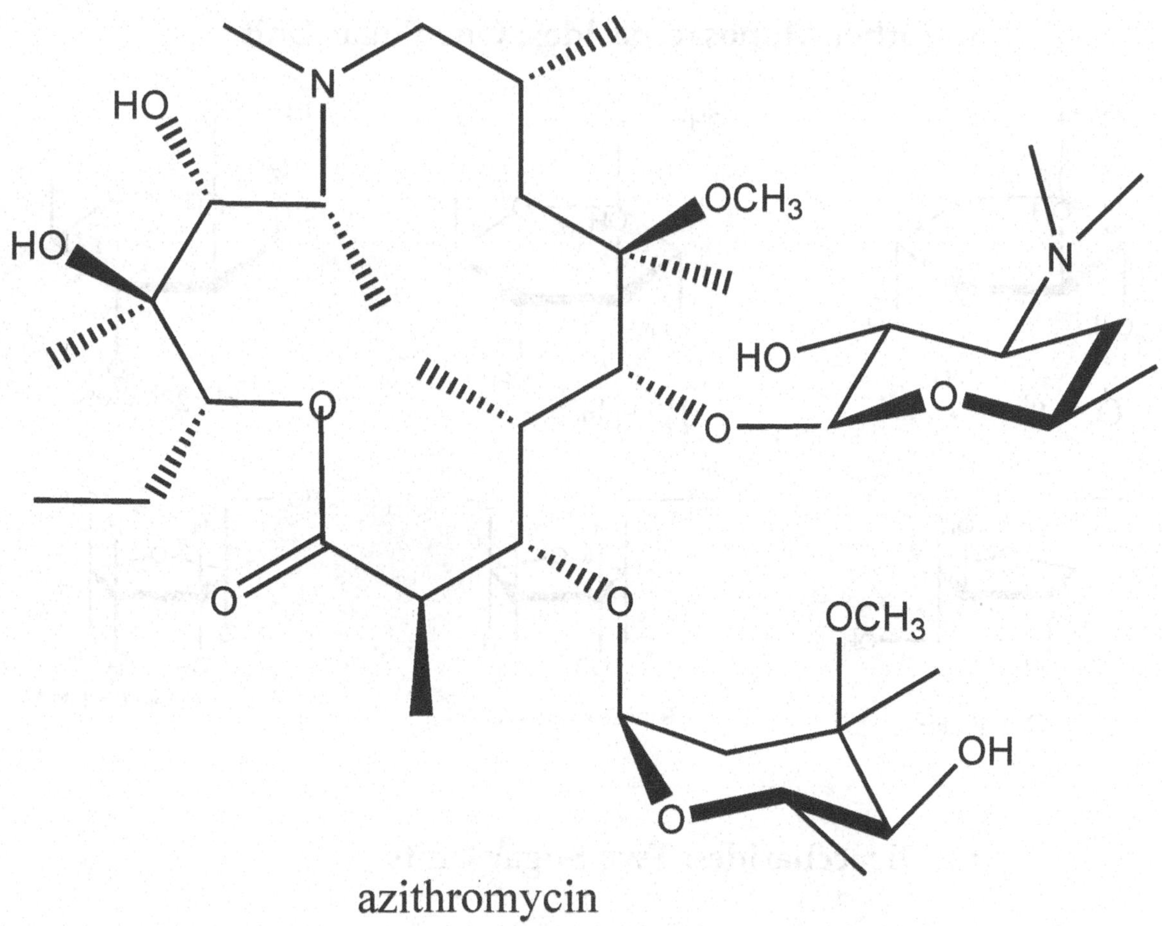

azithromycin

k. A Word about Reducing and non Reducing Sugars

- A sugar that has an **aldehyde, a ketone, a hemiacetal, or a hemiketal** is called a **reducing sugar**. Indeed reducing sugars **can reduce an oxidizing agent such as B_{r2}.**

Ex: glucose, fructose

- On the other hand, a **non reducing** sugar does not contain any of the functional groups aforementioned. Therefore, **it cannot reduce Br_2.**

Ex: Acetals, Ketals, **α** and **β** glycosides

l. Other Monosaccharides: One Sugar Unit

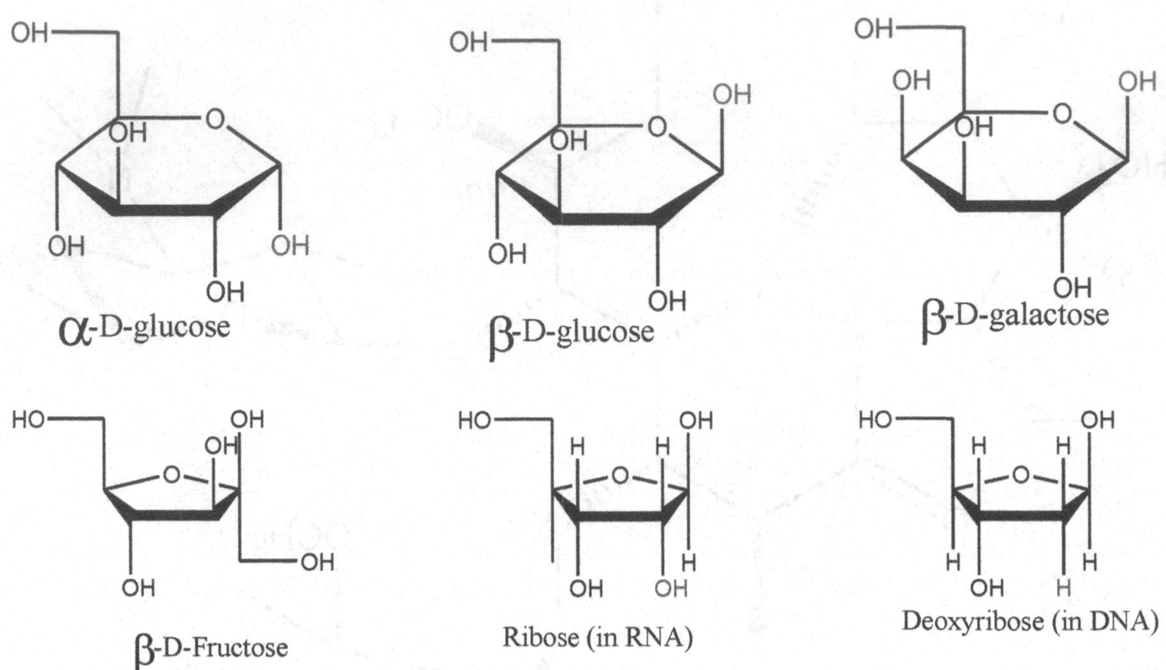

m. Disaccharides: Two Sugar Units

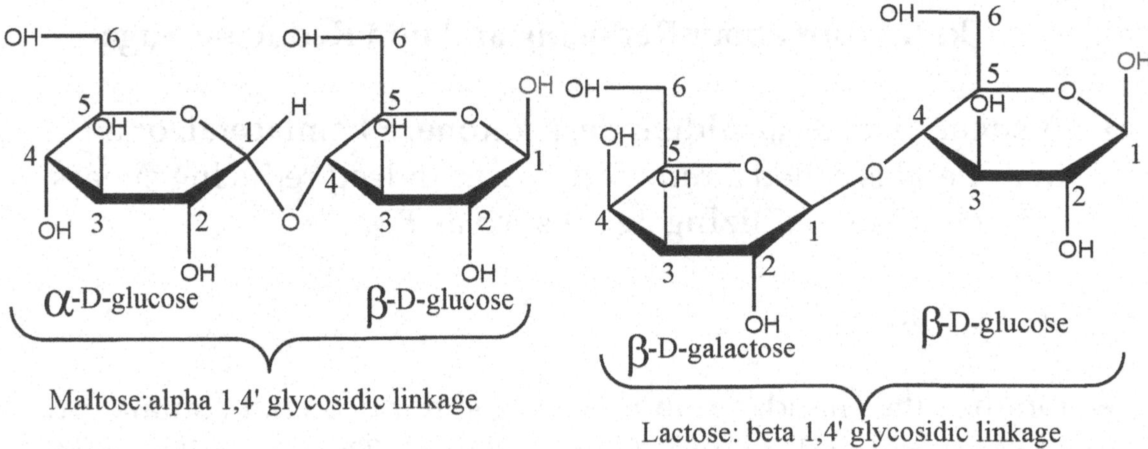

n. Polysaccharides: Several Sugar Units

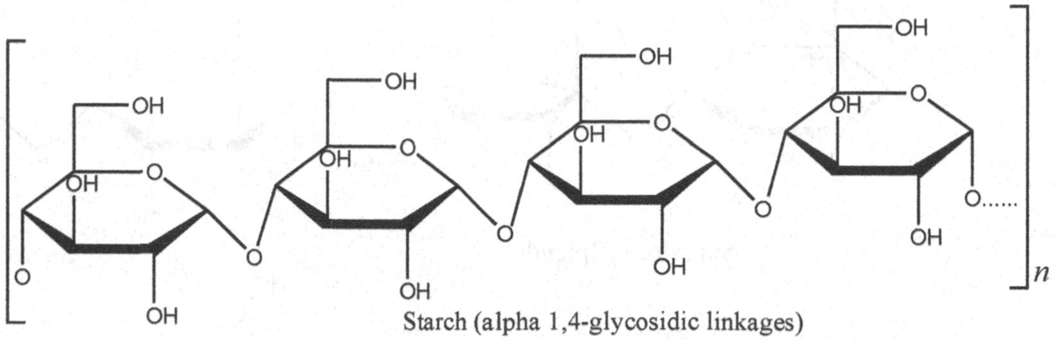

Starch (alpha 1,4-glycosidic linkages)

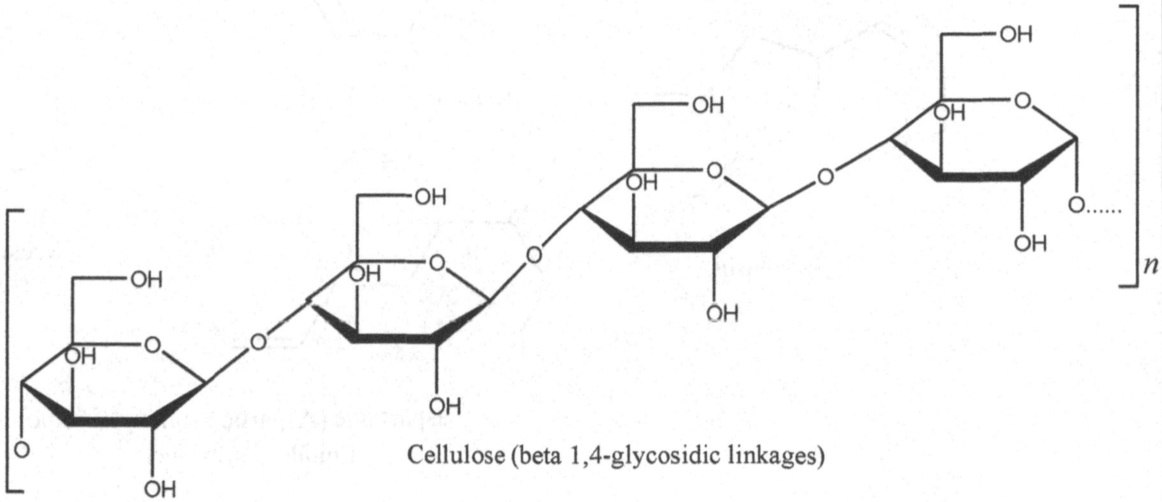

Cellulose (beta 1,4-glycosidic linkages)

o. Sucrose and Artificial Sweeteners

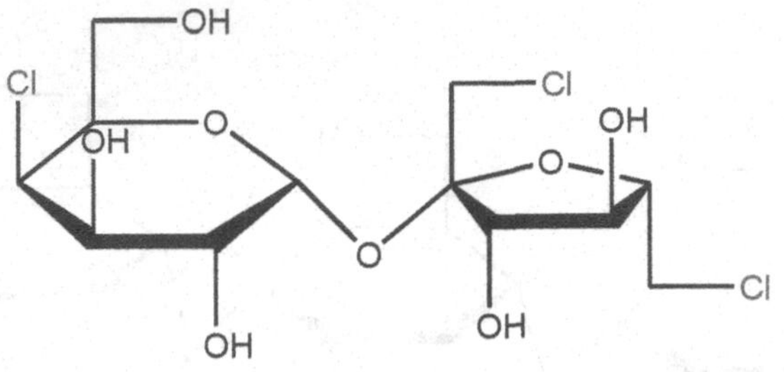

Sucralose (Splenda)

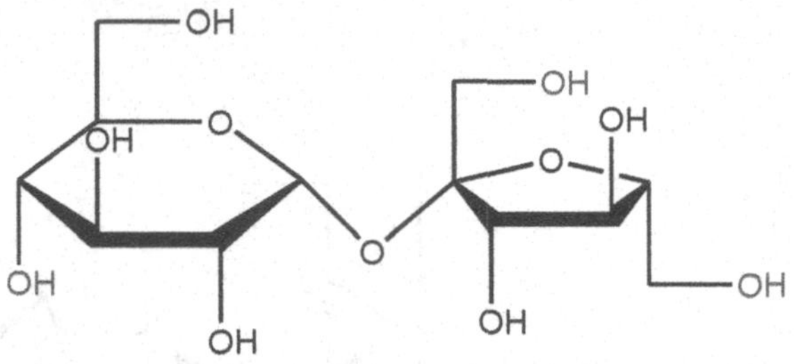

Sucrose

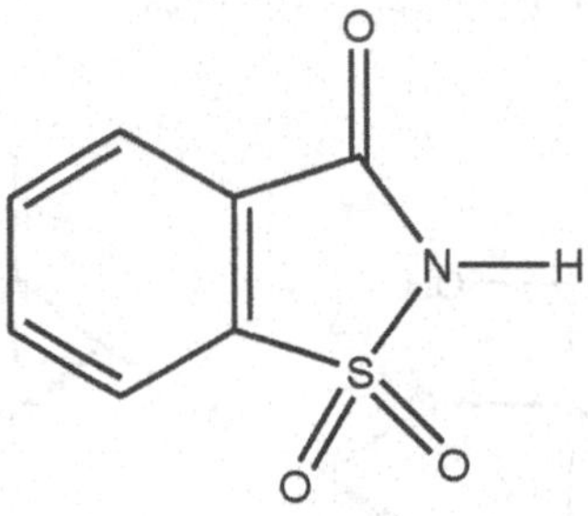

Saccharin

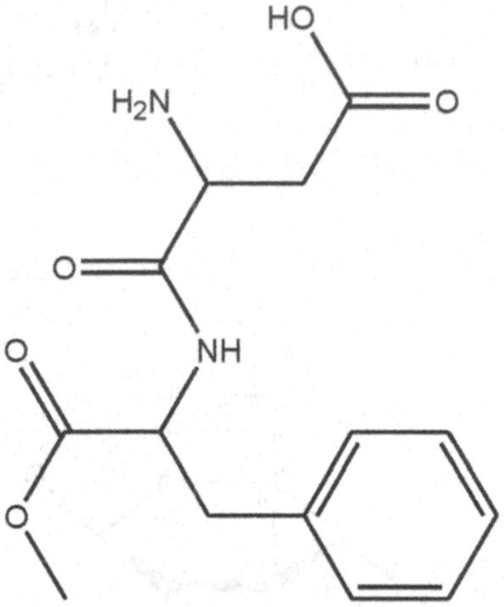

aspartame (Aspartic + phenylalanine)
(Equal, Nutrasweet)

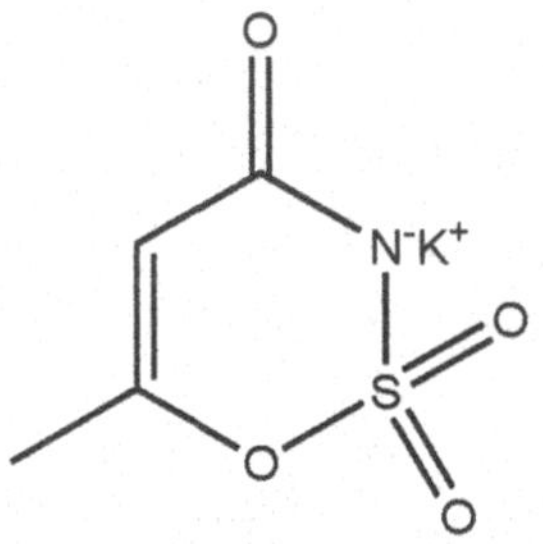

Acesulfame-K
(Sunette)

p. Carbohydrates in Blood Groups and Blood Types

i. Sugars in Blood groups

- There are **4 sugars** that are found in Blood Types: D-galactose, L-Fucose, N-Acetylglucosamine, and N-Acetylgalactosamine.

D-galactose

L-Fucose

Galactose

NH → Amine

Acetyl group

N-Acetylgalactosamine

Glucose

N-Acetylglucosamine

ii. Group Types: 4: A, B, AB, O

- Discovered by Austrian **Karl Landsteiner** (Nobel Prize in Medicine in 1930) in 1901. The following sketch summarizes it all.

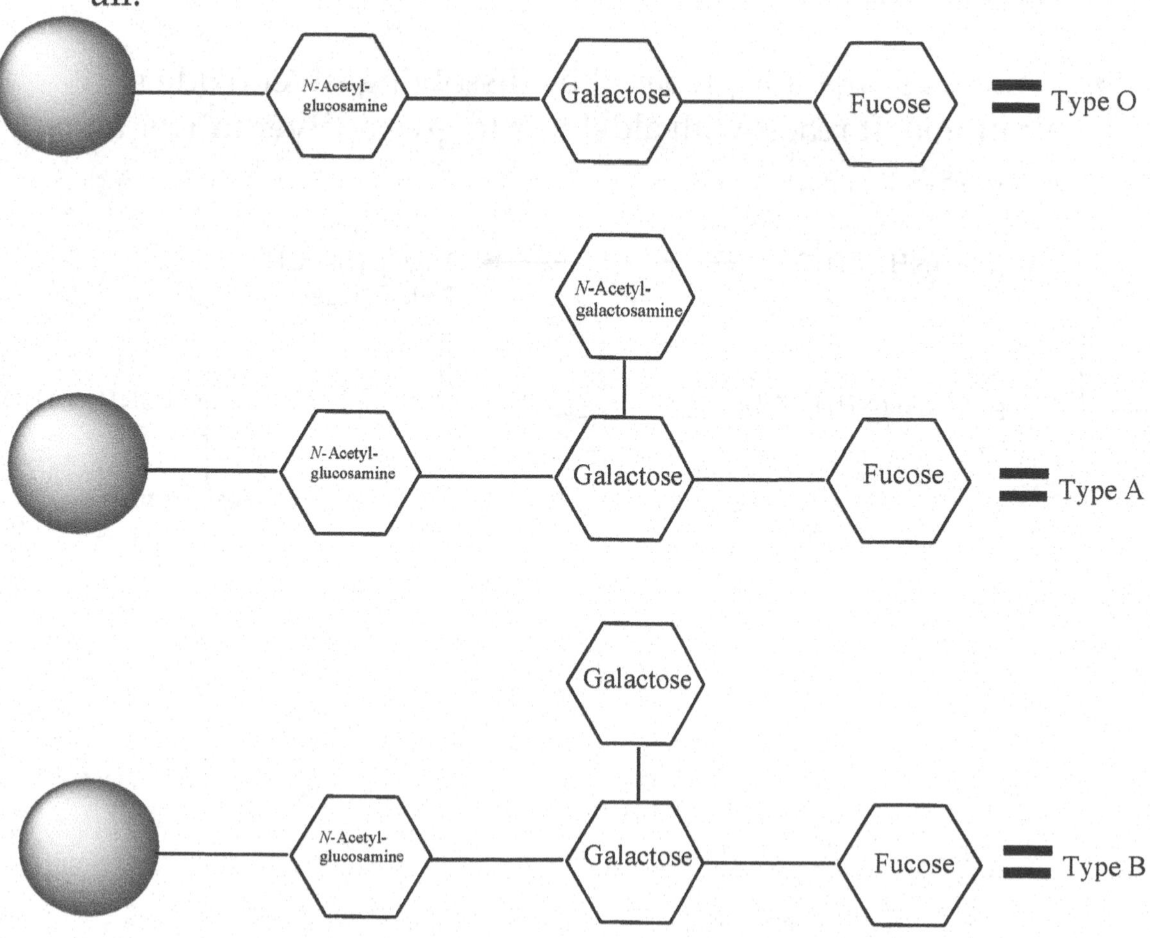

- **Note: The AB type blood contains both and A B types = Universal receiver.**

iii. Group Types and Blood Transfusion

Blood Group	Can receive from	Can donate to
A	A and O	A and AB
B	B and O	B and AB
AB	AB, A, B, and O	AB
O	O	A, B, AB, O

- **Note: O = Universal donor; AB = Universal acceptor.**

H. TESTING FOR ALDEHYDES AND KETONES

1. THE TOLLEN'S SILVER MIRROR TEST (ALDEHYDES ONLY): AN OXIDATION REACTION

- Tollen's reagent is prepared by dissolving silver oxide in ammonia. It reacts with aldehyde to give a silver mirror as follows:

$$Ag_2O + 4NH_3 + H_2O \longrightarrow \underbrace{2Ag(NH_3)_2^+ OH^-}_{\text{Tollen's reagent}}$$

$$R{-}\overset{\overset{\displaystyle O}{\|}}{C}{-}H + 2Ag(NH_3)_2^+ OH^- \longrightarrow R{-}\overset{\overset{\displaystyle O}{\|}}{C}{-}O^- NH_4^+ + \underset{\text{A silver mirror}}{\underset{\uparrow}{Ag^o}} + 3NH_3 + H_2O$$

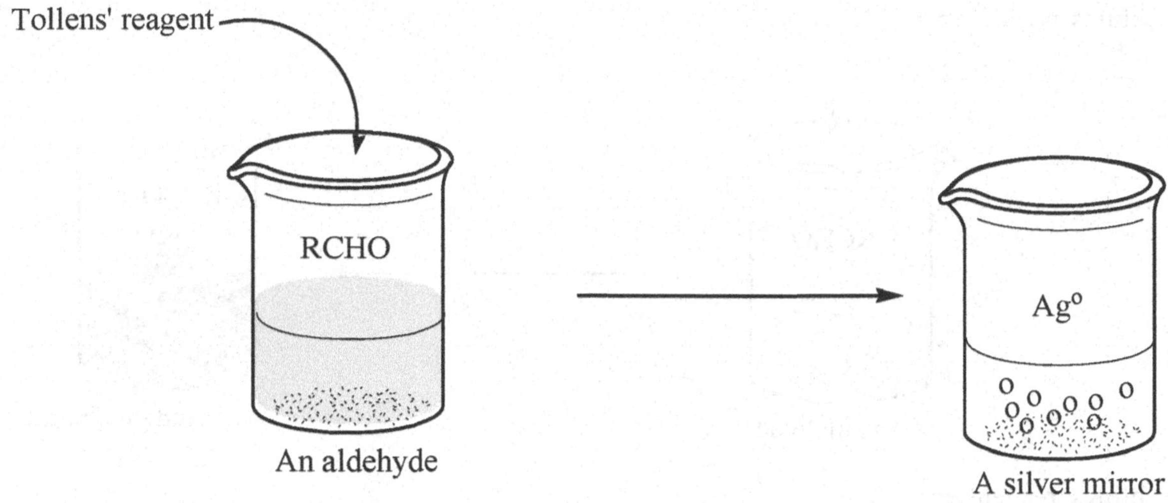

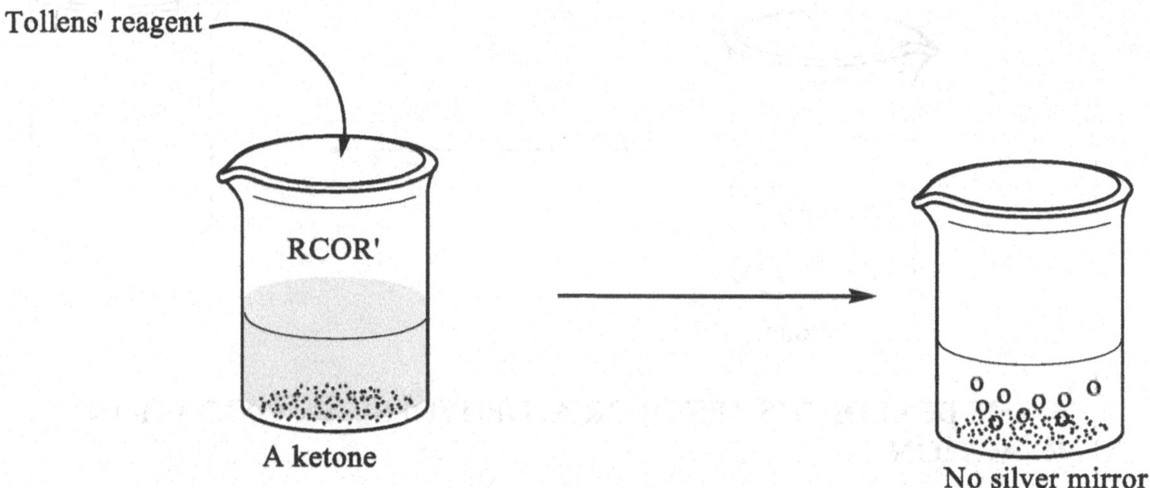

2. THE FEHLING'S TEST (FOR ALDEHYDES): AN OXIDATION REACTION

- **In the Fehling's test, a Cu^{2+}-tartrate complex in alkaline solution is used to oxidize an aldehyde to a carboxylic acid. The by-product Cu^{+} combines with O to give a Cu_2O brick red precipitate.**

$$\underset{}{RCHO} + \underset{\text{in tartariacid}}{Cu^{2+}} \longrightarrow RCOOH + \underset{\text{Brick red}}{Cu_2O(s)}$$

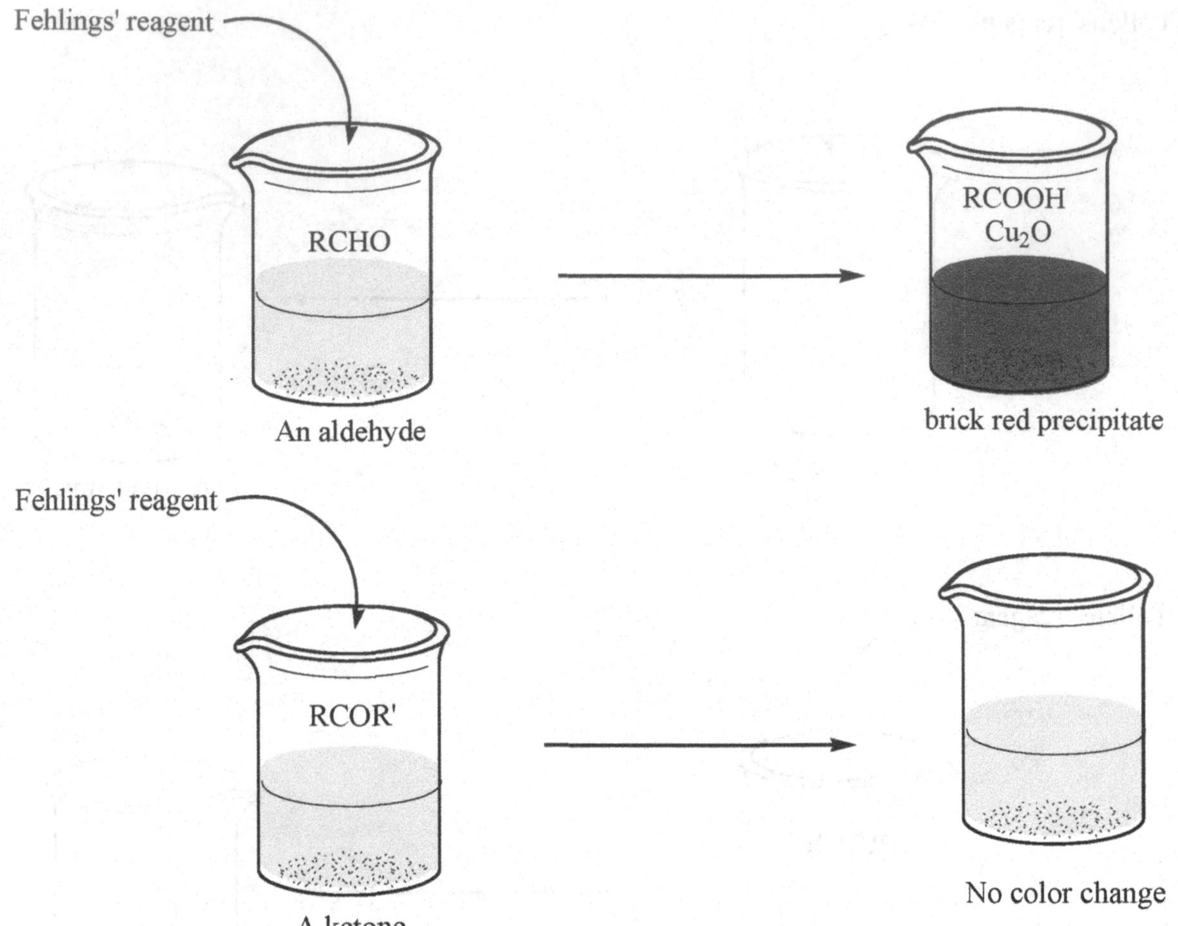

3. THE BENEDICT'S TEST (FOR ALDEHYDES): AN OXIDATION REACTION

- **The Benedict's reagent is similar to the Fehling's reagent, except citric acid is used instead of tartaric acid.**

$$\underset{\text{}}{RCHO} + \underset{\text{in citric acid}}{Cu^{2+}} \longrightarrow RCOOH + \underset{\text{Brick red}}{Cu_2O(s)}$$

4. THE BISULFITE TEST (FOR ALDEHYDES AND KETONES): AN ADDITION REACTION

- **In this test, a small amount of 20% aqueous sodium bisulfite solution is added to an aldehyde or a ketone. The reaction is:**

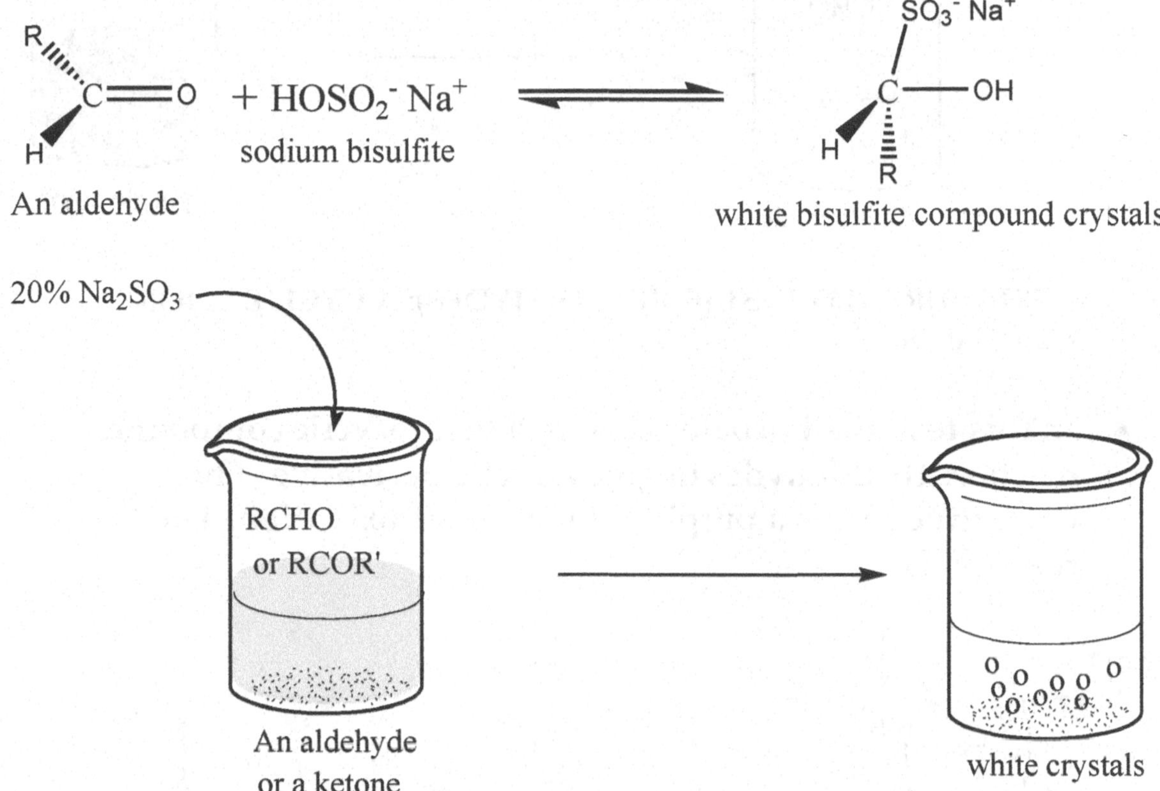

5. THE SCHIFF'S FUCHSIN-ALDEHYDE TEST (FOR ALDEHYDES ONLY): A COMPLEX ADDITION REACTION

- **In this test, Fuchsin, a faintly pink dye, is decolorized by sulfur dioxide. This reagent turns purple in the presence of an aldehyde.**

$NHSO_2H$

$^-Cl\ H_3\overset{+}{N}$— C — $NHSO_2H$

SO_3H

O

C

H

$NHSO_2$-CHR-OH

$^-Cl\ H_2\overset{+}{N}$= C

$NHSO_2$-CHR-OH

Fucsin

purple product

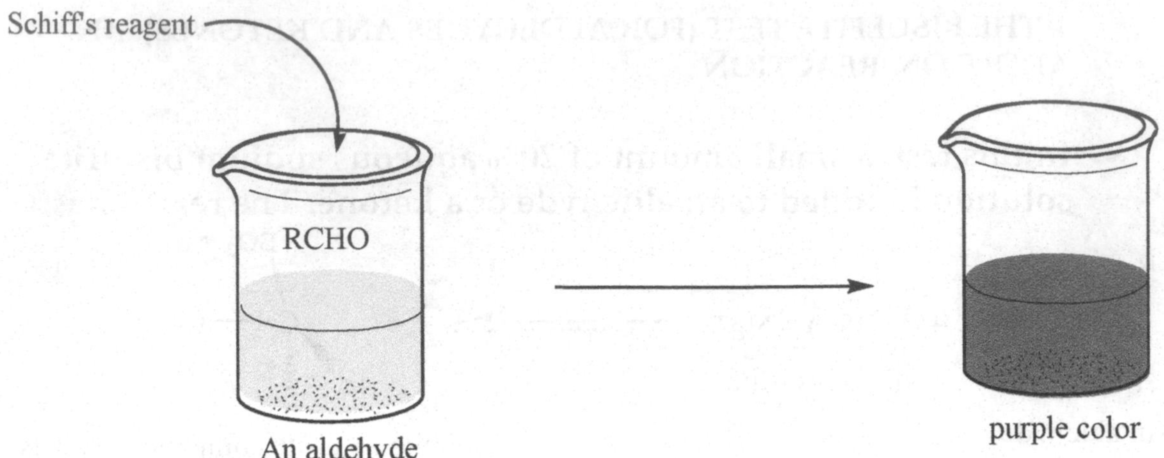

6. THE PURPALD TEST (FOR ALDEHYDES): A CYCLIZATION REACTION

- **In this test, the Purpald reagent, a heterocyclic compound, reacts with aldehydes to give a cyclic derivative. This derivative forms a purple color after air oxidation. The reaction is:**

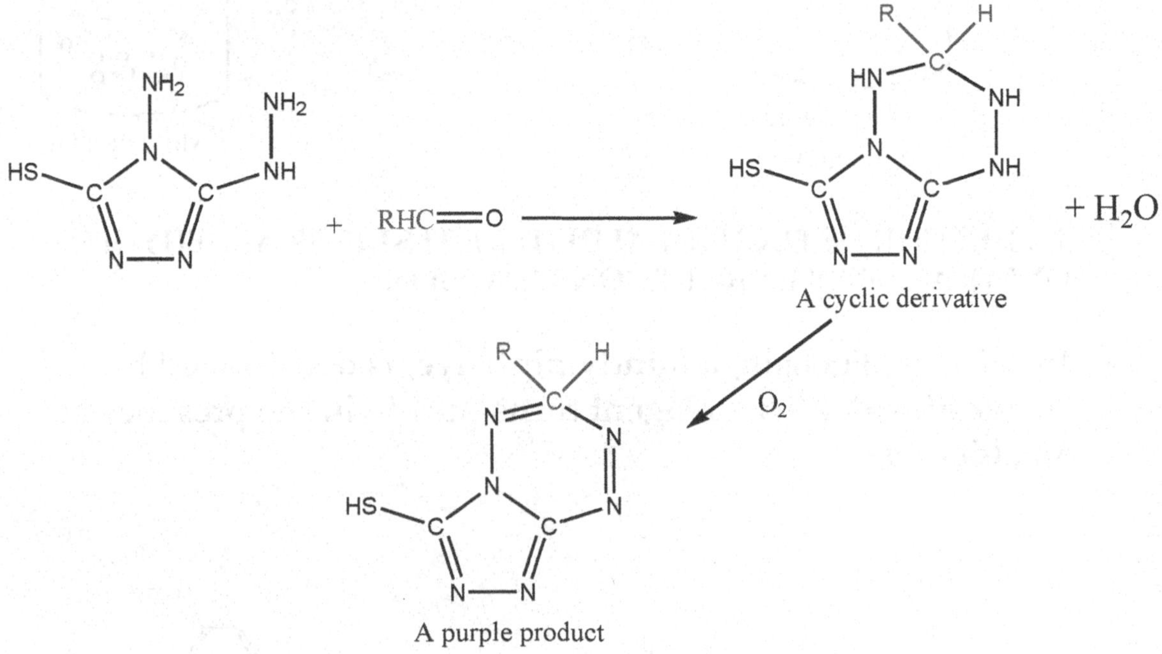

HS, NH_2, NH_2, NH + RHC=O → HS, HN, NH, NH, R, H + H_2O

A cyclic derivative

O_2 →

HS, N, N, R, H

A purple product

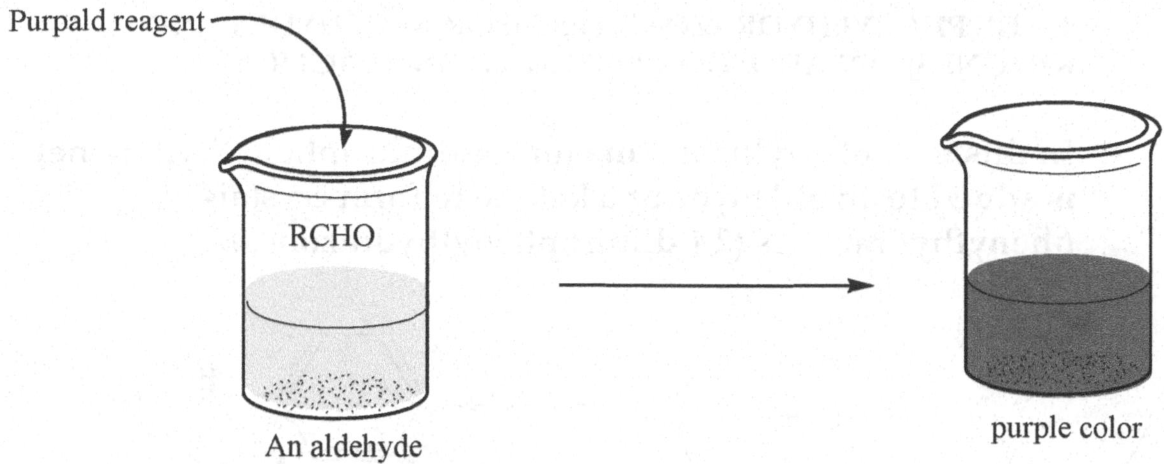

7. THE OXIME TEST (FOR ALDEHYDES AND KETONES): AN ADDITION/ELIMINATION REACTION

- **Aldehydes and ketones react with hydroxylamine(NH_2OH) to give crystalline oximes. The reaction is:**

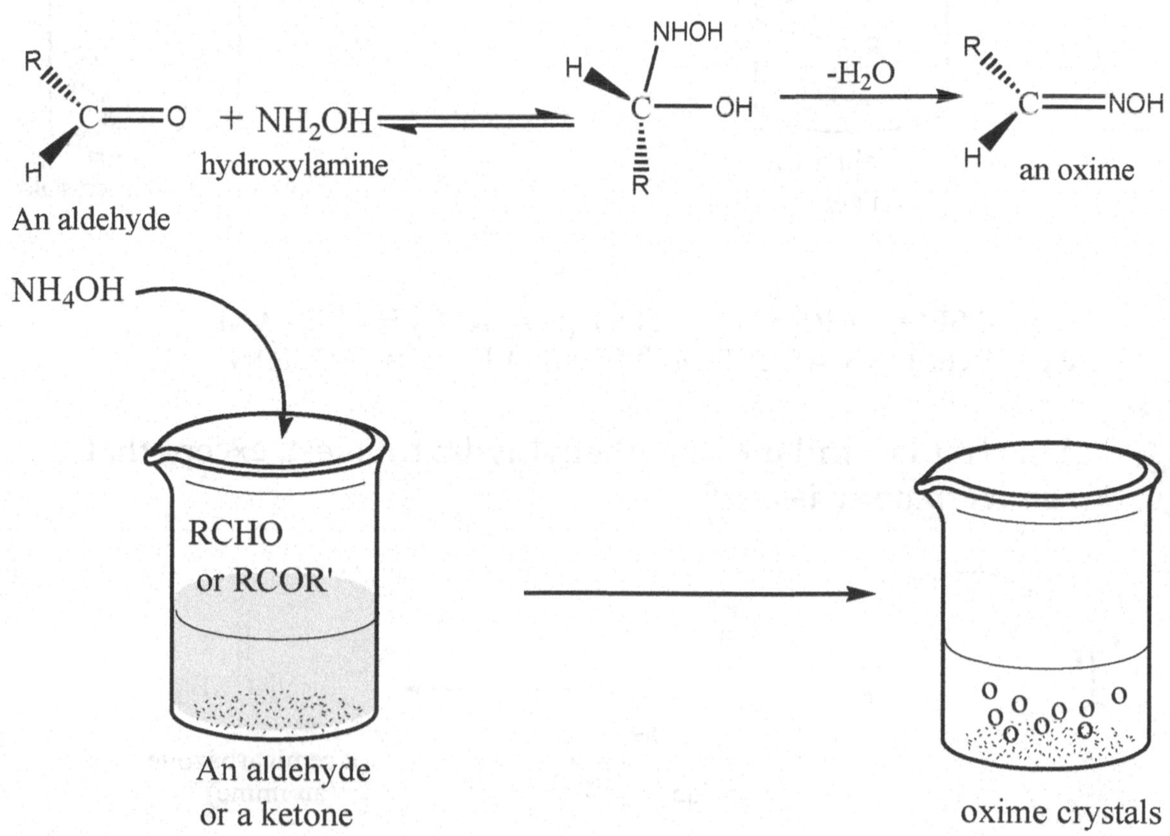

8. THE PHENYLHYDRAZONE TEST (FOR ALDEHYDES AND KETONES): AN ADDITION/ELIMINATION REACTION

- **In this test, phenylhydrazine (or 2,4-dinitrophenylhydrazine) is added to an aldehyde or a ketone to form crystals, phenylhydrazones (2,4-dinitrophenylhydrazones).**

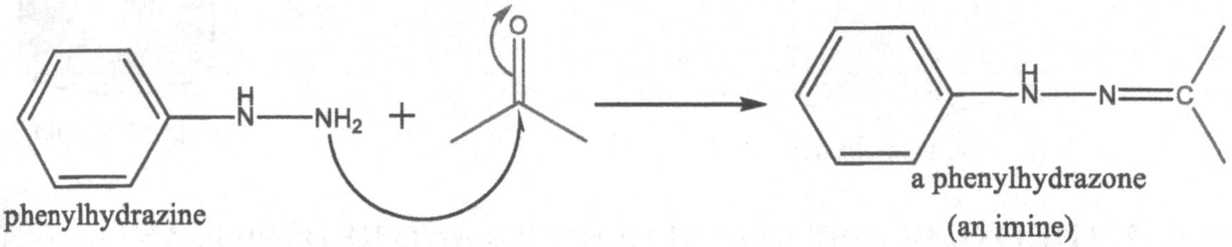

phenylhydrazine

a phenylhydrazone
(an imine)

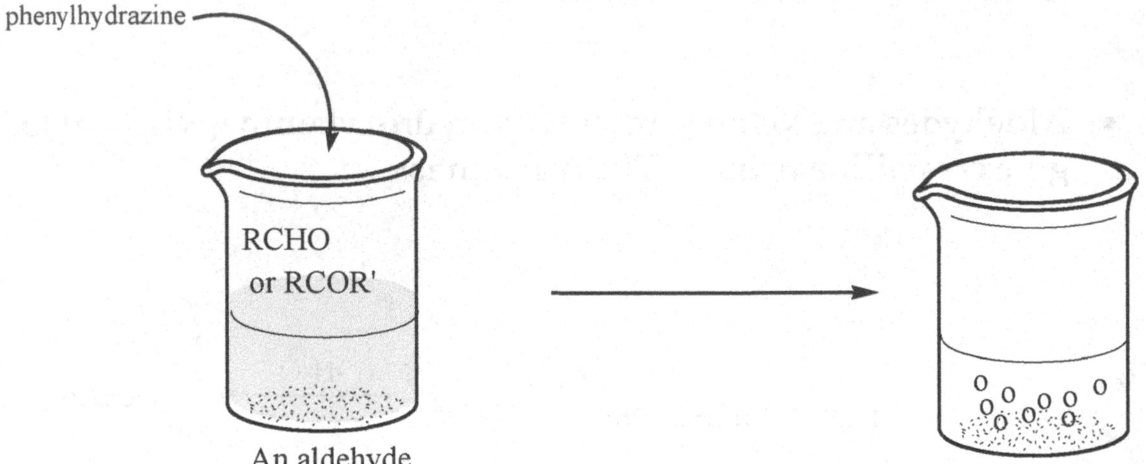

An aldehyde
or a ketone

phenylhydrazone crystals

9. THE SEMICARBAZONE TEST (FOR ALDEHYDES AND KETONES): AN ADDITION/ELIMINATION REACTION

- **This test is similar to the phenylhydrazine test, except that semicarbazide is used.**

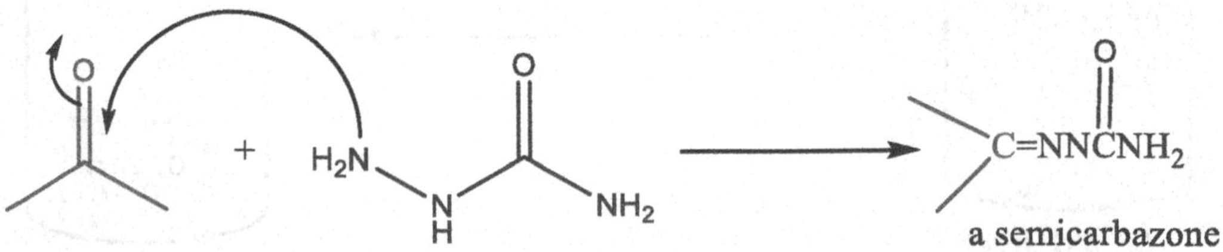

semicarbazide

a semicarbazone
(an imine)

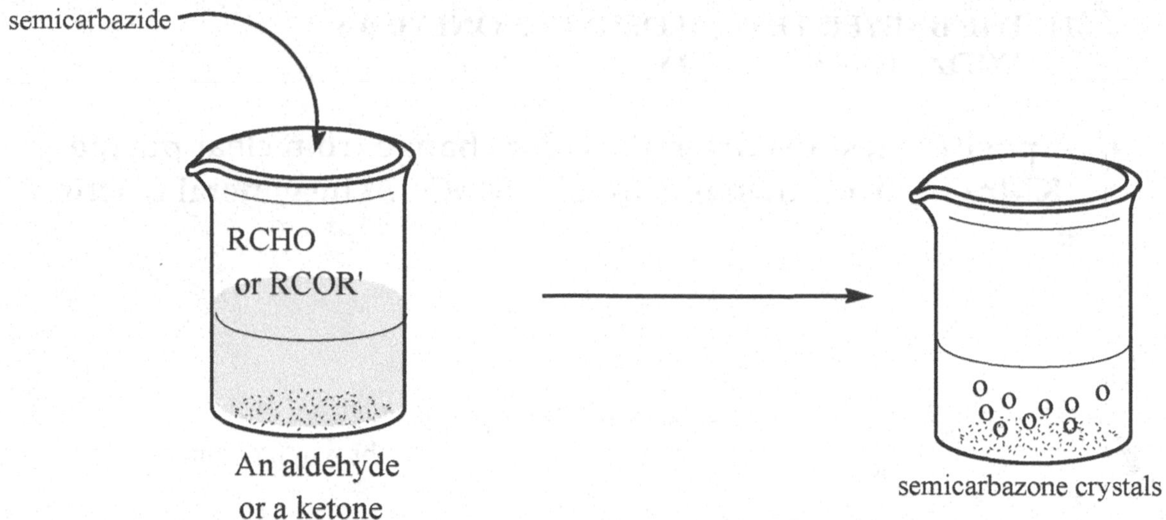

10. THE IDOFORM TEST (METHYL KETONES ONLY): REACTION

- **Methyl ketones react with iodine in alkaline media to give yellow iodoform crystals according to the reaction:**

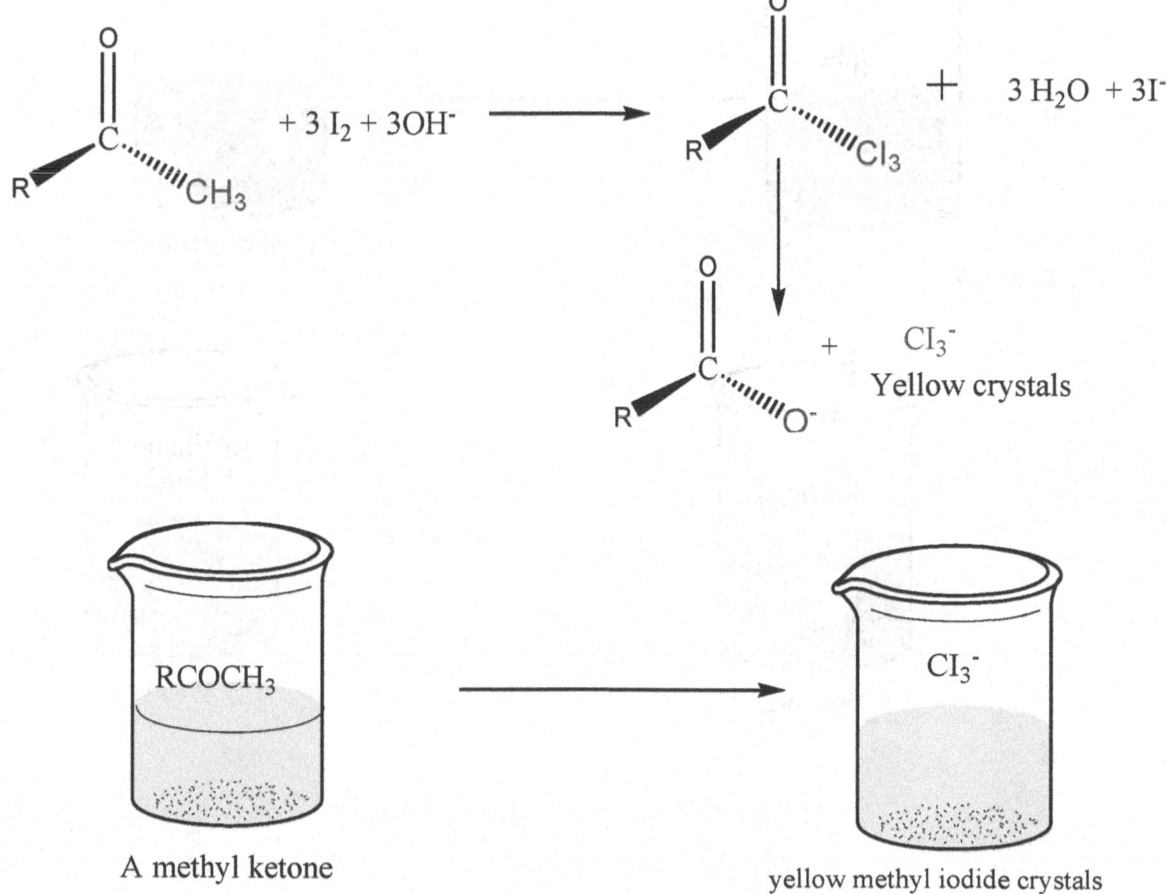

11. THE BAEYER TEST (ALDEHYDES ONLY): AN OXIDATION REACTION

- **A positive test results in the color change from clear purple ($KMnO_4$) to a brown precipitate (MnO_2). The general reaction is:**

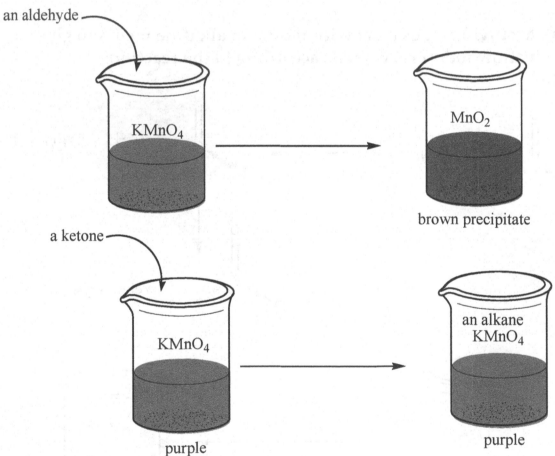

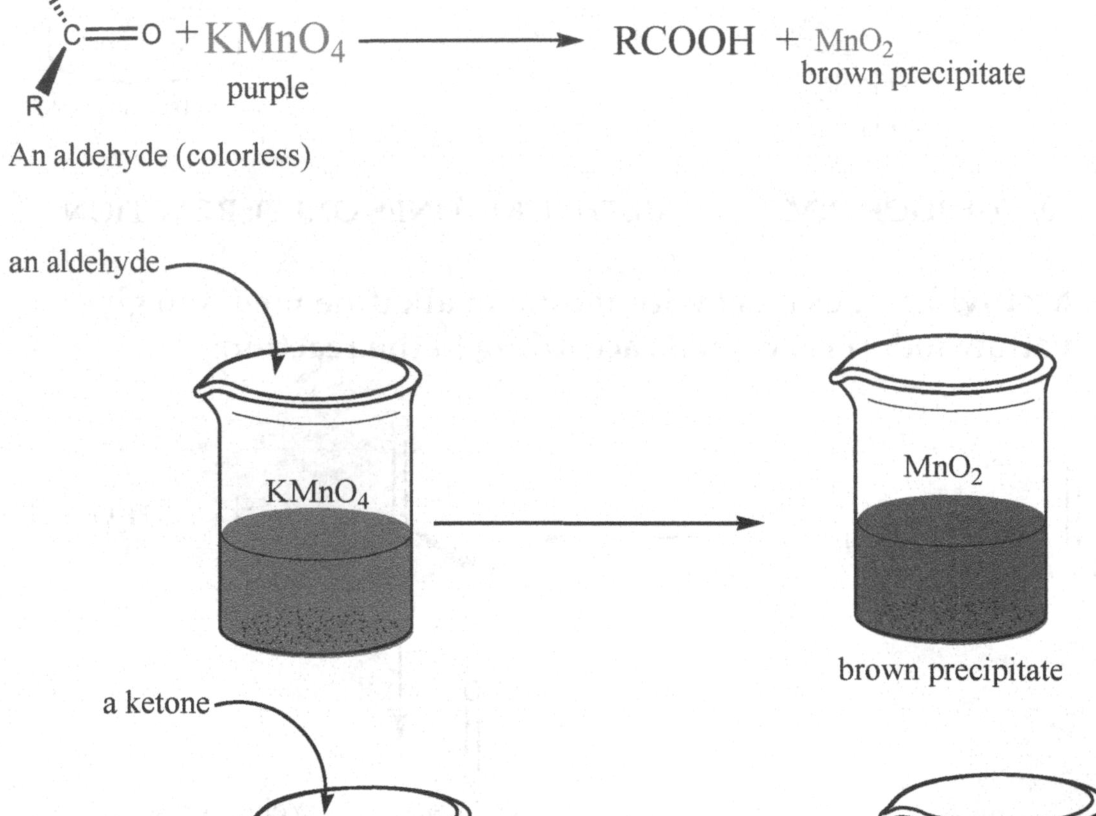

I. REACTIONS OF ALDEHYDES: A SUMMARY

$R-C(=O)-H$

- $NaBH_4/CH_3OH$ → $R-CH(OH)-H$ — a 1° alcohol
- 1. $LiALH_4$ 2. H_2O → $R-CH(OH)-H$ — a 1° alcohol
- 1. R'MgX 2. H_2O → $R-CH(OH)-R'$ — a 2° alcohol
- NaCN, HCl → $R-CH(OH)-CN$ — a cyanohydrin
- $R'NH_2$, mild acid → $R-CH=NR'$ — an imine
- R'_2NH, mild acid → $H-C(NR_2')=C<$ — an enamine
- H_2O, H^+ or OH^- (⇌) → $R-CH(OH)-OH$ — a gem-diol
- 2R'OH, TsOH (⇌) → $R-CH(OR')-OR'$ — an acetal
- $Ph_3P=CH_2$ → $R(HR)C=CH_2$ (Wittig) — an alkene

Cannizzaro's Reaction: A Disproportionation reaction

2 RCHO (an aldehyde (no α carbon)) —1. :HO⁻ 2. H_2O→ RCOOH (a carboxylic acid (oxidation product)) + RCH_2OH (anl alcohol (reduction product))

J. REACTIONS OF KETONES: A SUMMARY

$R-C(=O)-R'$

- $NaBH_4/CH_3OH$ → $R-C(OH)(H)-R'$, a 2° alcohol
- 1. $LiALH_4$ 2. H_2O → $R-C(OH)(H)-R'$, a 2° alcohol
- 1. $R''MgX$ 2. H_2O → $R-C(OH)(R'')-R'$, a 3° alcohol
- NaCN, HCl → $R-C(OH)(CN)-R'$, a cyanohydrin
- $R''NH_2$, mild acid → $R(R')C{=}NR''$, an imine
- R''_2NH, mild acid → $R-C(NR_2'')=C<$, an enamine
- H_2O, H^+ or OH^- ⇌ $R-C(OH)(OH)-R'$, a gem-diol
- $2R''OH$, TsOH ⇌ $R-C(OR'')(OR'')-R'$, an acetal
- $Ph_3P{=}CH_2$ → $R(R')C{=}CH_2$ (Wittig), an alkene

- See Key Concepts on pages _____ - _____.

OCHEM II UNIT 12: CARBOXYLIC ACIDS AND THEIR DERIVATIVES: NUCLEOPHILIC ACYL SUBSTITUTION REACTIONS

A. INTRODUCTION

1. COMMON CARBOXYLIC ACID DERIVATIVES

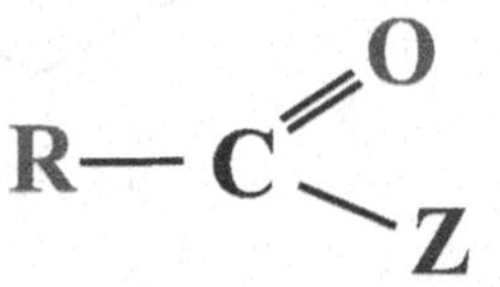

$R-C(=O)-Z$

a carboxylic acid derivative, z an electronegative atom

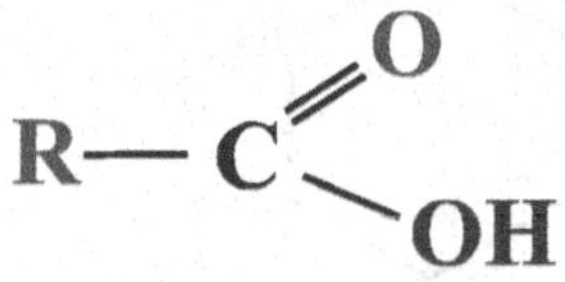

$R-C(=O)-OH$

a carboxylic acid

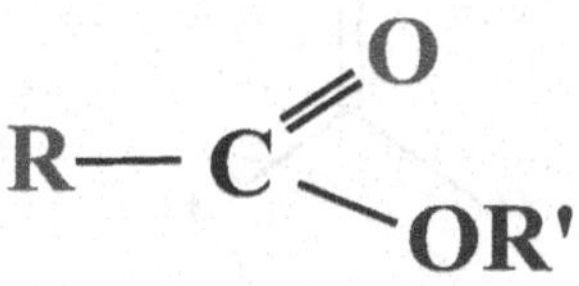

$R-C(=O)-OR'$

an ester

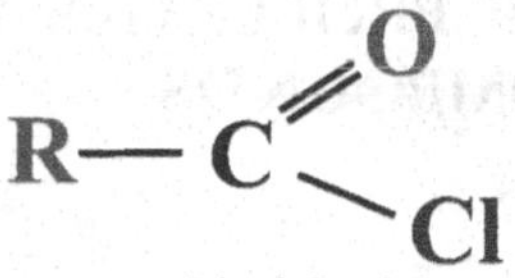

$R-C(=O)-Cl$

an acid chloride

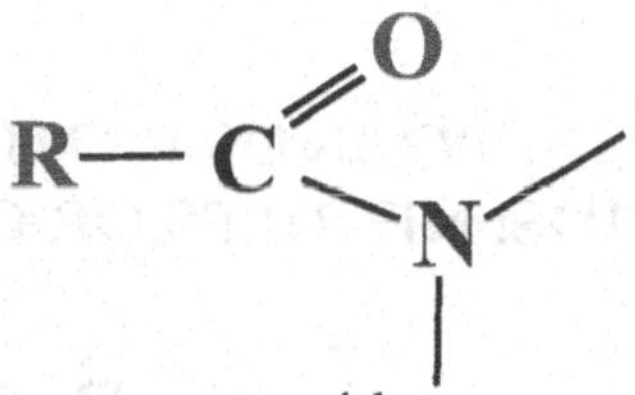

$R-C(=O)-N(CH_3)_2$

an amide

$R-C(=O)-O-C(=O)-R$

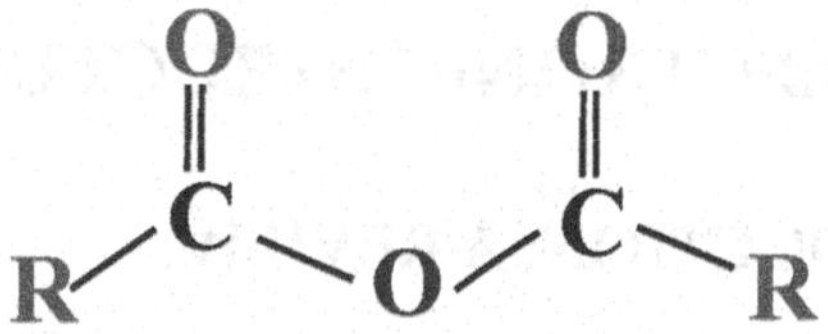

acid anhydride

- See page ______.
- Read pages _______ - _____.

2. SPECIAL CARBOXYLIC ACID DERIVATIVES

a. Lactones: Cyclic Esters

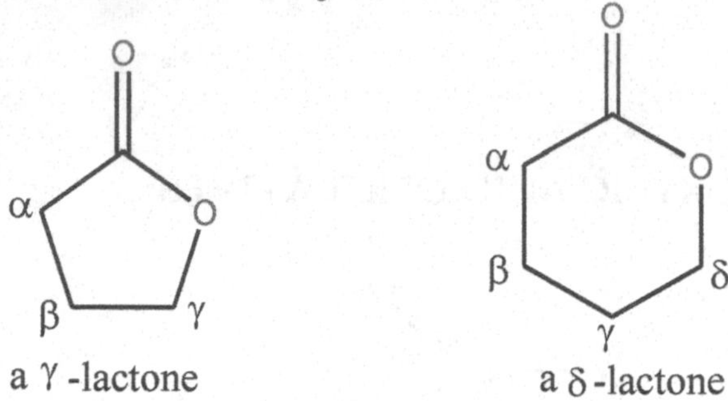

b. Lactams: Cyclic Amides

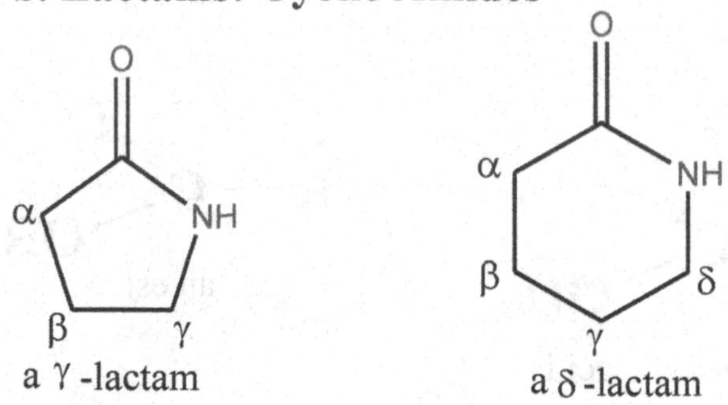

3. COMPOUNDS WITH PARALLEL PROPERTIES TO CARBOXYLIC ACIDS: NITRILES OR CYANOCOMPOUNDS

$$R\text{-}C\equiv N:$$

B. GENERAL ELECTRONIC STRUCTURE

1. INTRODUCTION: A REVIEW

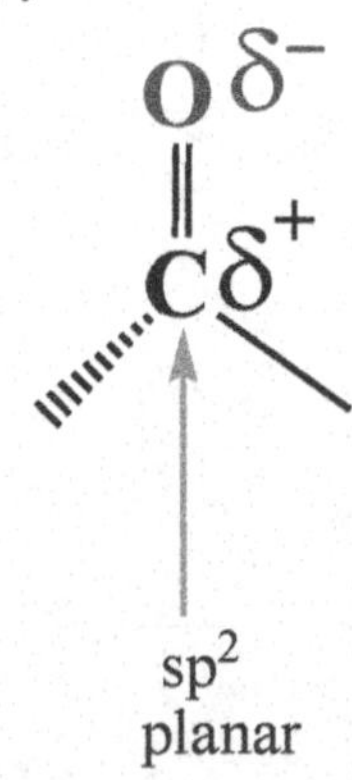

electrophilic and exposed and accessible to nucleophilic attack.

2. RESONANCE STRUCTURES OF CARBOXYLIC ACID DERIVATIVES

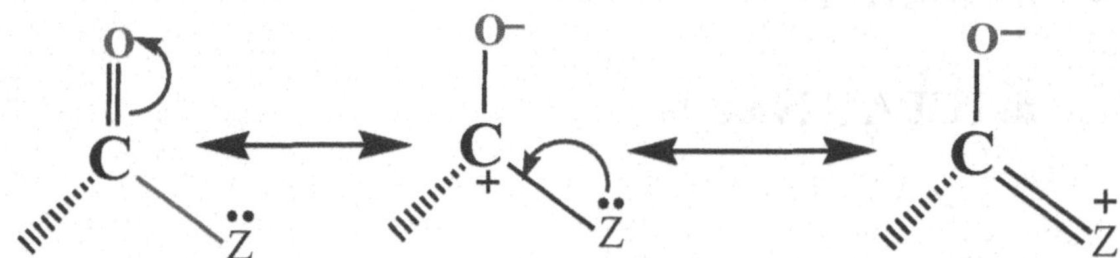

3. BASICITY OF RCOZ

- The order of basicity depends on Z. The increasing order of basicity of Z is:

  $Cl^- < RCO_2^- < OH^- \approx {}^-OR < {}^-NR_2$

- Therefore, the increasing order of basicity of RCOZ is:

  $RCOCl < RCO_2COR' < RCOOH \approx RCOOR' < RCONR_2$

- Read pages ______ - ______.

- See pKas of corresponding conjugate acids in Table _____, page _____.

- Note: The weak base whose conjugate acid has the highest pKa is the strongest.

- Do problems on pages ______ - _____.

4. A WORD ABOUT NITRILES

General Structure:

R-C≡N:

C. NOMENCLATURE OF CARBOXYLIC ACID DERIVATIVES

1. NAMING ACID CHLORIDES

a. IUPAC Names

i. Naming Acyclic Acid Chlorides

- **Recall: Acid chlorides come from carboxylic acids. So their names derive from those of carboxylic acids.**

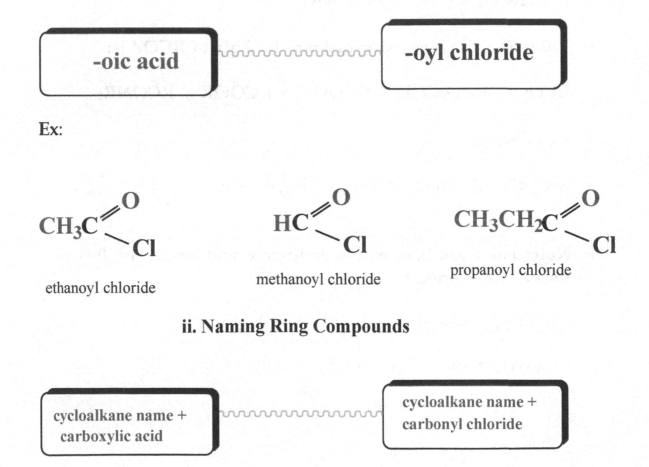

-oic acid ∿∿∿ **-oyl chloride**

Ex:

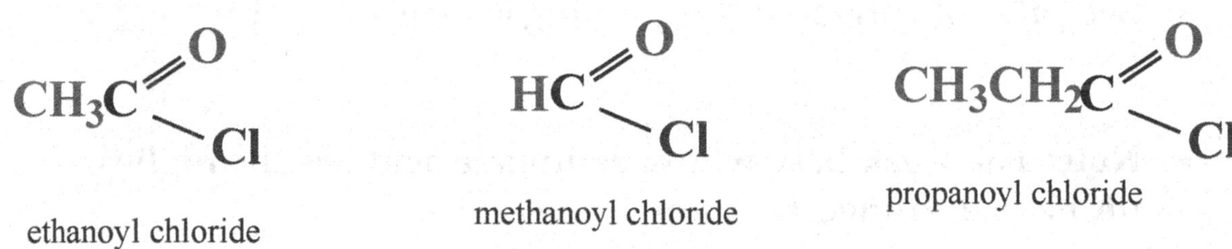

ii. Naming Ring Compounds

Ex:

Cyclohexanecarbonyl chloride

- See page _______.

b. Common Names

- **Review common names prefixes:**

Number of carbons	root
1	form-
2	acet-
3	propion-
4	butyr-
5	valer-
6	capro-

- Acid chlorides are named as:

Ex:

$CH_3C(=O)Cl$ $\quad$ $HC(=O)Cl$ $\quad$ $CH_3CH_2C(=O)Cl$

acetyl chloride

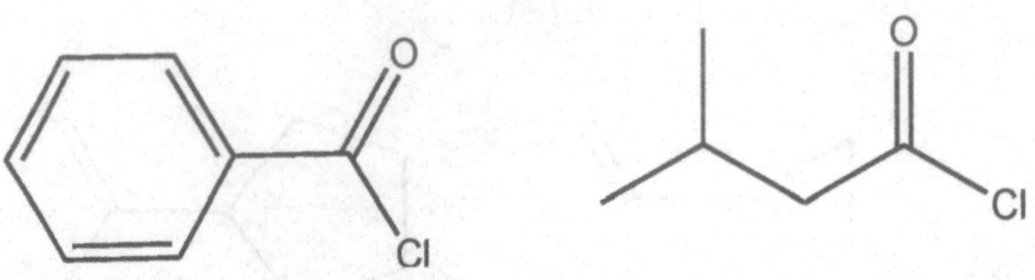

2. NAMING ACID ANHYDRIDES

a. Introduction

Acid anhydrides derive from two carboxylic acids.

b. Naming Symmetrical Anhydrides: RCOOCOR

Ex:

Acetic anhydride

c. Naming Mixed Anhydrides: RCOOCOR′

Name prefixes alphabetically.

names of two carboxylic acids

two prefixes + anhydride (alphabetically)

Ex:

Acetic benzoic anhydride

3. NAMING ESTERS

a. Introduction

- Esters are made of an **acyl group** and an **alkyl group.**

alkyl group

acyl group

b. General Nomenclature: IUPAC and Common Names

- **Name of R′ + Root of Name of Acid + -oate or -ate**

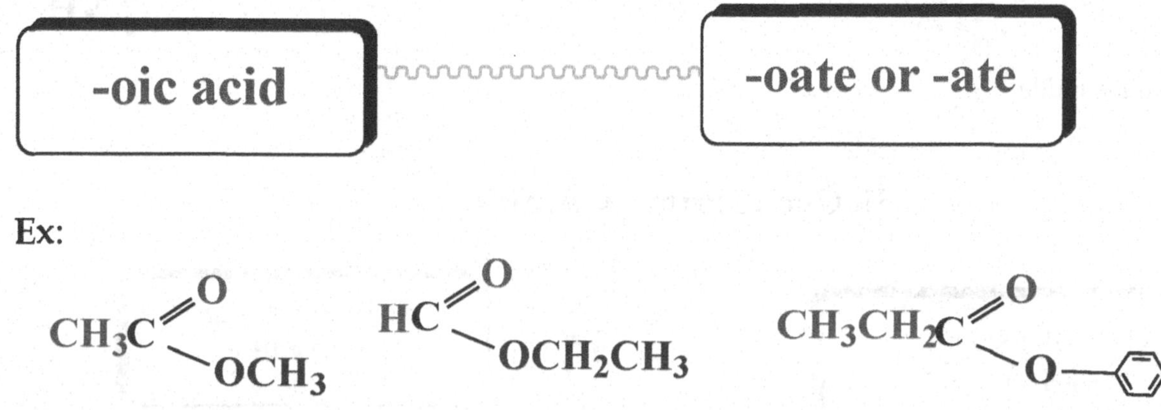

IUPAC: methyl ethanoate
common: methyl acetate

c. Special Ring Compounds: Some Examples

Name of R′ + Name of Cycloalkane Ring + Carboxylate

OCH_3 OCH_2CH_3 OCH_2CH_2CH

Methyl cyclohexanecarboxylate

4. NAMING AMIDES

a. IUPAC Names of 1° Amides

i. Acyclic Primary Amides

- Names end in **amide or carboxamide.**

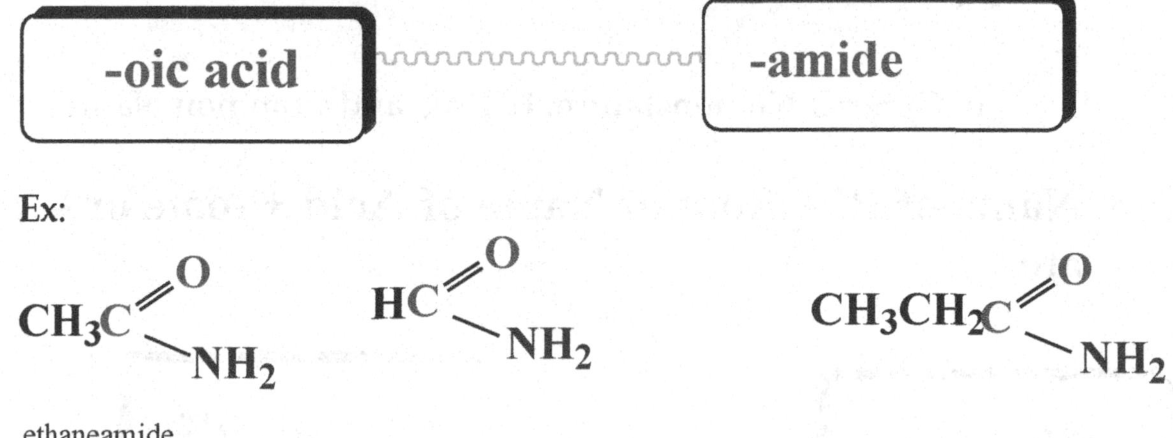

ii. Cyclic Primary Amides

Ex:

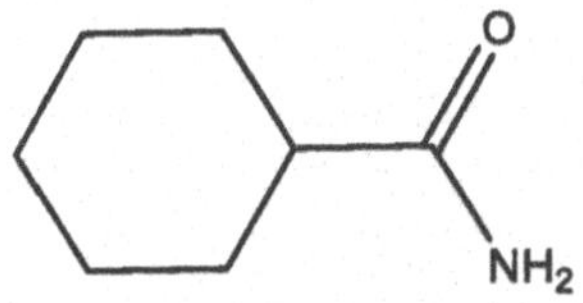

Cyclohexanecarboxamide

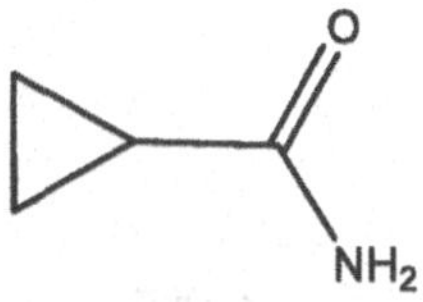

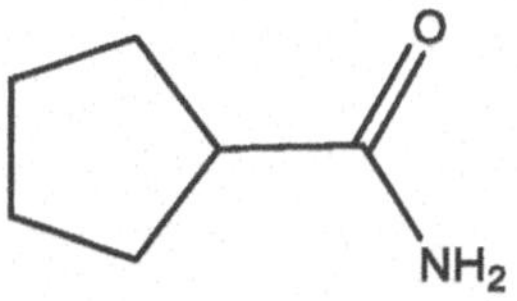

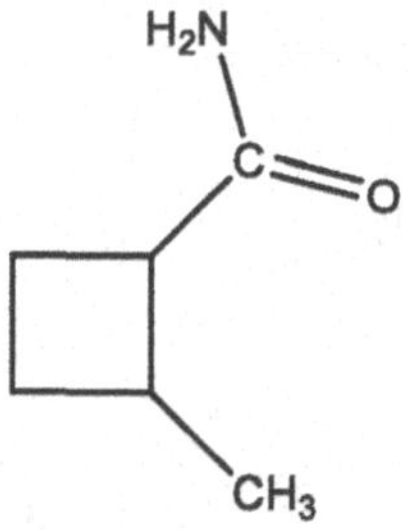

b. Common Names

prefix + -ic acid ∿∿∿∿∿ prefix + -amide

Ex:

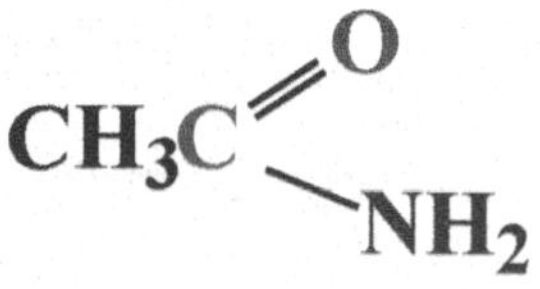

$CH_3C(=O)NH_2$

acetamide

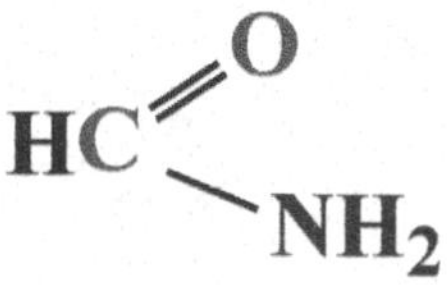

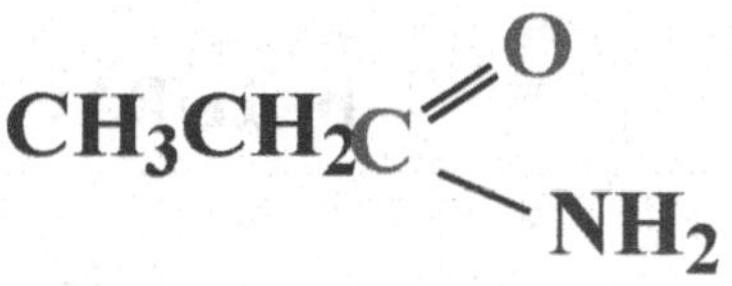

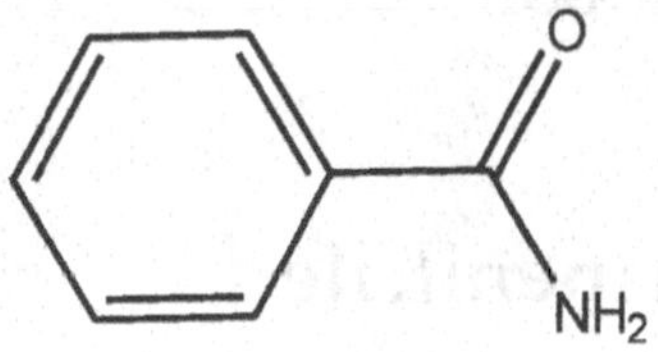

$HC(=O)NH_2$

$CH_3CH_2C(=O)NH_2$

c. Naming 2° and 3° Amides

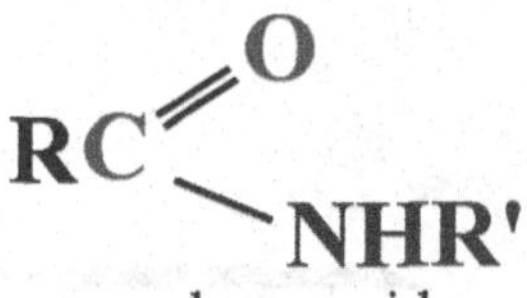

$RC(=O)NHR'$

a secondary amide

$RC(=O)NR_2'$

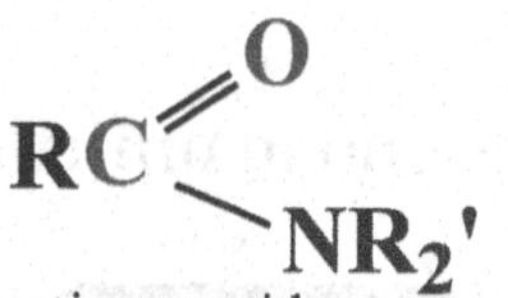

a tertiary amide

- **Named by using N or N- prefix+, N-prefix before each R' group**

Ex:

$$CH_3CH_2C(=O)NHCH_3 \qquad CH_3C(=O)N(CH_3)_2$$

5. NAMING NITRILES

a. Introduction

- Names end in -nitrile.

b. IUPAC Names

- -Named as **alkanenitriles.**
- -The parent chain is **the longest carbon chain that contains the -C≡N: group**.
- -The C≡N carbon is carbon # **1**.

Ex:

CH_3CN = Ethanenitrile

$CH_3CH_2CH_2CN$ =

c. Common Names

- Common names end in **onitrile.**

-common name prefix	+	**-onitrile**

Ex:

CH_3CN =

$CH_3CH_2CH_2CN$ =

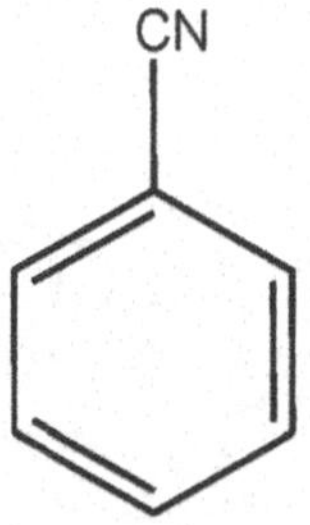

CN

benzonitrile

d. –CN as a Substituent: Cyano Group

Ex:

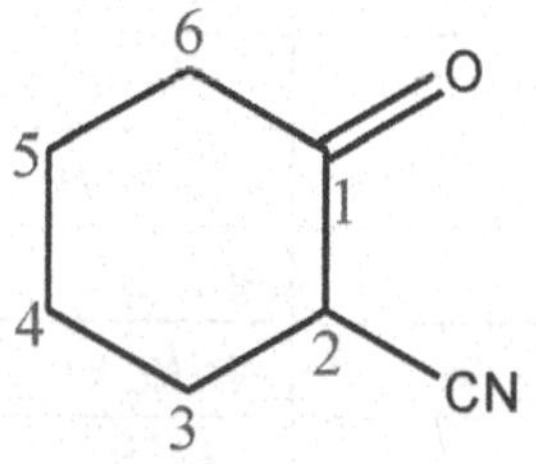

6 O 5 1 4 2 CN 3

2-Cyanocyclohexanone

6. NAMING POLYFUNCTIONAL CARBOXYLIC ACID DERIVATIVES: ESTERS AND AMIDES

a. Group Priorities: Déjà Vue

- When two or more different functional groups are present in a compound, functional group priorities are used. Group priorities are assigned based on the following Table (decreasing order of priority).
- **See Unit 9.**

Priority order	Group	Ending of name as a priority	Name as a non **Priority group**
Carboxylic acid	**RCOOH**	*-oic acid*	*-carboxy*
Ester	**RCOOR′**	*-oate*	*-alkoxycarbonyl*
Amide	**$RCONH_2$**	*-amide*	*-amido*
Nitrile	**RCN**	*-nitrile*	*-cyano*
Aldehyde	**RCHO**	*-al*	*-oxo(=O) or formyl(-CHO)*
Ketone	**RCOR′**	*-one*	*-oxo*
Alcohol	**ROH**	*-ol*	*-hydroxy*
Amine	**RNH_2**	*-amine*	*-amino*
Alkene	**-C=C-**	*-ene*	*-alkenyl*
Alkyne	**-C≡C-**	*-yne*	*-alkynyl*
Alkane	**-C-C-**	*-ane*	*-alkyl*
Ether	**ROR′**	*-none*	*-alkoxy*
Halide	**R-X**	*-none*	*-halo*

- **Note: Esters and amides have a higher priority than nitriles, alcohols, ketones, etc.**

b. Some Examples

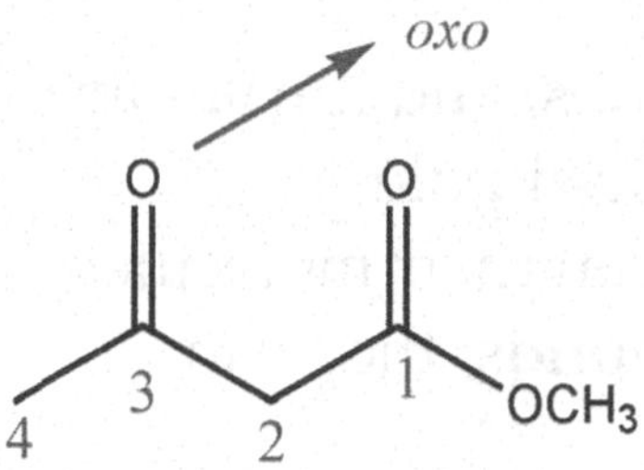

oxo

O O

4 3 2 1 OCH_3

methyl 3-*oxobutanoate*

OH O

NH_2

NH_2 O

O OCH_2CH_3

OCH_3 O

NH_2

O NH_2

CHO

OH O

NH_2

CHO O

OCH_2CH_3

$OCH_2CH_2CH_3$

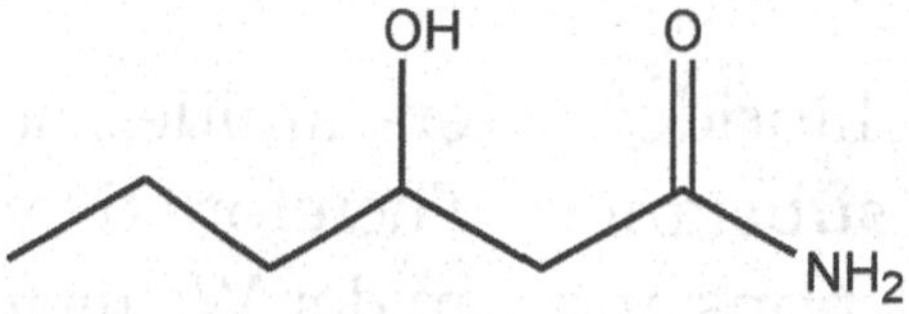

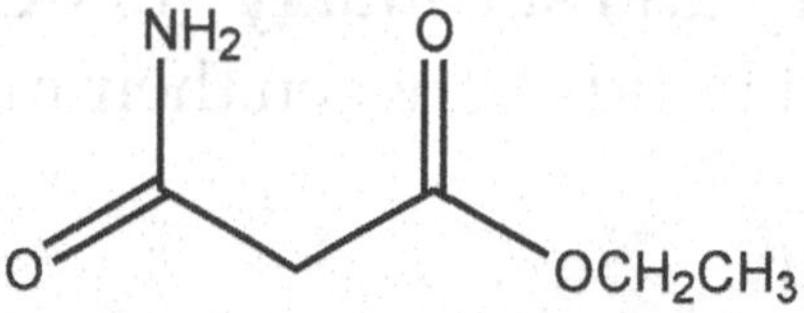

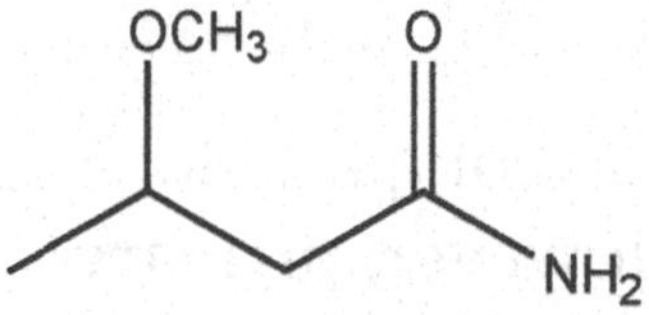

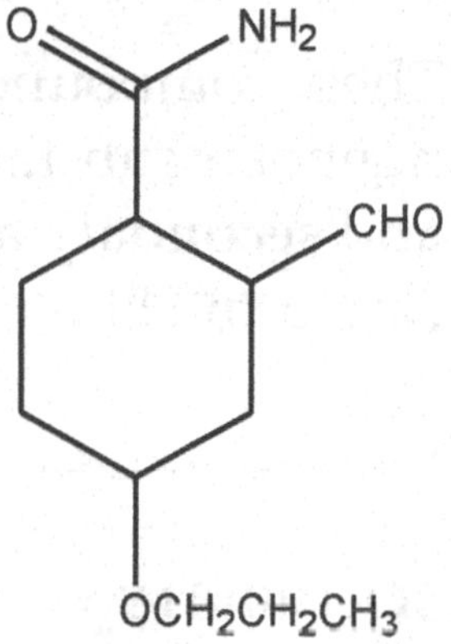

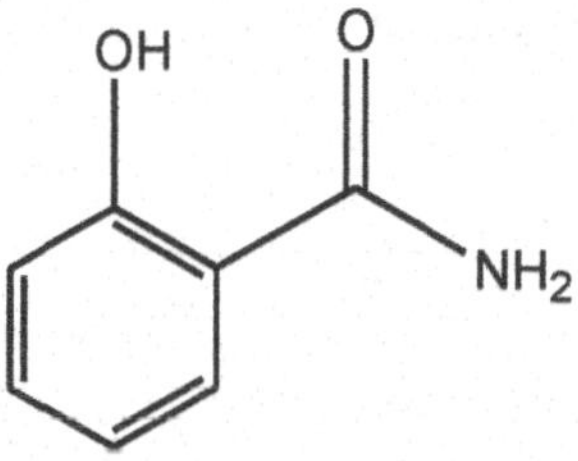

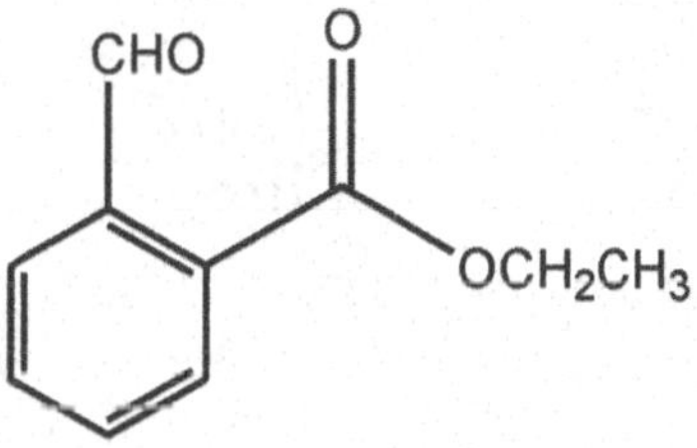

- **See Table _______, page 876: A Summary.**

- **Read pages ______ - ______.**

- **Do problems on pages ______ - ____.**

D. PHYSICAL PROPERTIES OF CARBOXYLIC ACIDS AND DERIVATIVES

1. INTRODUCTION

- Acid chlorides, esters, amides, acid anhydrides, and nitriles are **polar substances**. Therefore they have dipole-dipole interactions and van der Waals forces between their molecules. **Primary and secondary amides** have **N-H bonds**; they can have H bonds between their molecules.

2. BP + MP

- These compounds have higher BP and MP than comparable nonpolar substances of similar molar mass. However, **primary and secondary amides** have higher MP and BP because they can form H bonds.

- **See examples on page _____, Fig. _____.**

3. SOLUBILITY

- All sizes of carboxylic acids and derivatives are soluble in organic solvents.
- Carboxylic acids and derivatives with a number of carbons ≤ 5 are soluble in water because of their ability to **H bond** with water through the **-C=O group**.
- Carboxylic acids and derivatives with a number of carbons > 5 are insoluble in water because of the nonpolar, hydrophobic alkyl group attached to the -C=O group.

- **See Table ______, page _______. Read about solubility.**

- **Do Problem _______, page ______.**

E. SPECTROSCOPIC PROPERTIES

1. IR

- The **-C=O** of carboxylic acids and derivatives appears as a **strong band** between 1600 and 1850 cm^{-1}.
- The N-H stretch is between 3200 and 3400 cm^{-1}.
- The -C≡N band is at 2250 cm^{-1}.
- See Table _______, page ______.
- **Recall: Conjugated carbonyl groups absorb at lower frequencies. In other words, conjugation shifts the -C=O band to lower frequencies. Indeed for conjugated systems, frequency decreases with increasing conjugation because the -C=O in the conjugated system has a partial single bond character due to resonance (See Units 6 and 11).**

Ex:

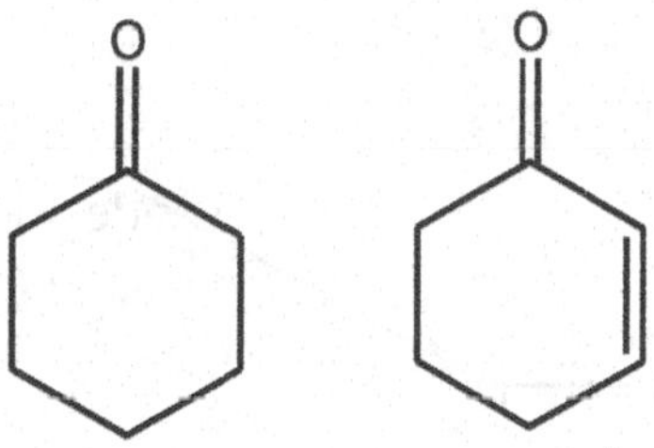

1715 cm^{-1}

1685 cm^{-1}

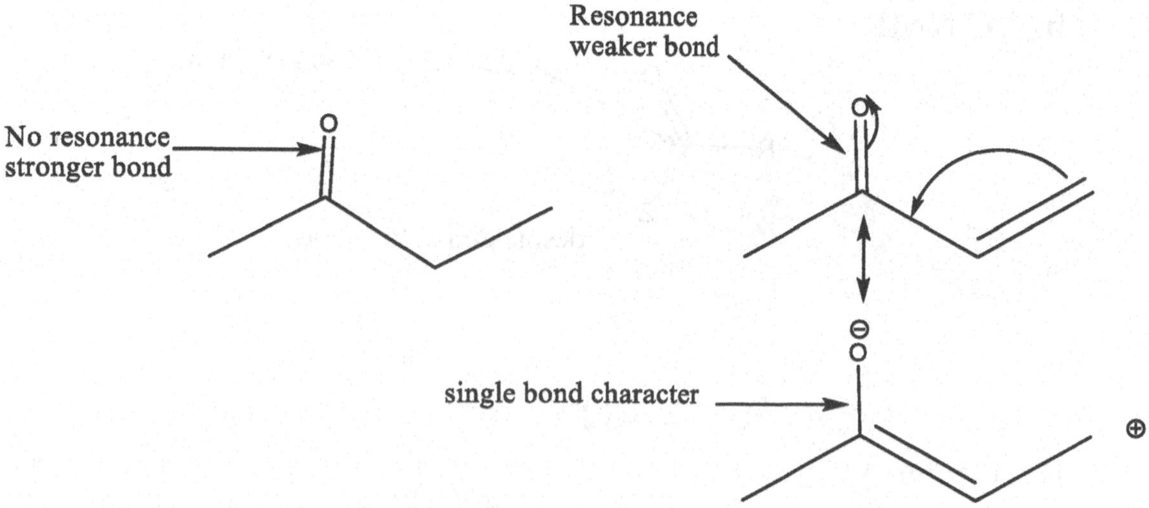

Resonance
weaker bond

No resonance
stronger bond

single bond character

- **Recall: As the ring size decreases, carbonyl groups absorb at higher frequencies. In other words, decreasing the size of the ring in cyclic lactones, lactams, and anhydrides increases the -C=O frequency because of ring strain. (See Units 6 and 11)**

Ex:

1715 cm^{-1} 1745 cm^{-1} 1780 cm^{-1}

2. NMR

a. ^{1}H NMR

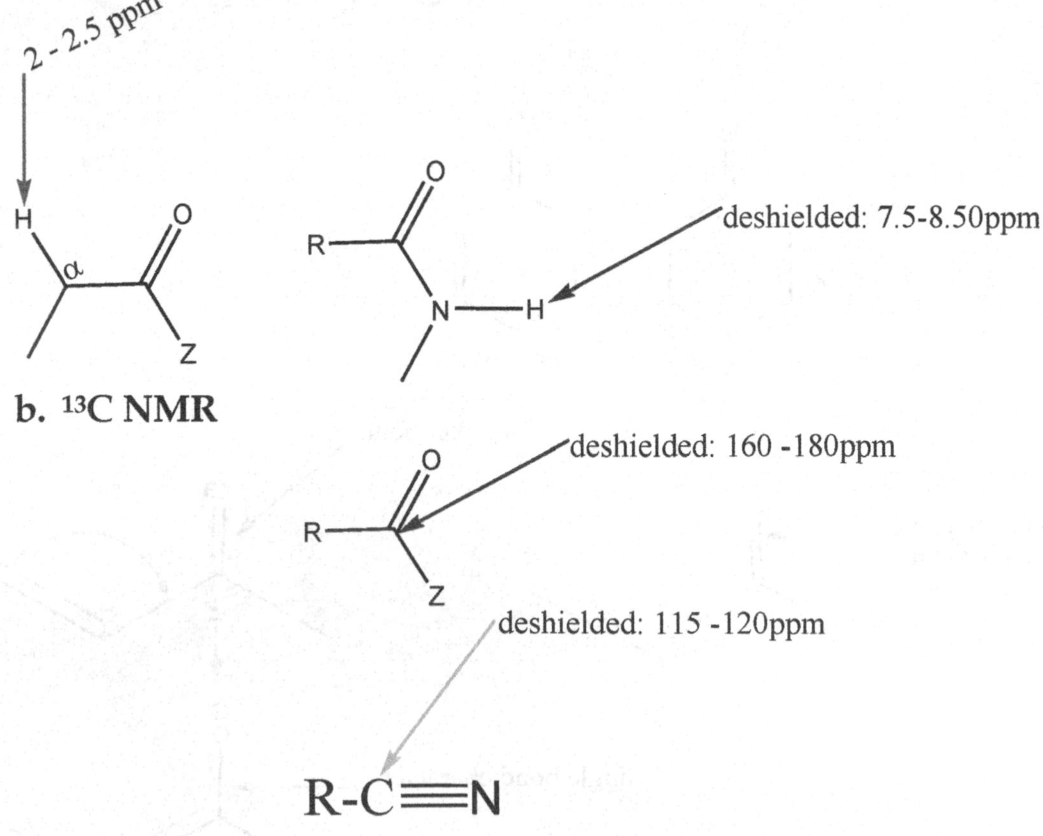

b. ^{13}C NMR

- **Read page ________.**

F. SOME INTERESTING ESTERS AND AMIDES

1. ESTERS

- **Esters have pleasant odors: fruits and perfumes.**
- **Isoamyl acetate (banana oil)**

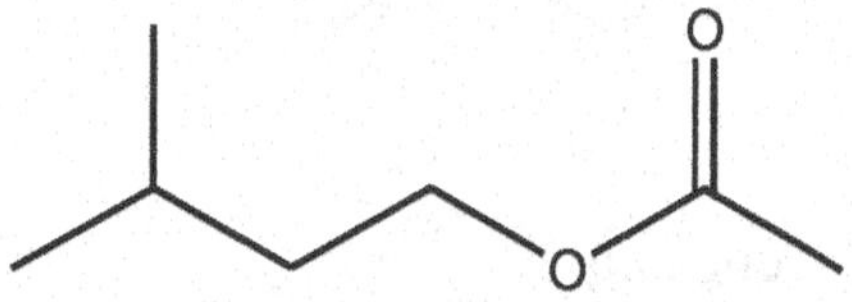

isoamyl acetate

- **Methyl α-methylbutyrate (pineapple oil) or Methyl 2-methylbutanoate** 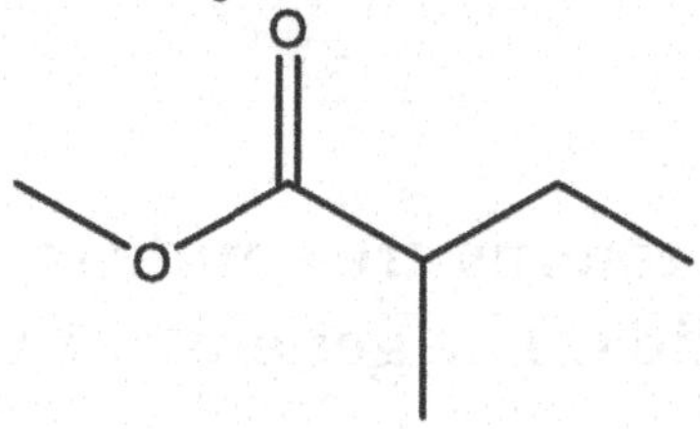
- **Ethyl butanoate (mango oil)**

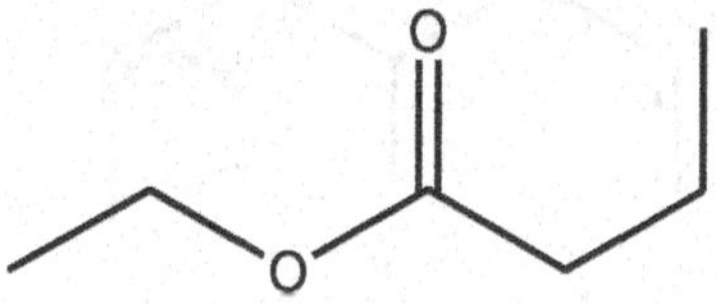

Ethyl butanoate

- **Ascorbic acid (vitamin C): An ester**

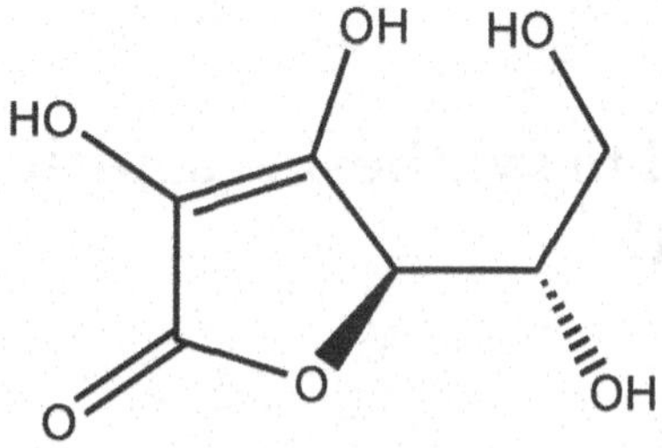

ascorbic acid

- **Cocaine: a diester. Can you locate the ester groups?**

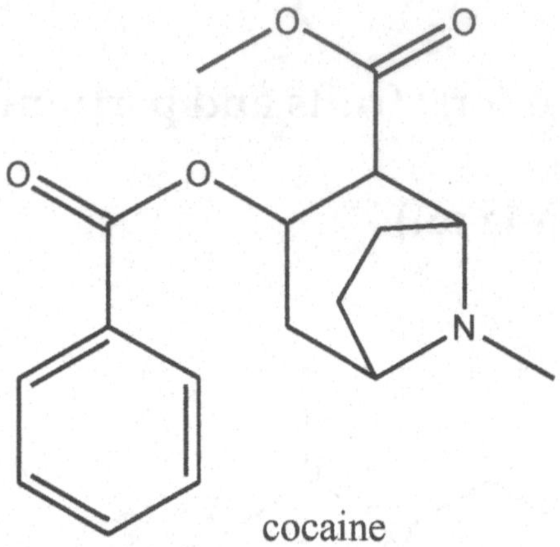

cocaine

- **Read pages ______ - ______.**

2. AMIDES

- **Peptides, polypeptides, proteins are amides = the latter two are polymers of amino acids. The general structure of a polypeptide is:**

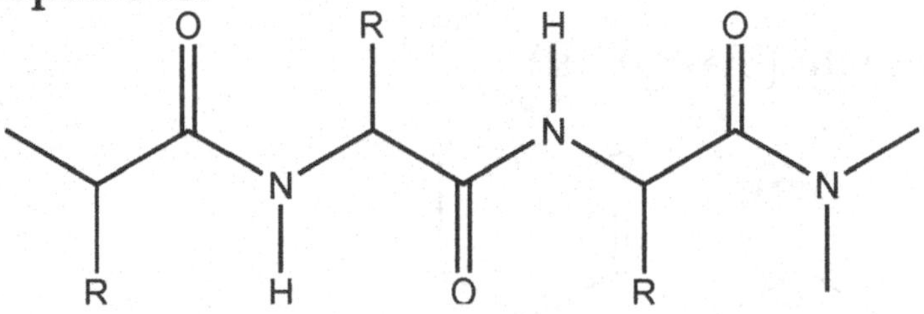

- **Read pages ______ - ______.**

- **Penicilins are β-lactams. Read pages ______ and _____.**

- **Acetic anhydride is used to synthesize aspirin (will be covered later in this unit).**

G. INTRODUCTION TO NUCLEOPHILIC ACYL SUBSTITUTION REACTIONS

1. INTRODUCTION

The general reaction is:

$$R-\overset{\overset{O}{\|}}{C}-Z \xrightarrow[\text{or HNu}]{:Nu^-} R-\overset{\overset{O}{\|}}{C}-Nu \quad + \quad Z^- \text{ or HZ}$$

Z = Cl, OCOR, OH, OR, NR'_2

2. MECHANISM REVISITED: 2 STEPS

$$R-\overset{\overset{O}{\|}}{C}-Z \quad :Nu^- \longrightarrow$$

Ex: Synthesis of Amides

$$R-\overset{\overset{O}{\|}}{C}-Cl \xrightarrow{:NH_3} R-\overset{\overset{O}{\|}}{C}-NH_2 \quad + \quad HCl$$

Ex: Synthesis of Esters: Esterification

$$R-\overset{\overset{O}{\|}}{C}-Cl \xrightarrow[H^+]{R'OH} R-\overset{\overset{O}{\|}}{C}-OR' \quad + \quad HCl$$

- **Do Problems on page _______.**

3. RELATIVE REACTIVITY OF RCOOH AND DERIVATIVES IN ACYL SUBSTITUTION REACTIONS

a. Introduction

- **Recall: The general structure of these compounds is:**

$$R-\overset{\overset{\large O}{\|}}{C}-Z$$

- **The better the leaving group Z, the faster the reaction. The increasing order of reactivity of the leaving group is:**

$$-NH_2 < -OH \sim -OR' < RCOO^- < Cl^-$$

- As a result, the increasing order of reactivity is:

$$RC(=O)NH_2 < RC(=O)OH \sim RC(=O)OR' < RC(=O)-O-C(=O)R < RC(=O)Cl$$

- **Note: More reactive derivatives can be used to synthesize less reactive ones.**

b. Occurrence of an Acyl Substitution Reaction

- **Recall:**

$$R-\overset{\overset{\large O}{\|}}{C}-Z \xrightarrow{:Nu^-} R-\overset{\overset{\large O}{\|}}{C}-Nu + Z^-$$

- **If Z is a better leaving group than Nu^-, the reaction occurs. However, if Z is a poor leaving group, no reaction occurs. Ex:**

$$H_3C-C(=O)-Cl + CH_3O^- \longrightarrow H_3C-C(=O)-OCH_3 + Cl^-$$

A good leaving group (Cl)

$$CH_3C(=O)NH_2 + Cl^- \longrightarrow \text{NR}$$

NH_2 ↓ A poor leaving group

- Do problem on page _______.

4. SOME EXAMPLES OF NAS REACTIONS WITH RCOZ

- **Synthesis of Acid Anhydrides**

$$R{-}C(=O)Z + {}^-O{-}C(=O){-}R \longrightarrow RC(=O){-}O{-}C(=O)R + Z^-$$

- **Synthesis of Carboxylic Acids**

$$R{-}C(=O)Z + H_2O \text{ or } OH^- \longrightarrow RC(=O)OH + Z^-$$

- **Synthesis of Esters: Fisher Esterification Revisited**

$$R{-}C(=O)Z + R'OH \longrightarrow RC(=O)OR' + Z^-$$

- **Synthesis of Amides Revisited**

$$R{-}C(=O)Z + R'_2NH \longrightarrow RC(=O)NR'_2 + Z^-$$

- See Preview of Reactions on page _______.

- **Note: Since Cl is the best leaving group among these compounds, acid chlorides are difficult to prepare through acyl nucleophilic substitution reactions.**

H. REACTIONS OF ACID CHLORIDES

1. GENERAL REACTION

$$R\text{—}C(=O)Cl + H\text{-}Nu \longrightarrow RC(=O)Nu + HCl$$

- **Pyridine is added to remove the HCl formed**

$$C_5H_5N + H^+Cl^- \longrightarrow C_5H_5N^+\text{—}H + Cl^-$$

Ex:

$$H_3C\text{—}C(=O)Cl + CH_3OH \xrightarrow{\text{Pyridine}} H_3C\text{—}C(=O)OCH_3 + HCl$$

An ester

2. REACTION OF ACID CHLORIDES WITH CARBOXYLATE IONS

$$R\text{—}C(=O)Cl + {}^-O\text{—}C(=O)R' \longrightarrow RC(=O)\text{—}O\text{—}C(=O)R' + Cl^-$$

acid chloride, carboxylate ion, acid anhydride

Ex:

$$C_3H_5\text{—}C(=O)Cl + {}^-O\text{—}C(=O)CH_3 \longrightarrow C_3H_5\text{—}C(=O)\text{—}O\text{—}C(=O)CH_3 + Cl^-$$

acid chloride (cyclopropyl), carboxylate ion, acid anhydride

- **Mechanism: see page ______.**

3. REACTION OF ACID CHLORIDES WITH WATER

$$R-C(=O)Cl + H_2O \xrightarrow{\text{pyridine}} RC(=O)OH + C_5H_5NH^+ Cl^-$$

a carboxylic acid

Ex:

$$C_4H_7-C(=O)Cl + H_2O \xrightarrow{\text{pyridine}} C_4H_7-C(=O)OH + C_5H_5NH^+ Cl^-$$

a carboxylic acid

- Mechanism: See page ________.

4. REACTION OF ACID CHLORIDES WITH ALCOHOLS

$$R-C(=O)Cl + R'OH \xrightarrow{\text{pyridine}} RC(=O)OR' + C_5H_5NH^+ Cl^-$$

an ester

Ex:

$$CH_3-C(=O)Cl + C_5H_9-OH \xrightarrow{\text{pyridine}} CH_3-C(=O)O-C_5H_9 + C_5H_5NH^+ Cl^-$$

an ester

5. REACTION OF ACID CHLORIDES WITH AMINES

$$R-C(=O)Cl + 2R'_2NH \longrightarrow RC(=O)NR'_2 + HCl$$

an amide, R' = H, alkyl group

- The HCl is removed by excess amine as follows:

$$R'_2NH + H^+Cl^- \longrightarrow R'_2\overset{+}{N}H_2\ Cl^-$$

Ex:

$$CH_3-C(=O)Cl + 2(CH_3CH_2)_2NH \longrightarrow CH_3C(=O)N(CH_2CH_3)_2 + (CH_3CH_2)_2\overset{+}{N}H_2\ Cl^-$$

- See the synthesis of N, N-diethyl-*m*-toluamide or DEET on page _______.
- Read pages _____ - ______.
- Do Problem _______, page _____.

I. REACTIONS OF ACID ANHYDRIDES

1. INTRODUCTION

- They are less reactive than acid chlorides. The general reaction is:

$$\underset{\text{acid anhydride}}{RC(=O)-O-C(=O)R'} + H\text{-}Nu \longrightarrow RC(=O)Nu + \underset{\text{a carboxylic acid}}{R'C(=O)OH}$$

2. REACTION OF ACID ANHYDRIDES WITH WATER

$$\underset{\text{acid anhydride}}{RC(=O)-O-C(=O)R'} + H_2O \longrightarrow \underset{\text{two carboxylic acids}}{RC(=O)OH + R'C(=O)OH}$$

Ex:

$$\underset{\text{acid anhydride}}{CH_3C(=O)-O-C(=O)H} + H_2O \longrightarrow \underset{\text{two carboxylic acids}}{CH_3C(=O)OH + HC(=O)OH}$$

3. REACTION OF ANHYDRIDES WITH ALCOHOLS

$$RC(=O)-O-C(=O)R' + R''OH \longrightarrow RC(=O)OR'' + R'C(=O)OH$$

acid anhydride — an ester — a carboxylic acid

Ex:

$$CH_3C(=O)-O-C(=O)H + CH_3OH \longrightarrow CH_3C(=O)OCH_3 + HC(=O)OH$$

acid anhydride — an ester — a carboxylic acid

- **Application: Synthesis of Aspirin From Salicylic Acid : Acetylation**

4. REACTIONS OF ANHYDRIDES WITH AMINES

$$RC(=O)-O-C(=O)R' + 2R''_2NH \longrightarrow RC(=O)NR''_2 + R'C(=O)O^- + R''_2\overset{+}{N}H_2$$

acid anhydride — an amide

Ex:

$$CH_3C(=O)-O-C(=O)H + 2(CH_3)_2NH \longrightarrow CH_3C(=O)N(CH_3)_2 + HC(=O)O^- + (CH_3)_2\overset{+}{N}H_2$$

acid anhydride

- **Conversion of an Anhydride to an Amide.**

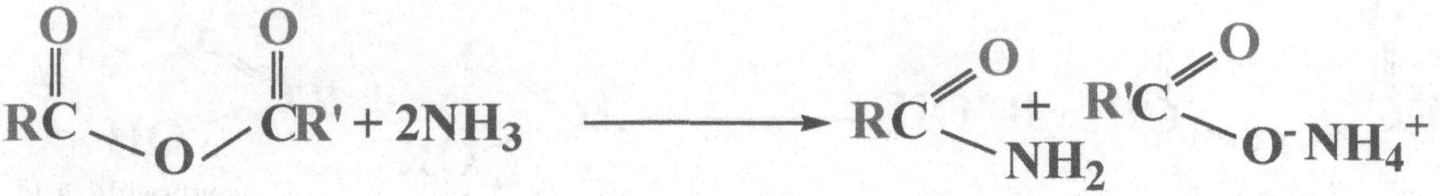

acid anhydride — an amide

Ex:

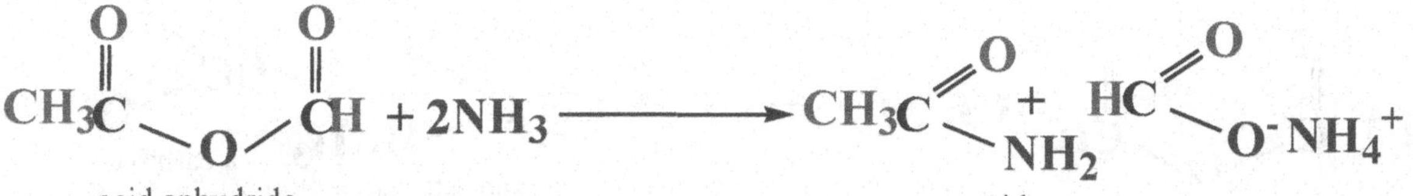

acid anhydride — an amide

- **Mechanism: See page ______.**

- **Application: Synthesis of Acetaminophen from PAP: Tylenol Synthesis: Acetylation**

- **See Synthesis of Heroin from Morphine on page ______.**

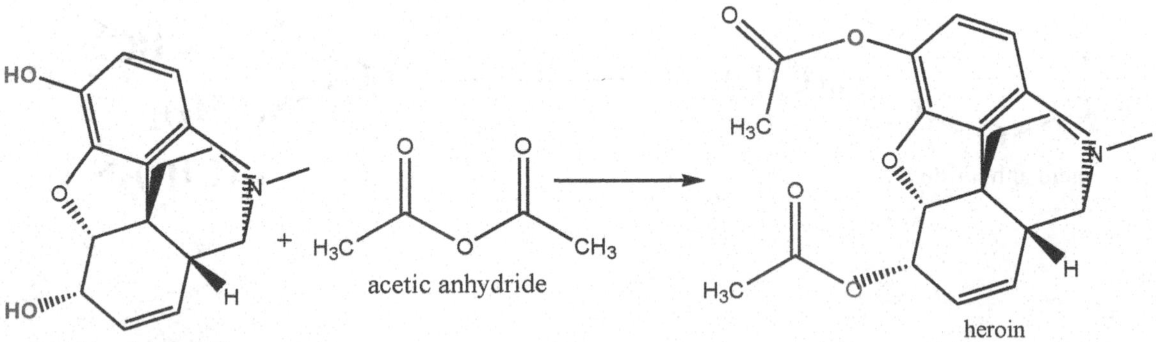

morphine (opium; from poppy seed) — heroin

- **Do Problem ______, page ______.**

J. REACTIONS OF CARBOXYLIC ACIDS

1. INTRODUCTION

- See Unit 9, Chapter _____.

- Carboxylic acids are strong organic acids. They can undergo acid-base reactions and nucleophilic reactions. Acid-base reactions are **faster** than nucleophilic ones.

- **The general reactions:**
  - **Acid-base reactions.**

$$RC(=O)-O-H + :Nu^- \longrightarrow RC(=O)-O^- + H\text{-}Nu$$

  - **Nucleophilic addition reactions.**

$$RC(=O)-O-H + :Nu^- \longrightarrow RC(O^-)(OH)-Nu$$

- See Reactions on page _______.
-

2. CONVERSION OF RCOOH TO ACID CHLORIDES REVISITED

$$RC(=O)OH + SOCl_2 \longrightarrow RC(=O)Cl + SO_2 + HCl$$

Ex:

$$CH_3C(=O)OH + SOCl_2 \longrightarrow CH_3C(=O)Cl + SO_2 + HCl$$

- See Mechanism on page ______.

- Do Problem _______ on page ______.

3. CONVERSION OF DICARBOXYLIC ACIDS TO CYCLIC ANHYDRIDES

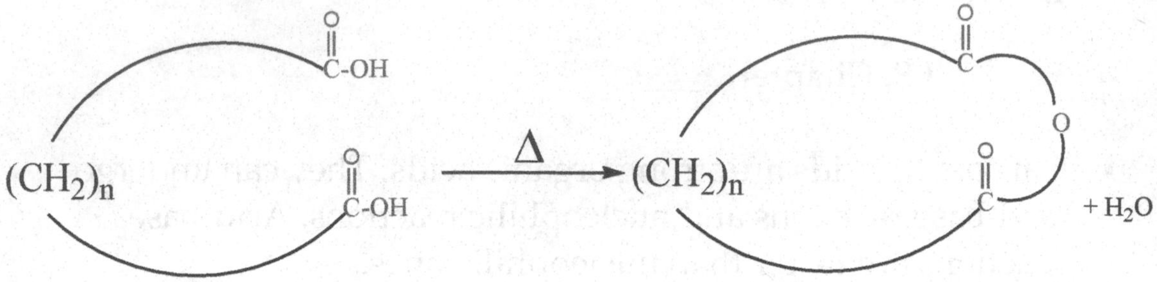

$(CH_2)_n$ (C(=O)-OH)(C(=O)-OH) $\xrightarrow{\Delta}$ $(CH_2)_n$ (C(=O)-O-C(=O)) $+ H_2O$

Ex:

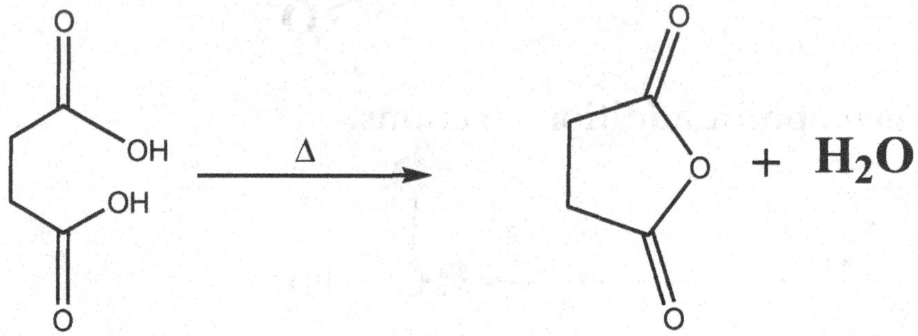

OH OH O O $\xrightarrow{\Delta}$ O O O $+ \ H_2O$

Ex:

O C OH OH C O $\xrightarrow{\Delta}$

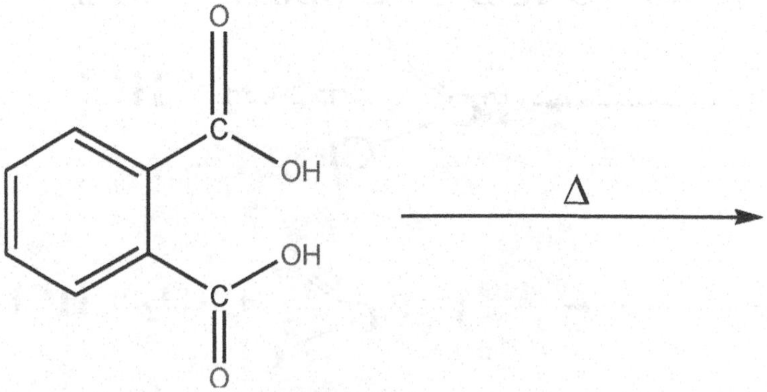

4. CONVERSION OF CARBOXYLIC ACIDS TO ESTERS: FISHER ESTERIFICATION

a. The General Reaction

$$R-C(=O)OH + R'OH \xrightleftharpoons{H_2SO_4} RC(=O)OR' + H_2O$$

an ester

- See Mechanism on page ______.

Ex:

$$CH_3-C(=O)OH + CH_3CH_2OH \xrightleftharpoons{H_2SO_4} CH_3C(=O)OCH_2CH_3 + H_2O$$

an ester

- Note: Since the reaction is reversible, it can be driven to the right either by removing the water byproduct or by using an excess of alcohol (application of Le Chatelier's principle).

- Read page ______.

b. Intramolecular Cyclization of γ- and δ- Hydroxy Carboxylic Acids

- γ and δ lactones can be prepared as follows: Internal Esterification

Ex:

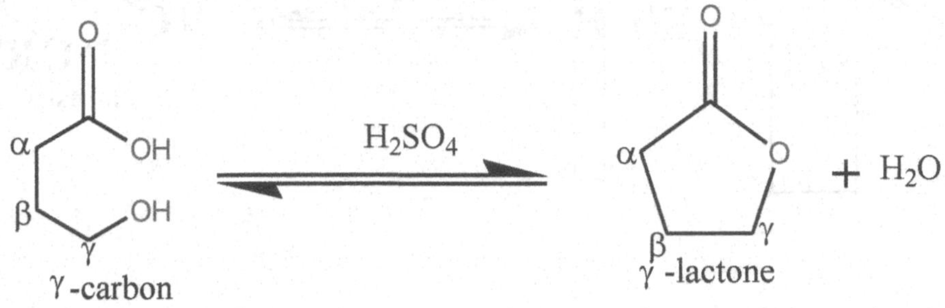

Ex:

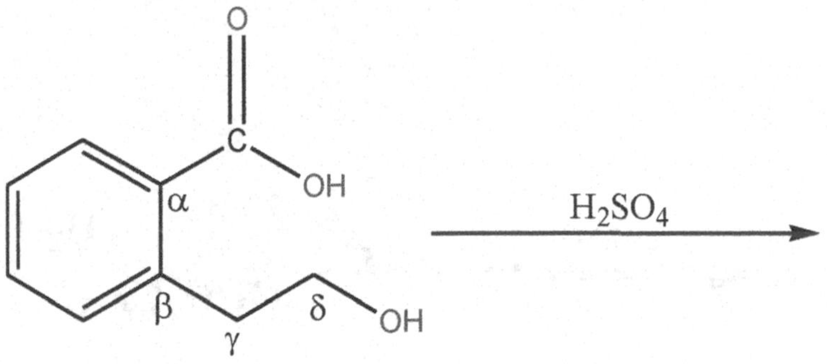

- Read page ______.
- Do problems on page______.

5. CONVERSION OF CARBOXYLIC ACIDS TO AMIDES

a. General Reaction

- **Dicyclohexyl carbodiimide (DCC), a dehydrating agent is added to make the reaction go faster. Dicyclohexylurea is the byproduct.**

- Read page_______.

$$R-C(=O)OH + R'NH_2 \xrightarrow{DCC} RC(=O)NR'H + \text{dicyclohexylurea}$$

an amide

dicyclohexylurea

Ex:

$$\text{cyclobutyl}-C(=O)OH + CH_3NH_2 \xrightarrow{DCC} \text{cyclobutyl}-C(=O)NHCH_3 + \text{dicyclohexylurea}$$

an amide

dicyclohexylurea

- See Mechanism on page ______.

b. Conversion of Carboxylic Acids to Primary Amides

$$R-C(=O)OH \xrightarrow[2.\ \Delta]{1.\ NH_3} RC(=O)NH_2 + H_2O$$

an amide

Ex:

$$C_6H_5-C(=O)OH \xrightarrow[2.\ \Delta]{1.\ NH_3} C_6H_5-C(=O)NH_2 + H_2O$$

Benzoic acid

benzamide

- **See Mechanism on page ______.**

c. Formation of the peptide bond between 2 aa

- **General Reaction:**

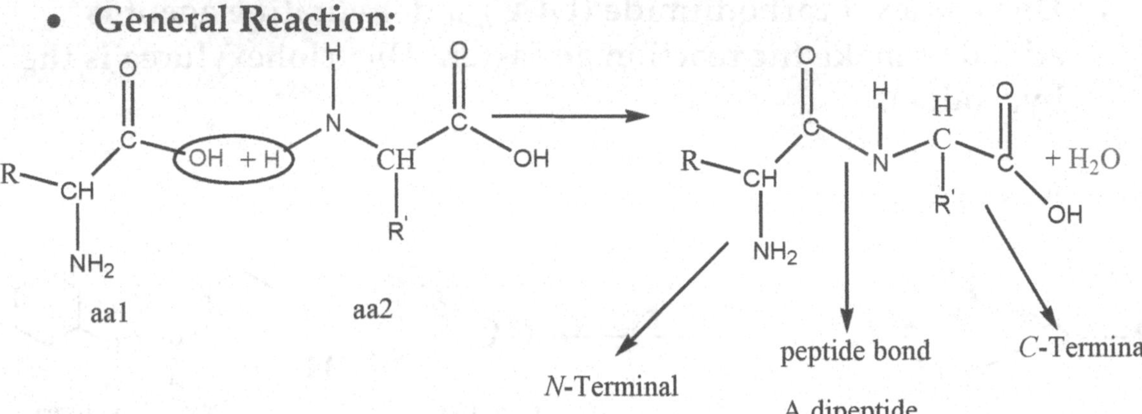

Ex:

$CH_3-CH(NH_2)-C(=O)-$ (OH + H) $-NH-CH_2-C(=O)-OH \longrightarrow CH_3-CH(NH_2)-C(=O)-NH-CH_2-C(=O)-OH + H_2O$

alanine glycine peptide bond

K. REACTIONS OF ESTERS

1. INTRODUCTION

- **Esters undergo 3 types of reactions:**
  - **Hydrolysis in acid.**
  - **Hydrolysis in base.**
  - **Reactions with amines.**

2. ESTER HYDROLYSIS IN ACID

$$\underset{\text{an ester}}{RC(=O)OR'} + H_2O \underset{}{\overset{H_2SO_4}{\rightleftharpoons}} R-C(=O)OH + R'OH$$

Ex:

$$CH_3C(=O)OCH_2CH_3 + H_2O \underset{}{\overset{H_2SO_4}{\rightleftharpoons}} CH_3{-}C(=O)OH + CH_3CH_2OH$$

an ester

- See Mechanism on page _____.

- **Note: Since the reaction is reversible, a large excess of water is used to drive the reaction to the right (application of Le Chatelier's principle).**

- Do problems on page _____.

3. ESTER HYDROLYSIS IN BASE

$$RC(=O)OR' \xrightarrow[H_2O]{OH^-} R{-}C(=O)O^- + R'OH$$

an ester — a carboxylate ion

Ex:

$$CH_3C(=O)OCH_2CH_3 \xrightarrow[H_2O]{OH^-} CH_3{-}C(=O)O^- + CH_3CH_2OH$$

an ester

- See Mechanism on page _____.

- **Note: The carboxylate ion can be converted to a carboxylic acid with a strong acid.** See page _____.

Ex.

$$CH_3{-}C(=O)O^- \xrightarrow{H_3O^+} CH_3{-}C(=O)OH$$

- Do problems on page _____.

- **Application of Ester Hydrolysis:**

- **Enzyme lipase catalyzes hydrolysis of triacylglycerols or lipids (polyesters) in cells, releasing a lot of fatty acids that can be stored as fats, contributing to obesity. The reaction is:**

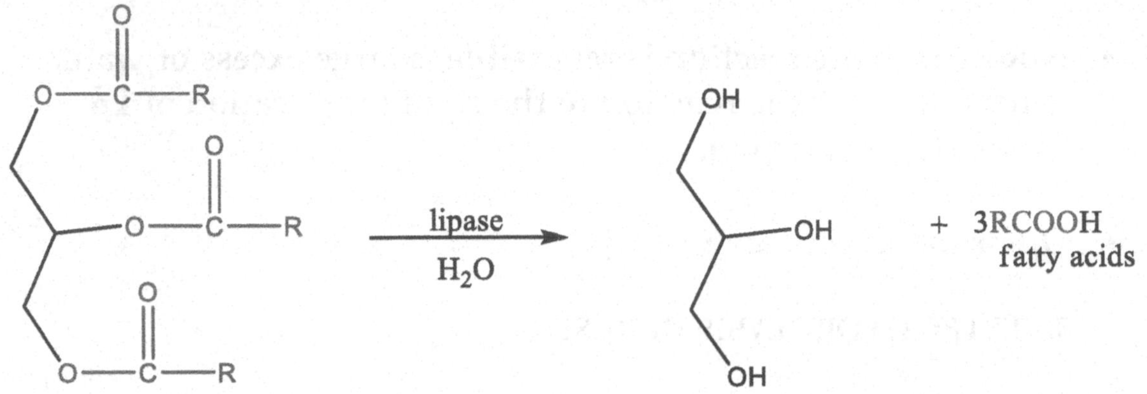

A triacylglycerol (fat or lipid) $\xrightarrow[H_2O]{\text{lipase}}$ Glycerol + 3RCOOH (fatty acids)

- **Olestra is a polyester of sucrose (synthetic fat) that is difficult to hydrolyze. It passes through the digestive system without hydrolysis.**

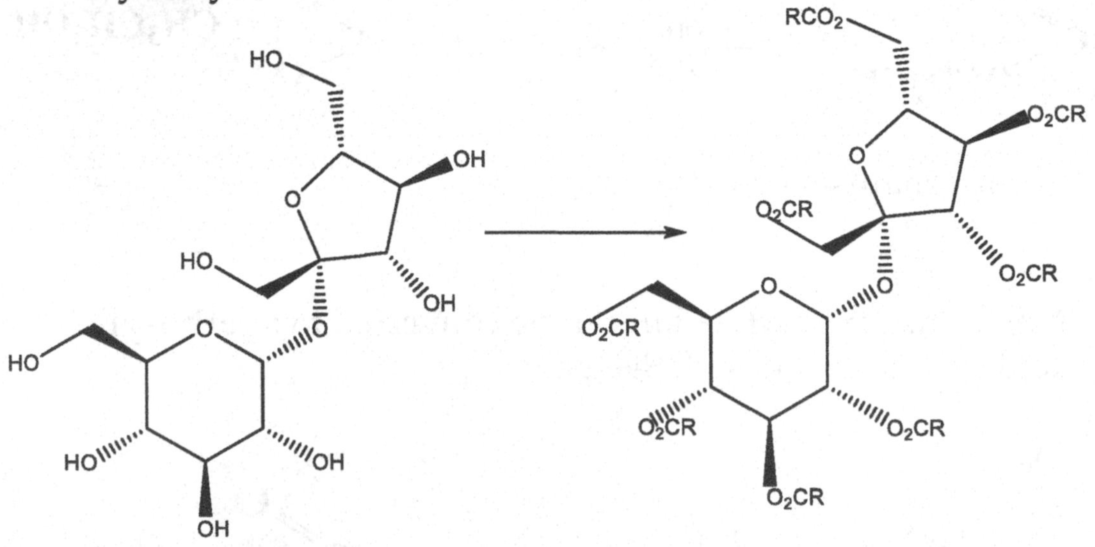

sucrose → olestra

- Read pages _____ - _____.

- Do problem _____, page _____.

- **Soap Making Revisited: Base Hydrolysis of Triacylglycerols: See Unit 9 (Saponification).**

$$\text{A triacylglycerol (fat or lipid)} + 3NaOH \longrightarrow \text{Glycerol} + 3RCOO^-Na^+ \text{ (soaps)}$$

A triacylglycerol
(fat or lipid)

Glycerol

4. REACTIONS OF ESTERS WITH AMINES

$$R{-}C(=O){-}OR' \xrightarrow{R''_2NH} RC(=O){-}NR''_2 + R'OH$$

an amide, R' = H, alkyl

Ex:

$$CH_3{-}C(=O){-}OCH_2CH_3 \xrightarrow{(CH_3)_2NH} CH_3C(=O){-}N(CH_3)_2 + CH_3CH_2OH$$

L. REACTIONS OF AMIDES

1. INTRODUCTION

- **Recall: -NH_2 is a poor leaving group. Therefore, amides are the least reactive of all the compounds covered in this unit. However, they do hydrolyze under strenuous conditions in acid or base.**

- See page ______.

2. HYDROLYSIS OF AMIDES IN ACIDS

$$RC(=O)NR_2' \xrightarrow[H^+]{H_2O} R-C(=O)OH + R'_2\overset{\oplus}{N}H_2$$

an amide → a carboxylic acid + an ammonium ion

Ex:

$$CH_3C(=O)N(CH_3)_2 \xrightarrow[H^+]{H_2O} CH_3-C(=O)OH + (CH_3)_2\overset{\oplus}{N}H_2$$

3. HYDROLYSIS OF AMIDES IN BASE

$$RC(=O)NR_2' \xrightarrow[OH^-]{H_2O} R-C(=O)O^- + R'_2NH$$

an amide → a carboxylate ion + an amine

Ex:

$$CH_3C(=O)N(CH_3)_2 \xrightarrow[OH^-]{H_2O} CH_3-C(=O)O^- + (CH_3)_2NH$$

- See Mechanism on page ______.
- **Note: The lack of reactivity of amides makes proteins very stable in the body.**
- Read page _______.
- Do Problems ______ and _____, page ______.

- **Read about β-lactam antibiotics and their mode of action: penicillins, on page ______. Penicillins stop cell wall construction as follows:**

- **See Summary of Nucleophilic Acyl Substitution reactions on page ______.**

- **See Table ______, page ______.**

- **Read about synthetic fibers (polyamides or nylon) on pages _______ - ______.**

- **Read about cholesteryl esters: LDL and HDL, pages _____ - _______.**

M. NITRILES

1. INTRODUCTION

- **Recall: General Structure**

$$\underset{\delta^+}{\text{R-C}}\equiv\underset{\delta^-}{\text{N:}}$$

- **Nitriles do not have leaving groups. Therefore, they do not undergo nucleophilic substitution reactions. However, they undergo nucleophilic addition reactions at the electrophilic carbon.**

- See General Reactions on page ______.

2. PREPARATION OF NITRILES: SN2

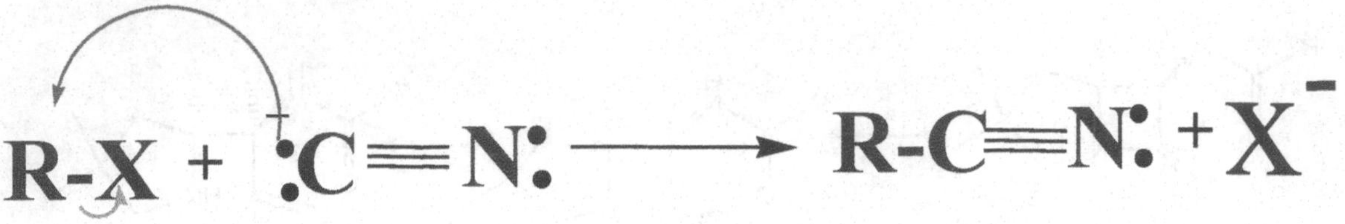

$$R\text{-}X + {}^{-}:C{\equiv}N: \longrightarrow R\text{-}C{\equiv}N: + X^{-}$$

Ex

$$CH_3CH_2\text{-}Br + {}^{-}:C{\equiv}N: \longrightarrow$$

- Do problems page ______.

3. REACTIONS OF NITRILES

a. Hydrolysis of Nitriles

i. In Acid

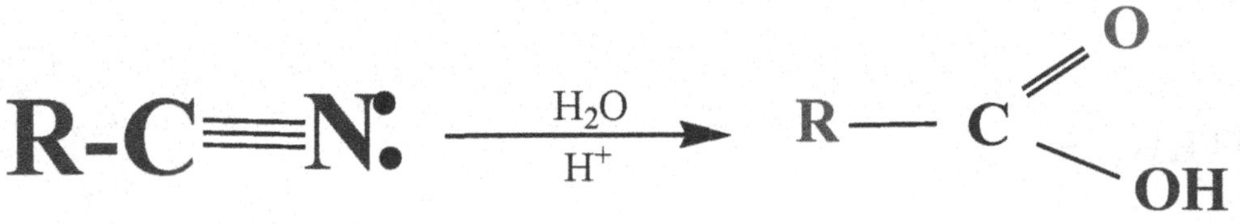

$$R\text{-}C{\equiv}N: \xrightarrow[H^+]{H_2O} R\text{—}C(=O)OH$$

a carboxylic acid

Ex:

$$CH_3CH_2\text{-}C{\equiv}N: \xrightarrow[H^+]{H_2O}$$

ii. In Base

$$R\text{-}C{\equiv}N\!: \xrightarrow[OH^-]{H_2O} R\text{—}C(=O)O^-$$

a nitrile

a carboxylate ion

Ex:

$$CH_3CH_2\text{-}C{\equiv}N\!: \xrightarrow[OH^-]{H_2O}$$

- See page _______ for mechanism.

b. Reduction of Nitriles

i. Reduction of Nitriles with $LiAlH_4/H_2O$

$$R\text{-}C{\equiv}N\!: \xrightarrow[2.\ H_2O]{1. LiAlH_4} RCH_2NH_2$$

a nitrile

amine

Ex:

$$CH_3CH_2\text{-}C{\equiv}N\!: \xrightarrow[2.\ H_2O]{1.\ LiAlH_4}$$

- See page ______ for mechanism.
- Do problems on pages ______ - _____.

ii. Reduction of Nitriles with DIBAL-H/H_2O

$$R\text{-}C{\equiv}N\!: \xrightarrow[2.\ H_2O]{1.\ DIBAL\text{-}H} R\text{—}C(=O)H$$

a nitrile

an aldehyde

Ex:

$$CH_3CH_2\text{-}C\equiv N: \xrightarrow[\text{2. } H_2O]{\text{1. DIBAL-H}}$$

- See Summary on page ______.
- Do problems on ______.

c. Reaction of Nitriles with Organometallic Reagents

The general reaction:

$$R\text{-}C\equiv N: \xrightarrow[\text{2. } H_2O]{\text{1. R'M}} R\text{-}C(=O)\text{-}R'$$

a nitrile — a ketone

R'M = R'MgX, R'Li

Ex:

$$CH_3CH_2\text{-}C\equiv N: \xrightarrow[\text{2. } H_2O]{\text{1. } CH_3CH_2MgBr}$$

- Do problems on page ______.
- See Key Concepts on pages _____ - _____.

SOME SELECTED REACTIONS OF CARBOXYLIC ACIDS AND DERIVATIVES

1. GENERAL REACTIONS

$$R{-}C({=}O){-}Z \xrightarrow{^{-}O{-}C({=}O){-}R'} R{-}C({=}O){-}O{-}C({=}O){-}R'$$

$$R{-}C({=}O){-}Z \xrightarrow{H_2O \text{ or } OH^-} R{-}C({=}O){-}OH$$

$$R{-}C({=}O){-}Z \xrightarrow{R'OH} R{-}C({=}O){-}OR'$$

$$R{-}C({=}O){-}Z \xrightarrow{R'_2NH} R{-}C({=}O){-}NR'_2$$

2. REACTIONS OF ACID CHLORIDES

$$R{-}C({=}O){-}Cl \xrightarrow[Py]{HNu} R{-}C({=}O){-}Nu$$

$$R{-}C({=}O){-}Cl \xrightarrow{^{-}O{-}C({=}O){-}R'} R{-}C({=}O){-}O{-}C({=}O){-}R'$$

$$R{-}C({=}O){-}Cl \xrightarrow{H_2O \text{ or } OH^-} R{-}C({=}O){-}OH$$

$$R{-}C({=}O){-}Cl \xrightarrow[Py]{R'OH} R{-}C({=}O){-}OR'$$

$$R{-}C({=}O){-}Cl \xrightarrow{2R'_2NH} R{-}C({=}O){-}NR'_2$$

$$R{-}C({=}O){-}Cl \xrightarrow[2.\ H_2O]{1.\ DIBAL\text{-}H} R{-}C({=}O){-}H$$

$$R{-}C({=}O){-}Cl \xrightarrow[2.\ H_2O]{1.\ LiAlH_4} R{-}CH_2OH$$

3. REACTIONS OF CARBOXYLIC ACIDS

$R{-}COOH \xrightarrow{:B^-} R{-}COO^-$

$R{-}COOH \xrightarrow{:Nu^-} R{-}C(O^-)(OH){-}Nu$

$R{-}COOH \xrightarrow{SOCl_2} R{-}COCl$

$R{-}COOH \xrightarrow[H^+]{R'OH} R{-}COOR' + H_2O$

$R{-}COOH \xrightarrow[DCC]{R'NH_2} R{-}CONR'H$

$R{-}COOH \xrightarrow[2.\ H_2O]{1. NH_3} R{-}CONH_2$

$R{-}COOH \xrightarrow[2.\ H_2O]{1.\ LiAlH_4} R{-}CH_2OH$

4. REACTIONS OF ACID ANHYDRIDES

$RC(=O){-}O{-}C(=O)R' \xrightarrow{HNu} R{-}C(=O){-}Nu + R'C(=O){-}OH$

$RC(=O){-}O{-}C(=O)R' \xrightarrow{H_2O} RC(=O){-}OH + R'{-}C(=O){-}OH$

$RC(=O){-}O{-}C(=O)R' \xrightarrow{R''OH} R{-}C(=O){-}OR'' + R'{-}C(=O){-}OH$

$RC(=O){-}O{-}C(=O)R' \xrightarrow{2R''_2NH} R{-}C(=O){-}NR_2'' + R'{-}C(=O){-}O^-$

$RC(=O){-}O{-}C(=O)R' \xrightarrow{NH_3} R{-}C(=O){-}NH_2$

5. REACTIONS OF ESTERS

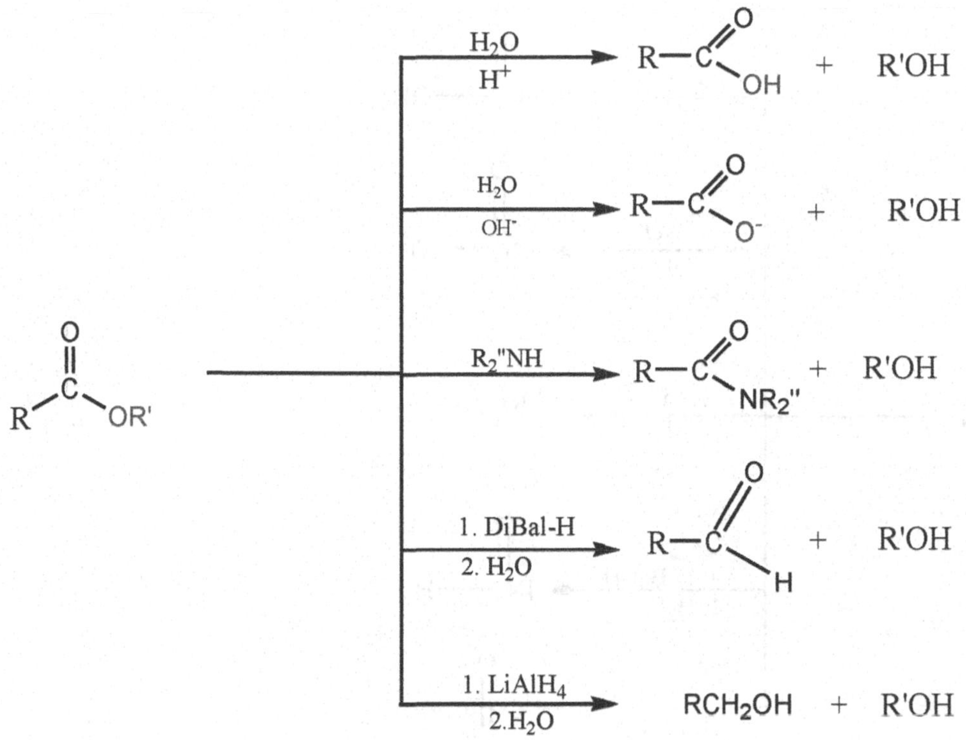

$RCOOR'$ →

- H_2O, H^+ → $R-C(=O)OH$ + R'OH
- H_2O, OH^- → $R-C(=O)O^-$ + R'OH
- $R_2''NH$ → $R-C(=O)NR_2''$ + R'OH
- 1. DiBal-H 2. H_2O → $R-C(=O)H$ + R'OH
- 1. $LiAlH_4$ 2. H_2O → RCH_2OH + R'OH

6. REACTIONS OF AMIDES

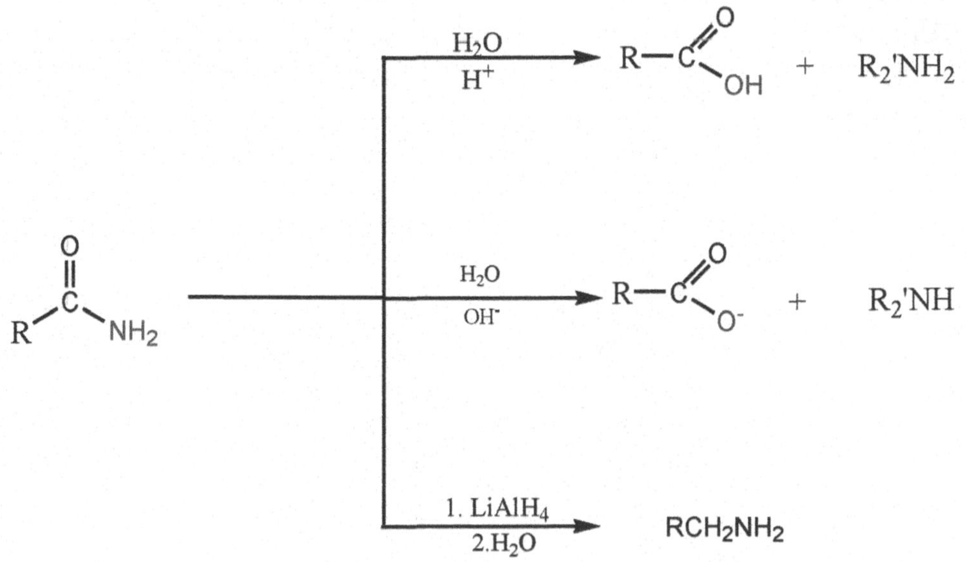

$RCONH_2$ →

- H_2O, H^+ → $R-C(=O)OH$ + $R_2'NH_2$
- H_2O, OH^- → $R-C(=O)O^-$ + $R_2'NH$
- 1. $LiAlH_4$ 2. H_2O → RCH_2NH_2

7. REACTIONS OF NITRILES

$$RC{\equiv}N \xrightarrow[H^+]{H_2O} RC(=O)-OH$$

$$RC{\equiv}N \xrightarrow[OH^-]{H_2O} RC(=O)-O^-$$

$$RC{\equiv}N \xrightarrow[2.\ H_2O]{1.\ LiAlH_4} RCH_2NH_2$$

$$RC{\equiv}N \xrightarrow[2.\ H_2O]{1.\ DiBal\text{-}H} RC(=O)-H$$

$$RC{\equiv}N \xrightarrow[2.\ H_2O]{1.\ R'MgX} R-C(=O)-R'$$

OCHEM II UNIT 13: AMINES

A. INTRODUCTION

1. GENERAL STRUCTURE

- Amines can be thought of as derivatives of ammonia.

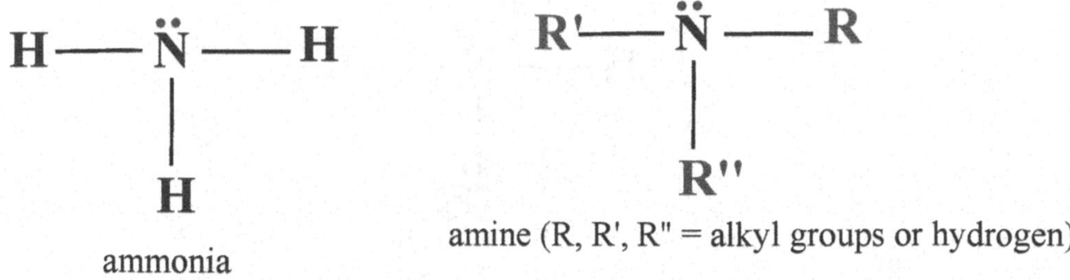

ammonia

amine (R, R', R" = alkyl groups or hydrogen)

2. TYPES OF AMINES: 4 TYPES

a. Primary (1°) Amines: One R Group

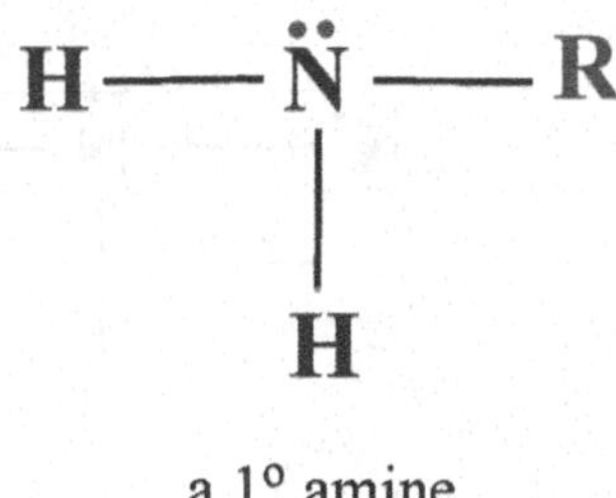

a 1° amine

Ex:

$$H-\ddot{N}(H)-CH_2CH_3$$

a 1° amine

b. Secondary (2°) Amines: Two R Groups

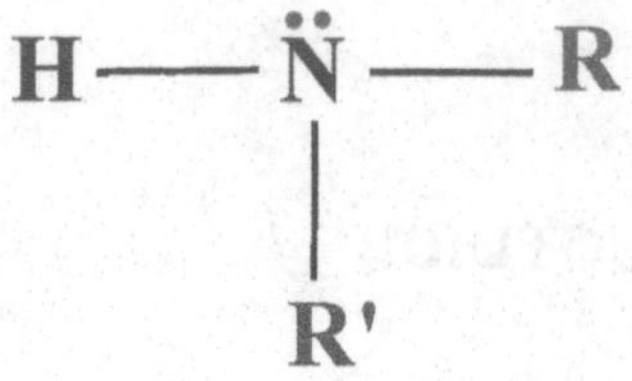

$$H-\ddot{N}(R')-R$$

a 2° amine

Ex:

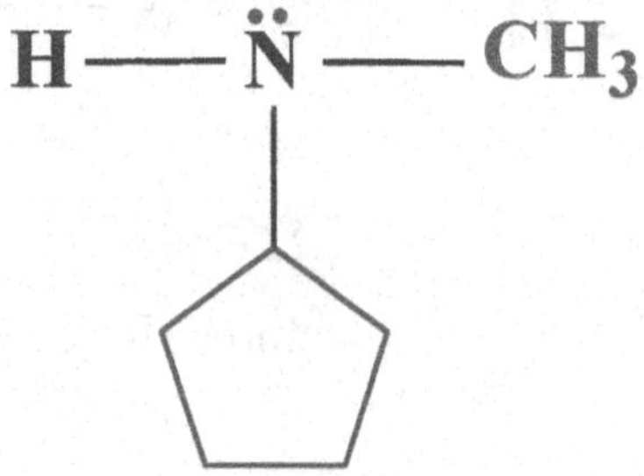

$$H-\ddot{N}(\text{cyclopentyl})-CH_3$$

a 2° amine

c. Tertiary (3°) Amines: Three R Groups

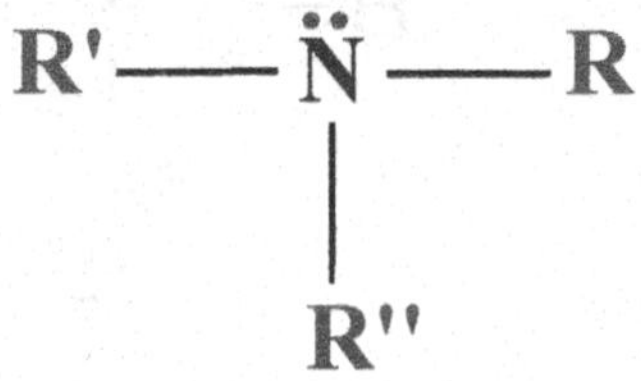

$$R'-\ddot{N}(R'')-R$$

Ex:

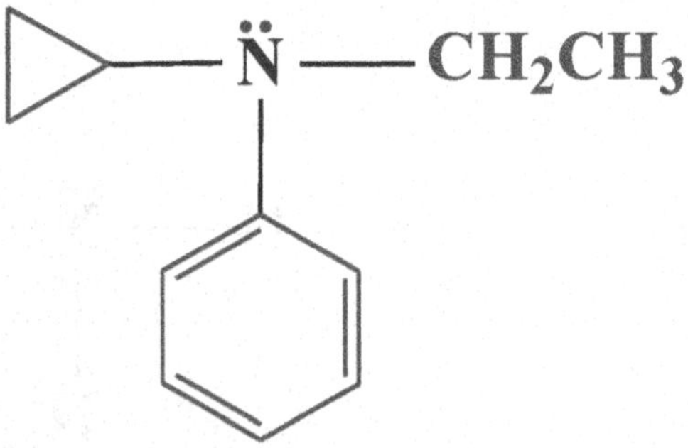

$$(\text{cyclopropyl})-\ddot{N}(\text{phenyl})-CH_2CH_3$$

a 3° amine

d. Quaternary (4°) Ammonium Salts: Four R Groups

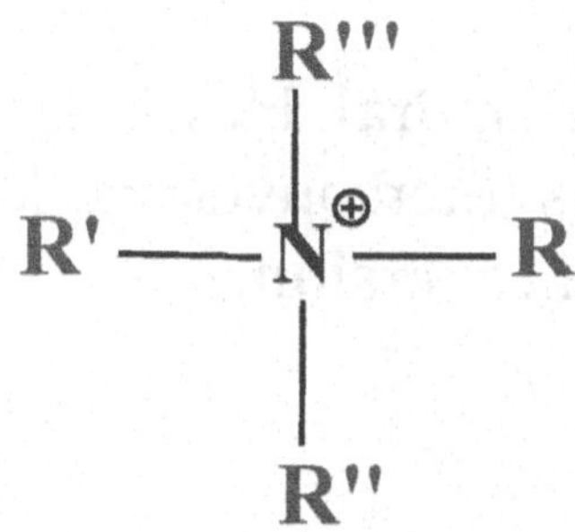

Ex:

CH_3

N

CH_2CH_3

- Do Problem ______ page ______.

3. **BONDING IN AMINES**

- The N in amines is sp^3 hybridized.

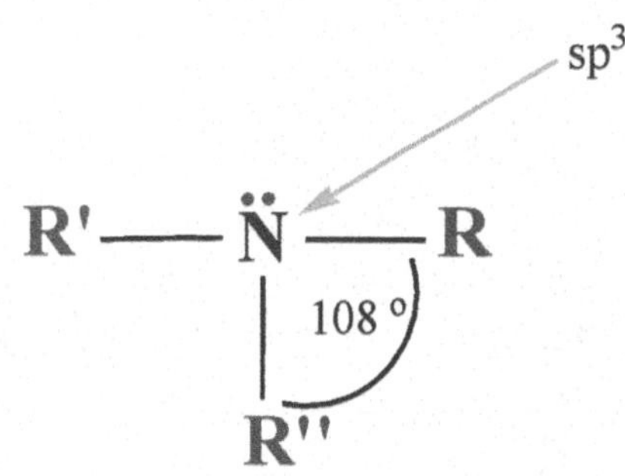

4. CHIRALITY OF AMINES

a. Simple Amines

- Simple amines are **chiral**. However, chirality can be ignored as the 2 enantiomers interconvert rapidly at room temperature thru **a pyramidal inversion.**

- See page ______.

b. Quaternary Ammonium Salts

- Tetrahedral ammonium salts with 4 different R groups are chiral and chirality **cannot** be ignored.

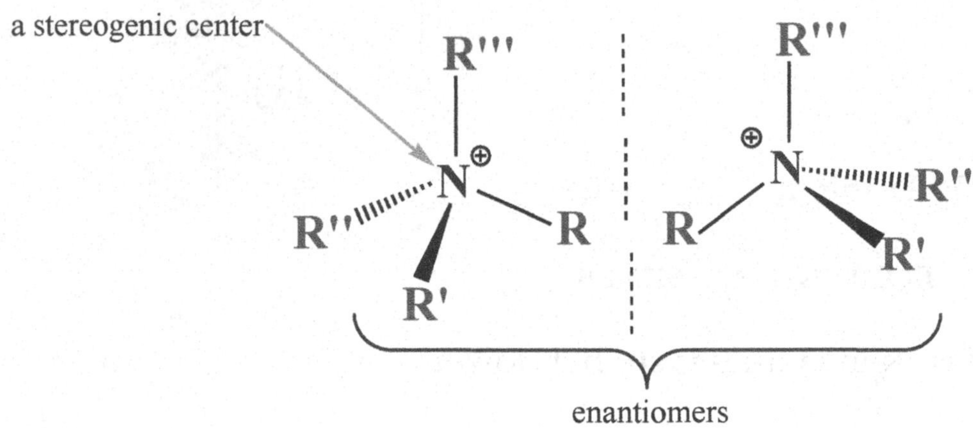

- **Read pages _____ - _____.**

- **Do Problems ______ and _____, page _____.**

B. NOMENCLATURE OF AMINES

1. NAMING 1° AMINES

- **Primary amines are named as *alkanamines.***
- **The parent chain is the longest carbon chain that contains the amine group.**

name of alkane ~~ -e ~~ **alkanamine**

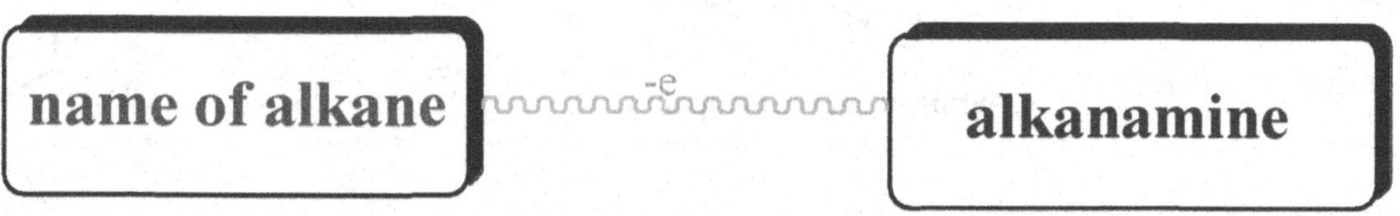

Ex:

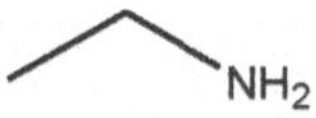

ethanamine

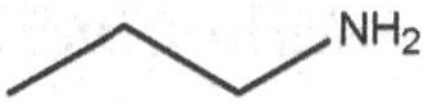

propanamine

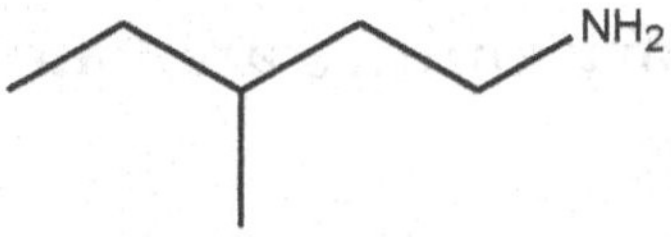

3-methyl-1-pentanamine

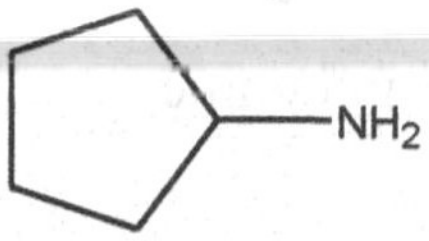

cyclopentanamine

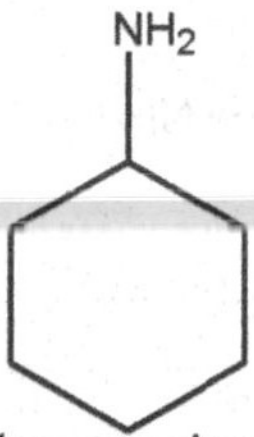

cyclohexanamine

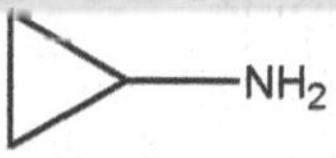

cyclopropanamine

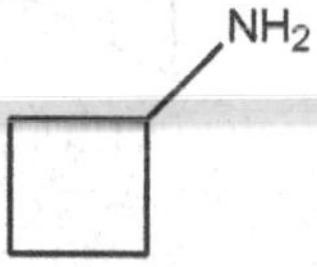

cyclobutanamine

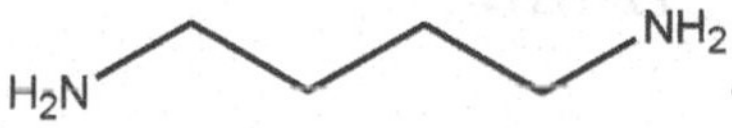

butanediamine

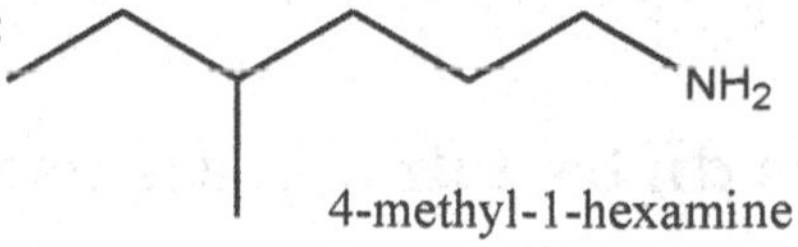

4-methyl-1-hexamine

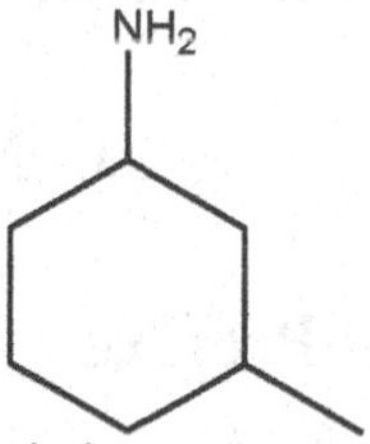

3-methyl-cyclohexanamine

2. NAMING 1° AMINES WITH MORE THAN ONE FUNCTIONAL GROUP

- **The $-NH_2$ group is called *amino*.**

Ex:

2-amino-1-ethanol

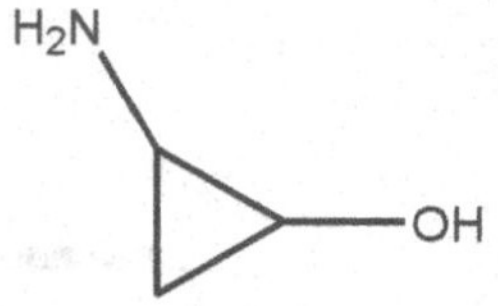

2-amino-1-cyclopropanol

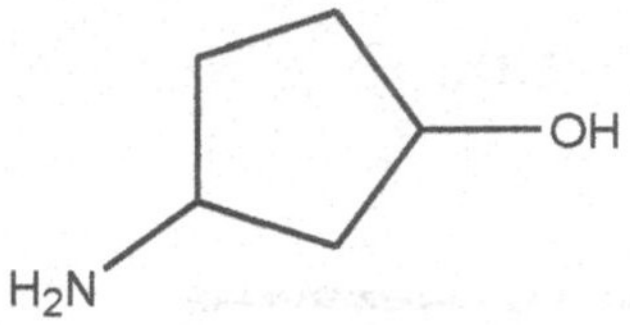

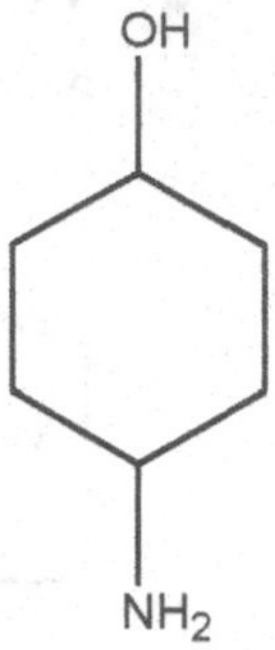

3. COMMON NAMES OF 1° AMINES

- **Primary amines are named as *alkylamines*.**

Ex:

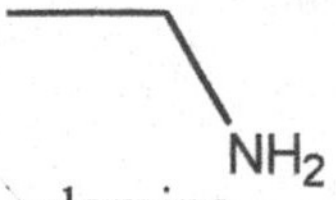

e..ylamine

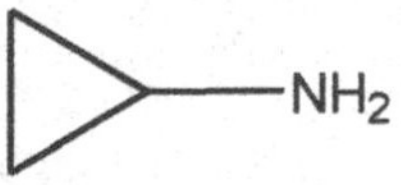

cyclopropylamine

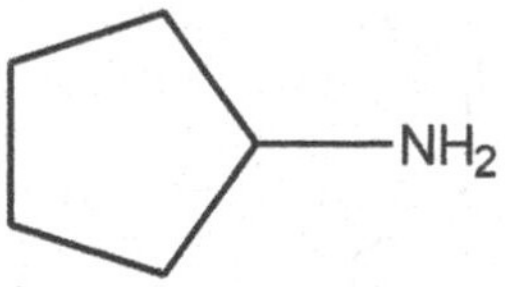

cyclopentylamine

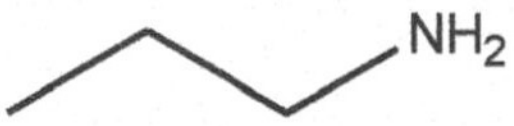

propylamine

4. NAMING 2° AND 3° AMINES

a. Naming Symmetrical 2° and 3° Amines

- **Prefixes such as di, tri, tetra, ... are used.**

Ex:

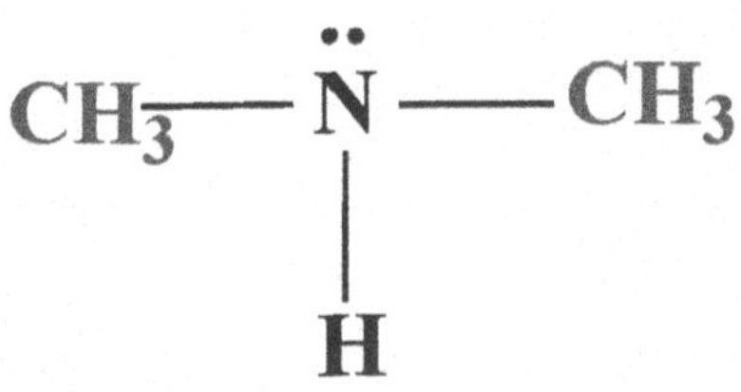

$CH_3-\ddot{N}(H)-CH_3$

Dimethylamine

$CH_3-\ddot{N}(CH_3)-CH_3$

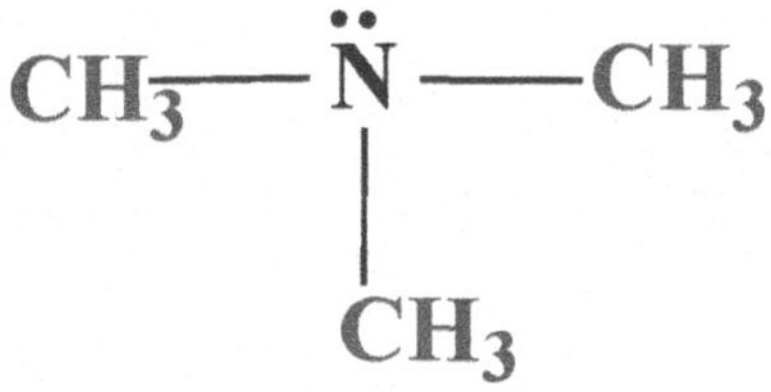

Trimethylamine

b. Naming Unsymmetrical 2° and 3° Amines

- **They are named as N- common names of 1° amines.**

Ex:

CH_3–N(H)—CH_2CH_3

parent chain

***N*-methylethylamine (common name)**
***N*-methylethanamine (IUPAC name)**

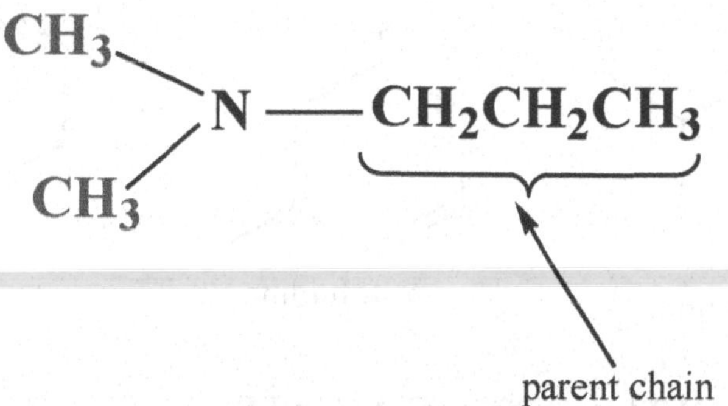

***N,N*-dimethylpropylamine (common name)**
***N,N*-dimethylpropanamine (IUPAC name)**

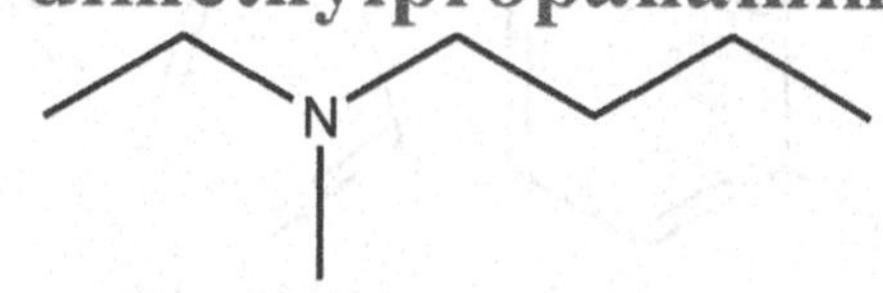

N-ethyl-N-methylbutylamine (common name)
N-ethyl-N-methylbutanamine (IUPAC name)

- **Read pages ______ - ______.**
- **Do Problems on page ______.**

5. NAMING AROMATIC AMINES

- **Named as derivatives of aniline.**

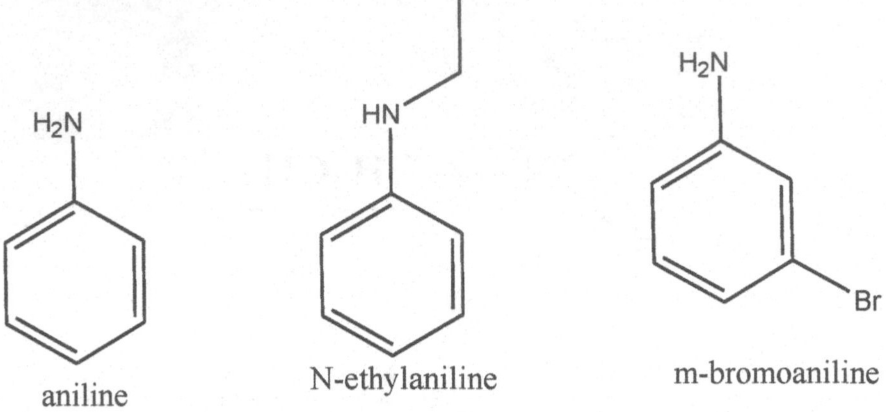

aniline N-ethylaniline m-bromoaniline

- Exceptions:

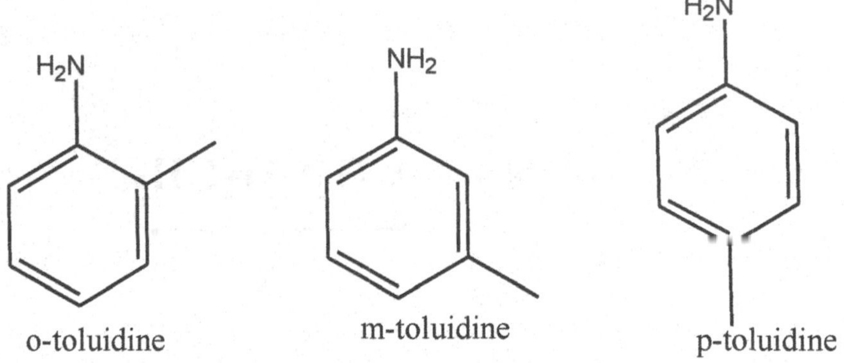

o-toluidine m-toluidine p-toluidine

6. NAMING HETEROCYCLIC AMINES

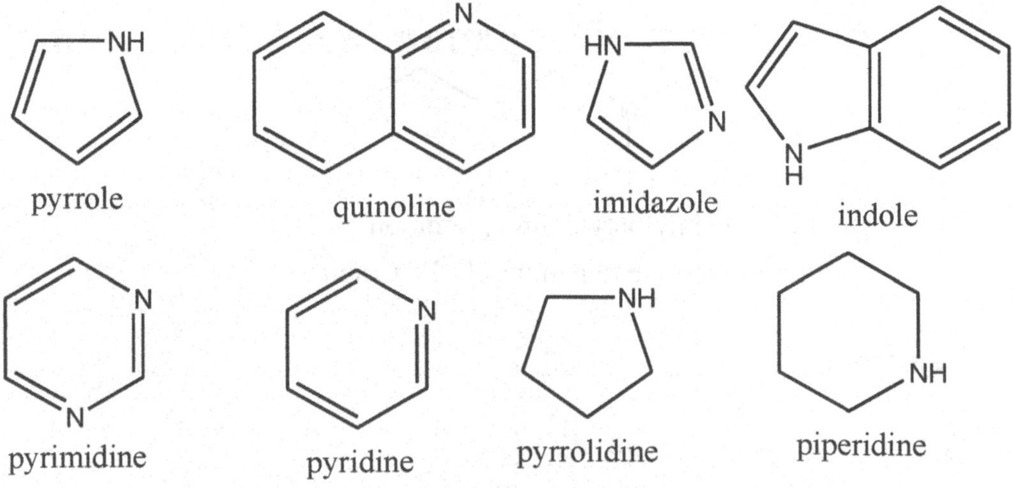

pyrrole quinoline imidazole indole

pyrimidine pyridine pyrrolidine piperidine

- **Read page _____. Do Problems _____ and ____, page _____.**

7. NAMING POLYFUNCTIONAL AMINES

a. Group Priorities: Déjà Vue

- When two or more different functional groups are present in a compound, functional group priorities are used. Group priorities are assigned based on the following Table (decreasing order of priority).
- **See Unit 9.**

Priority order	Group	Ending of name as a priority	Name as a non **Priority group**
Carboxylic acid	**RCOOH**	*-oic acid*	*-carboxy*
Ester	**RCOOR′**	*-oate*	*-alkoxycarbonyl*
Amide	**$RCONH_2$**	*-amide*	*-amido*
Nitrile	**RCN**	*-nitrile*	*-cyano*
Aldehyde	**RCHO**	*-al*	*-oxo(=O) or formyl(-CHO)*
Ketone	**RCOR′**	*-one*	*-oxo*
Alcohol	**ROH**	*-ol*	*-hydroxy*
Amine	**RNH_2**	*-amine*	*-amino*
Alkene	**-C=C-**	*-ene*	*-alkenyl*
Alkyne	**-C≡C-**	*-yne*	*-alkynyl*
Alkane	**-C-C-**	*-ane*	*-alkyl*
Ether	**ROR′**	*-none*	*-alkoxy*
Halide	**R-X**	*-none*	*-halo*

- **Note: Amines have a higher priority than alkenes, alkynes, ethers, and halides.**

b. Some Examples

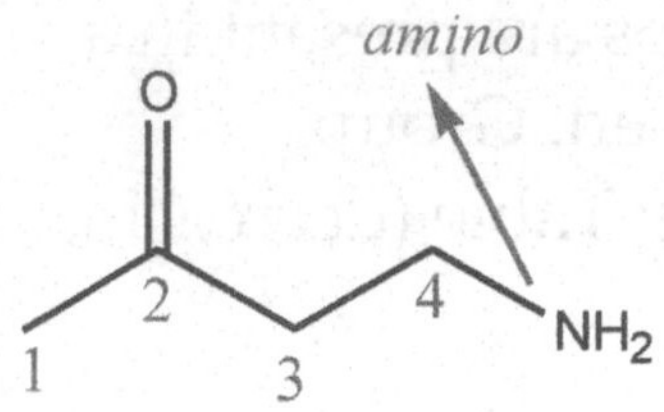

amino

4-*aminobutanone*

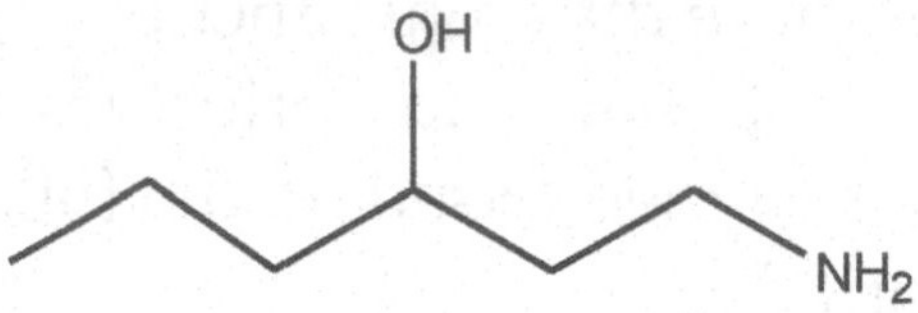

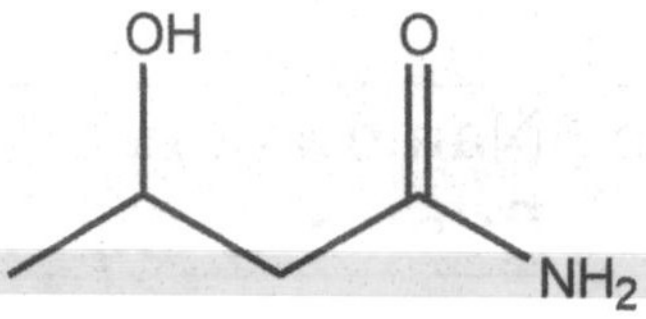

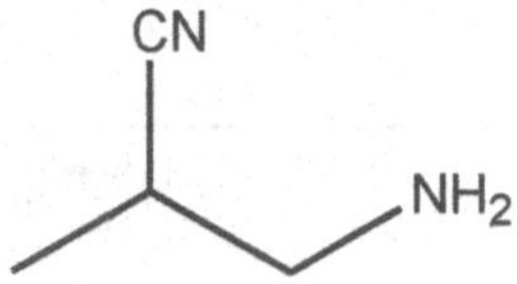

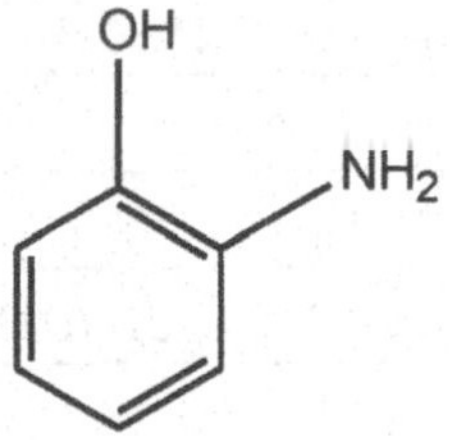

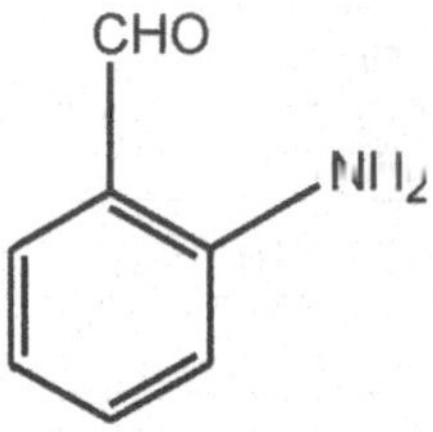

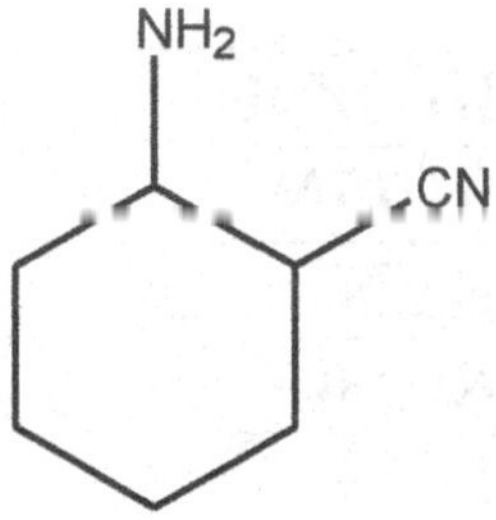

C. PHYSICAL PROPERTIES AND SOURCES OF AMINES

1. MP + BP

- **Primary and secondary amines can H bond. Therefore they have higher MP + BP than comparable substances. However, they have lower MP + BP than alcohols and carboxylic acids.**

Ex:

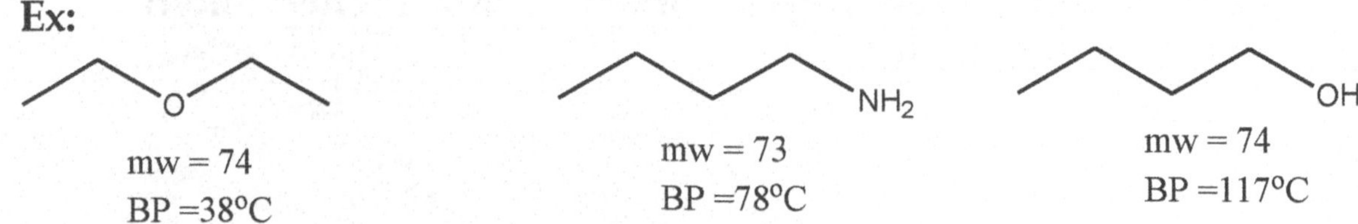

- **Note: Tertiary amines have lower BP + MP than primary amines.**

2. SOLUBILITY

- See Table _____, page ______.

- All amines are soluble in organic solvents.
- Amines having a number of carbons ≤ 5 are soluble in water because of their ability to H bond with water molecules.
- Amines with a number of carbons > 5 are insoluble in water.

3. ODORS

- Low-molecular weight amines have fishlike odors.
- Diamines smell like cadavers.

Ex: Cadaverine

H_2N NH_2

cadaverine

D. SPECTROSCOPIC PROPERTIES

1. MASS SPECT

- Amines with odd number of nitrogens have odd molecular weights.

- See Fig. ______, page ______.

2. IR

- 1° amines have two N-H bands: 3300 – 3500 cm^{-1}.
- 2° amines have one N-H band: 3300 – 3500 cm^{-1}.
- 3° amines have no N-H band between 3300 – 3500 cm^{-1}.

- See Fig. ______, page ____. Do Problem _____, page _____.

3. NMR

a. ^1H NMR

- **The N-H proton is deshielded: 2.2 -3.0 ppm.**
- **Does not split signals of adjacent hydrogens.**

b. ^{13}C NMR

- **The C of C-N is deshielded: 30–50 ppm.**

- **Read page _____.**

- **Do problem on page ______.**

E. IMPORTANT AMINES

1. INTRODUCTION

- **Although amines have foul odors, they are very important.**

2. SIMPLE AMINES AND ALKALOIDS

- **Putrescine: 1,4-butanediamine or 1,4-diaminobutane in dead animals.**

$$NH_2CH_2CH_2CH_2CH_2\ NH_2$$

- **Cadaverine: 1,5-pentanediamine or 1,5-diaminopentane in dead animals.**

$$NH_2CH_2CH_2CH_2CH_2CH_2NH_2$$

- **Trimethylamine: found in dead fish.**

$$(CH_3)_3N:$$

- **Alkaloids: complex basic amines derived from plants.**

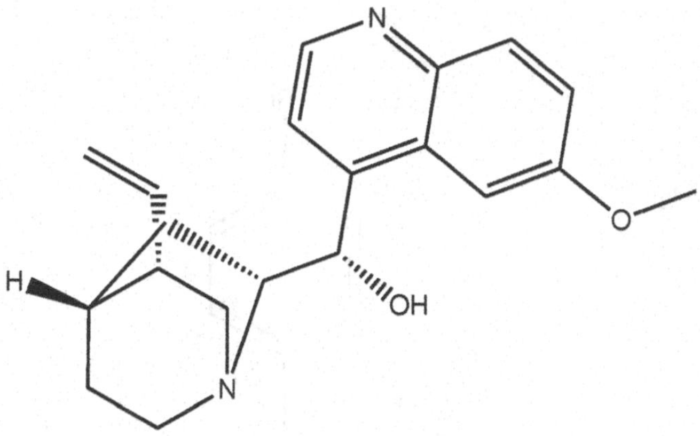

quinine

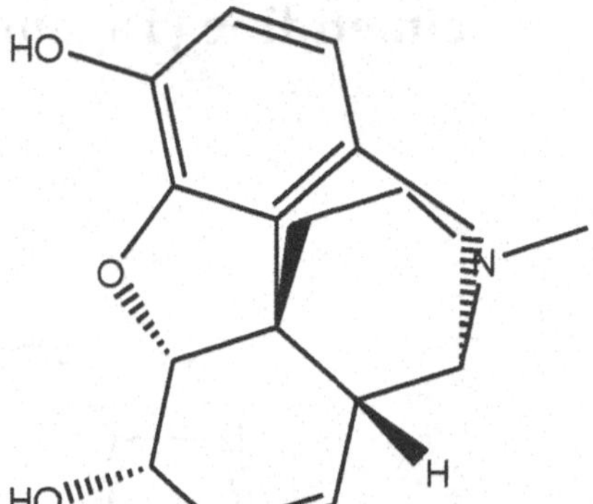

morphine

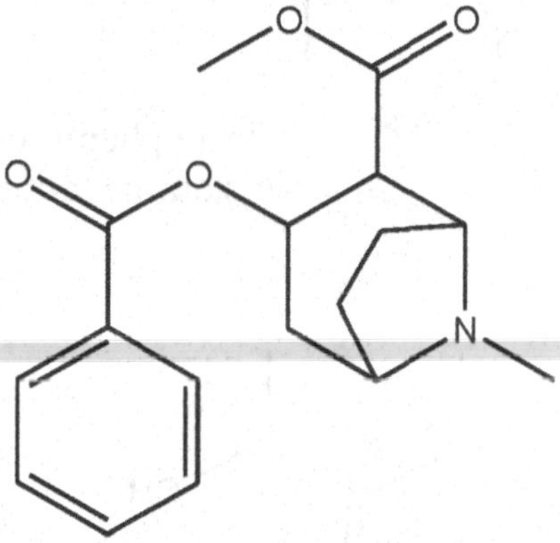

cocaine

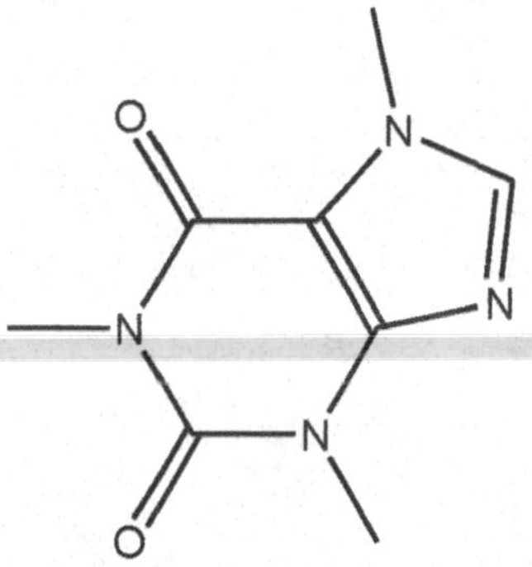

caffeine

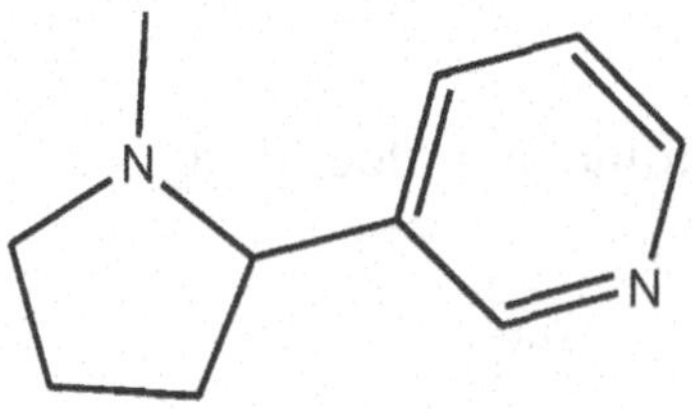

nicotine

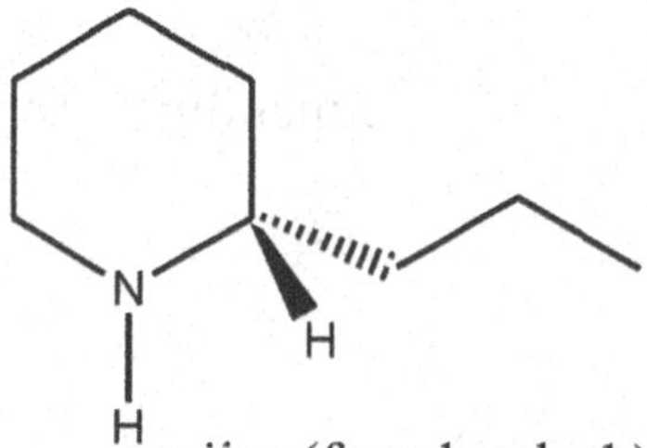

coniine (from hemlock)

3. HISTAMINE AND ANTIHISTAMINES

- **Histamine is a compound that dilates capillaries to help blood flow.**
- **Antihistamines work against it; brompheniramine and cimetidine (Tagamet)**

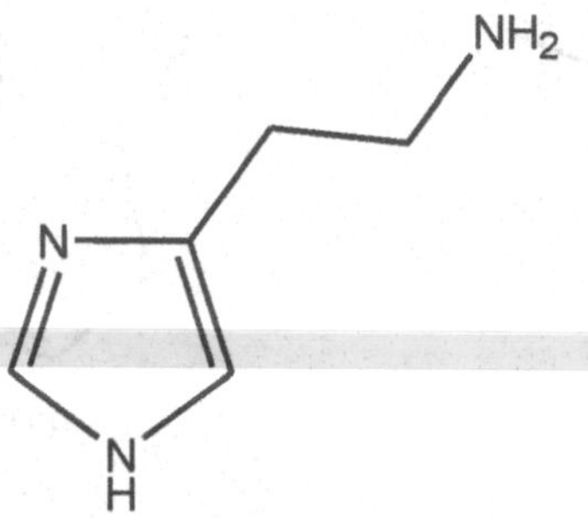

NH_2
N
N
H

histamine: cold causing agent

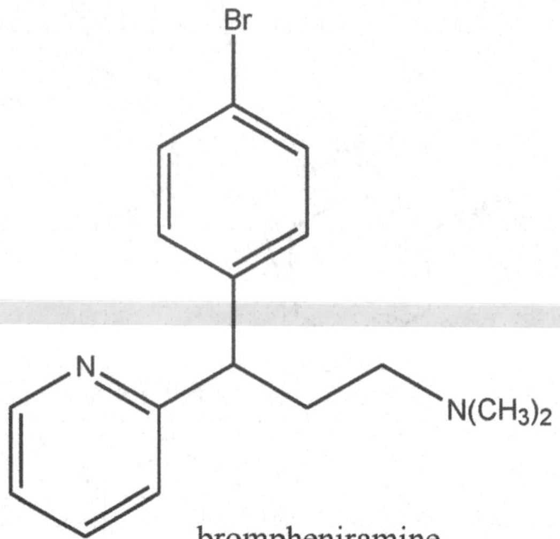

Br
N
$N(CH_3)_2$

brompheniramine
an antihistamine: anticold agent

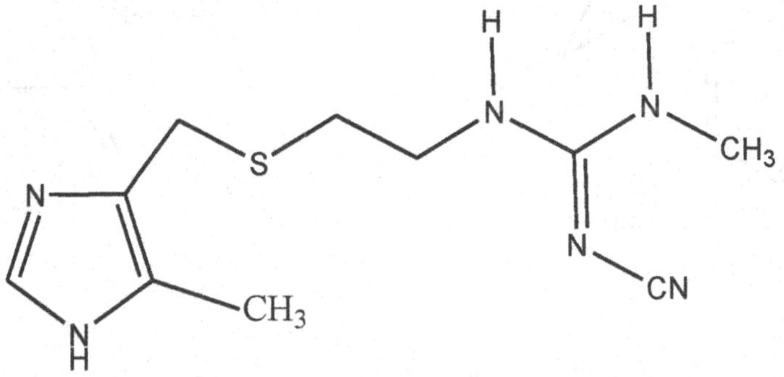

H H
N N
S CH_3
N
N
N CH_3 CN
H

cimetidine:(Tagamet) antihistidine: antiulcer drug

- **Read pages ______ - _____.**

4. SOME WELL KNOWN LEGAL AMINE DRUGS

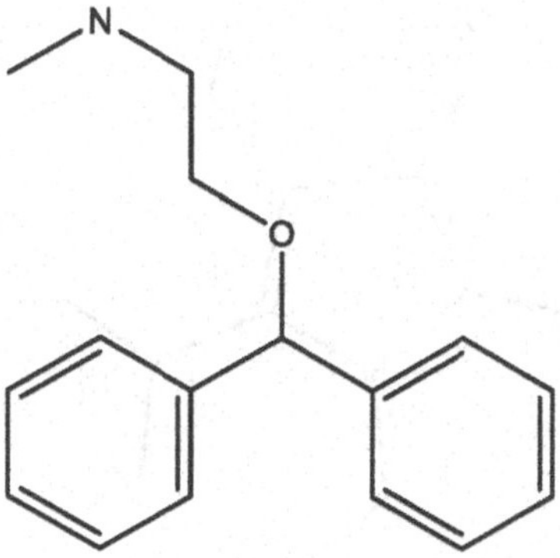

diphenhydramine (Benadryl)

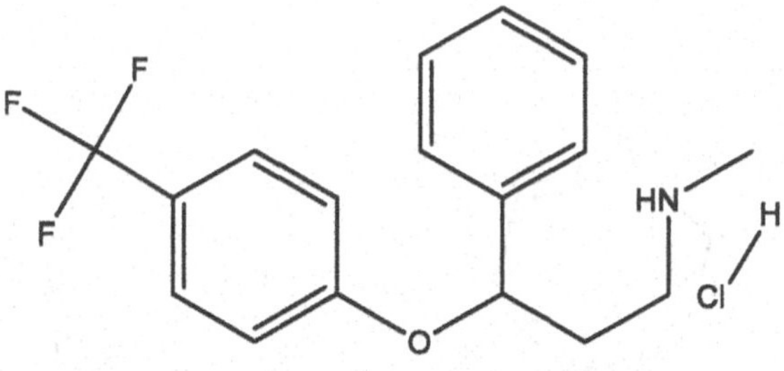

prozac

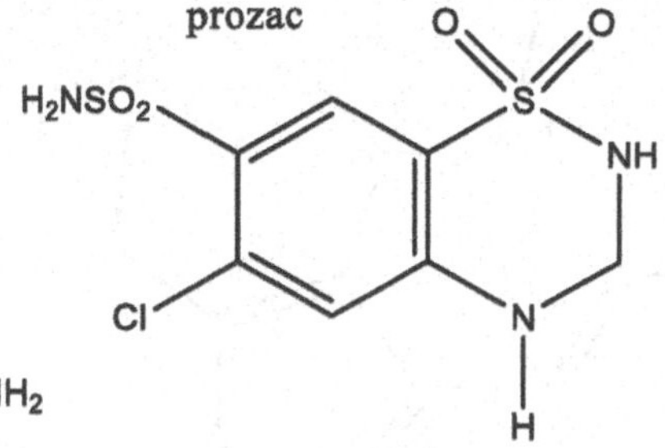

hydrochlorothiazide

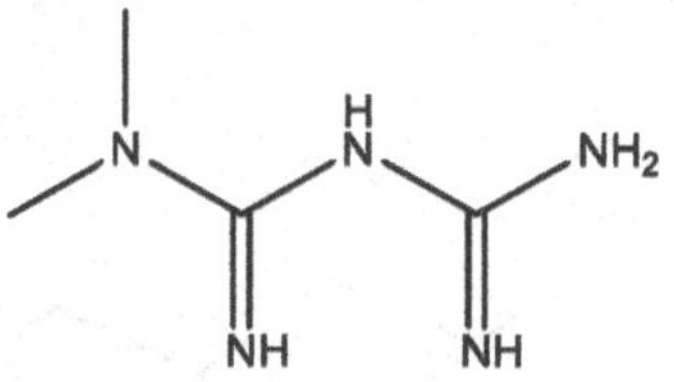

metformin {glucophage)

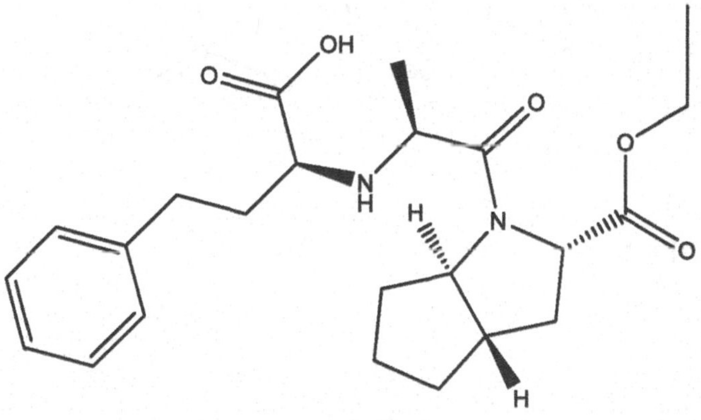

ramipril

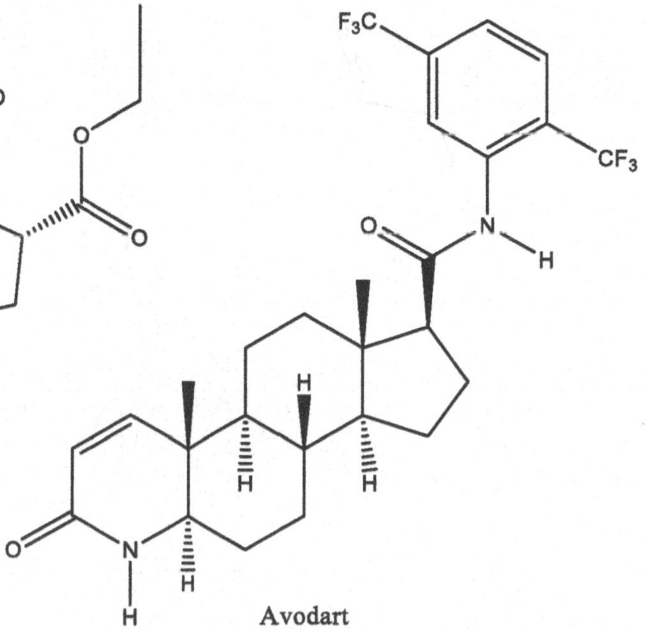

Avodart

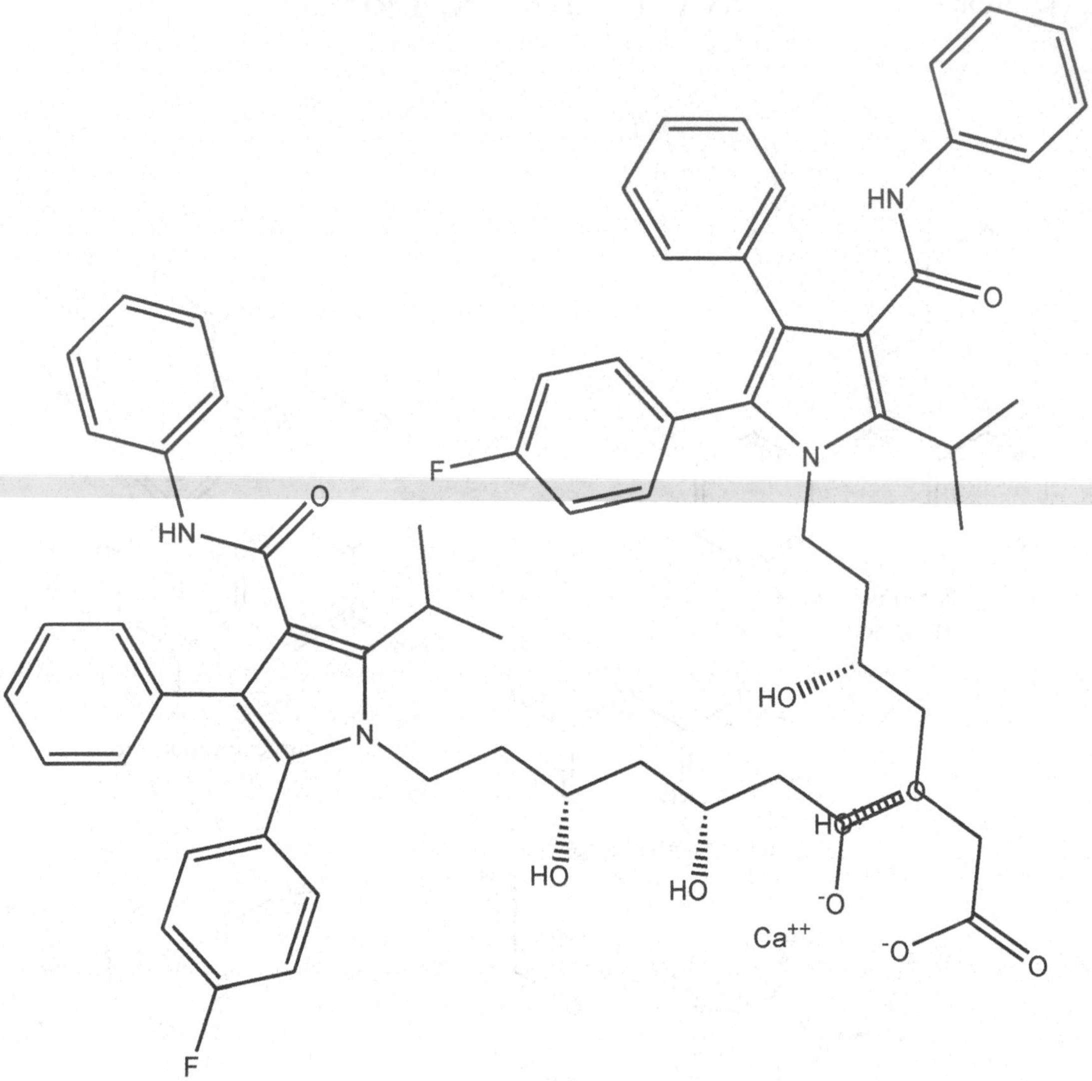

lipitor

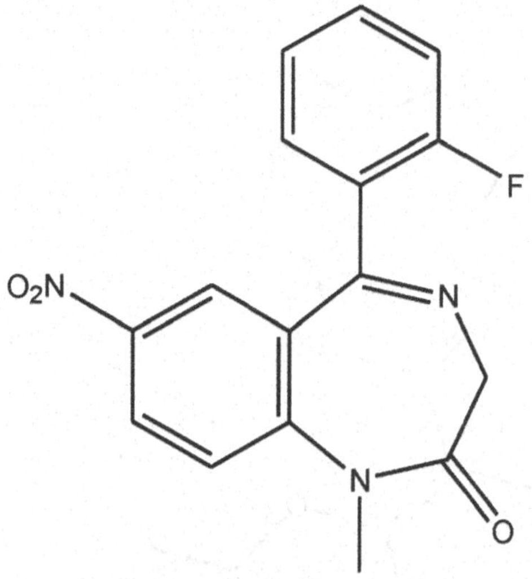

ROHYPNOL = ROOFIES

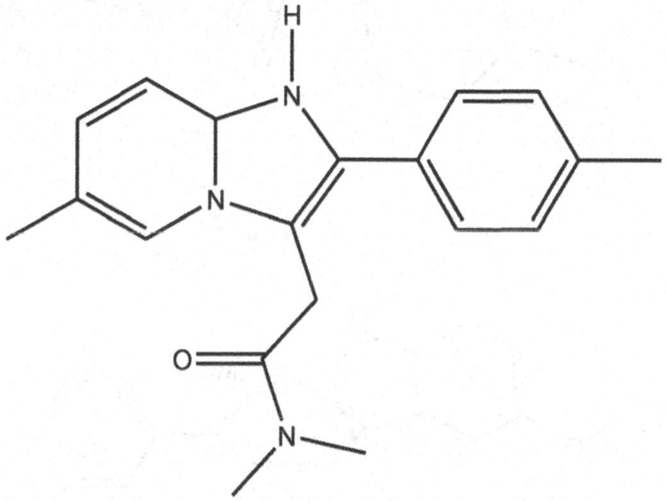

Ambien

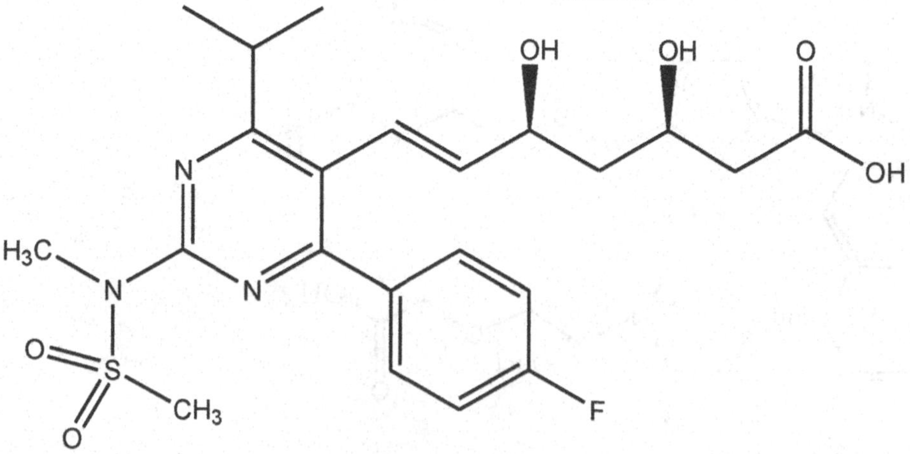

Crestor (Rosuvastatin)

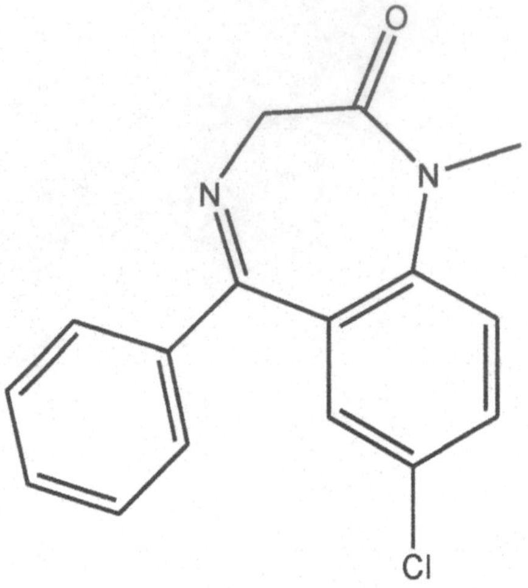

valium

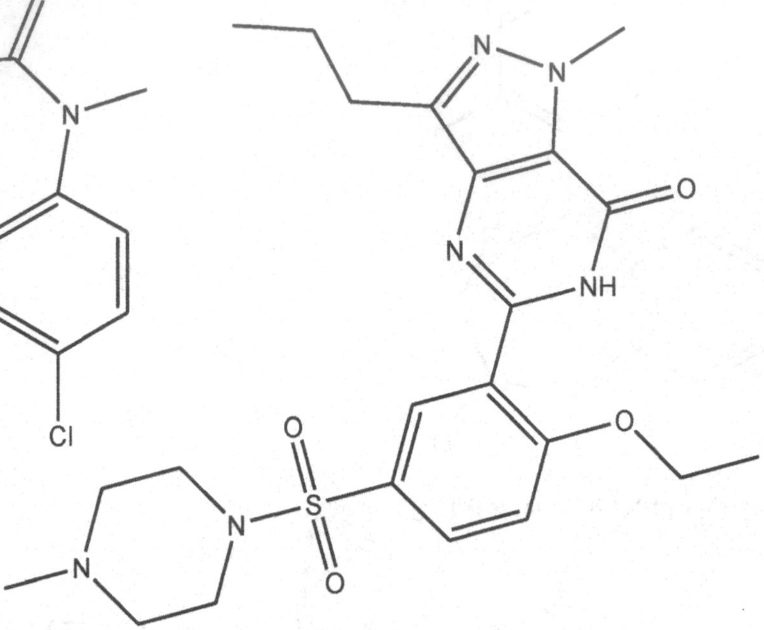

viagra

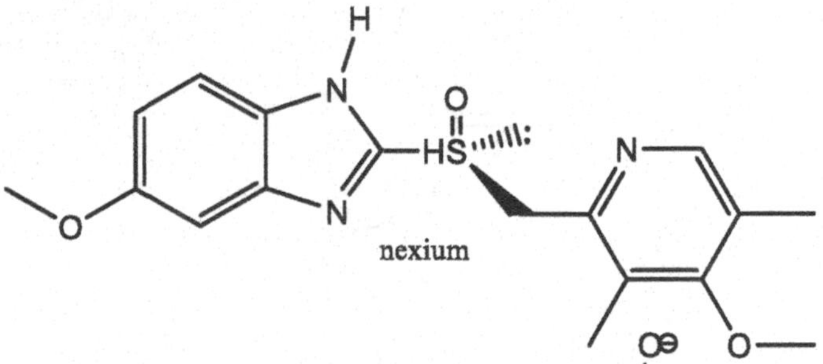

nexium

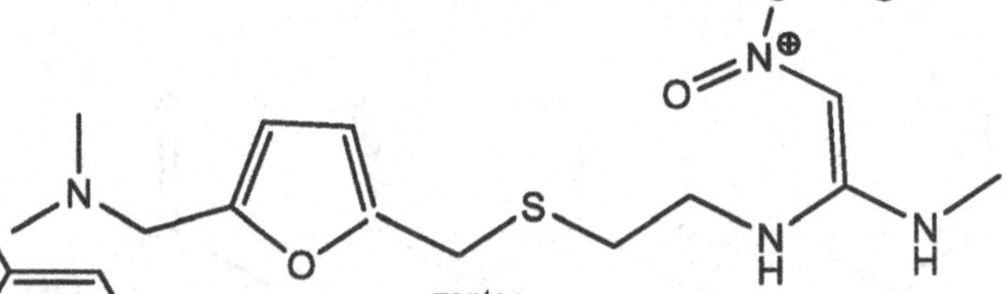

zantac

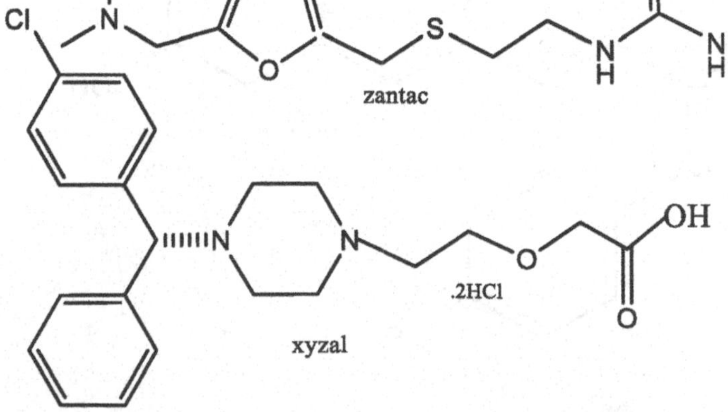

.2HCl

xyzal

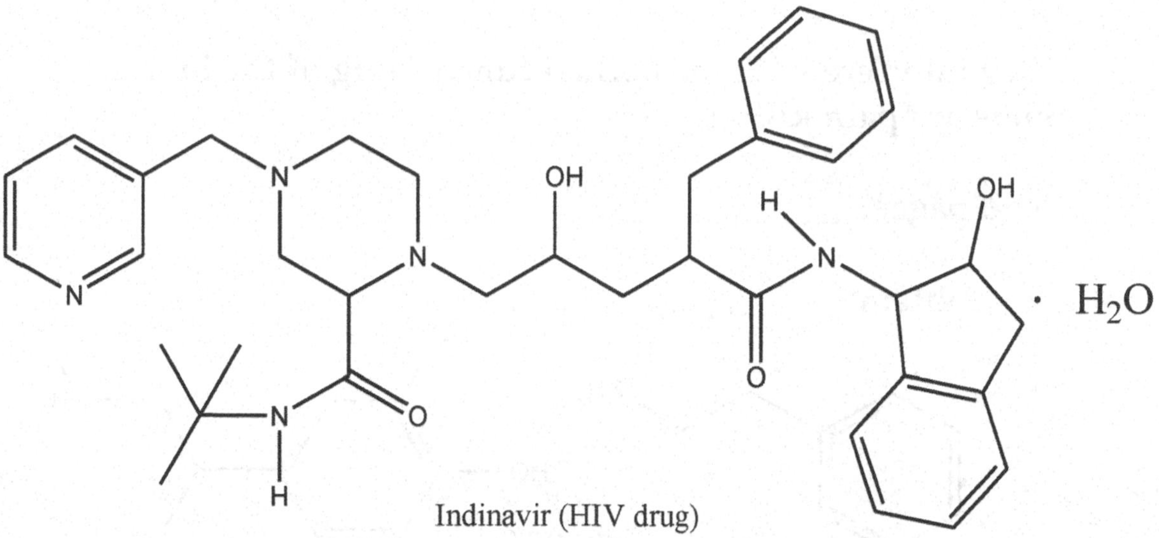

Indinavir (HIV drug)

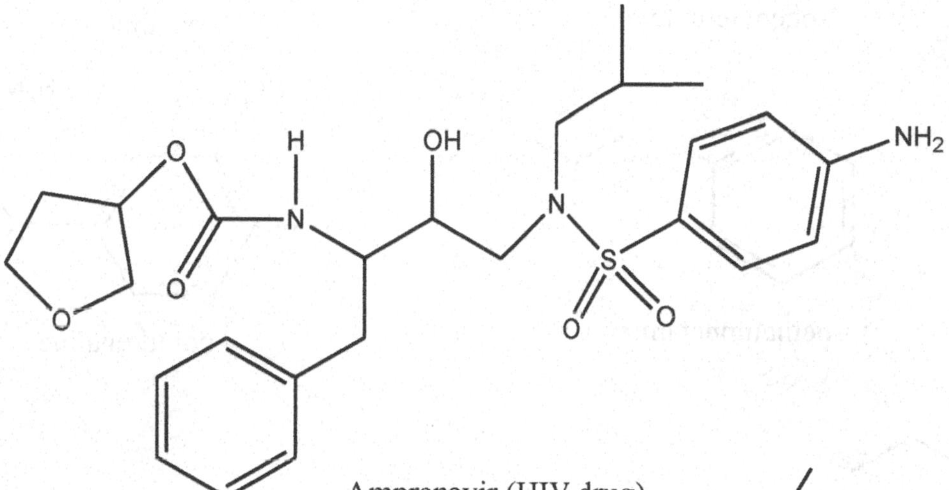

Amprenavir (HIV drug)

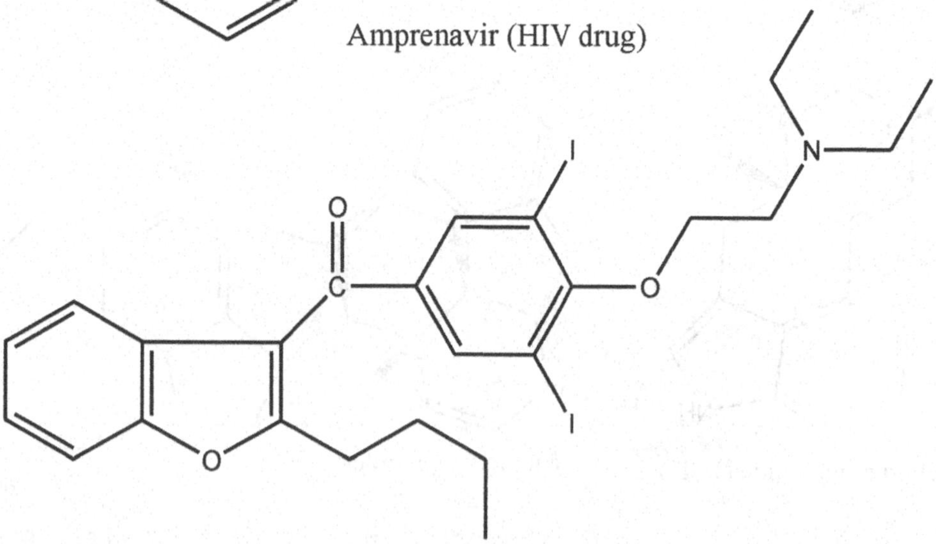

Cordarone (amiodarone) (atrial fibrillation drug)

5. 2-PHENYLETHYLAMINE AND DERIVATIVES

- **They interfere with the normal functioning of the brain. Some are pain killers.**
- **Read pages ______ - ______.**
- **Do problem ______, page ______.**

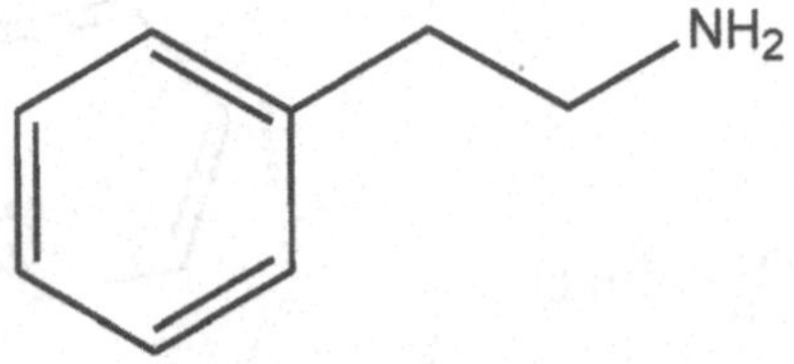

2-phenylethylamine

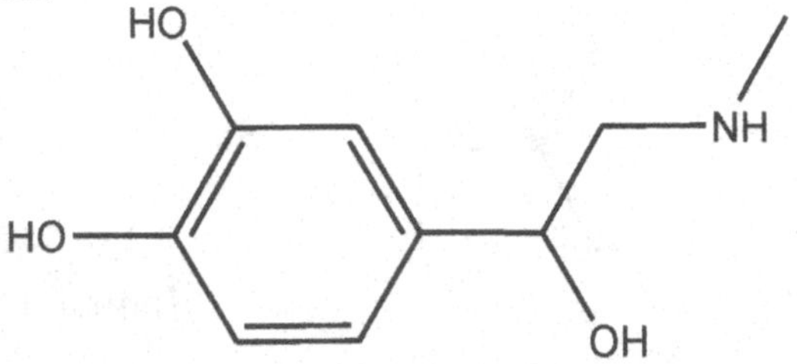

adrenaline

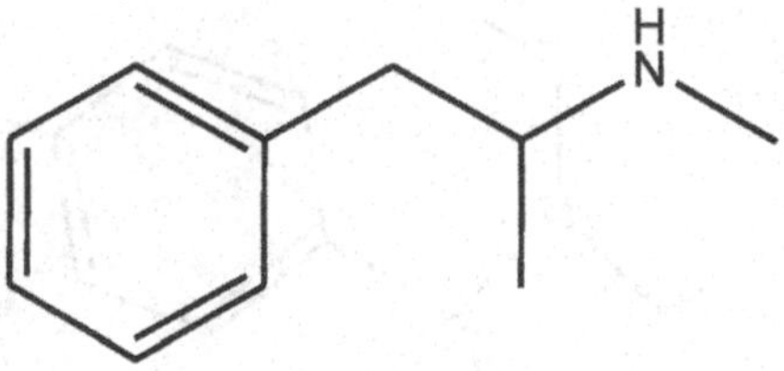

methamphetamine

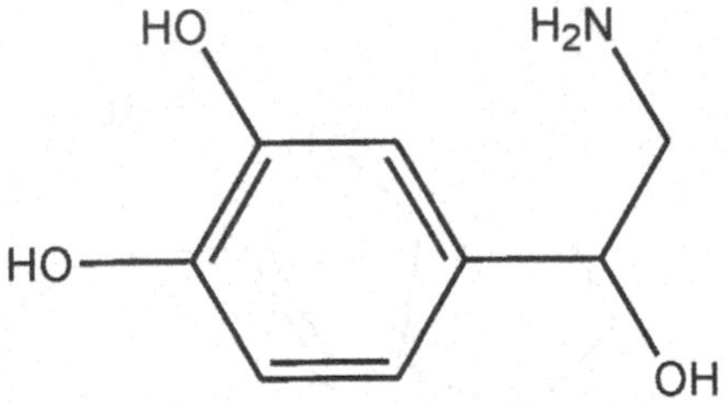

noradrenaline

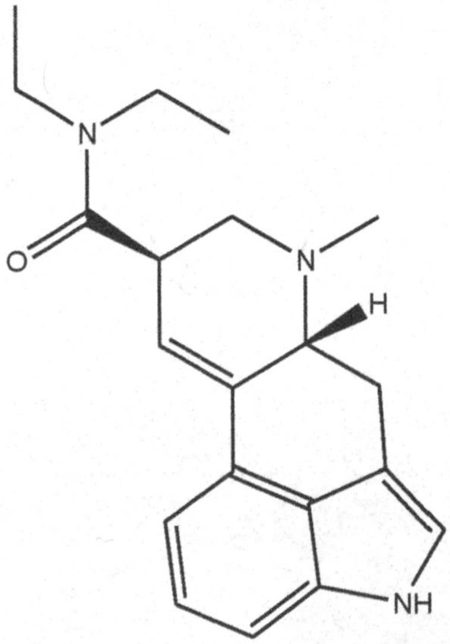

lysergic acid diethyl amide (LSD)

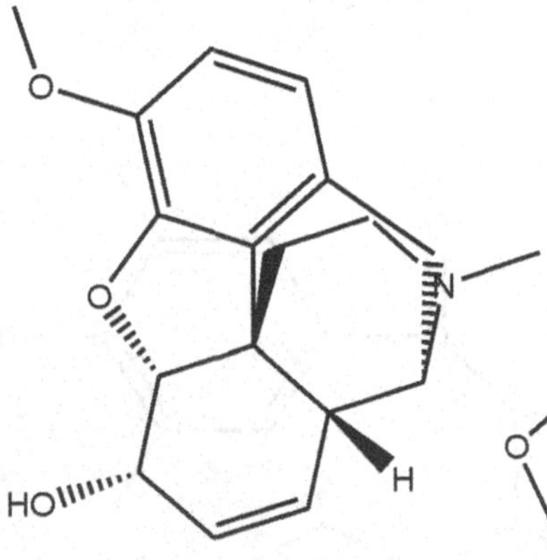

codeine: pain killer

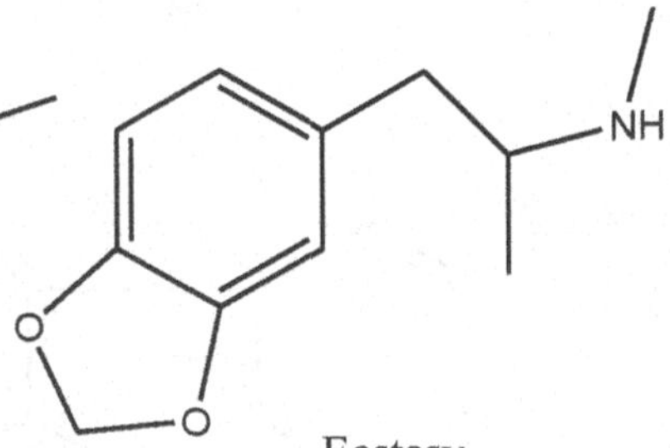

Ecstasy

F. PREPARATION OF AMINES

1. INTRODUCTION

- There are 3 ways:
  - Nucleophilic (SN2) substitution with nitrogen nucleophiles
  - Reduction of nitrogen containing functional groups
  - Reductive amination of aldehydes and ketones

2. NUCLEOPHILIC SUBSTITUTION: SN2

a. Primary Amines

$$R\text{-}X + 2\ddot{N}H_3 \longrightarrow R\text{-}\ddot{N}H_2 + NH_4^+X^-$$

excess ammonia

Ex:

$$CH_3\text{-}Br + 2\ddot{N}H_3 \longrightarrow CH_3\text{-}\ddot{N}H_2 + NH_4^+Br^-$$

excess ammonia

- Note: for 2° and 3° amines, $R'NH_2$ and R'_2NH are used, respectively.

b. Quaternary Ammonium Salts

$$R'\text{-}X + R'_3\ddot{N} \longrightarrow R'_4\overset{+}{N}\,X^-$$

excess alkyl halide

Ex:

$$CH_3CH_2\text{-}Br + (CH_3)_3\ddot{N} \longrightarrow (CH_3)_3\overset{+}{N}(CH_2CH_3)\,Br^-$$

excess alkyl halide

- Do problem on page ______.

c. The Gabriel Synthesis of 1° Amines

R-X + phthalimide $\xrightarrow[H_2O]{OH^-}$ RNH_2 + $C_6H_4(CO_2^-)_2$ + X^-

phthalimide

Ex:

Br-cyclopentane + phtalimide $\xrightarrow[H_2O]{OH^-}$ cyclopentyl-NH_2 + Br^-

- **Read pages ______ - ______.**

3. REDUCTION OF COMPOUNDS CONTAINING N

a. Reduction of Nitro Compounds

$$R\text{-}NO_2 \xrightarrow[\text{or Fe/HCl or Sn/HCl}]{H_2/Pd} R\text{-}\ddot{N}H_2$$

a 1° amine

Ex:

$$CH_3CH_2\text{-}NO_2 \xrightarrow{H_2/Pd} CH_3CH_2\text{-}\ddot{N}H_2$$

a 1° amine

C_6H_5-NO_2 $\xrightarrow{Fe/HCl}$ C_6H_5-NH_2

cyclobutyl-NO_2 $\xrightarrow{Sn/HCl}$ cyclobutyl-NH_2

b. Reduction of Nitriles: R-C≡N: See Unit 12

$$R\text{-}C{\equiv}N: \xrightarrow[2.\ H_2O]{1.\ LiAlH_4} R\text{-}CH_2\ddot{N}H_2$$

a 1° amine

Ex:

$$\text{cyclopentyl}\text{-}C{\equiv}N \xrightarrow[2.\ H_2O]{1.\ LiAlH_4} \text{cyclopentyl}\text{-}CH_2NH_2$$

c. Reduction of Amides: R-CONH₂

$$R\text{-}C(=O)NH_2 \xrightarrow[2.\ H_2O]{1.\ LiAlH_4} RCH_2\ddot{N}H_2$$

an amide → a 1° amine

or

$$R\text{-}C(=O)NR'_2 \xrightarrow[2.\ H_2O]{1.\ LiAlH_4} RCH_2\ddot{N}R'_2$$

an amide → a 3° amine

Ex:

$$\text{cyclopentyl}\text{-}C(=O)NH_2 \xrightarrow[2.\ H_2O]{1.\ LiAlH_4} \text{cyclopentyl}\text{-}CH_2NH_2$$

- **Read page ______.**
- **Do problems on page ______.**

4. REDUCTIVE AMINATION OF ALDEHYDES AND KETONES

- The catalyst is **sodium cyanoborohydride, $NaBH_3CN$. The general reaction for aldehydes is:**

$$RR'C{=}O \;(\text{aldehyde or ketone}) + R_2''NH \xrightarrow{NaBH_3CN} R'\text{—}CH(R)\text{—}NR_2''$$

aldehyde or ketone

1°, 2°, 3° amines

Ex:

$$(H_3CH_2C)(H_3C)C{=}O + (CH_3)_2NH \xrightarrow{NaBH_3CN} (H_3CH_2C)(H_3C)CH\text{—}N(CH_3)_2$$

- Do problems on pages ______ - ______.

G. REACTIONS OF AMINES

1. INTRODUCTION

- **Amines use the lone pair of electrons on N to react as bases or nucleophiles.**

2. REACTIONS AS BASES

- **The general reaction:**

$$R\text{-}\ddot{N}H_2 + H\text{-}A \rightleftharpoons R\text{-}\overset{+}{N}H_3 + :A^-$$

pKa of HA should be less than 10

pKa~10-11

Ex:

$$C_5H_9\text{—}\ddot{N}H_2 + H\text{-}Cl \longrightarrow C_5H_9\text{—}NH_3^+\,Cl^-$$

- Do problems on page ______.

- **Read about the use of acid-base properties of amines to extract them as aqueous ammonium salts. See Fig. ______, page ______.**

Ex:

$$R\text{-}\ddot{N}H_2 + H\text{-}Cl \longrightarrow R\text{-}\overset{+}{N}H_3\ Cl^-$$

insoluble in water, but soluble in CH_2Cl_2 .

soluble in water, but insoluble in CH_2Cl_2

3. RELATIVE BASICITY OF AMINES

- The best way of assessing the strength of a basic amine is to use the pKa of its conjugate acid, **RNH_3^+. The higher the pKa of the conjugate acid, the stronger the amine as a weak base, and vice versa (See Unit 12).**

Ex:

NH $\xrightarrow{+H^+}$ NH_2^+

pKa = 11.1

$NH_3 \xrightarrow{+H^+} NH_4^+$

pKa = 9.3

- **Conclusion: NH_4^+ is a stronger acid; therefore, NH_3 is a weaker base.**

- **Do problems ______, ______, ______, pages ______ - ______.**

4. FACTORS AFFECTING AMINE BASICITY

a. Comparing Alkyl Amines and Ammonia

- **Alkyl amines are more basic than NH_3 since alkyl groups are electron donating groups (EDG).**

Ex:

$$CH_3NH_2 > NH_3$$

b. Comparing Alkyl Amines and Aniline ($PhNH_2$)

- **Alkyl Amines are more basic than aniline because the lone pair of electrons on the N is delocalized in Aniline through resonance.**

Ex:

$$C_6H_5NH_2 \xrightarrow{+H^+} C_6H_5NH_3^+$$

pKa = 4.6

$$CH_3CH_2NH_2 \xrightarrow{+H^+} CH_3CH_2N_3^+$$

pKa = 10.8

- **Conclusion:**

c. Comparing Aniline and Substituted Anilines

- **EDG ($-NH_2$, -OH, -OR, -NHCOR, -R) make arylamines more basic than aniline.**

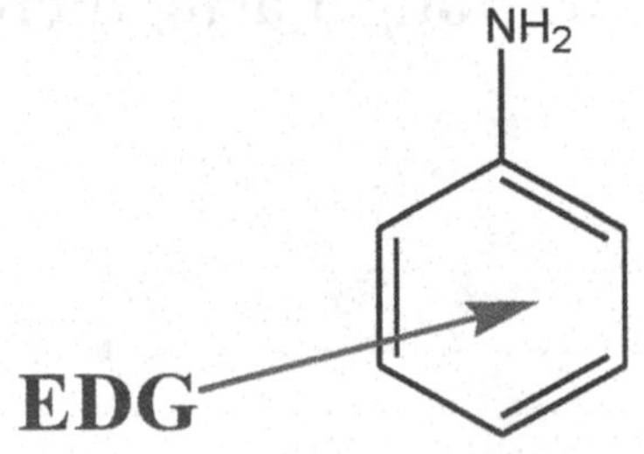

- EWG (-X, -CHO, -COR, -COOR, -COOH, -CN, $-SO_3$, $-NO_2$, $-NR_3^+$) make arylamines less basic than aniline.

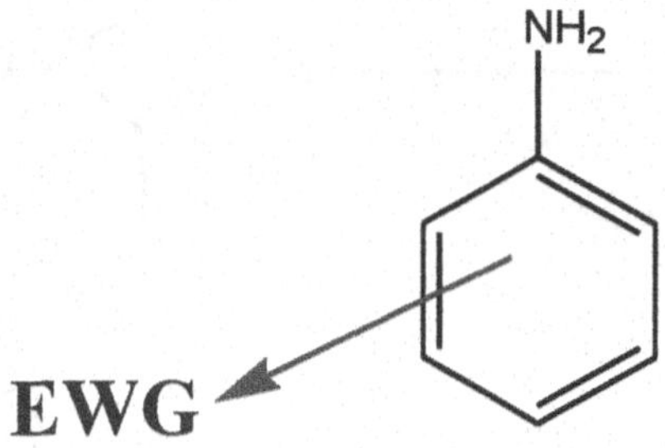

NH_2

EWG

Ex: Rank the following compounds in increasing order of basicity.

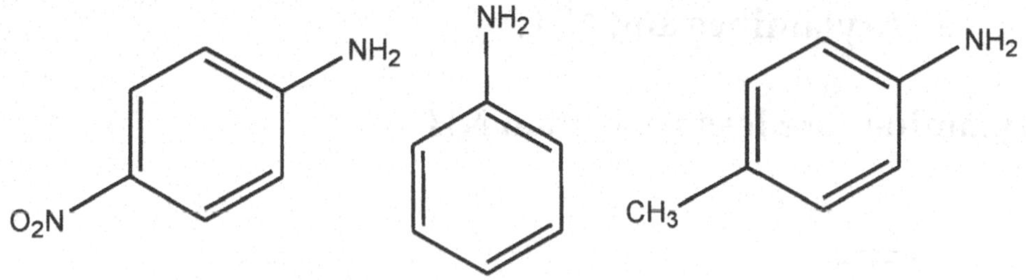

NH_2 NH_2 NH_2

O_2N CH_3

- Read pages _____ - _____.

d. Comparing Alkylamines and Amides

- Alkylamines are more basic than amides since protonation at the -C=O group leads to a resonance stabilized structure in amides. As a result, the lone pair on the N in amides is delocalized and not as readily available as in alkylamines.

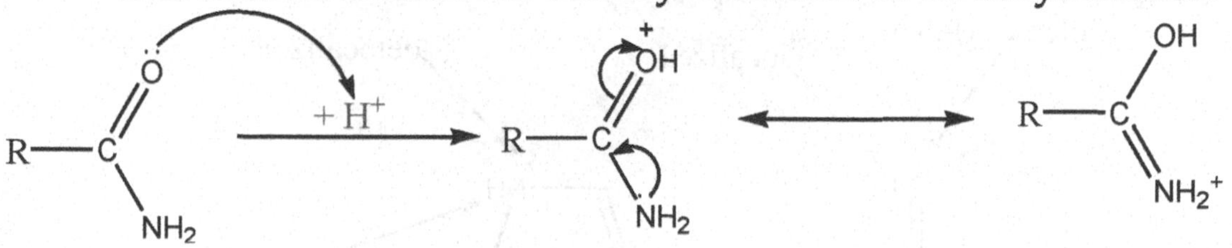

R—C(=O)—NH_2 $\xrightarrow{+H^+}$ R—C(=$\overset{+}{O}$H)—NH_2 $\longleftrightarrow$ R—C(OH)=NH_2^+

- See page _____.

Ex:

$$\text{(CH}_3\text{CH}_2)_2\text{NH} \xrightarrow{+H^+} (\text{CH}_3\text{CH}_2)_2\text{NH}_2^+$$

pKa = 11.1

$$\text{R–C(=O)–NH}_2 \xrightarrow{+H^+} \text{R–C(=O)–NH}_3^+$$

pKa = -1

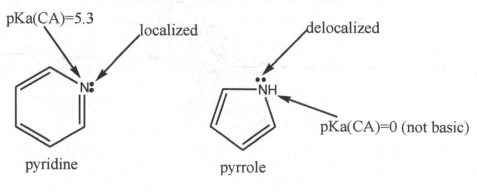

e. Arylamines and NH3

- **Arylamines are less basic than NH3.**

Ex:

$$\text{C}_6\text{H}_5\text{–NH}_2 \xrightarrow{+H^+} \text{C}_6\text{H}_5\text{–NH}_3^+$$

pKa = 4.6

$$NH_3 \xrightarrow{+H^+} NH_4^+$$

pKa = 9.3

- **Conclusion: Ammonia is more basic than aniline.**

f. Heterocyclic Aromatic Amines: Pyridine and Pyrolle

pKa(CA)=5.3

localized

pyridine

delocalized

pKa(CA)=0 (not basic)

pyrrole

CA = conjugate acid

g. Hybridization Effects: Pyridine and Piperidine

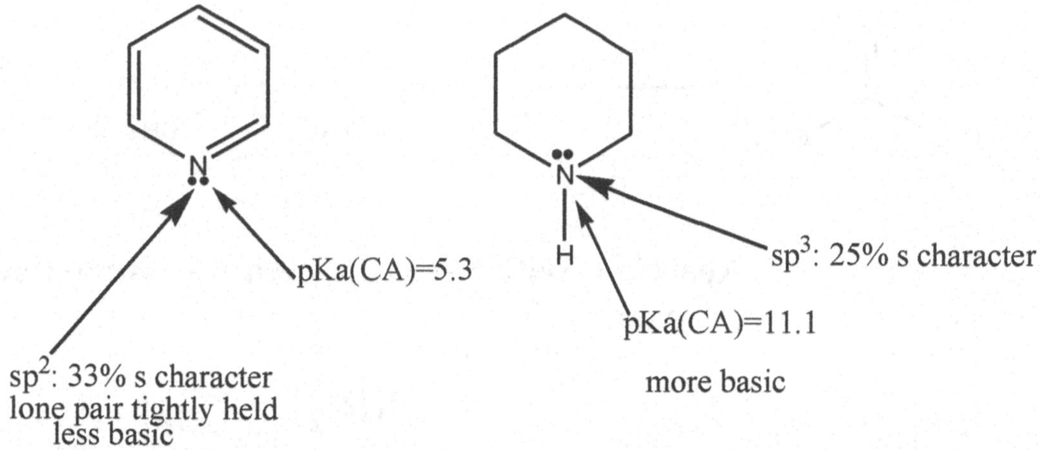

pKa(CA)=5.3

sp[2]: 33% s character
lone pair tightly held
less basic

sp[3]: 25% s character

pKa(CA)=11.1

more basic

- Do problem on page ______.
- See Table _____, page _____. See pKa values on page ______.
- Read about LSD: an amine on page ______.

5. REACTIONS OF AMINES AS NUCLEOPHILES

a. Introduction

- Amines react with carbonyl groups to give substitution or addition compounds.

b. Reactions of 1° and 2° Amines with Aldehydes and Ketones

i. Reactions with Primary Amines Revisited: See Unit 11

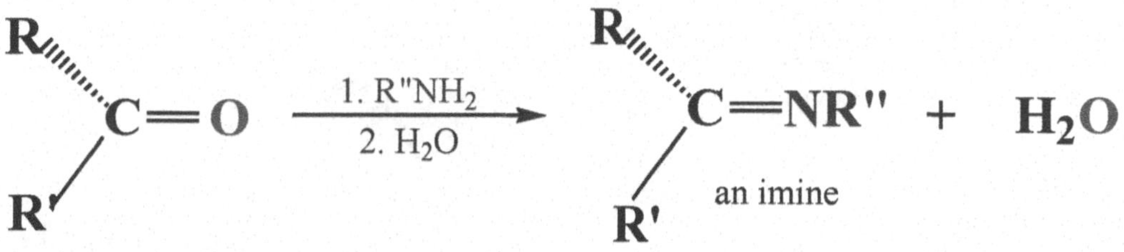

$$RR'C{=}O \xrightarrow[2.\ H_2O]{1.\ R''NH_2} RR'C{=}NR'' + H_2O$$

R' = H or alkyl group

an imine

Ex:

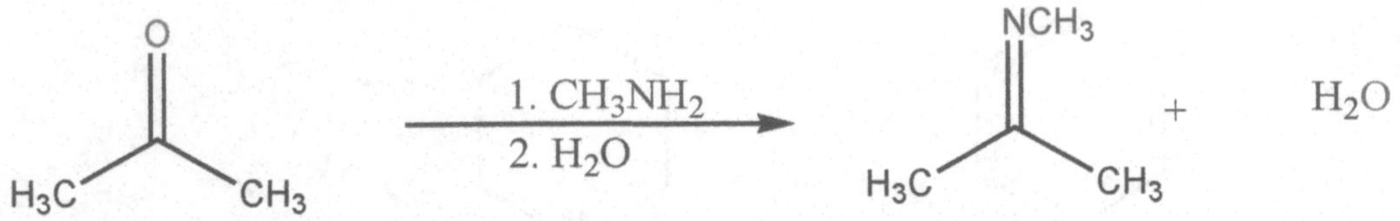

$$CH_3COCH_3 \xrightarrow[2.\ H_2O]{1.\ CH_3NH_2} (CH_3)_2C{=}NCH_3 + H_2O$$

ii. Reactions with Secondary Amines Revisited: See Unit 12

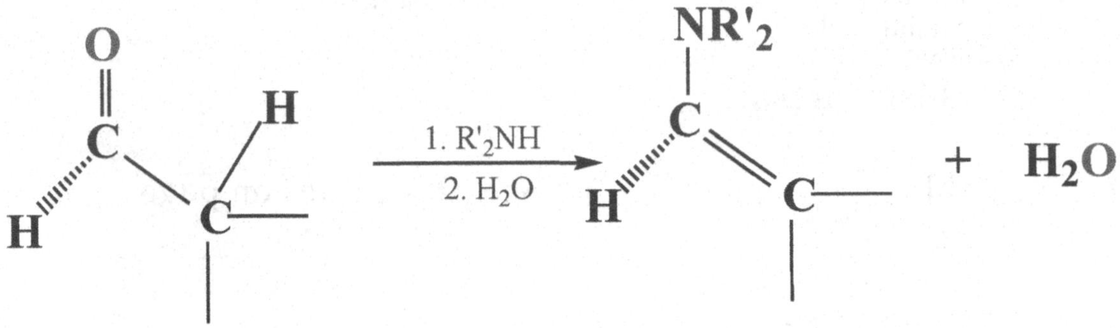

$$HC(=O)-C(H)< \xrightarrow[2.\ H_2O]{1.\ R'_2NH} HC(NR'_2){=}C< + H_2O$$

an enamine

Ex:

$$CH_3COCH_3 \xrightarrow[2.\ H_2O]{1.\ (CH_3)_2NH} H_3C{-}C(N(CH_3)_2){=}CH2 + H_2O$$

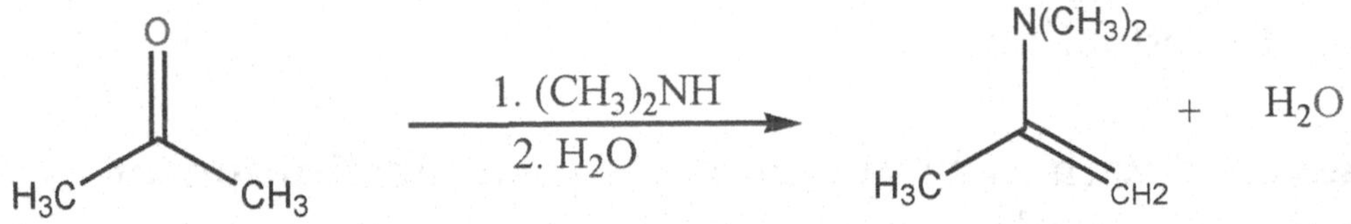

c. Reactions of NH_3, 1° and 2° Amines with RCOZ

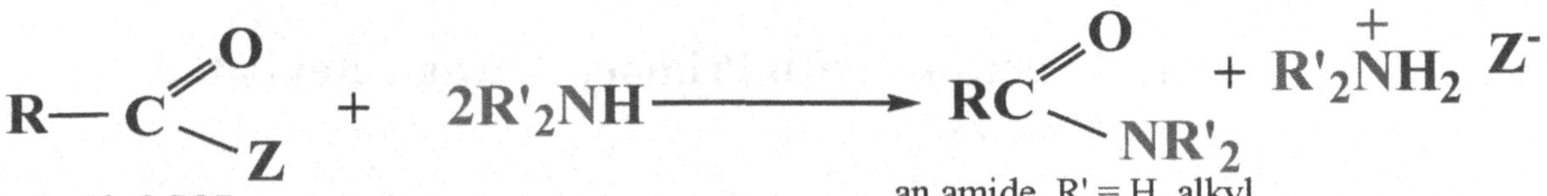

$$R{-}C({=}O){-}Z + 2R'_2NH \longrightarrow RC({=}O)NR'_2 + R'_2\overset{+}{N}H_2\ Z^-$$

Z= Cl, OCOR

an amide, R' = H, alkyl

Ex:

$$\text{cyclopropyl–C(=O)Cl} + 2(CH_3)_2NH \longrightarrow \text{cyclopropyl–C(=O)N}(CH_3)_2 + (CH_3)_2\overset{+}{N}H_2\,Cl^-$$

- **Read about the protection of $-NH_3$ in aniline undergoing Friedel-Crafts reaction on page ______.**

- **Do Problem _______, page ______.**

d. Hoffman Elimination Reaction

- **This reaction converts first an amine to a quaternary ammonium salt; the salt is then converted to an alkene.**

$$\text{H–C}_\beta\text{–C}_\alpha\text{–NH}_2 \xrightarrow[\text{2. } Ag_2O \quad \text{3. } \Delta]{\text{1. excess } CH_3I} \text{C=C (an alkene)}$$

β

H NH$_2$

α

an alkene

Ex:

$$\text{1-methylcyclopentylamine (CH}_3\text{, NH}_2\text{)} \xrightarrow[\text{2. } Ag_2O \quad \text{3. } \Delta]{\text{1. excess } CH_3I} \text{1-methylcyclopentene (–CH}_3\text{)} + \text{methylenecyclopentane (=CH2)}$$

minor product

major product

- **Note: The least substituted alkene is the major product (non-Zaisev).**

- **See Mechanism on page _____.**

- **Read pages ______ - ______ for details.**

- **Do problems on pages _______ - ______.**

e. Reactions of Amines with HNO_2

i. Primary Amines

$$R\text{-}\ddot{N}H_2 \xrightarrow[\text{HCl (or } HNO_2)]{NaNO_2} \underbrace{R\text{-}\overset{+}{N}{\equiv}N{:}\ Cl^-}_{\text{alkyl diazonium salt}} \text{ or } R\text{-}\overset{+}{N_2}\ Cl^-$$

Ex:

$$CH_3CH_2CH_2CH_2NH_2 \xrightarrow[HCl]{NaNO_2} CH_3CH_2CH_2CH_2N_2^+Cl^-$$

Ex: Aryl Diazonium salts

$$C_6H_5\text{-}NH_2 \xrightarrow[\text{HCl (or } HNO_2)]{NaNO_2} \underbrace{C_6H_5\text{-}\overset{+}{N}{\equiv}N{:}\ Cl^-}_{\text{An aryl diazonium salt}} \text{ or } C_6H_5\text{-}\overset{+}{N_2}\ Cl^-$$

ii. Secondary Amines

$$R\text{-}\ddot{N}(R')\text{-}H \xrightarrow[HCl]{NaNO_2} \underbrace{R\text{-}\ddot{N}(R')\text{-}\ddot{N}{=}\ddot{O}}_{\text{N-Nitrosamine}}$$

Ex:

$$CH_3\text{-}NH\text{-}CH_2CH_2CH_3 \xrightarrow[HCl]{NaNO_2} CH_3\text{-}N(N{=}O)\text{-}CH_2CH_2CH_3$$

- Do Problem ______, page ______. Read pages ______ - ______.

f. Substitution Reactions with Aryl Diazonium Salts

i. The general reaction is:

a good leaving group

N_2^+ —Z→ Z + N_2 + Cl^-

aryl diazonium salt

ii. Some examples:

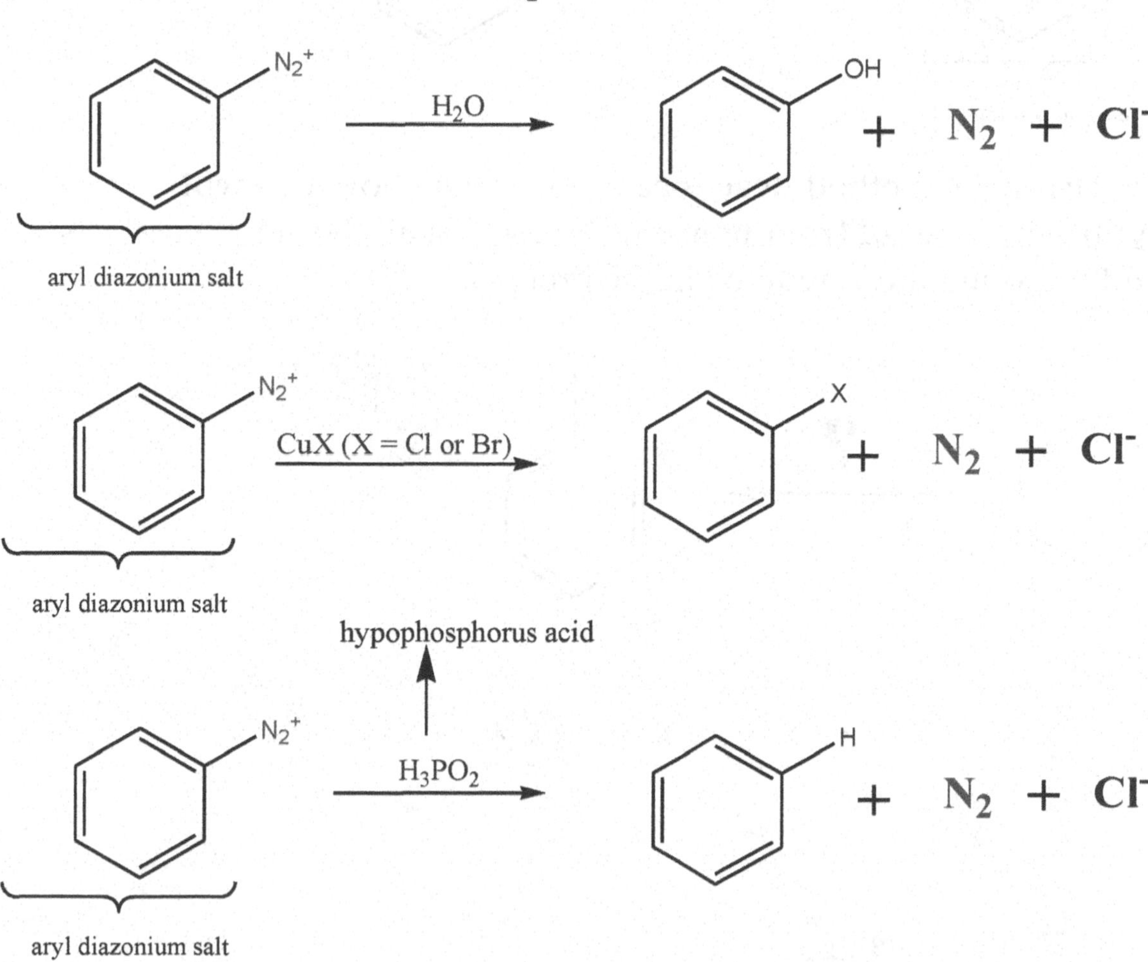

N_2^+ (aryl diazonium salt) $\xrightarrow{HBF_4}$ fluorobenzene (F) + N_2 + Cl^-

aryl diazonium salt

N_2^+ (aryl diazonium salt) $\xrightarrow{\text{NaI or KI}}$ iodobenzene (I) + N_2 + Cl^-

aryl diazonium salt

N_2^+ (aryl diazonium salt) $\xrightarrow{CuCN}$ benzonitrile (CN) + N_2 + Cl^-

aryl diazonium salt

Ex: Using the method **described above**, show how you will synthesize **phenol from benzene**. Please, list all the steps, catalysts, and intermediates involved in the process.

benzene $\xrightarrow{?}$ phenol (OH)

- **Read pages _____ - ______.**
- **Do all problems on page ______.**

iii. Coupling Reactions of Aryl Diazonium Salts

- **The general reaction is:**

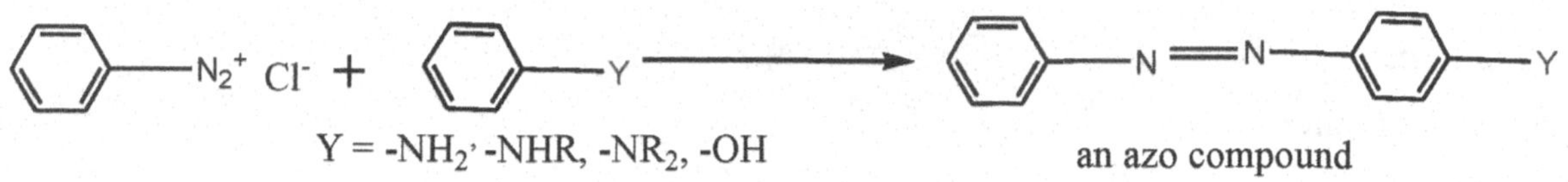

$$C_6H_5-N_2^+\ Cl^- + C_6H_5-Y \longrightarrow C_6H_5-N{=}N-C_6H_4-Y$$

Y = $-NH_2$, $-NHR$, $-NR_2$, $-OH$

an azo compound

- **Note: Y is a very strong electron donor.**

Ex:

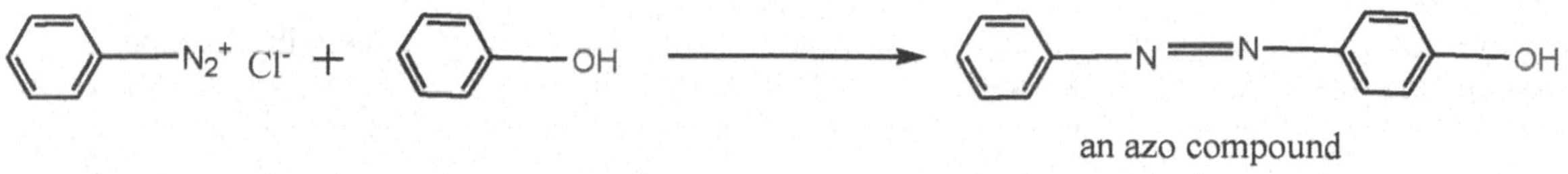

$$C_6H_5-N_2^+\ Cl^- + C_6H_5-OH \longrightarrow C_6H_5-N{=}N-C_6H_4-OH$$

an azo compound

- **Read pages ________ - ______ about Synthesis of Dyes and Sulfa Drugs.**

- **Do problems on page ______.**

6. PHENYLHYDRAZONES AND SEMICARBAZONES (SEE SECTION 5)

a. Phenylhydrazones

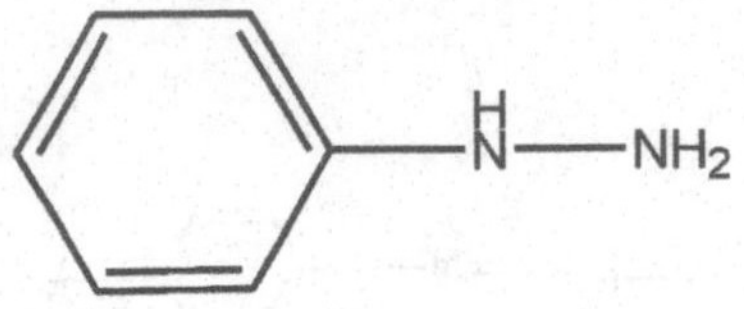

phenylhydrazine

- **Phenyl hydrazine** and **2,4-dinitrophenylhydrazine (2,4-DNP) (primary amines)** react with aldehydes and ketones to give crystalline **phenylhydrazones (imines)** with distinctive MP as follows:

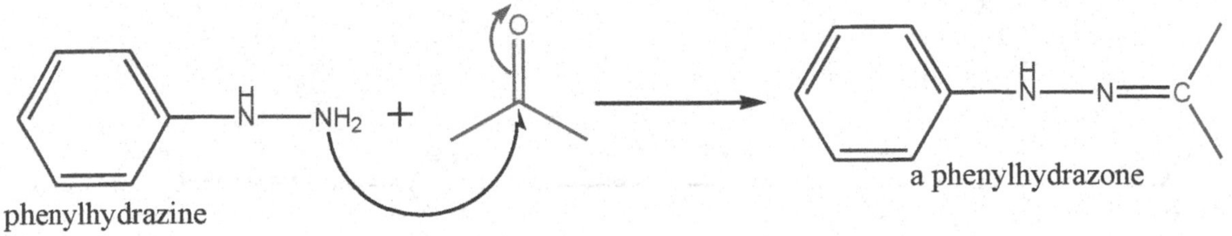

phenylhydrazine

a phenylhydrazone

Ex:

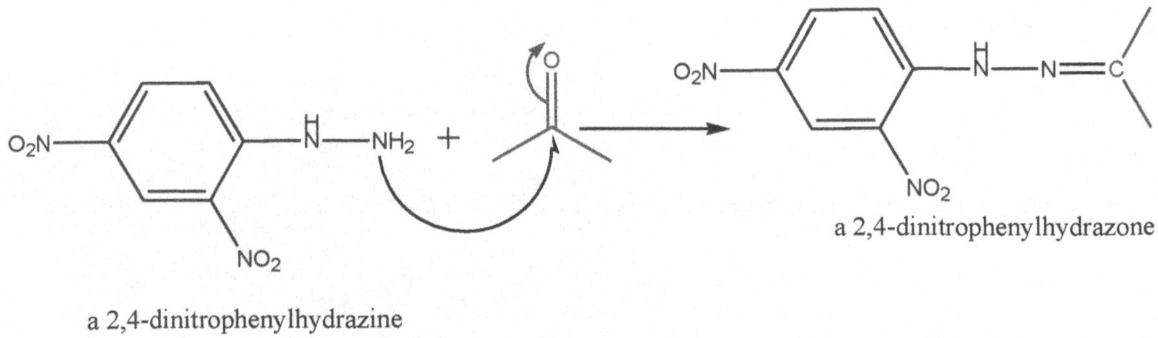

a 2,4-dinitrophenylhydrazine

a 2,4-dinitrophenylhydrazone

b. Semicarbazones

H_2N–NH–C(=O)–NH_2

semicarbazide

- **Smicarbazide (a primary amine) reacts with** aldehydes and ketones to give **crystalline semicarbazones (imines)** as follows.

+ H_2N–NH–C(=O)–NH_2 ⟶ C=NNCNH$_2$ (with O on C)

semicarbazide

a semicarbazone

+ H$_2$N–NH–C(=O)–NH$_2$ ⟶ Ph–CH=CH–CH=N–NH–C(=O)–NH$_2$

semicarbazide

a semicarbazone

cinnamaldehyde

- **Note: 2,4-DNP and semicarbazone can be used to identify aldehydes and ketones.**

H. NITROGEN CONTAINING LIFE MOLECULES: NUCLEIC ACIDS

1. PURINE AND PYRIMIDINE BASES

a. Pyrimidine Bases: 3

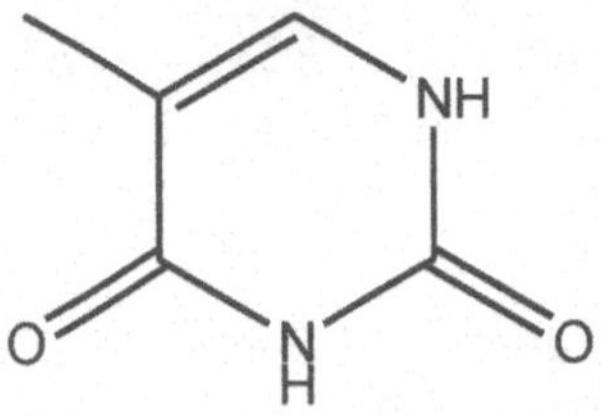

Thymine

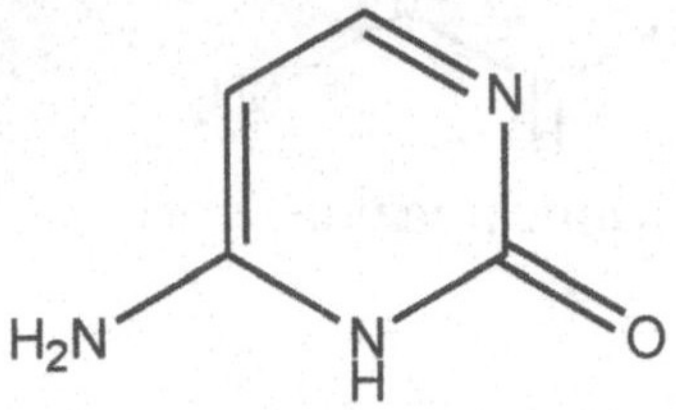

Cytosine

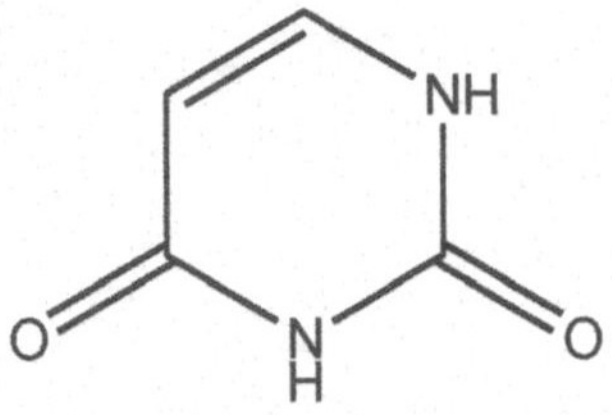

uracil

b. Purine Bases: 2

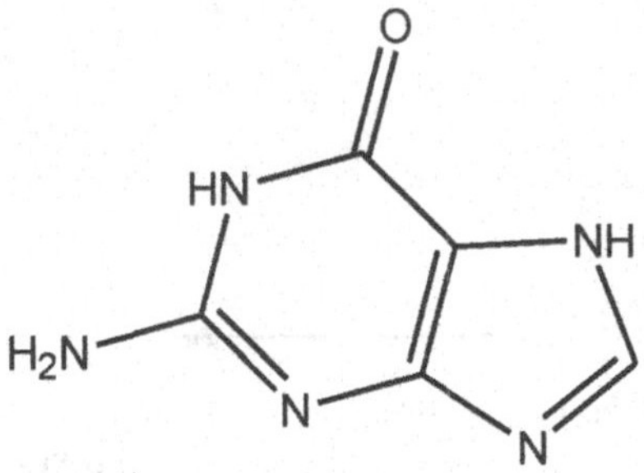

Guanine

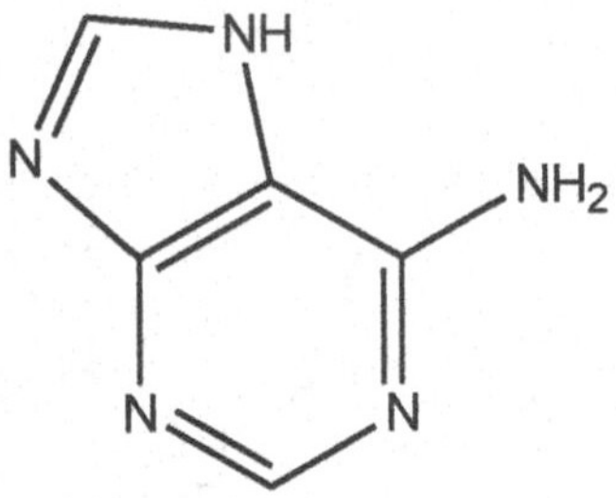

Adenine

c. Hydrogen Bonding and Base Pairs in DNA

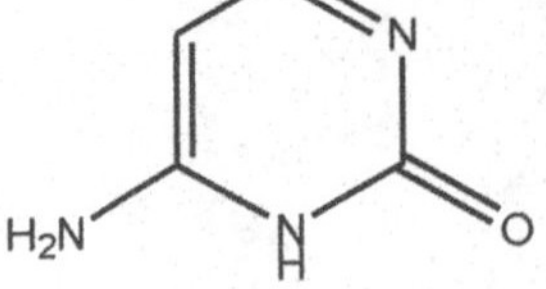

Cytosine

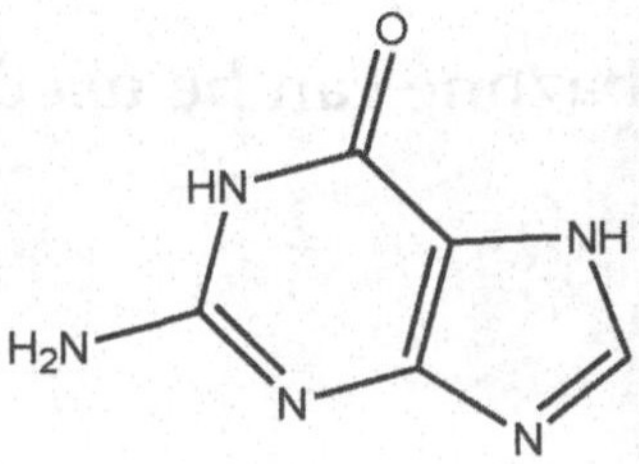

Guanine

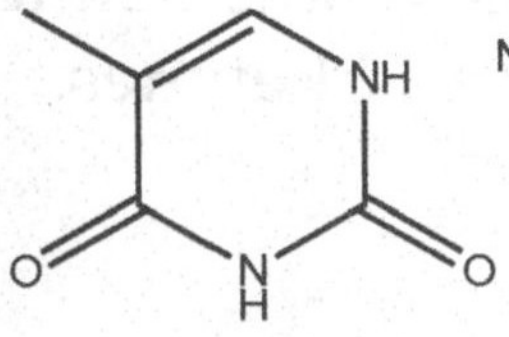

Thymine (Uracil for RNA)

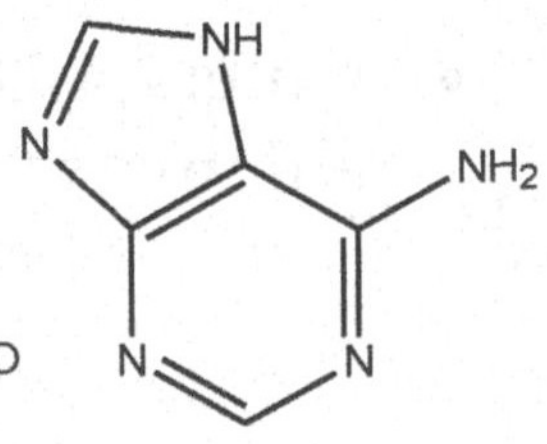

Adenine

Hydrogen Bonds

Cytosine ::::::::: Guanine

C≡G

Hydrogen Bonds

Thymine :::::::: Adenine

T=A

2. THE SUGARS IN DNA AND RNA

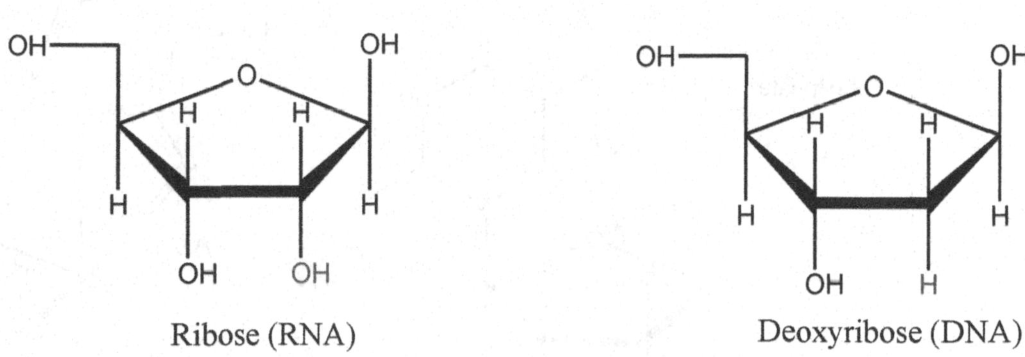

Ribose (RNA) Deoxyribose (DNA)

3. THE PHOSPHATE GROUP

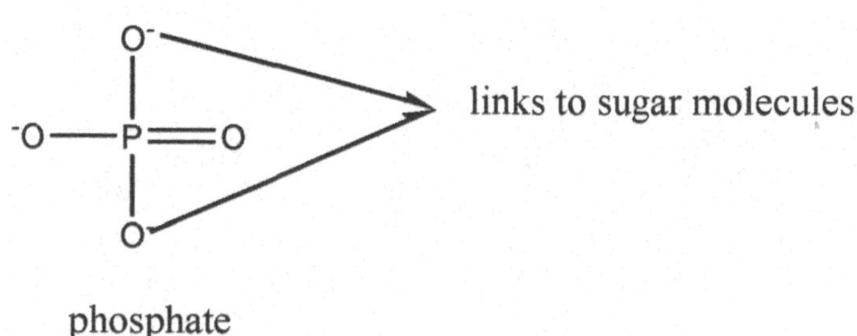

phosphate

4. FORMING NUCLEOTIDE UNITS

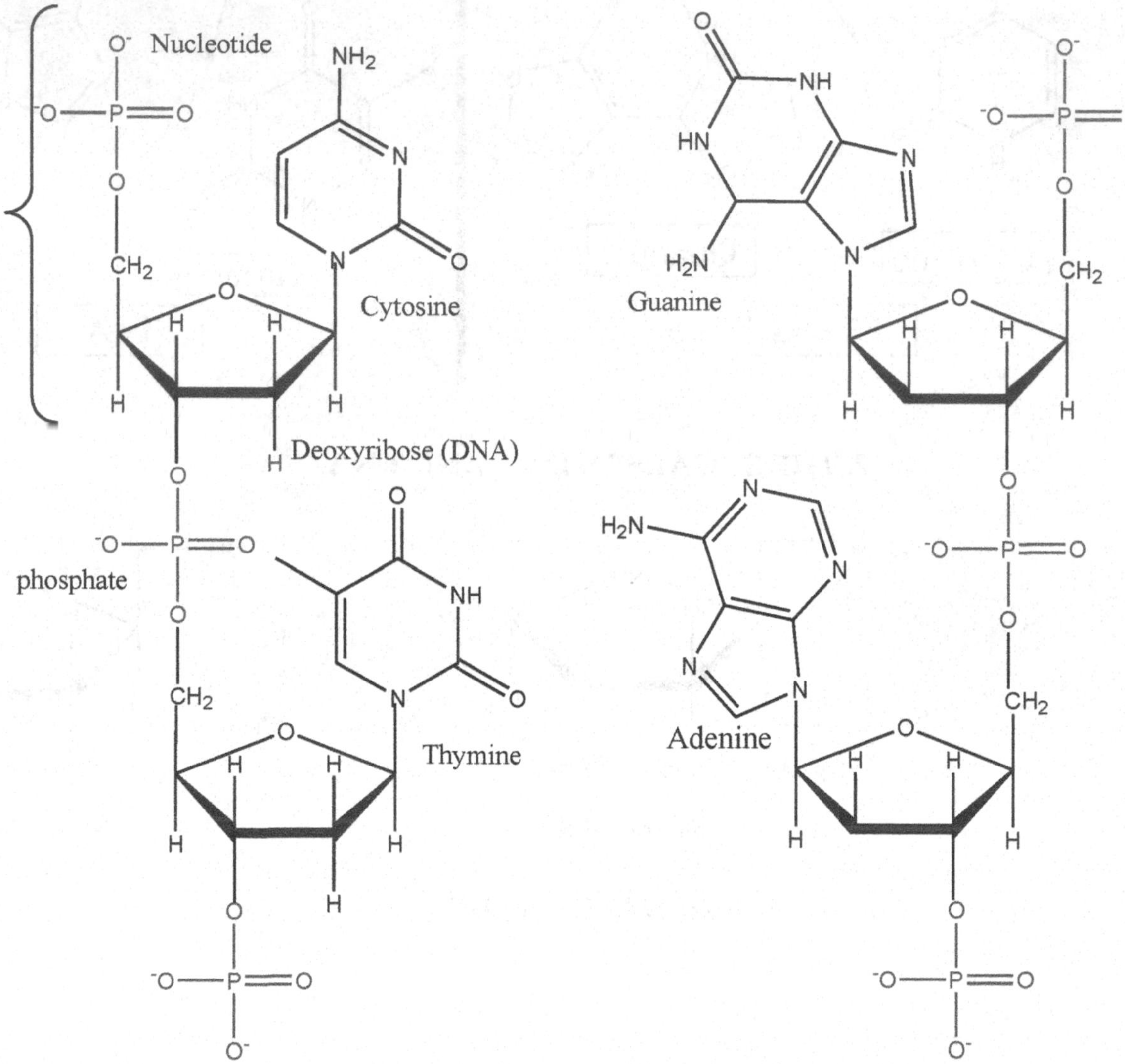

5. PUTTING ALL TOGETHER: A DNA SEGMENT

1. A DNA Segment

2. A DNA Ladder

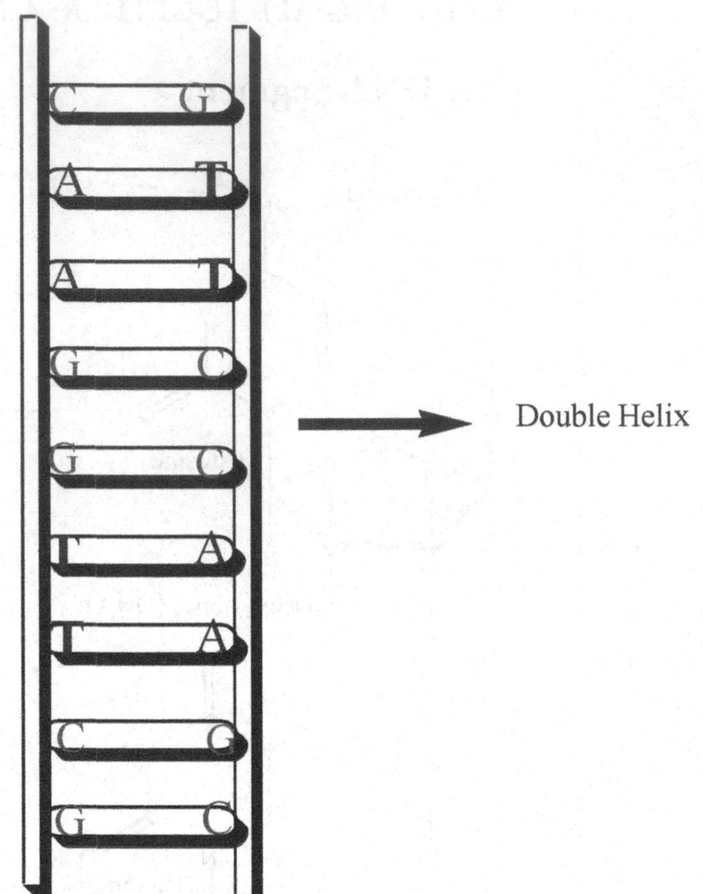

3. DNA vs. RNA

	DNA	RNA
Base	A, T, C, G	A, U, C, G
Sugar	2-Deoxyribose	D-Ribose
Structure	Helical, double-stranded	Single-stranded

I. A WORD ABOUT OPIATES AND OPIODS CONTAINING

1. INTRODUCTION

- **Opiates** and **opioids** are painkillers that have similar chemical structures. They are all nitrogen compounds (**amines**). They are prescription drugs used to **alleviate pain**. Basically, they act on the **nervous system**. However, they can become addictive when overused.

2. OPIATES

- They are opium **alkaloids** extracted from opium poppy. In other word, opiates are painkillers that occur **naturally**. Some are codeine, morphine, heroin, opium, etc.

3. OPIOIDS

- These are **synthetic** (prescription) drugs used to alleviate pain. They are lab made painkillers that have **similar** chemical structures as opiates (amines). Oxycontin (percocet, oxydone), hydrocodone (Lortab), Demerol (pethidine), hydromorphone (Dilaudid), and fentanyl (Duragesic) are some examples.

- **Please, see the chemical structures of some opiates and opioids below.**

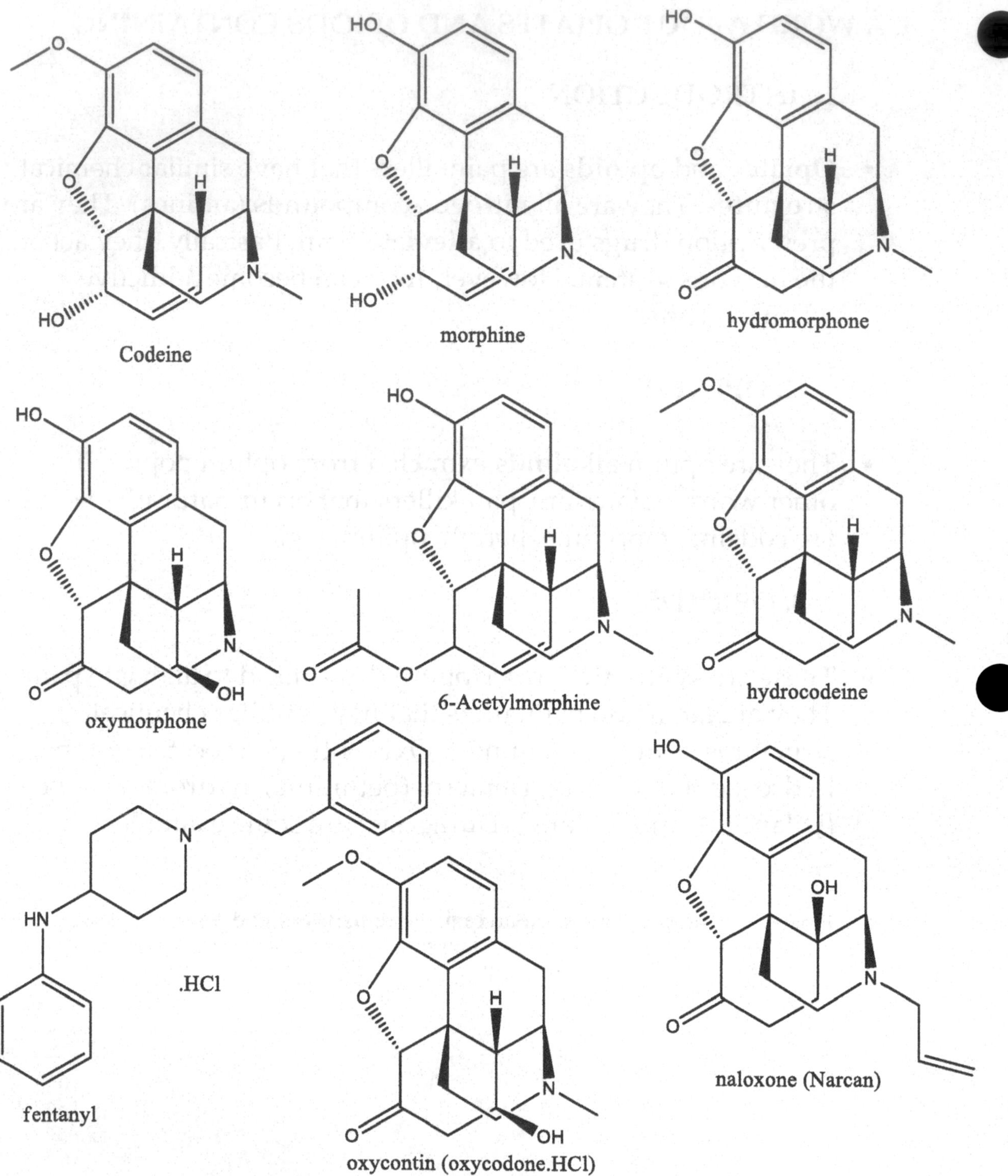

Codeine

morphine

hydromorphone

oxymorphone

6-Acetylmorphine

hydrocodeine

fentanyl

oxycontin (oxycodone.HCl)

naloxone (Narcan)

- **See Key Concepts on pages ______ – ______.**

SOME SELECTED REACTIONS FROM UNIT 13

1. PREPARATION OF THE AMINES

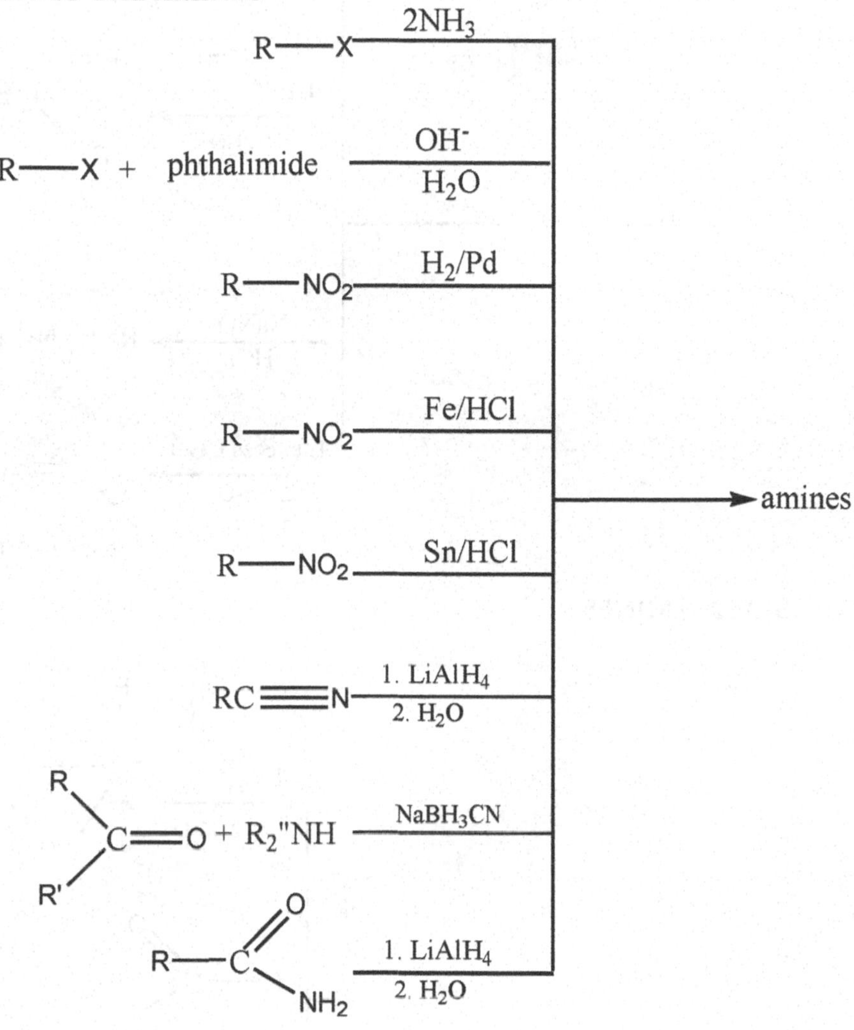

2. REACTIONS OF 1° AMINES

R—NH_2 →
- HA → $RNH_3^+ + A^-$
- 1. R'(R")C=O 2. H_2O → R'(R")C=NR
- $NaNO_2$, HCl → R—N_2^+ Cl^-
- 1. excess CH_3I 2. Ag_2O 3. Δ → C=C

a non-Zaisev product

3. REACTIONS OF 2° AMINES

R_2"NH →
- 1. H—C(=O)—C(H)< 2. H_2O → H—C(NR"$_2$)=C<
- R—C(=O)Z → R—C(=O)Z
- R'(R)C=O, $NaBH_3CN$ → R'(R)CH—NR"$_2$
- $NaNO_2$, HCl → R—N(R')—N=O

4. REACTIONS OF DIAZONIUM SALTS

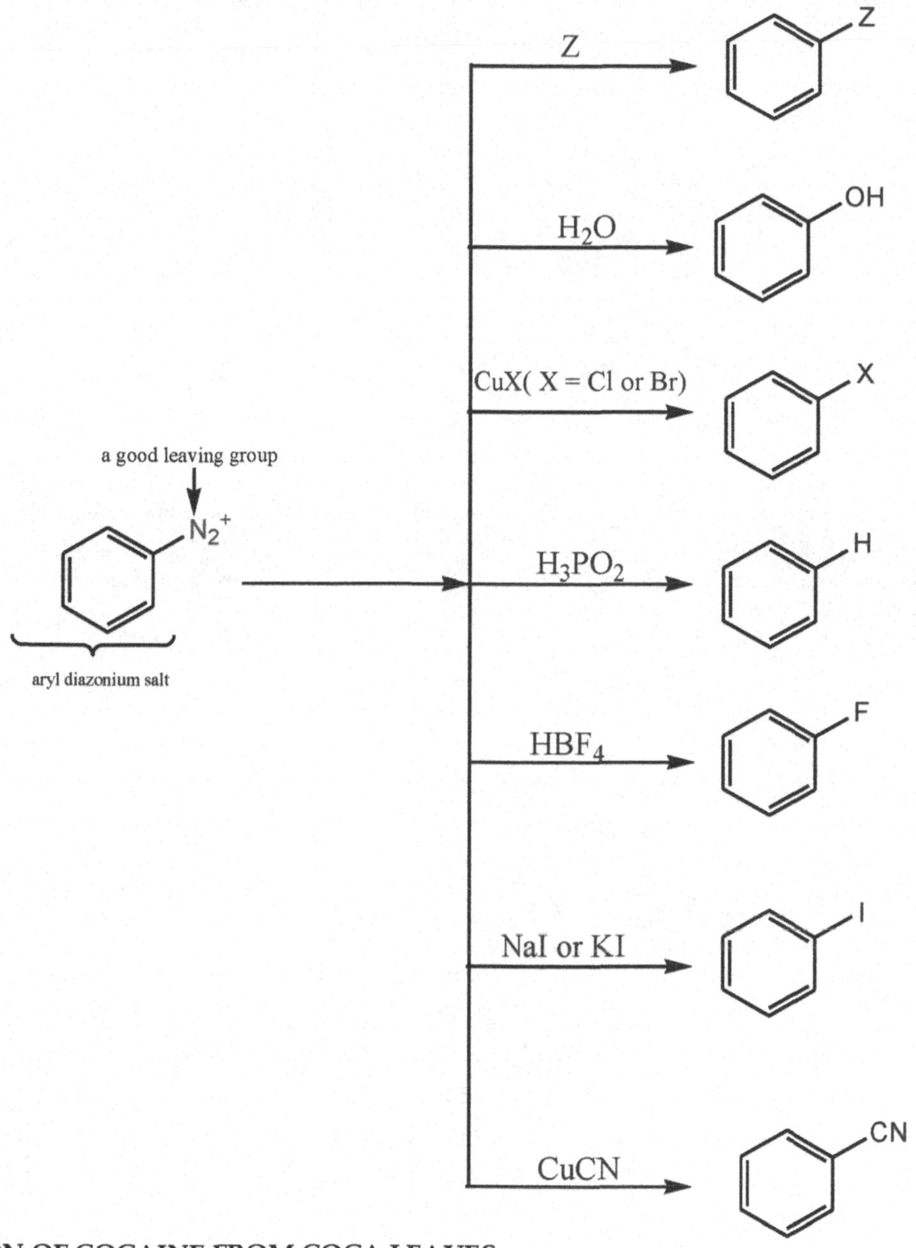

a good leaving group

N_2^+

aryl diazonium salt

Z

H_2O

CuX(X = Cl or Br)

H_3PO_2

HBF_4

NaI or KI

CuCN

Z

OH

X

H

F

I

CN

5. EXTRACTION OF COCAINE FROM COCA LEAVES

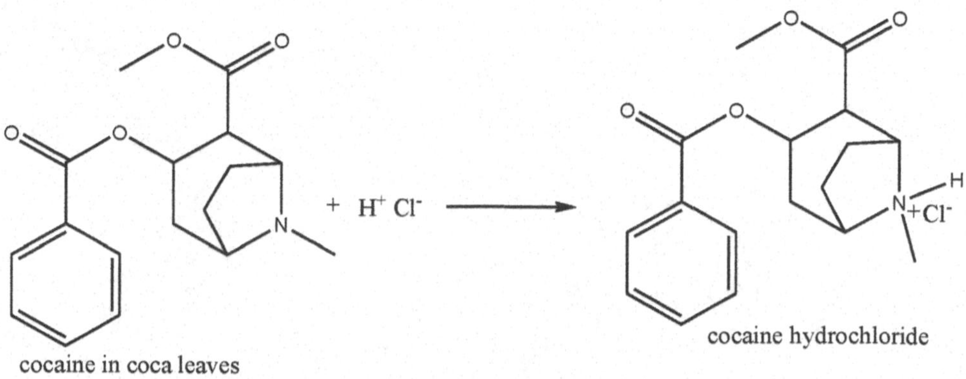

+ $H^+ Cl^-$

cocaine in coca leaves

cocaine hydrochloride

OCHEM II UNIT 14: SUBSTITUTION REACTIONS OF CARBONYL COMPOUNDS AT THE α CARBON

A. INTRODUCTION

- **Carbonyl compounds** can undergo **3 kinds of reactions:**
  - **Nucleophilic Addition Reactions: occur with carbonyl compounds that do not have an electronegative atom on the -C=O group. In other words, there is no leaving group Z on the CO.** See OCHEM II Units 10 and 11.
  - **The general reaction is:**

$$\mathrm{R{-}C({=}O){-}R'} \xrightarrow[\text{2. } H_2O]{\text{1. } Nu^-} \mathrm{R{-}C(OH)(R'){-}Nu}$$

  - **Nucleophilic Acyl Substitution (NAS) Reactions: There is an electronegative atom attached to the -C=O. There is a leaving group Z.** See OCHEM II Unit 12.
  - **The general reaction is:**

$$\mathrm{R{-}C({=}O){-}Z} \xrightarrow{:Nu^-} \mathrm{R{-}C({=}O){-}Nu} + Z^-$$

Z = OH, Cl, OR, NH_2

- **Substitution Reactions at the α carbon through enols and enolates.**

- **Enolate and enol formation:**

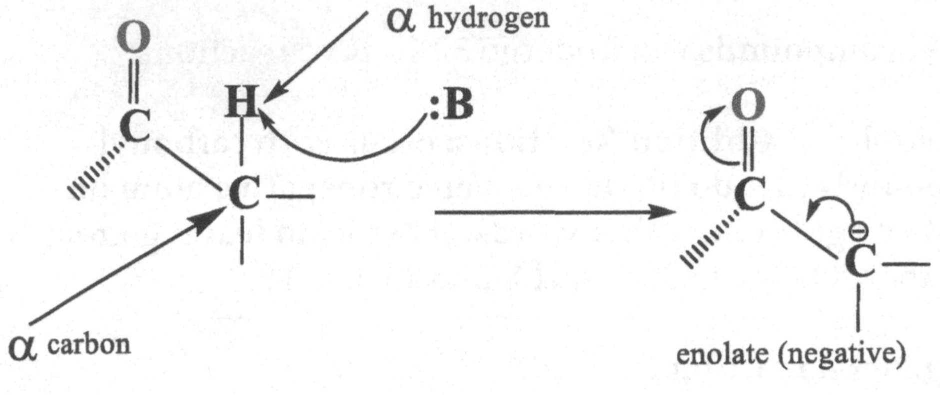

α hydrogen

:B

α carbon

enolate (negative)

or

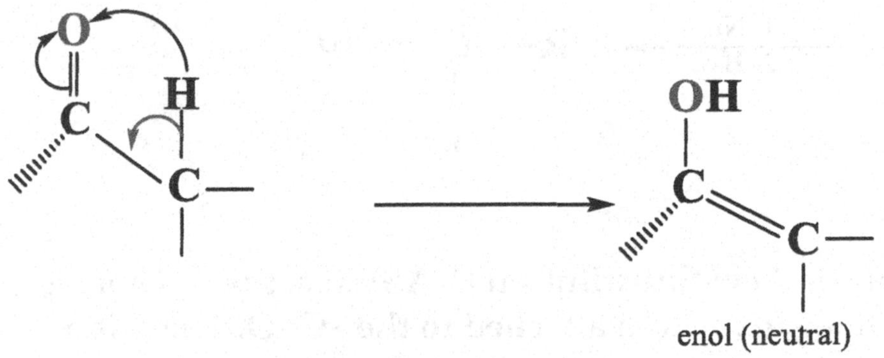

enol (neutral)

- **Note: The enolate or enol can react with an electrophile since enols and enolates are electron rich (a nucleophile = Nu^-).**

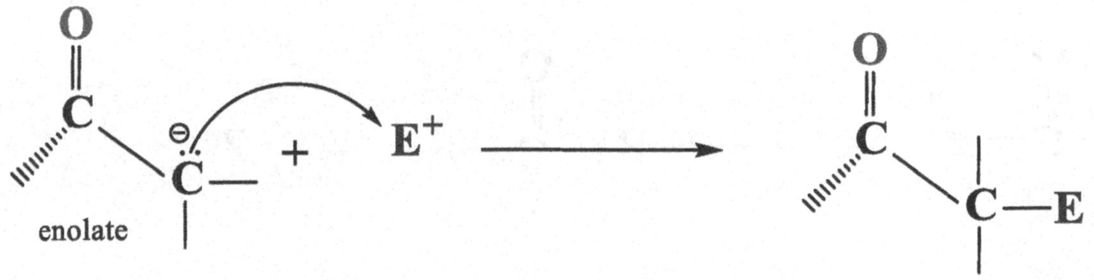

E^+

enolate

or

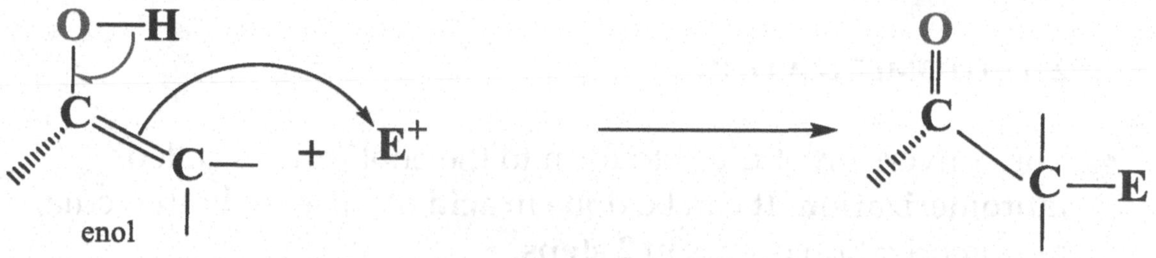

O—H

C=C— + E^+ ⟶ O=C—C—E

enol

- Read pages ______ and ______.
- Do problems on pages _____ - _____.

B. ENOLS

1. KETO FORM VS. ENOL FORM = TAUTOMERS

- **The 2 forms are at equilibrium as follows:**

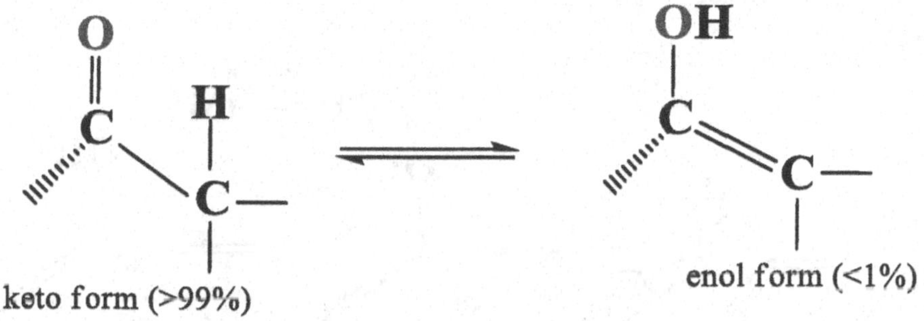

keto form (>99%) ⇌ enol form (<1%)

- **Note: With β-dicarbonyl compounds, the concentration of the enol form exceeds that of the keto form because the enol form is stabilized by conjugation and formation of intramolecular H bonds.**

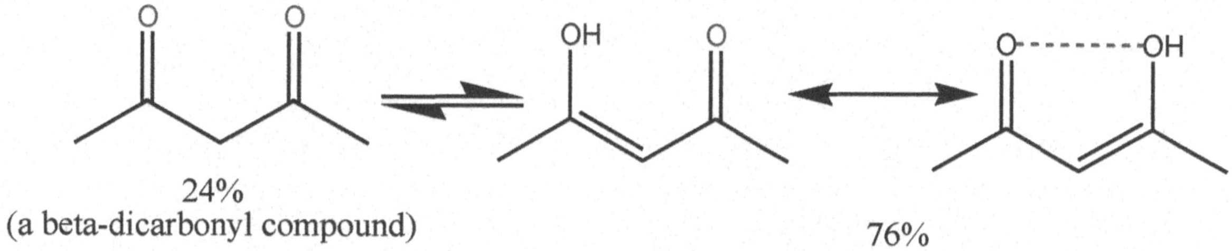

O O ⇌ OH O ⟷ O-------OH

24%
(a beta-dicarbonyl compound)

76%

- Read page ______. Do problems on page ______.

2. TAUTOMERIZATION

- The conversion of the keto form to the enol form is called **tautomerization**. It can be done in **acid** or **base**. In both media, tautomerization occurs in **2 steps**.

- **In acid:**

:O: H-OH_2^+ $^+$O H OH H_2O: OH C H C H C H C C C C C

keto form enol form

- **In base:**

H OH O :OH^- $^-$O: OH C H O C C C C C C C

keto form enol form

- **Read pages _____ - ______.**

- **Do Problems on pages _____ - _____.**

3. ENOLS AS NUCLEOPHILES: ELECTRON RICH

- The general reaction is:

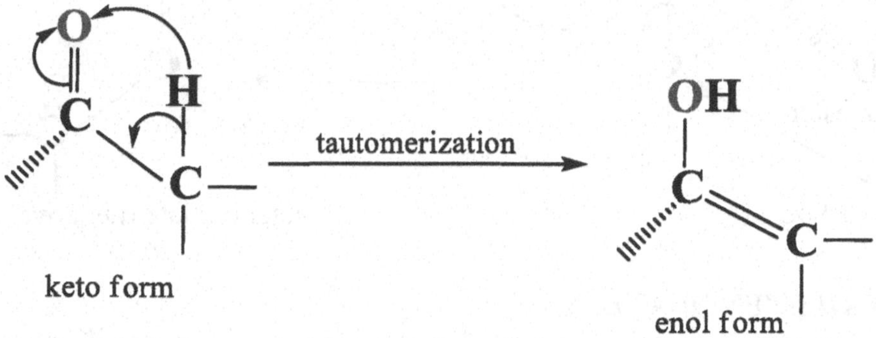

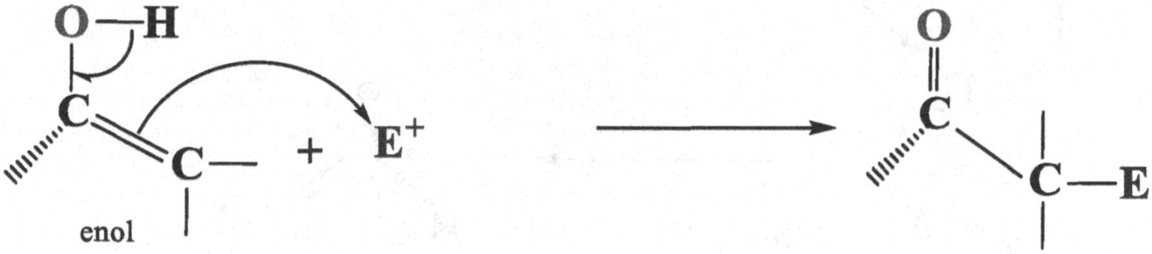

keto form

tautomerization

enol form

enol

+ E^+

- Note: The α hydrogen is replaced by the electrophile.

C. ENOLATES

1. ENOLATE FORMATION: KETO FORM VS. ENOL ATE FORM

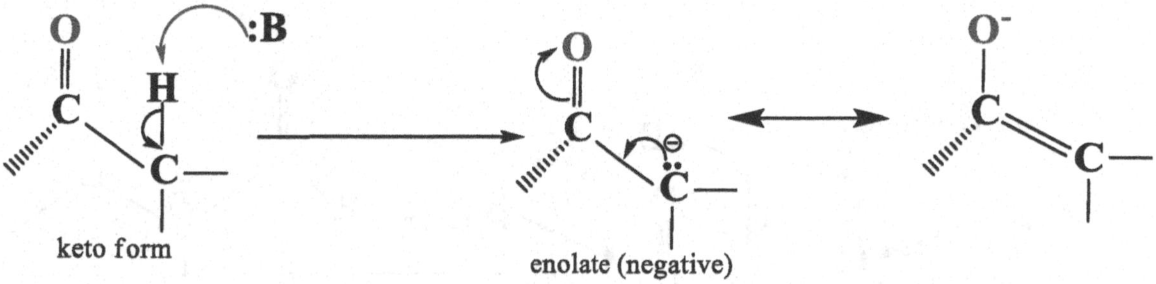

:B

keto form

enolate (negative)

- Note:
  - The enolate ion is resonance stabilized.
  - The α hydrogen is more acidic than a normal sp^3 alkane H.

- See examples on pages ______ - ______.

2. ESTER ENOLATES

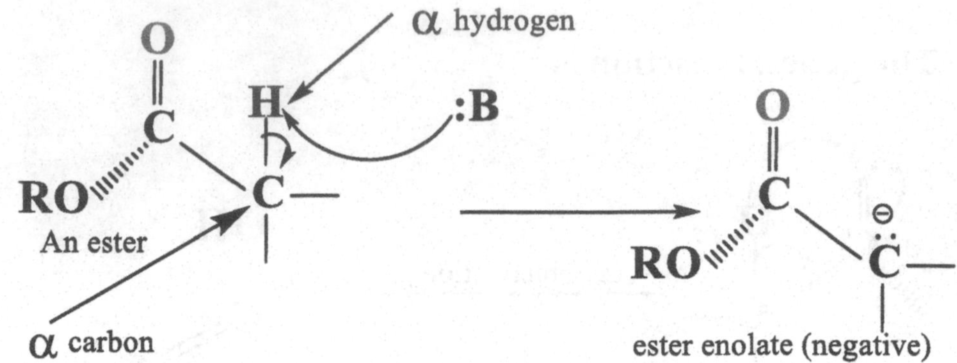

3. NITRILE "ENOLATES

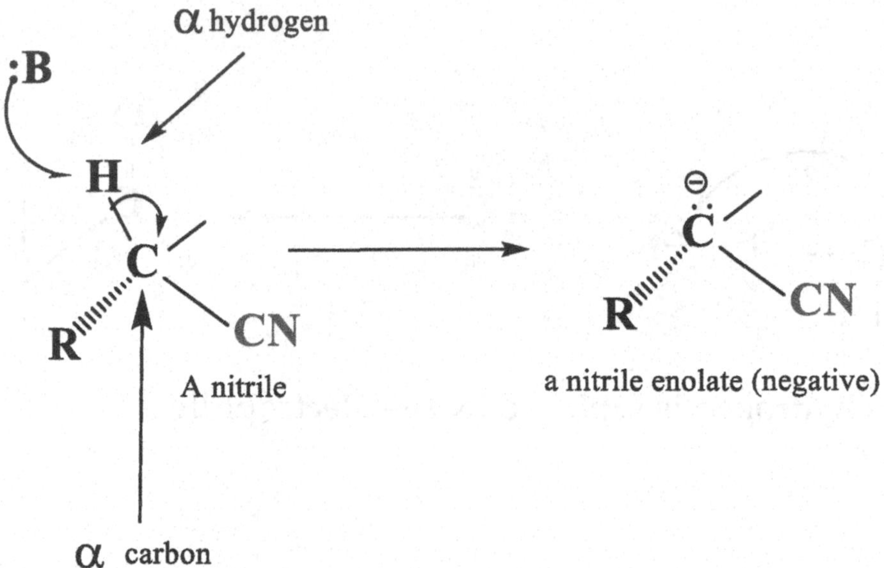

4. β-DICARBONYL ENOLATES

α hydrogen

:B

β carbon

β carbon

a β dicarbonyl enolate (negative)

α carbon

- Do problems on page ______.
- See Table ______, page ______ for acidity.

5. THE BASE AND ENOLATE FORMATION

Recall:

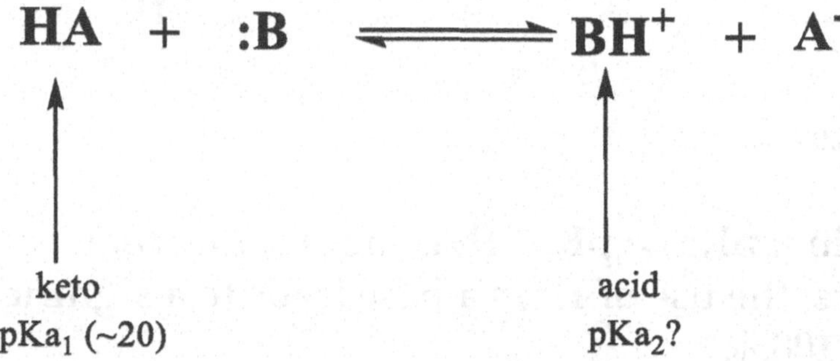

- **If pKa2 > pKa1, A^- is more stable than HA. As a result, a lot of A^- is produced.**
- **If $pKa2 < pKa1$, A^- is less stable than HA. Therefore, a small A^- yield is expected ($<1\%$).**
- **Let's apply this concept to enolate reaction with bases:**

O, C, H, C, :B ⇌ O, C, C + BH^+

acid 1

acid 2

Ex:

$$CH_3C(=O)CH_2\text{-}H + :H^- \rightleftharpoons CH_3C(=O)\ddot{C}^{\ominus}H_2 + H_2$$

$pKa_1 = 20$ $pKa_2 = 35$

- **Conclusion: $pKa_2 > pKa_1$. H_2 is more stable than acetone. Therefore, the use of H^- as a base leads to a significant enolate yield : ~ 100%.**

$$CH_3C(=O)CH_2\text{-}H + :OH^- \rightleftharpoons CH_3C(=O)\ddot{C}^{\ominus}H_2 + H_2O$$

$pKa_1 = 20$ $pKa_2 = 15.7$

- **Conclusion: $pKa_2 < pKa_1$. H_2O is less stable than acetone. Therefore, using OH^- as a base leads to a small enolate yield of enolate: ~ 1%.**

- Read pages _______ - _____.

- See Table ______, page ______.

- **Using Lithium diisopropylamide, LDA: Li^{+} $^{-}N[CH(CH_3)_2]_2$. The actual structure of LDA is:**

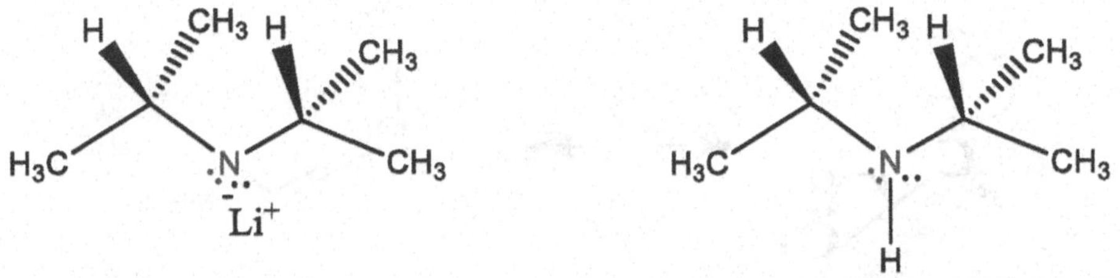

Lithium diisopropylamide or LDA

Conjugate acid of LDA: pKa = 40

- **The reaction is:**

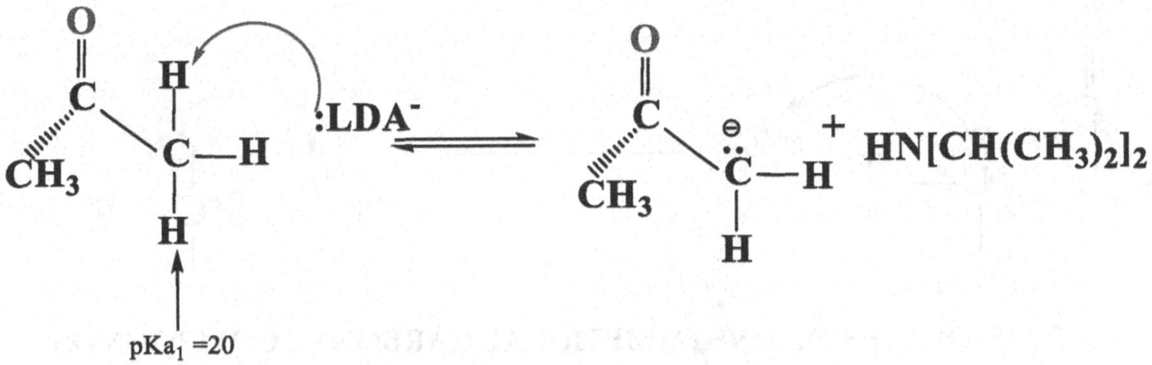

CH_3COCH_3 + :LDA$^-$ ⇌ $CH_3COCH_2^{-}$ + $HN[CH(CH_3)_2]_2$

pKa_1 =20

- **Conclusion: $pKa_2 > pKa_1$. The conjugate acid of LDA is more stable than acetone. Therefore, the use of LDA as a base leads to a significant enolate yield: ~ 100%.**

- **Note: The stronger the base (large pKa of its conjugate acid), the more enolate is produced. The ideal base is LDA.**

- **Do Problems on page ______.**

6. **GENERAL REACTIONS OF ENOLATES**

- **Enolates are:**
    - **electron rich**
    - **nucleophiles**
- **They have two sites that can react: the CO carbon and the negative α carbon. They are said to be ambident nucleophiles. Theoretically, the two sites can react with electrophiles.**

- **Recall: Enolates are stabilized by resonance.**

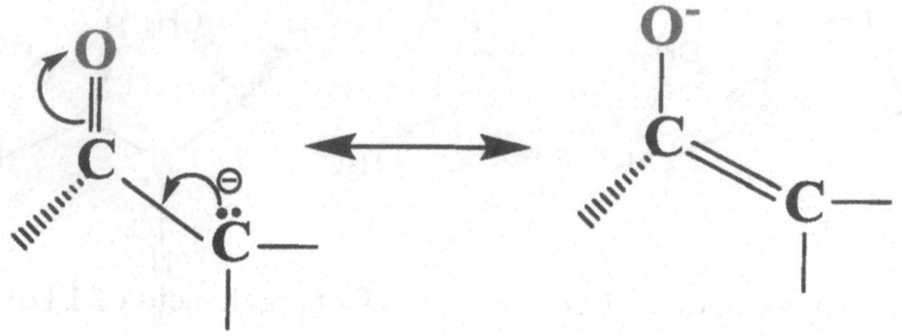

- **In practice, enolate reactions occur at the most nucleophilic α carbon site, not at the O of the -CO.**

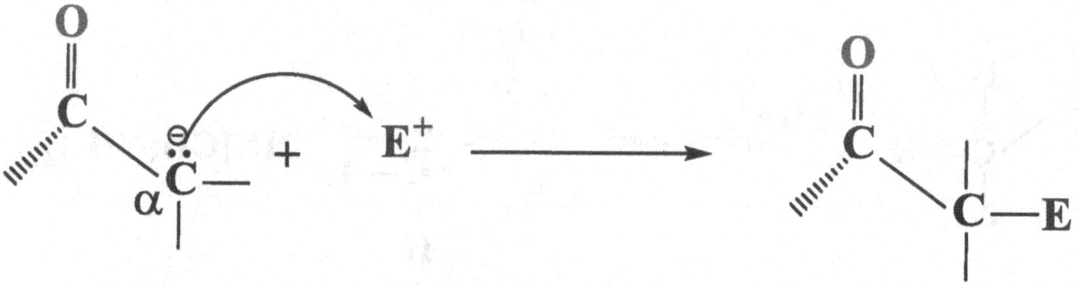

7. ENOLATES OF UNSYMMETRICAL CARBONYL COMPOUNDS

a. Introduction

- For unsymmetrical carbonyl compounds, there are **2 possible enolates. The more substituted enolate is the most stable.**

Ex. 1: a thermodynamic enolate

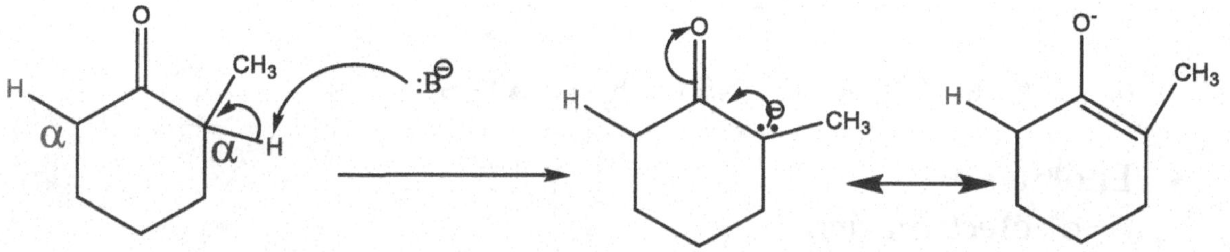

Note: The more substituted enolate has a lower energy. It is called the thermodynamic enolate.

Ex. 2: a kinetic enolate

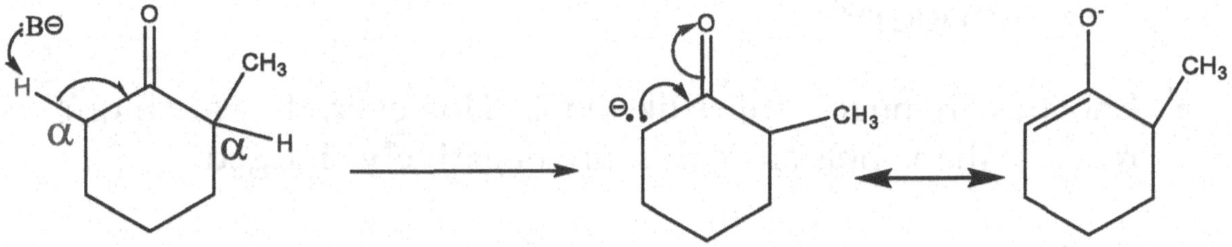

- **Note: This less substituted enolate forms faster. But it is the least stable. It is called the kinetic enolate.**

b. Kinetic vs. Thermodynamic Enolates (Review Unit 6, Section H)

- **A kinetic enolate has a higher energy. It forms faster under milder conditions. It does not reach equilibrium. Its formation is favored by strong bases such as LDA in polar aprotic solvent (LDA in THF) at -78 °C. It is the least substituted.**

- **A thermodynamic enolate has a lower energy and forms more slowly under room temperature. It is the most substituted. Its formation is favored by strong bases in polar protic solvents at higher temperatures ($NaOCH_2CH_3$ in CH_3CH_2OH at room temperature).**

- Read pages ______ - ______. Do problem ______, page ______.

D. RACEMIZATION AT THE α CARBON

- **Since all atoms of the enolate are sp^2 hybridized, attack can occur above and below the α carbon, leading to racemization.**

- Read page ______. Do problem on page ______.

E. REACTIONS AT THE α CARBON

1. INTRODUCTION

- Enolates are nucleophiles like enols. However, they are more reactive than enols since they are **negatively charged**.

- Enolates can undergo **two types of reactions:**

    - **Substitution reactions**
    - **Reactions with carbonyl compounds: will cover these reactions in OCHEM II UNIT 15.**

- **Read page ______ for Preview of Reactions.**

2. HALOGENATION AT THE α CARBON

a. Introduction

- Enolates react readily with Cl_2, Br_2, and I_2 **to give α-halo ketones or aldehydes. The reaction can be carried out in acid or base.**

- **The general reaction is:**

keto form $\xrightarrow[H^+ \text{ or } OH^-]{X_2}$ α-halo ketone or aldehyde + HX

Ex:

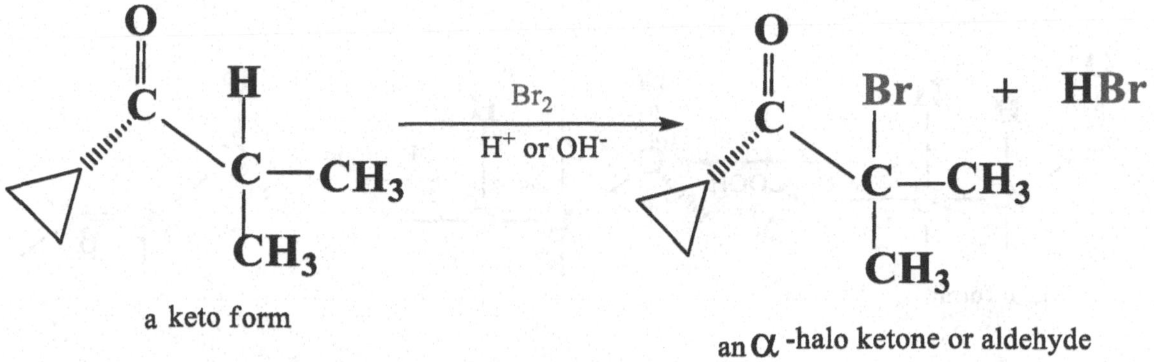

O
‖
C   H
     |
   C—CH_3
     |
   CH_3

a keto form

$\xrightarrow[\text{H}^+ \text{ or OH}^-]{\text{Br}_2}$

O
‖
C   Br   +   HBr
     |
   C—CH_3
     |
   CH_3

an α -halo ketone or aldehyde

b. Halogenation in Acid: Acetic Acid

- Acetic acid plays the role of a solvent and a catalysis.

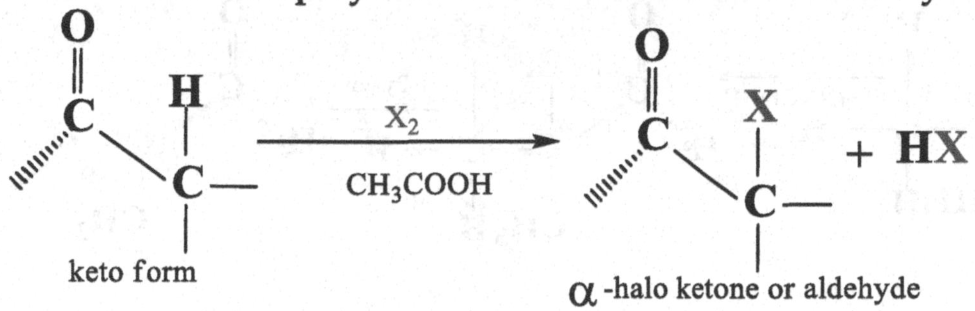

O
‖
C   H
     |
   C—
     |

keto form

$\xrightarrow[\text{CH}_3\text{COOH}]{\text{X}_2}$

O
‖
C   X
     | + HX
   C—
     |

α -halo ketone or aldehyde

- See Mechanism on page _______.

2 steps:

1. Tautomerization to enol.
2. Nucleophilic reaction of enol with X_2.

c. Conversion of α-halo Compounds to α, β-Unsaturated Compounds

2 steps:

1. Halogenation.
2. Elimination of HX.

- **The general reaction is:**

$$\text{C(=O)–CH–CH (keto form)} \xrightarrow[CH_3COOH]{X_2} \text{C(=O)–CX–CH} \xrightarrow[\Delta]{\text{Pyridine}} \text{C(=O)–C}_\alpha\text{=C}_\beta$$

keto form

Ex:

$$HC(=O)–CH(CH_3)–CH_3 \xrightarrow[CH_3COOH]{Br_2} HC(=O)–CBr(CH_3)–CH_3 \xrightarrow[\Delta]{\text{Pyridine}} HC(=O)–C_\alpha(CH_3)=C_\beta$$

or

$$\text{C(=O)–CH–CH (keto form)} \xrightarrow[CH_3COOH]{X_2} \text{C(=O)–CX–CH} \xrightarrow[\text{LiBr, DMF}]{Li_2CO_3} \text{C(=O)–C}_\alpha\text{=C}_\beta$$

keto form

Ex:

$$HC(=O)–CH(CH_3)–CH_3 \xrightarrow[CH_3COOH]{Br_2} HC(=O)–CBr(CH_3)–CH_3 \xrightarrow{Li_2CO_3,\ LiBr/THF} HC(=O)–C_\alpha(CH_3)=C_\beta H_2$$

- **Read pages _____ - ______. Do problem on page _____.**

d. Halogenation in Base

- In this case, polysubstitution occurs at the α carbon. Since the reaction cannot be stopped, a dihalide is produced. This makes this reaction less useful.

- The general reaction is:

keto form

$$\text{keto form} \xrightarrow[\text{OH}^- \; (X_2 = Cl_2,\, Br_2,\, I_2)]{X_2 \text{ (excess)}} \text{C(=O)–CX}_2\text{–} + X^-$$

Ex:

$$CH_3CH_2C(=O)CH(H)CH_2CH_3 \xrightarrow[\text{OH}^-]{Br_2 \text{ (excess)}} CH_3CH_2C(=O)CBr_2CH_2CH_3 + Br^-$$

- See Mechanism on page ______.

e. Haloform Formation: Halogenation of Methyl Ketones

- CHX_3 = haloform
- CHI_3 = iodoform
- $CHBr_3$ = bromoform
- $CHCl_3$ = chloroform

- An excess X_2 is used with a methyl ketone.

- **The general reaction is:**

$$R-C(=O)-CH_3 \xrightarrow[OH^-]{X_2\ (excess)} R-C(=O)-O^- + HCX_3$$

methyl ketone — a carboxylate ion — a haloform

- **The iodoform test: used to detect methyl ketones. HCI_3 = yellow crystals.**

Ex:

$$H_3C-C(=O)-CH_3 \xrightarrow[OH^-]{I_2\ (excess)} H_3C-C(=O)-O^- + HCI_3$$

methyl ketone — a carboxylate ion — iodoform (yellow crystals)

- **The leaving group is CI_3.**
- **See Mechanism on page ______.**

f. Summary on α halogenation:

- **In acid, get α- halo ketones or aldehydes ➔ α, β- unsaturated compounds**
- **In base, get polysubstitution.**
- **In base, with methyl ketones, get carboxylates and haloforms.**

- **For Summary, see Fig. ______, page ______.**
- **Do Problem ______, page ______.**

X_2
H^- or OH^-
+HX

X_2
CH_3COOH

1. X_2, CH_3COOH
2. pyridine

1. X_2, CH_3COOH
2. Li_2CO_3, LiBr, DMF

X_2 (excess)
OH^-
+ X^-

X_2 (excess)
OH^-
+ HCX
a haloform

a methyl ketone

3. ENOLATE ALKYLATION OF SYMMETRICAL KETONES: THE DIRECT WAY

a. Introduction

- In this reaction an α H is replaced with R, an alkyl group.
- The general reaction is:

$$\text{keto form} \xrightarrow[\text{2. RX}]{\text{1. :B}} \text{an alkyl ketone} + X^- + HB^+$$

Ex:

$$H_3C-CO-CH_2-H \text{ (keto form)} \xrightarrow[\text{2. } CH_3CH_2Br]{\text{1. LDA/THF}} H_3C-CO-CH_2CH_2CH_3 \text{ (an alkyl ketone)} + Br^- + HLDA^+$$

b. Reaction Conditions

- A nonnucleophilic base like LDA is used in THF at low temperatures (-78 °C).
- LDA leads to an enolate which attacks the RX in an S_N2 reaction.

Ex:

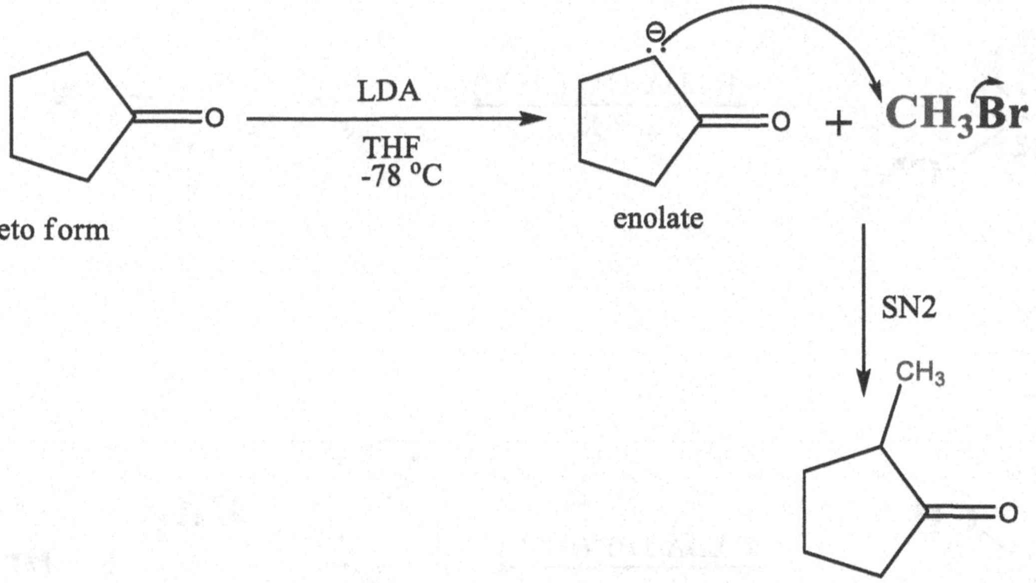

keto form $\xrightarrow[\text{THF, -78 °C}]{\text{LDA}}$ enolate + CH_3Br $\xrightarrow{\text{SN2}}$ 2-methylcyclopentanone (CH_3)

- **Note: The reaction can be performed with esters and nitriles too.**
- See examples on pages ______ - _____.
- **The general reaction with esters is:**

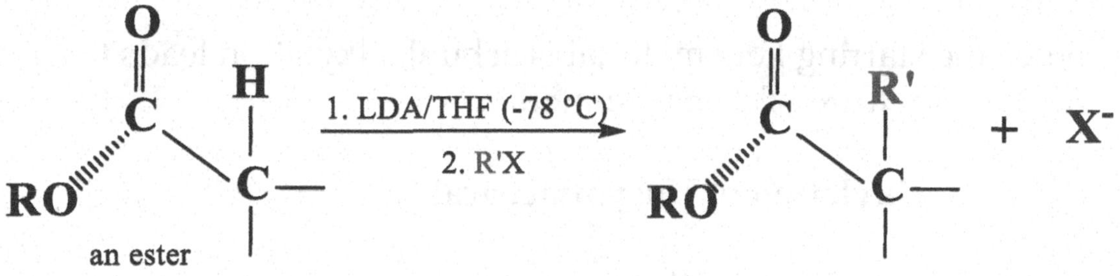

$$RO-C(=O)-C(H)< \xrightarrow[\text{2. R'X}]{\text{1. LDA/THF (-78 °C)}} RO-C(=O)-C(R')< + X^-$$

an ester

Ex:

$$CH_3O-C(=O)-C(H)(CH_3)-CH_3 \xrightarrow[\text{2. } CH_3CH_2Br]{\text{1. LDA/THF (-78 °C)}} CH_3O-C(=O)-C(CH_2CH_3)(CH_3)-CH_3 + Br^-$$

an ester

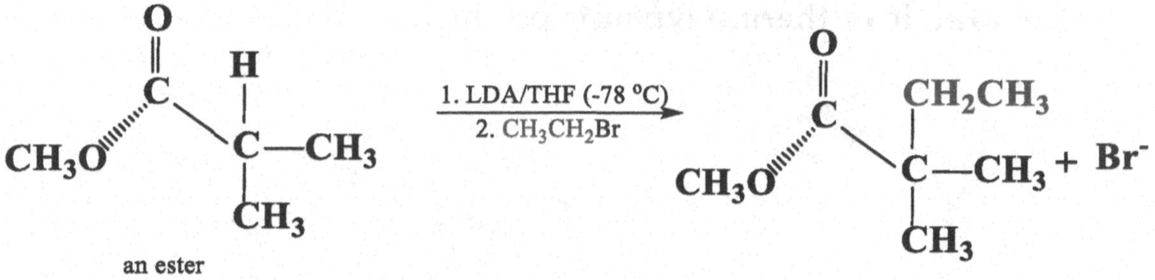

- **The general reaction with nitriles is:**

$$R(H)C_{\alpha}H\text{-}CN \xrightarrow[\text{2. R'X}]{\text{1. LDA/THF (-78°C)}} R(R')C\text{-}CN + X^-$$

a nitrile

Ex:

$$CH_3(H)C_{\alpha}(CH_3)\text{-}CN \xrightarrow[\text{2. } CH_3CH_2Br]{\text{1. LDA/THF (-78°C)}} CH_3(CH_2CH_3)C(CH_3)\text{-}CN + Br^-$$

a nitrile

- **Do problems on page ________.**

c. Stereochemistry of Enolate Alkylation

- Since the starting keto material is achiral, alkylation leads to a **racemic mixture of 2 enantiomers.**

d. Alkylation of Unsymmetrical Ketones

i. Introduction

- Reaction conditions can be controlled to produce exclusively **the kinetic or thermodynamic** product.

ii. Reaction under Kinetic Control

- **The less substituted enolate intermediate forms faster.**
- **The general reaction is:**

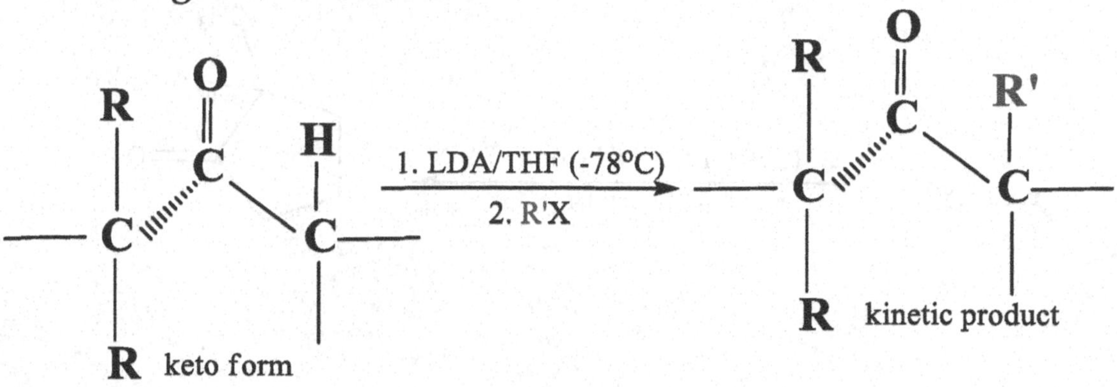

R
O
C
H
C
C
R keto form
1. LDA/THF (-78°C)
2. R'X
R
O
C
R'
C
C
R kinetic product

Ex:

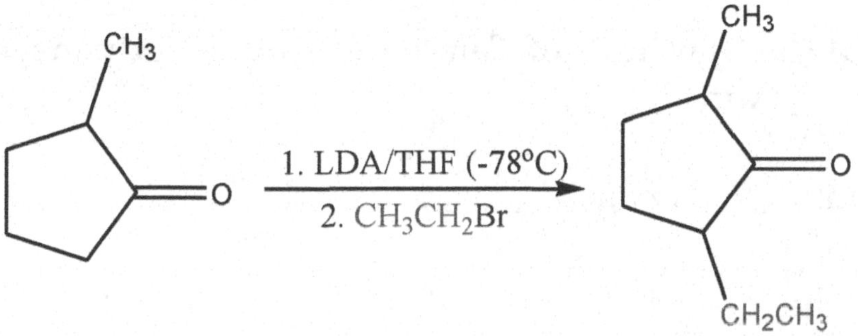

CH3
O
1. LDA/THF (-78°C)
2. CH3CH2Br
CH3
O
CH2CH3

iii. Reaction Under Thermodynamic Control

- **The more substituted enolate intermediate forms more slowly. The general reaction is:**

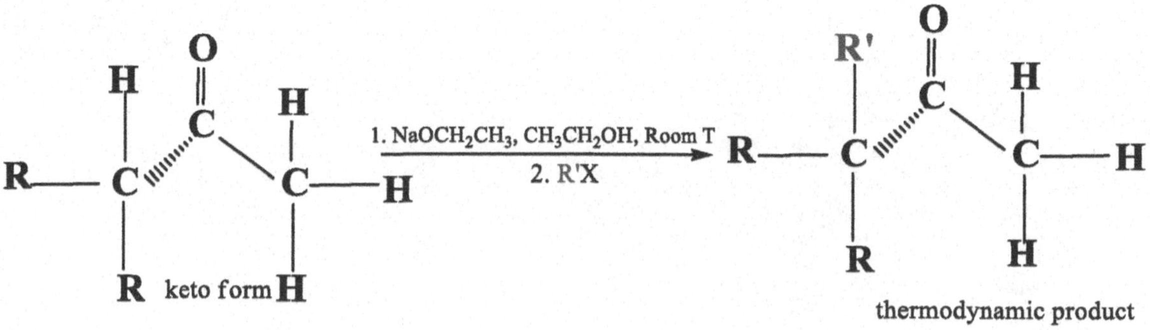

H
O
C
H
R
C
C
H
R keto form H
1. NaOCH2CH3, CH3CH2OH, Room T
2. R'X
R'
O
C
H
R
C
C
H
R
H
thermodynamic product

Ex:

CH_3 (2-methylcyclopentanone) → [1. NaOEt, EtOH (room T°); 2. CH_3CH_2Br] → 2-ethyl-2-methylcyclopentanone (CH_3, CH_2CH_3, =O)

- **Read about the Synthesis of Tamoxifen and β-Vetivone, pages _______ and ______.**
- **Do Problem ______, page _______.**

e. Enolate (direct) Alkylation of Ketones: A Summary

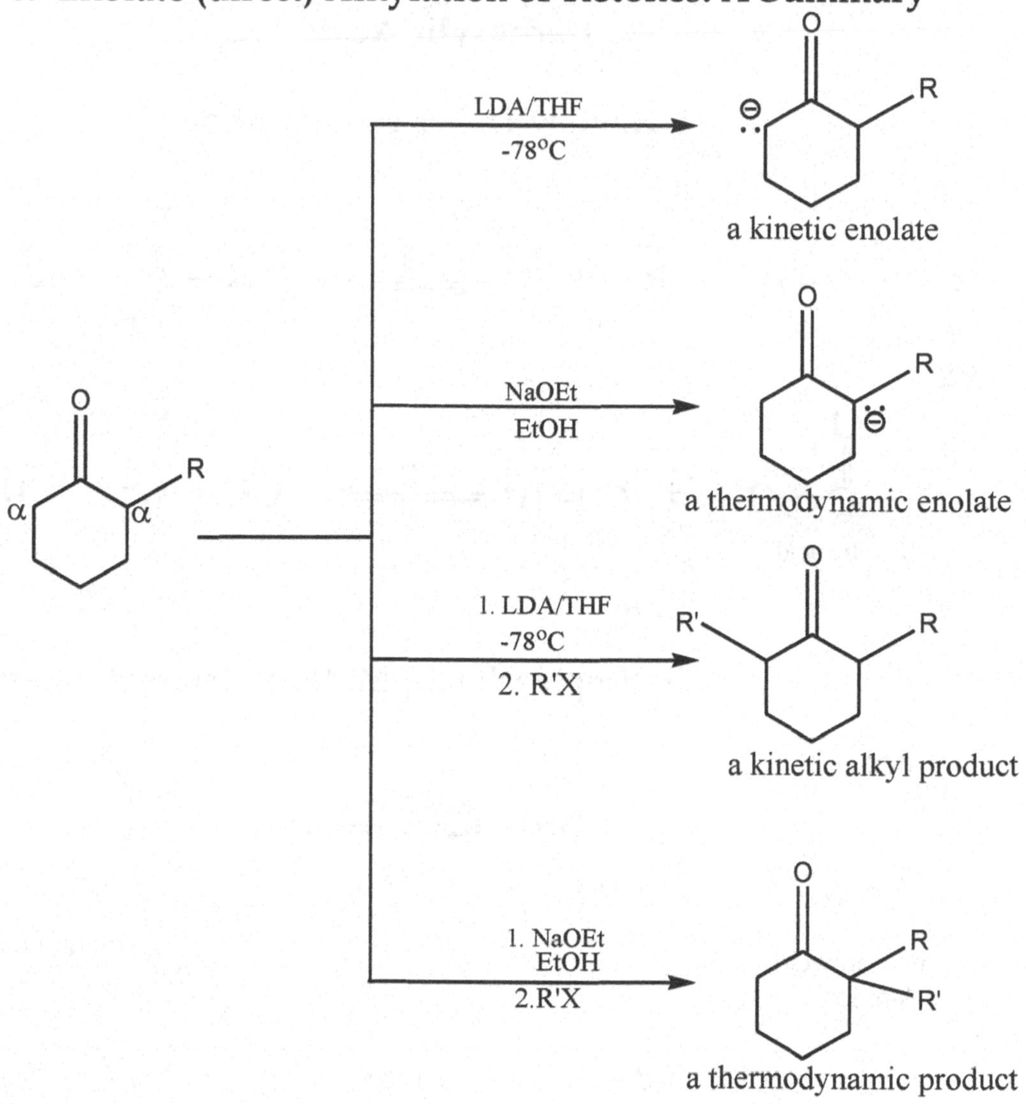

4. MALONIC ESTER SYNTHESIS: INDIRECT ALKYLATION

a. Introduction

- **This reaction is an indirect way of putting an alkyl group on the α carbon. There are two types of reactions:**
  - **Malonic synthesis of α substituted carboxylic acids**
  - **Acetoacetic ester synthesis of α substituted ketones**

b. The Malonic Ester Synthesis of α Substituted Carboxylic Acids

i. Review Fisher Esterification:

$$R-\overset{\overset{O}{\|}}{C}-OH + R'OH \xrightleftharpoons{H^+} R-\overset{\overset{O}{\|}}{C}-OR' + H_2O$$

carboxylic acid, alcohol, Ester

Ex:

$$CH_3-\overset{\overset{O}{\|}}{C}-OH + CH_3OH \xrightleftharpoons{H^+} CH_3-\overset{\overset{O}{\|}}{C}-OCH_3 + H_2O$$

carboxylic acid, alcohol, Ester

ii. Preparation of Malonic Esters: Malonates

$$H_2C(COOH)_2 + 2CH_3CH_2OH \xrightleftharpoons{\text{esterification}} H_2C(COOCH_2CH_3)_2 + 2H_2O$$

malonic acid
a β-diacid

ethanol

diethyl malonate

- **Abbreviated diethyl malonate**

$$H_2C(\underbrace{COOCH_2CH_3}_{COO\,Et})_2 \quad \text{or} \quad H_2C(COOEt)_2$$

diethyl malonate

iii. Hydrolysis of Diethyl Malonate

$$H_2C(COOEt)_2 + 2H_2O \xrightleftharpoons{\text{hydrolysis}} H_2C(COOH)_2 + 2EtOH$$

β -diacid

ethanol

iv. Decarboxylation of β-Diacids

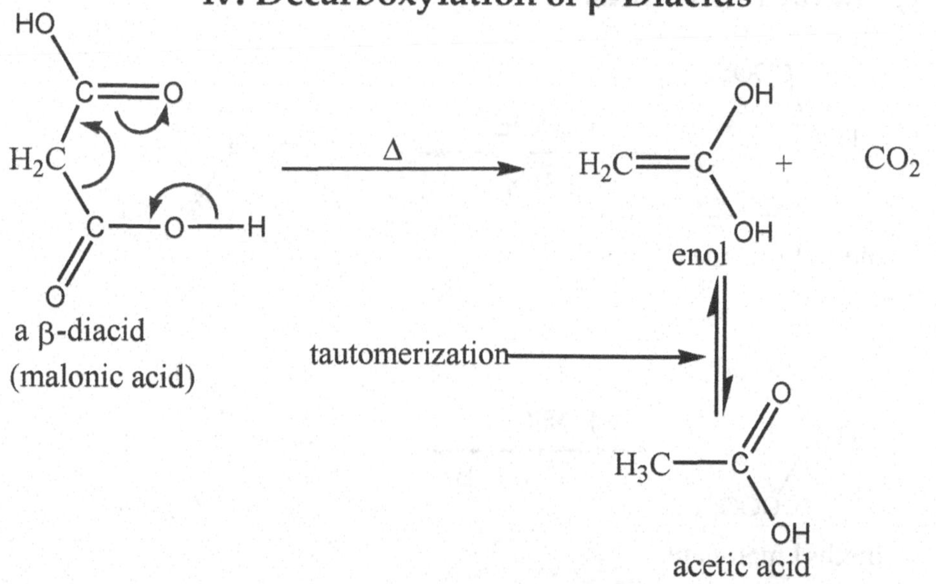

HO

C=O

H_2C

C—O—H

O

a β-diacid
(malonic acid)

Δ

$H_2C{=}C(OH)_2$ + CO_2

enol

tautomerization

$H_3C{-}C(=O)OH$

acetic acid

- **Note: Anytime there is a -COOH on the α carbon, decarboxylation can occur.**

Ex:

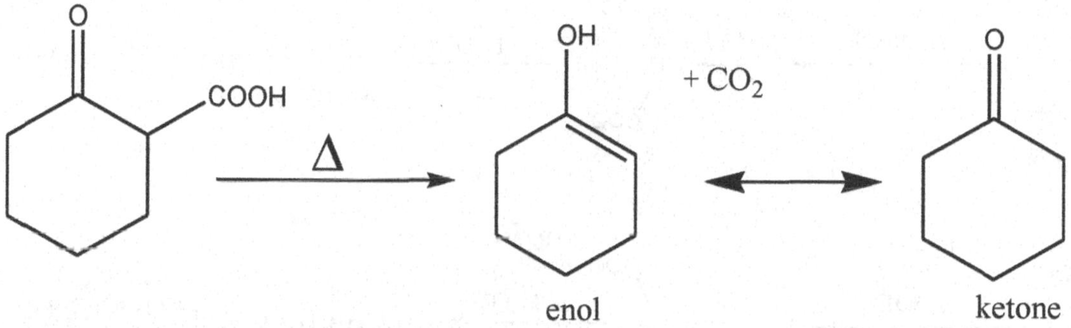

COOH

Δ

OH

+ CO_2

enol

ketone

- **See steps of reaction on page ______.**

- **Do Problems on pages ______ - ______.**

v. Introducing an Alkyl Group, R, to the Malonic Ester

- **The reaction uses sodium ethoxide (CH_3CH_2ONa = NaOEt), a strong base.**

- **The general reaction is:**

$$H_2C(COOEt)_2 \xrightarrow[\text{2. R'X}]{\text{1. NaOEt}} R'\text{—}CH(COOEt)_2$$

diethyl malonate

Ex:

$$H_2C(COOEt)_2 \xrightarrow[\text{2. } CH_3CH_2Br]{\text{1. NaOEt}} CH_3CH_2\text{—}CH(COOEt)_2$$

diethyl malonate

Note: The overall reaction leads to a carboxylic acid.

$$H_2C(COOEt)_2 \xrightarrow[\text{2. R'X}]{\text{1. NaOEt}} R'\text{—}CH(COOEt)_2 \xrightarrow[\Delta]{H_3O^+} R'CH_2COOH + CO_2 + EtOH$$

diethyl malonate; a carboxylic acid

Ex:

$$H_2C(COOEt)_2 \xrightarrow[\text{2. } CH_3CH_2Br]{\text{1. NaOEt}} CH_3CH_2\text{—}CH(COOEt)_2 \xrightarrow[\Delta]{H_3O^+} CH_3CH_2CH_2COOH + CO_2 + EtOH$$

diethyl malonate; a carboxylic acid

vi. Introducing 2 R Groups: Repeat the First Step

$$H_2C(COOEt)_2 \xrightarrow[\text{2. RX}]{\text{1. NaOEt}} R\text{—}CH(COOEt)_2 \xrightarrow[\text{2. R'X}]{\text{1. NaOEt}} R(R')C(COOEt)_2 \xrightarrow[\Delta]{H_3O^+} R(R')CHCOOH + CO_2 + EtOH$$

diethyl malonate

- **Note: This is a powerful method of synthesizing carboxylic acids.**

Ex:

$H_2C(COOEt)_2$ (diethyl malonate) $\xrightarrow[\text{2. } CH_3CH_2Br]{\text{1. NaOEt}}$ $CH_3CH_2-CH(COOEt)_2$ $\xrightarrow[\text{2. } CH_3Br]{\text{1. NaOEt}}$ $CH_3CH_2-C(CH_3)(COOEt)_2$ $\xrightarrow[\Delta]{H_3O^+}$ $CH_3CH_2-CH(CH_3)-CH_2COOH$ $+ CO_2 + EtOH$

diethyl malonate

vii. Intramolecular Malonic Ester Synthesis

- **Formation of an enolate:**

$H_2C(COOEt)_2$ $\xrightarrow{\text{NaOEt}}$ $^{-}HC(COOEt)_2$

diethyl malonate

- **Enolate attack**

Br–(CH$_2$)$_n$–Br + $^{-}HC(COOEt)_2$ $\longrightarrow$

- **Do Problems _______ and ______, page ______.**

viii. Retrosynthesis: An Example

- **Question: How would you synthesize the following compound from diethyl Malonate?** $CH_3CH_2CH_2COOH$

- **Read pages _____ - _____. Do Problems ______, ______, page _______.**

ix. A Summary on Diethyl malonate Syntheses

- **Introduction of one R group**

$$H_2C(COOEt)_2 \xrightarrow[\text{2. R'X} \;\; \text{3. } H_3O^+/\Delta]{\text{1. NaOEt}} R'\overset{\alpha}{C}H_2COOH + CO_2 + EtOH$$

diethyl malonate → a carboxylic acid

- **Introduction of two R groups**

$$H_2C(COOEt)_2 \xrightarrow[\text{2. RX}]{\text{1. NaOEt}} R{-}CH(COOEt)_2 \xrightarrow[\text{2. R'X} \;\; \text{3. } H_3O^+/\Delta]{\text{1. NaOEt}} RR'\overset{\alpha}{C}HCOOH + CO_2 + EtOH + X^-$$

diethyl malonate → a carboxylic acid

c. Acetoacetic Ester Synthesis of α Substituted Ketones

i. Introduction

- This synthesis method converts ethyl acetoacetate to a ketone with one or two R groups on the α carbon.

$$H_3C{-}C({=}O){-}CH_2COOEt \longrightarrow H_3C{-}C({=}O){-}CH_2{-}R \quad \text{or} \quad H_3C{-}C({=}O){-}CH(R){-}R'$$

ethyl acetoacetate → a ketone or a ketone

ii. Steps

- See page ______.
- Same steps as in malonate synthesis, except that ethyl acetoacetate is used.

iii. Using one R Group

- The general reaction is:

$$\underset{\text{ethyl acetoacetate}}{H_3C{-}C({=}O){-}CH_2COOEt} \xrightarrow[\text{2. RX} \quad \text{3. } H_3O^+/\Delta]{\text{1. NaOEt}} \underset{\text{a ketone}}{H_3C{-}C({=}O){-}\overset{\alpha}{CH_2}R} + CO_2 + EtOH$$

Ex:

$$\underset{\text{ethyl acetoacetate}}{H_3C{-}C({=}O){-}CH_2COOEt} \xrightarrow[\text{2. } CH_3Br \quad \text{3. } H_3O^+/\Delta]{\text{1. NaOEt}} \underset{\text{a ketone}}{H_3C{-}C({=}O){-}\overset{\alpha}{CH_2}CH_3} + CO_2 + EtOH$$

iv. Using two R Groups

$$H_3C{-}C(=O){-}CH_2COOEt \xrightarrow[\text{2. RX}]{\text{1. NaOEt}} H_3C{-}C(=O){-}CH(R){-}COOEt \xrightarrow[\text{2. R'X; 3. } H_3O^+/\Delta]{\text{1. NaOEt}} H_3C{-}C(=O){-}CH(R)(R') + CO_2 + EtOH$$

ethyl acetoacetate → a ketone (α carbon indicated)

Ex:

$$H_3C{-}C(=O){-}CH_2COOEt \xrightarrow[\text{2. } CH_3Br]{\text{1. NaOEt}} H_3C{-}C(=O){-}CH(CH_3){-}COOET \xrightarrow[\text{2. } CH_3CH_2Br\text{; 3. } H_3O^+/\Delta]{\text{1. NaOEt}} H_3C{-}C(=O){-}CH(CH_3)(CH_2CH_3)$$

ethyl acetoacetate → a ketone (α carbon indicated)

- Note: Can do retrosynthesis. See page ____.
- Do problems on pages ____ - ____.
- Read pages ____ - ____.

- Note: In order to introduce an R group, one can use either enolate alkylation or the malonic/acetoacetate route.

d. Diethyl Malonate and Ethyl Acetoacetate alkylations: A Summary

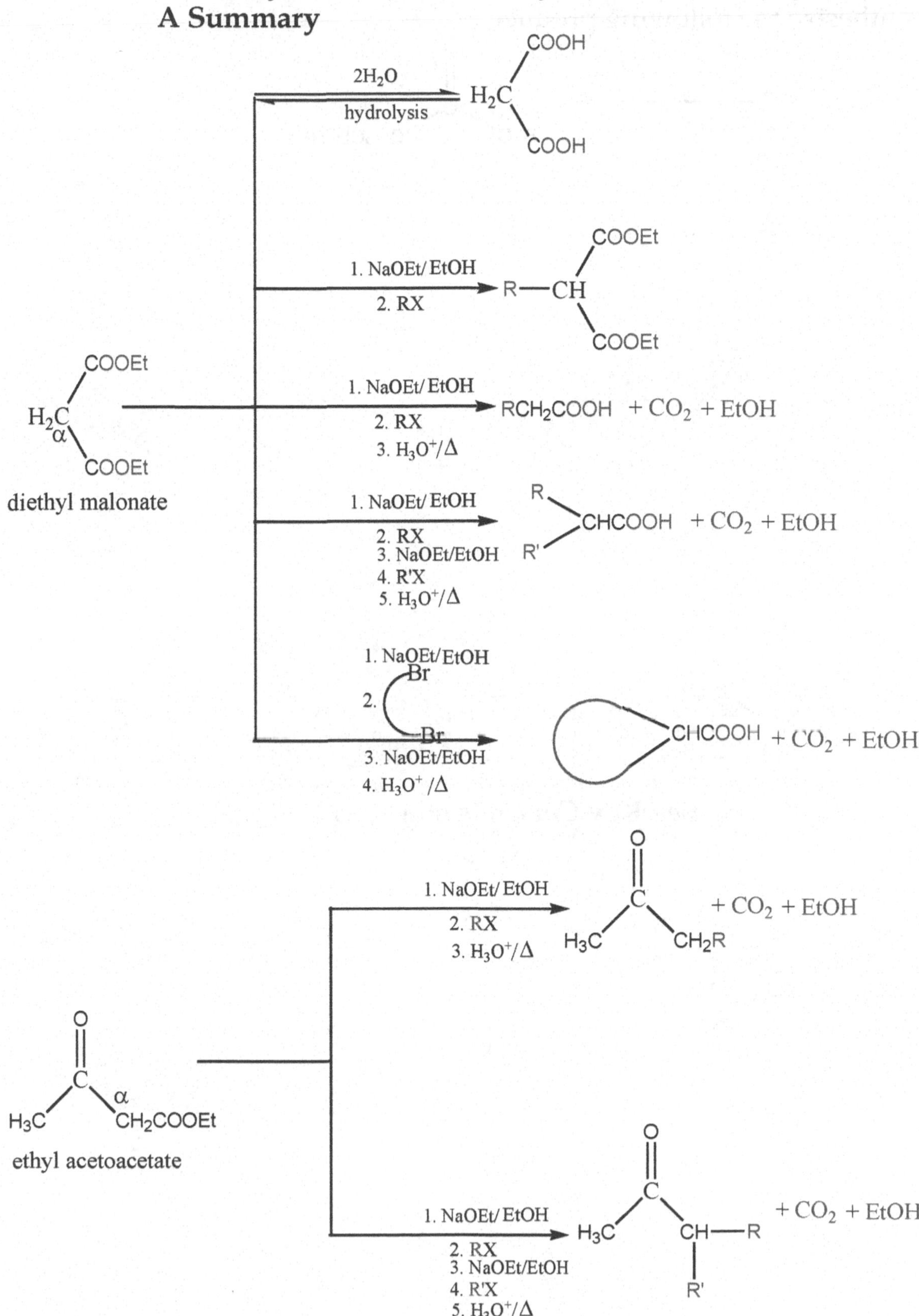

Ex: Use the **enolate (direct way)** and the **ethyl acetoacetate** routes to synthesize the following product:

$$? \longrightarrow H_3C-C(=O)-CH_2CH_2H_3$$

- **See Key Concepts on pages _____ - ______.**

OCHEM II UNIT 15: CARBONYL ALDOL CONDENSATION REACTIONS

A. THE ALDOL CONDENSATION REACTION

1. INTRODUCTION

- **In an aldol condensation reaction, two molecules of ketone or aldehyde react with each other in basic media to give a bigger compound.**

- **The general reaction is:**

$$2RCH_2CHO \underset{}{\overset{OH^-, H_2O}{\rightleftharpoons}} RCH_2CH(OH){-}CH(R){-}CHO$$

an aldehyde — a β-hydroxy aldehyde

Ex:

$$2CH_3CH_2CHO \underset{}{\overset{OH^-, H_2O}{\rightleftharpoons}} CH_3CH_2CH(OH){-}CH(CH_3){-}CHO$$

an aldehyde — a β-hydroxy aldehyde

- **See Mechanism on page ______.**

- **The reaction is an example of nucleophilic addition reaction.**

- **Recall:**

$$RCHO + :Nu^- \xrightarrow{H_2O} RCH(OH)Nu$$

enolate

- In the aldol reaction, the **enolate** is the nucleophile.

An example is:

$$H_3C{-}CHO + \overset{\ominus}{C}H_2{-}CHO \text{ (enolate)} \xrightarrow[H_2O]{OH^-} CH_3CH(OH){-}CH_2CHO$$

enolate

- **Read pages ______ - _____.**

- **Do Problems on page ______.**

2. DEHYDRATION OF ALDOL PRODUCT

- **β - hydroxy ketones and aldehydes undergo dehydration to give conjugated α, β unsaturated compounds. The reaction proceeds in two steps:**

- **Step 1: aldol reaction**

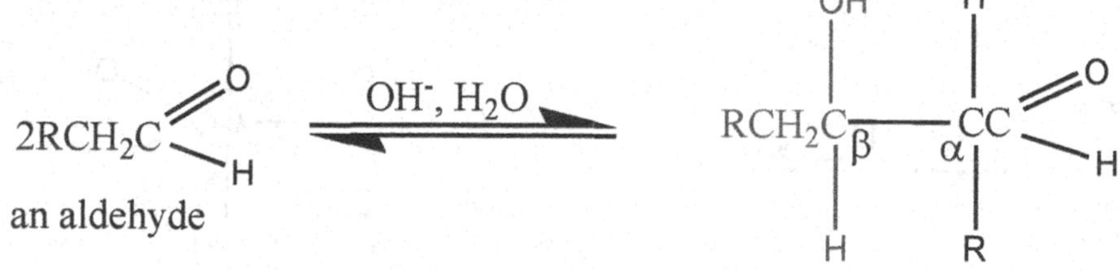

a β-hydroxy aldehyde

- **Step 2: Dehydration of β-hydroxy aldehydes or ketones**

$$RCH_2CH(OH)_{\beta}{-}CH(R)_{\alpha}CHO \underset{-H_2O}{\overset{OH^-}{\rightleftharpoons}} RCH_2CH{=}C(R)CHO$$

a β-hydroxy ketone

an α β -unsaturated conjugated aldehyde

- See examples on page ______.
- See Mechanism on page _____.
- Do Problems on page ______.

3. RETROSYNTHETIC ANALYSIS OF ALDOL PRODUCT

- Go from products to starting materials.

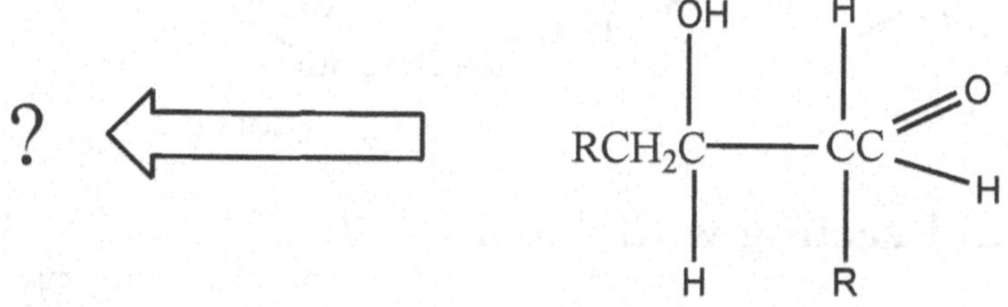

4. GENERAL CROSS ALDOL REACTIONS

a. Introduction

- A cross aldol reaction is an aldol reaction between two different aldehydes or ketones.

b. Both Aldehydes Have α H Atoms

- There are 4 possible products since two enolates can form.

Ex:

RCH_2CHO (aldehyde 1) $\underset{H_2O}{\overset{OH^-}{\rightleftharpoons}}$ $R\overset{\ominus}{C}HCHO$ (enolate 1)

H_3CH_2CCHO (aldehyde 1) $\underset{H_2O}{\overset{OH^-}{\rightleftharpoons}}$ $H_3C\overset{\ominus}{C}HCHO$ (enolate 1)

$$R'H_2C{-}C(=O){-}H \;\underset{H_2O}{\overset{OH^-}{\rightleftharpoons}}\; R'\ddot{H}C^{\ominus}{-}C(=O){-}H$$

aldehyde 2 → enolate 2

Ex:

$$H_3CH_2CH_2C{-}C(=O){-}H \;\underset{H_2O}{\overset{OH^-}{\rightleftharpoons}}\; H_3CH_2CH\ddot{C}^{\ominus}{-}C(=O){-}H$$

aldehyde 2 → enolate 2

1. Enolate 1 (Nu) Reacting with Aldehyde 1:

$$RH_2C{-}C(=O){-}H + R\ddot{H}C^{\ominus}{-}C(=O){-}H \longrightarrow HC(OH)(R){-}CC(H)(R)(=O)H$$

aldehyde 1 + enolate 1

2. Enolate 1 (Nu) Reacting with Aldehyde 2:

$$R'H_2C{-}C(=O){-}H + R\ddot{H}C^{\ominus}{-}C(=O){-}H \longrightarrow HC(OH)(R'){-}CC(H)(R)(=O)H$$

aldehyde 2 + enolate 1

3. Enolate 2 (Nu) Reacting with Aldehyde 1:

$$RH_2C{-}C(=O){-}H + R'\ddot{H}C^{\ominus}{-}C(=O){-}H \longrightarrow HC(OH)(R){-}CC(H)(R')(=O)H$$

aldehyde 1 + enolate 2

4. Enolate 2 (Nu) Reacting with Aldehyde 2:

$$R'H_2C{-}C(=O){-}H + R'\ddot{H}C^{\ominus}{-}C(=O){-}H \longrightarrow HC(OH)(R'){-}CC(H)(R')(=O)H$$

aldehyde 2 + enolate 2

- See example on page ______.

- Note: The cross aldol reaction previously described is not useful since a mixture of products is obtained.

5. USEFUL CROSS ALDOL REACTIONS

a. Only one Carbonyl Compound Has α H Atoms

- One product is obtained. Use formaldehyde or benzaldehyde.

- The general reaction is:

$$HCHO\ (\text{formaldehyde}) + RCH_2CHO \xrightarrow[H_2O]{OH^-} HOCH_2\text{–}CH(R)\text{–}CHO \xrightarrow{-H_2O} H_2C{=}C(R)\text{–}CHO$$

Ex:

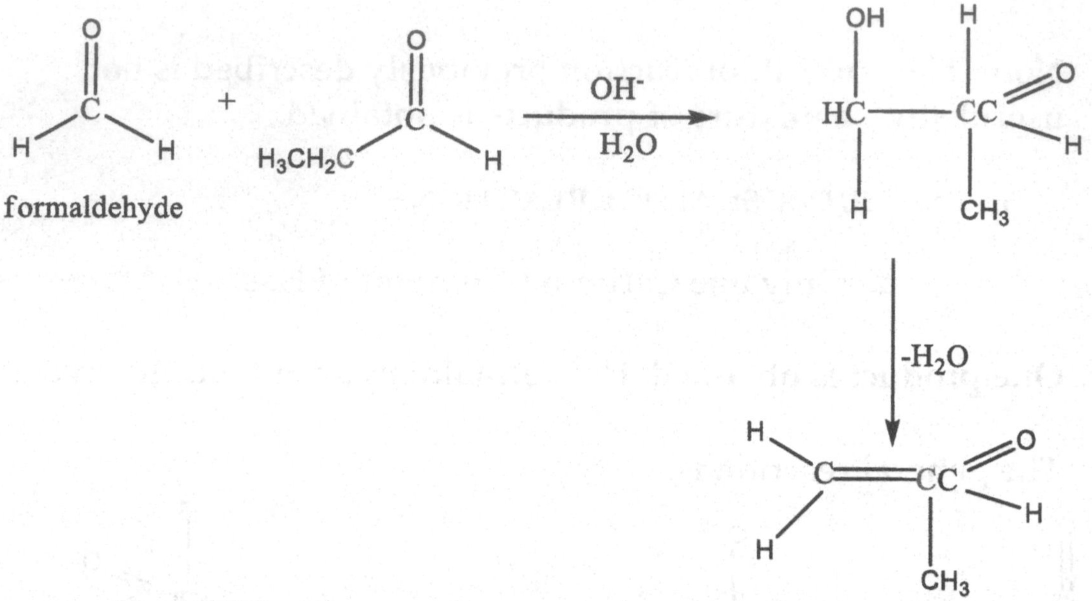

formaldehyde + H_3CH_2C–CHO $\xrightarrow[H_2O]{OH^-}$ HOCH$_2$–CH(CH$_3$)–CHO $\xrightarrow{-H_2O}$ H_2C=C(CH$_3$)–CHO

- **Note: In order to get a good yield, use excess unhindered carbonyl compounds.**
- **See example on page _______.**
- **Do problems on pages _______ - ______.**

b. Using Active Methylene Compounds: Y-CH$_2$-Z

- **Some examples:**

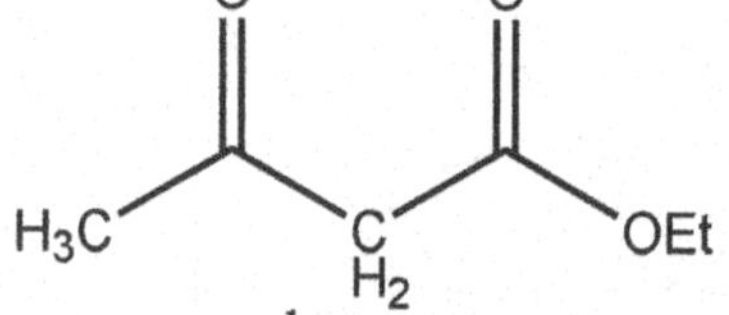

H_3C–CO–CH_2–CO–OEt
β keto ester

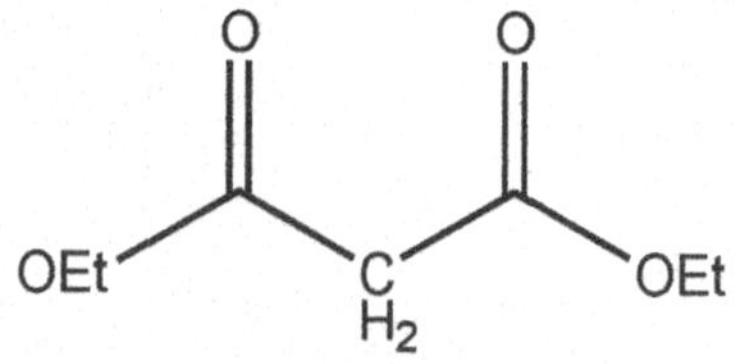

EtO–CO–CH_2–CO–OEt
β-diester

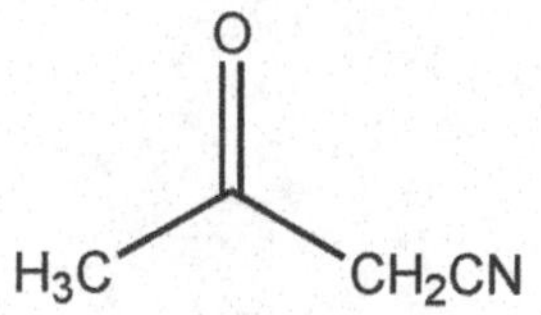

H_3C–CO–CH_2CN
α-cyanocarbonyl compound

N≡CCH$_2$C≡N
1,3-dinitrile

- **The general reaction is:**

$$R_2C{=}O + Y\text{-}CH_2\text{-}Z \xrightarrow[\text{EtOH}]{\text{NaOET}} R_2C{=}C(Y)(Z)$$

Ex: Use diethyl malonate

- **Recall: Diethyl malonate is:**

$H_2C(COOCH_2CH_3)_2$ (Et = CH_2CH_3) **or** $H_2C(COOEt)_2$ **or** $CH_2(COOET)_2$

diethyl malonate

Enolate formation revisited:

$$H_2C(COOEt)_2 \xrightarrow[\text{EtOH}]{\text{NaOET}} {}^{-}\ddot{H}C(COOEt)_2 \text{ or } {}^{-}CH(COOET)_2$$

diethyl malonate

$$\underset{\text{diethyl malonate}}{CH_2(COOEt)_2} \xrightarrow[\text{EtOH}]{\text{NaOEt}} \underset{\text{enolate}}{{}^{-}CH(COOEt)_2} + R_2C{=}O \longrightarrow R\text{-}C(O^{-})(R)\text{-}CH(COOEt)_2$$

$$R\text{-}C(O^{-})(R)\text{-}CH(COOEt)_2 \xrightarrow{\text{EtOH}} R\text{-}C(OH)(R)\text{-}CH(COOEt)_2 \xrightarrow{-H_2O} R_2C{=}C(COOEt)_2$$

Ex:

$CH_2(COOEt)_2$ diethyl malonate $\xrightarrow[EtOH]{NaOEt}$ $^-CH(COOEt)_2$ enolate + CH_3CHO → $H-C(O^-)(CH_3)-CH(COOEt)_2$ $\xrightarrow{EtOH}$ $H-C(OH)(CH_3)-CH(COOEt)_2$ $\xrightarrow{-H_2O}$ $(H)(H_3C)C=C(COOEt)(COOEt)$

- **A Summary:**

$RR'C{=}O + H_2C(COOEt)(COOET) \xrightarrow{-H_2O} RR'C{=}C(COOEt)(COOET)$

- See example on page _____, Fig. _____.

- Do Problems ______ and ______, page _____.

6. DIRECTED ALDOL REACTIONS

- The carbonyl compound that becomes the enolate nucleophile is clearly defined.
- There are three steps in these reactions. First, the enolate is made using LDA prior to the aldol reaction. Then the condensation reaction is carried out.

- Both carbonyl compounds can have α H atoms.

- Recall: LDA leads to the less substituted enolate (kinetic enolate).

- **The general reaction is:**
  - **Step 1: Enolate formation:**

RH_2C–C(=O)–C(H)(R')– $\xrightarrow{LDA/THF}$ $RHC^{\ominus}$–C(=O)–C(H)(R')–

  - **Step 2: Aldol condensation**

R''–C(=O)–H (aldehyde 2) + $RHC^{\ominus}$–C(=O)–C(H)(R')– ⟶ O^-–HC(R'')–C(R)(HO)C–C(H)(R')–

H_2O Step 3: H_2O

OH–HC(R'')–C(R)(HO)C–C(H)(R')–

- See example on page ______.

- See Fig. ______, page ______: synthesis of Periplanone B (cockroach).

- See examples on pages ______ - ______.
- Do Problems ______ and ______, page ______.

7. INTRAMOLECULAR ALDOL REACTIONS

- **In these reactions, 1,4 – and 1,5- dicarbonyl compounds are used to form ring compounds.**

Ex:

carbon
α β α α
NaOET
too strained
6 5 4 2 1 7 α CH_2 α
2 1 3 4 5 6 7 CH_2^-
EtOH
α H β OH
$-H_2O$
α β

- Read pages ______ - ______.

- See example on page ______.

- See Problems on pages ______ - ______.

- See Synthesis of Progesterone on page ______.
- Do problems on page ______.

B. THE CLAISEN REACTION

1. INTRODUCTION

- **This reaction involves enolates of esters.**
- **β-Keto esters substitution products are obtained.**
- **The leaving group is -OR.**

- **Recall: Nucleophilic substitution of esters**

$$R-\overset{\overset{\Large O}{\|}}{C}-OR' \xrightarrow{:Nu^-} R-\overset{\overset{\Large O}{\|}}{C}-Nu \quad + \quad {}^-OR'^-$$

- **Here the ester enolate is: Nu^-.**

- **The general reaction is:**

$$2\ RH_2C-\overset{\overset{O}{\|}}{C}-OR' \xrightarrow[H_3O^+]{NaOR'} RH_2C-\overset{\overset{O}{\|}}{C}-\underset{\underset{R}{|}}{CH}-\overset{\overset{O}{\|}}{C}-OR' \quad + \quad {}^-OR'$$

ester (2 moles) β-keto-ester

Ex:

$$2\ H_3C-\overset{\overset{O}{\|}}{C}-OEt \xrightarrow[H_3O^+]{NaOET} H_3C-\overset{\overset{O}{\|}}{C}-CH_2-\overset{\overset{O}{\|}}{C}-OEt \quad + \quad {}^-OEt$$

ester (2 moles) β-keto-ester

- **Read pages ______ - _____.**

- **See examples on pages _____ - _____.**

- **See Mechanism on page ______.**

- **Do problem on page ______.**

2. THE CROSSED CLAISEN AND RELATED REACTIONS

a. Introduction

- These are the most **useful** Claisen reactions.

b. Claisen with 2 Different Esters, But one has α Hydrogens

Common esters used: Ethyl formate (HCO_2Et) or ethyl benzoate ($C_6H_5CO_2Et$). Why? The leaving group is –OEt.

The general reaction:

H–C(=O)–OEt (ethyl formate) + RH_2C–C(=O)–OR $\xrightarrow[H_3O^+]{NaOET}$ H–C(=O)–CH(R)–C(=O)–OR (β-keto-ester) + -OEt

Ex:

H–C(=O)–OEt (ethyl formate) + H_3C–C(=O)–OEt (ester) $\xrightarrow[H_3O^+]{NaOET}$ H–C(=O)–CH_2–C(=O)–OEt (β-keto-ester) + -OEt

- See example on page ______.

c. Claisen with a Ketone and an Ester with the Ester Having No α H Atoms

- **The leaving group is -OEt.**
- **The General Reaction:**

H–C(=O)–OEt (ethyl formate) + RH_2C–C(=O)–R' (a ketone) $\xrightarrow[H_3O^+]{NaOET}$ H–C(=O)–CH(R)–C(=O)–R' (β-dicarbonyl compound) + -OEt

Ex:

H–C(=O)–OEt (ethyl formate) + H_3C–C(=O)–CH_3 (a ketone) $\xrightarrow[H_3O^+]{NaOET}$ H–C(=O)–CH_2–C(=O)–CH_3 (β-dicarbonyl compound) + -OEt

d. Synthesis of β-keto Esters Using Ethyl Chloroformate

- **-Cl (not OEt) is the leaving group. Why?**
- **The general reaction:**

Cl–C(=O)–OEt (chloroformate) + H_3C–C(=O)–R' (a ketone) $\xrightarrow[H_3O^+]{NaOET}$ EtO–C(=O)–CH_2–C(=O)–R' (β-keto-ester) + Cl^-

Ex:

Cl–C(=O)–OEt (chloroformate) + H_3C–C(=O)–cyclopropyl (a ketone) $\xrightarrow[H_3O^+]{NaOET}$ EtO–C(=O)–CH_2–C(=O)–cyclopropyl (β-keto-ester) + Cl^-

- **Read pages ______ - _____.**
- **Do all problems on pages _____ - _____.**

e. Synthesis of Diethyl Malonates Using Diethyl Carbonates: A Claisen Reaction

- **The leaving group is –OEt.**
- **The general reaction:**

EtO–C(=O)–OEt (diethyl carbonate) + RH_2C–C(=O)–OEt (ester) $\xrightarrow[H_3O^+]{NaOET}$ EtO–C(=O)–CH(R)–C(=O)–OET (diethyl malonate) + ^-OEt

Ex:

EtO–C(=O)–OEt (diethyl carbonate) + H_3C–C(=O)–OEt (ester) $\xrightarrow[H_3O^+]{NaOET}$ EtO–C(=O)–CH_2–C(=O)–OET (diethyl malonate) + ^-OEt

C. THE DICKMANN REACTION

1. INTRODUCTION

- This is an **intermolecular Claisen reaction** between **diesters** to give 5-and 6–member rings.

2. 1,6-DIESTER REACTION

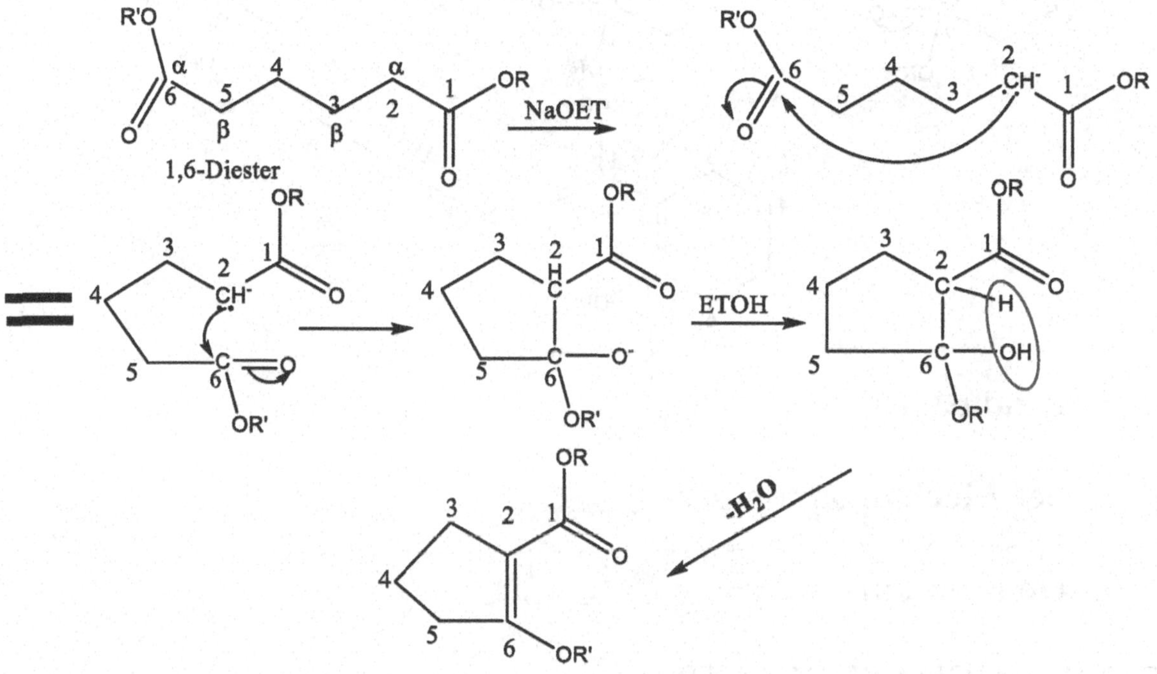

3. 1,7-DIESTER REACTION

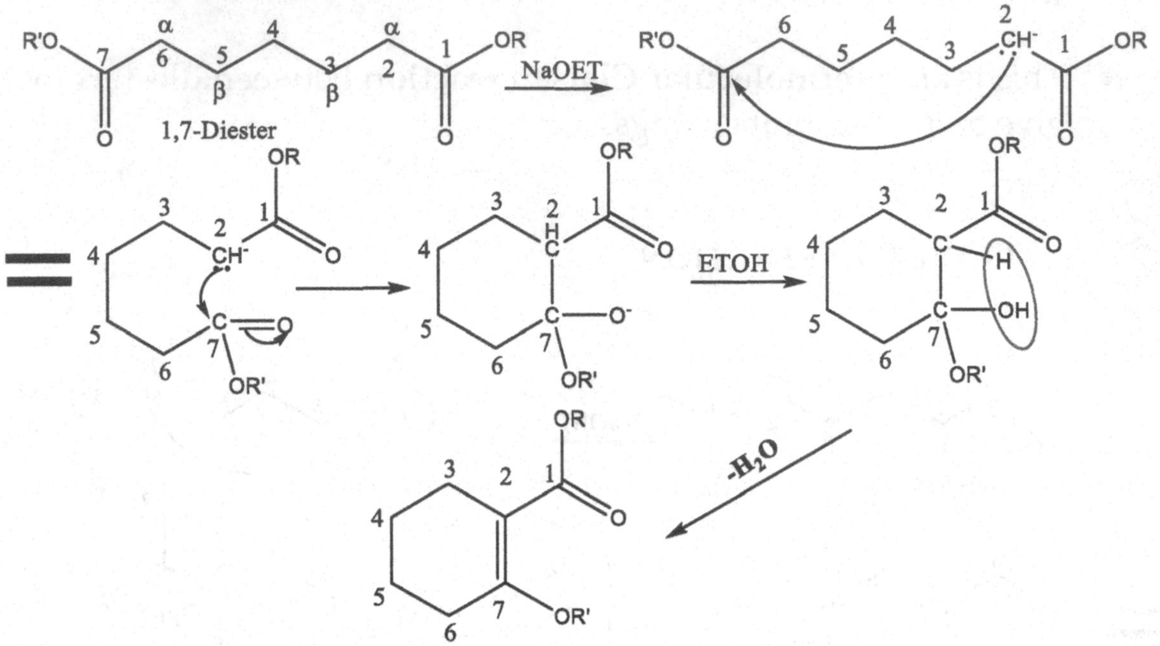

$$R'O-C(=O)-CH_2-CH_2-CH_2-CH_2-CH_2-C(=O)-OR \xrightarrow{NaOEt} \ldots$$

1,7-Diester

ETOH

$-H_2O$

- Read pages ______ - ______.
- See Mechanism on page ______.
- Do problem on page ______.

D. THE MICHAEL REACTION

1. INTRODUCTION

- The reaction involves an enolate and an α-, β -unsaturated carbonyl compound called a Michael acceptor.
- **The nucleophilic attack occurs at the β carbon.**

2. THE GENERAL REACTION

- **Recall:**

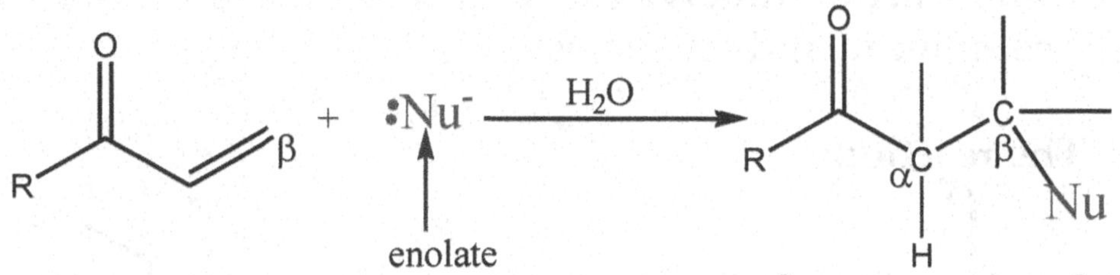

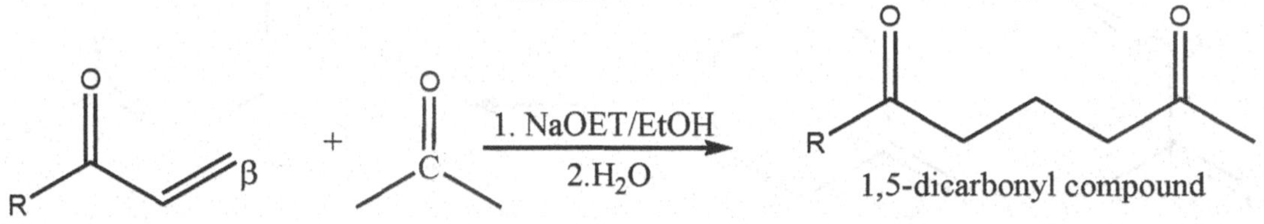

R–C(=O)–CH=CH₂ (β) + :Nu⁻ (enolate) —H_2O→ R–C(=O)–C(α)(H)–C(β)–Nu

α βunsaturated compound + ketone —1. NaOET/EtOH; 2.H_2O→ 1,5-dicarbonyl compound

- **See Mechanism on page ______.**

Ex:

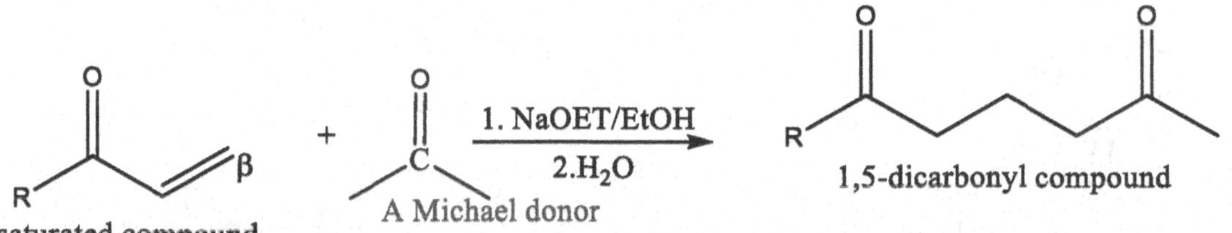

α βunsaturated compound (a Michael acceptor) + A Michael donor —1. NaOET/EtOH; 2.H_2O→ 1,5-dicarbonyl compound

- **Do problems on pages ______ - ______.**

E. THE ROBINSON ANNULATION

- The reaction involves an α-, β -unsaturated carbonyl compound and an enolate.

Ex: The reaction:

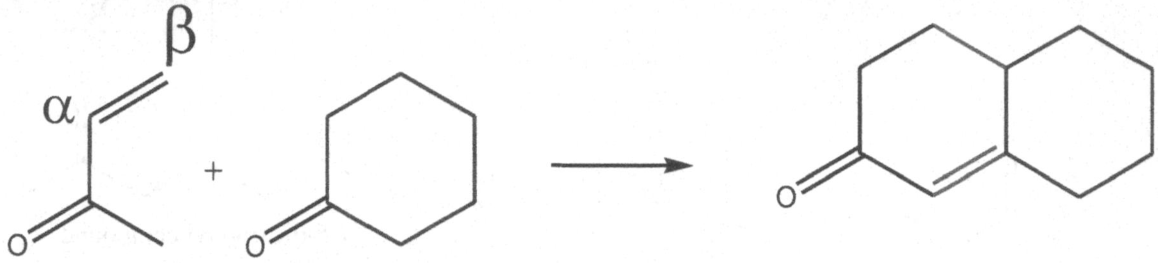

β

α

O + O → O

- Mechanism:

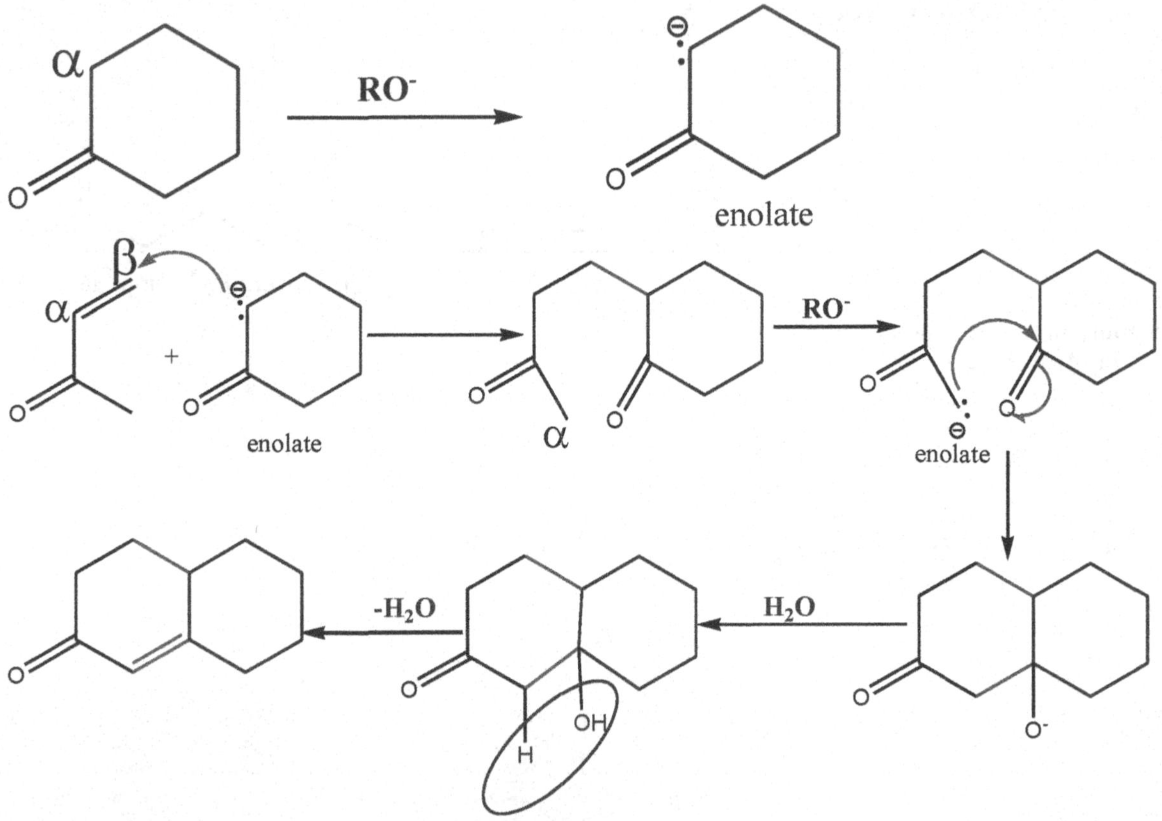

α

O —RO⁻→ O enolate

β α O + O enolate → O α O —RO⁻→ O O enolate ↓

O —$-H_2O$← O H OH ←H_2O— O O^-

- See Mechanism on pages _____ - ______.

- Read pages ______ - _____. Do related problems.

- See Key Concepts on pages _______ - ______.

- **Exercise 1: Please, write a stepwise mechanism for the following reaction:**

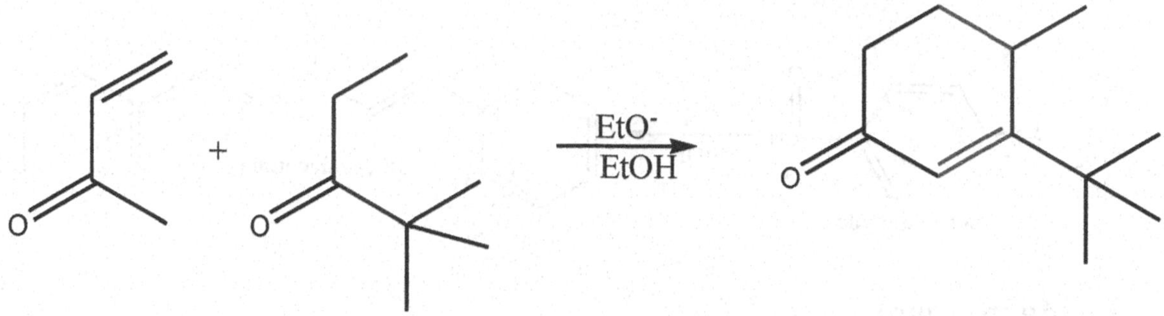

O + O $\xrightarrow[\text{EtOH}]{\text{EtO}^-}$ O

- **Mechanism:**

β

α α

O α + O $\xrightarrow[\text{EtOH}]{\text{EtO}^-}$

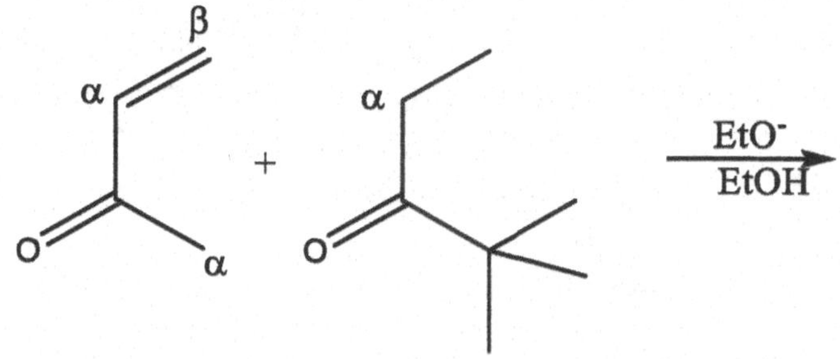

- **Exercise 2: Please, write a stepwise mechanism for the following reaction; you have to show the first 4 detailed steps with appropriate electron movements in order to get credit.**

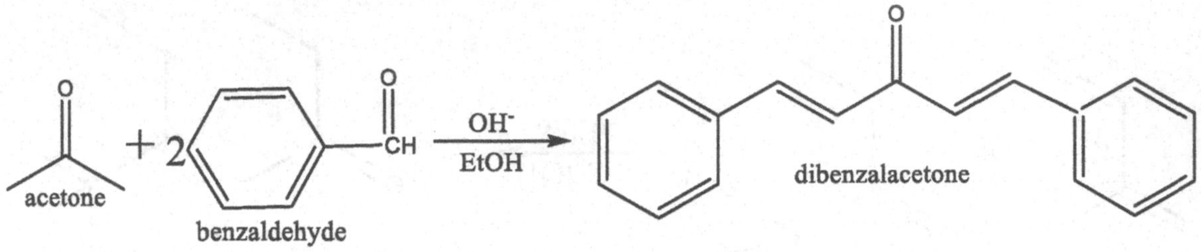

acetone + 2 benzaldehyde $\xrightarrow[\text{EtOH}]{\text{OH}^-}$ dibenzalacetone

- **Mechanism**

- SOME GLOSSARY OF CONDENSATION REACTIONS

- Aldol condensation reaction: a reaction in which a carbonyl compound (aldehyde or ketone) reacts with its enolate in alkaline (basic) media.

- cross-aldol condensation reaction: an aldol reaction in which a carbonyl compound reacts with the enolate of another carbonyl compound.

- directed cross aldol reaction: a cross aldol reaction in which the carbonyl compound that supplies the enolate (nucleophile) is clearly defined before the reaction is carried out.

- Claisen reaction: an aldol condensation reaction in which an ester and its enolate are used.

- cross-Claisen reaction: a Claisen reaction in which an ester (substrate) reacts with the enolate (nucleophile) of another ester.

- Dickman reaction: an intramolecular Claisen reaction that takes place within 1, 6 and 1, 7 diesters.

- Michael reaction: a cross aldol reaction that involve an enolate (a Michael donor) and an α,β-unsaturated compound (a Michael acceptor).

- Robinson annulation reaction: a cyclization reaction with an enolate and an α,β-unsaturated compound.